TRAVEL&LEISURE
ROAD ATLAS
THE COMPLETE TRAVEL PLANNING GUIDE

UNITED STATES
CANADA
MEXICO

Golden Press • New York
Western Publishing Company, Inc.

Created exclusively for American Express and Travel & Leisure
by General Drafting Co., Inc.

© MCMLXXXI, GENERAL DRAFTING CO., INC., Convent Station, N.J. 07961
All rights reserved. This work must not be copied in whole or in part.

Campground information © MCMLXXXI by Woodall Publishing Company,
500 Hyacinth Place, Highland Park, Illinois 60035.

Printed in U.S.A. by Western Publishing Company, Inc.
ISBN 0-307-48600-1

CONTENTS

	Page
Locator Map/Index to City Maps	3
Features of Your Road Atlas	4
How to Read the Maps	5
Planning Your Trip	6-7
Weather—What to Expect	8
Mileage Guide	9
North America	10-11
U.S. Road Map	12-13
U.S. State Maps	14-52
Places of Interest	53-61
Canada	62-69
Mexico	70-73

	Page
U.S. Cities	74-112
Canadian Cities	113-115
Mexican Cities	116
U.S. National Parks	117-124
National Park Areas	125-126
U.S. Accommodations	127-134
U.S. Campgrounds	135-140
Discount Coupons	141-148
Fun Along the Way	149-150
U.S. Facts and Figures	151-152
Index to Cities and Towns	153-160

Index to U.S./Canada/Mexico Maps and Places of Interest

	Road Map	Places of Interest
United States	14-52	53-61
Alabama	22	53
Alaska	52	53
Arizona	46	53
Arkansas	24	53
California	50-51	53
Colorado	45	54
Connecticut	14	54
Delaware	19	54
District of Columbia	19	54
Florida	25	54
Georgia	22-23	55
Hawaii	52	55
Idaho	42	55
Illinois	31	55
Indiana	26	55
Iowa	37	55
Kansas	38-39	55
Kentucky	32-33	56
Louisiana	24	56
Maine	15	56
Maryland	18-19	56
Massachusetts	14	56
Michigan	28-29	56
Minnesota	35	56
Mississippi	24	57
Missouri	37	57
Montana	42-43	57
Nebraska	36-37	57
Nevada	49	57
New Hampshire	14	57
New Jersey	19	57
New Mexico	47	57
New York	16-17	58
North Carolina	20-21	58

	Road Map	Places of Interest
North Dakota	34-35	59
Ohio	26-27	59
Oklahoma	38-39	59
Oregon	48	59
Pennsylvania	18-19	59
Rhode Island	14	59
South Carolina	23	59
South Dakota	34-35	60
Tennessee	32-33	60
Texas	40-41	60
Utah	44	60
Vermont	14	61
Virginia	20-21	61
Washington	48	61
West Virginia	18-19	61
Wisconsin	30	61
Wyoming	42-43	61
Canada	62-69	63-69
Alberta	63	63
British Columbia	62-63	63
Manitoba	65	65
New Brunswick	68	68
Newfoundland	69	69
Nova Scotia	68-69	68
Ontario	65-67	66
Prince Edward Island	68-69	69
Québec	67-69	66 & 68
Saskatchewan	64-65	65
Mexico	70-73	71
Mexico, Northern	70-71	71
Mexico, Southern	72-73	71

Locator Map

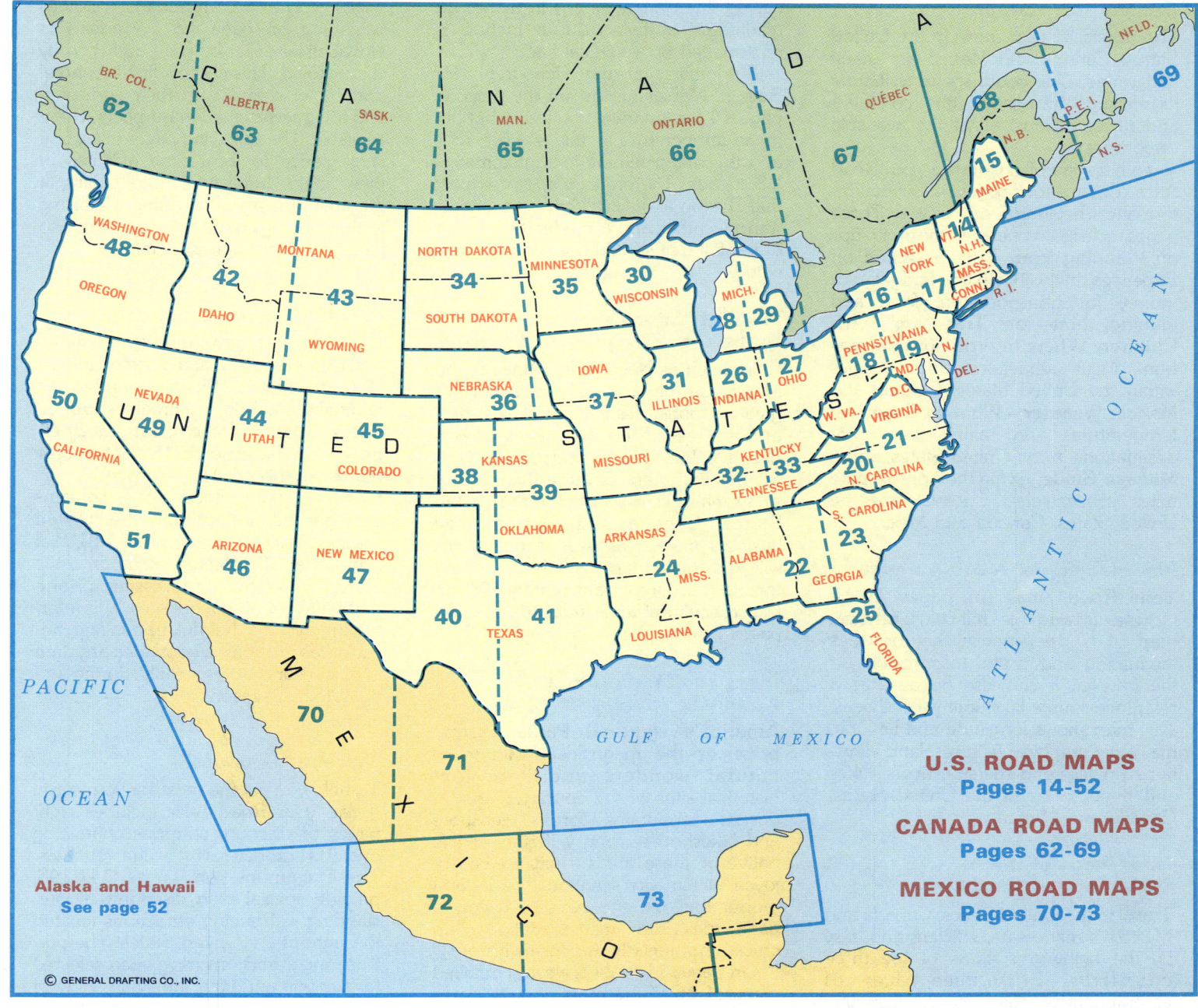

U.S. ROAD MAPS
Pages 14-52

CANADA ROAD MAPS
Pages 62-69

MEXICO ROAD MAPS
Pages 70-73

Alaska and Hawaii
See page 52

© GENERAL DRAFTING CO., INC.

Index to City Maps

United States 74-112

Akron, Ohio 75
Albany—Troy, N.Y. 75
Albuquerque, N.Mex. 74
Alexandria, La. 75
Allentown—Bethlehem, Pa. . . 74
Amarillo, Tex. 74
Annapolis, Md. 74
Asheville, N.C. 75
Atlanta, Ga. 74
Atlantic City, N.J. 75
Austin, Tex. 75
Baltimore, Md. 78
Baton Rouge, La. 76
Beaumont, Tex. 76
Billings, Mont. 77
Biloxi—Gulfport, Miss. 77
Birmingham, Ala. 76
Bismarck, N.Dak. 76
Boise, Idaho 77
Boston, Mass. 77
Buffalo—Niagara Falls, N.Y. . 76
Butte, Mont. 77
Camden, N.J. 80
Canton, Ohio 78
Carson City, Nev. 80
Casper, Wyo. 78
Charleston, S.C. 80
Charleston, W.Va. 80
Charlotte, N.C. 80
Chattanooga, Tenn. 78
Cheyenne, Wyo. 79
Chicago, Ill. 79

Cincinnati, Ohio 80
Cleveland, Ohio 77
Coeur D'Alene, Idaho 78
Colorado Springs, Colo..... 80
Columbia, S.C. 79
Columbus, Ga. 78
Columbus, Ohio 78
Corpus Christi, Tex. 78
Dallas, Tex. 83
Dayton, Ohio 81
Denver, Colo. 81
Detroit, Mich. 81
Duluth, Minn. 81
Durham, N.C. 81
El Paso, Tex. 82
Erie, Pa. 82
Fargo, N.Dak. 82
Flint, Mich. 83
Fort Smith, Ark. 82
Fort Wayne, Ind. 82
Fort Worth, Tex. 82
Galveston, Tex. 85
Grand Rapids, Mich. 84
Great Falls, Mont. 84
Greensboro, N.C. 83
Greenville, S.C. 84
Harrisburg, Pa. 84
Hartford, Conn. 84
Helena, Mont. 84
Houston, Tex. 85
Huntington, W.Va. 85
Idaho Falls, Idaho 85
Indianapolis, Ind. 86
Jackson, Miss. 86

Jacksonville, Fla. 86
Kalamazoo, Mich. 86
Kansas City, Kans.—Mo. 87
Knoxville, Tenn. 86
Lafayette, Ind. 90
Lansing, Mich. 87
Las Vegas, Nev. 91
Lexington, Ky. 91
Lincoln, Nebr. 87
Little Rock, Ark. 91
Los Angeles, Calif. 88-89
Los Angeles Area 90
Louisville, Ky. 90
Memphis, Tenn. 94
Miami to
    Fort Lauderdale, Fla. . . 92-93
Milwaukee, Wis. 94
Minneapolis—St. Paul, Minn. . 91
Mobile, Ala. 94
Monroe, La. 94
Montgomery, Ala. 94
Nashville, Tenn. 98
New Haven, Conn. 95
New Orleans, La. 95
New Orleans
    (Vieux Carré), La. 95
New York City, N.Y. 96-97
Norfolk, Va. 98
Oklahoma City, Okla. 98
Omaha, Nebr. 98
Orlando, Fla. 98
Philadelphia, Pa. 99
Phoenix, Ariz. 100
Pittsburgh, Pa. 101

Portland, Me. 101
Portland, Oreg. 100
Providence, R.I. 99
Raleigh, N.C. 102
Reno, Nev. 101
Richmond, Va. 101
Roanoke, Va. 101
Rochester, N.Y. 102
Sacramento, Calif. 103
St. Augustine, Fla. 102
St. Louis, Mo. 106
St. Petersburg, Fla. 106
Salt Lake City, Utah 107
San Antonio, Tex. 103
San Diego, Calif. 105
San Francisco, Calif. 104
San Francisco Area 104
Santa Fe, N.Mex. 103
Sarasota, Fla. 106
Savannah, Ga. 105
Scranton, Pa. 102
Seattle, Wash. 107
Shreveport, La. 103
Sioux Falls, S.Dak. 107
South Bend, Ind. 102
Spartanburg, S.C. 105
Spokane, Wash. 105
Springfield, Ill. 102
Springfield, Mass. 102
Syracuse, N.Y. 105
Tallahassee, Fla. 106
Tampa, Fla. 108
Terre Haute, Ind. 107
Toledo, Ohio 108

Topeka, Kans. 108
Trenton, N.J. 108
Tucson, Ariz. 108
Tulsa, Okla. 108
Waco, Tex. 109
Washington, D.C. 109
Washington, D.C.,
    Central 110-111
Wheeling, W.Va. 109
Wichita, Kans. 112
Wichita Falls, Tex. 109
Wilkes-Barre, Pa. 108
Williamsburg, Va. 109
Wilmington, Del. 112
Wilmington, N.C. 112
Winston-Salem, N.C. 112
Worcester, Mass. 112
Youngstown, Ohio 112

Canada 113-115

Calgary, Alta. 113
Edmonton, Alta. 113
Montréal, Qué. 113
Niagara Falls, Ont. 113
Ottawa, Ont. 113
Québec Central, Qué. 114
Regina, Sask. 115
Toronto (Metro.), Ont. ... 114
Vancouver, B.C. 115
Winnipeg, Man. 115

Mexico 116

Mexico City 116
Mexico City, Central 116

FEATURES OF YOUR ROAD ATLAS

General Information

The possibilities for travel in the United States, Canada and Mexico are many and varied. One could spend a lifetime journeying throughout this continent and never exhaust all of the available opportunities.

To help you plan your trip, your Road Atlas contains four pages of General Information beginning on page 6. These pages include an overall section with tips on **Planning Your Trip;** a section on **Expenses** that will help you budget your money for transportation, food and lodging; hints on **Traveling With Children; Where to Write for Information** about attractions and facilities within the United States, Canada and Mexico; **Weather—What to Expect** of temperatures and rainfall for major destinations in the United States; and a **Mileage Guide,** giving the approximate miles between selected points across the United States, Canada and Mexico.

Physical Map of North America

Your Road Atlas on pages 10-11 includes a two-page, full-color, physical map of North America that provides fascinating and useful information for the traveler. It can also be used as a reference source by students of all ages. This map shows longitude and latitude, major geographical features, land elevations (in feet and meters), rivers, lakes and major cities in the United States, Canada and Mexico.

Road Maps of the United States, Canada and Mexico

Travel in the United States, Canada and Mexico is made safe, efficient and easy by the highway systems of the three countries. Fifty- three pages of comprehensive road maps, beginning on page 12 of your Atlas, detail those systems state by state and province by province. All of the maps give you a great deal of useful information. See the **How to Read the Maps** section on page 5 for a clear interpretation of this information.

Each country offers the traveler a variety of sights to enjoy and interesting places to explore. A list of selected **Places of Interest,** arranged alphabetically, can be found on pages 53-71. Use the grid reference within parentheses after the entry to locate the attraction on the appropriate map.

Climographs

On many maps throughout your Road Atlas, climographs are provided for representative cities and destinations. These show the monthly temperature and rainfall derived from past records for each month of the year. Mean temperature in Fahrenheit, in gradations of twenty degrees, is represented by a curved line. Mean rainfall in inches, in gradations of two or four inches, is represented by a vertical bar.

By looking at the climograph for Boston, Massachusetts, on the map on page 14 for instance, you learn that the mean temperature in July is about 73°F and the mean rainfall is approximately 2.8 inches. Therefore, you can expect that Boston in July will have reasonably warm weather and a moderate amount of rainfall that should not dampen your visit.

Maps of Major Urban Centers in the U.S., Canada and Mexico

Cities such as New York, Montreal and Mexico City are among North America's greatest tourist destinations. The variety and richness of this continent's cities is dazzling. Included in your Road Atlas are detailed maps of more than 150 of these major cities located in the United States, Canada and Mexico. The business traveler as well as the tourist will find these maps useful in planning trips to these important centers of commerce, culture and industry. The 43 pages of maps begin on page 74.

Maps and Descriptions of National Park System Areas

America's National Parks System preserves the magnificent wilderness, natural wonders and important historical sites of the country. Glaciers, deserts, mountains, forests, seashore and lakeshore—along with the wildlife native to these areas—may all be enjoyed at America's national parks and monuments. At the nation's historical parks the visitor may step back into crucial moments in American history.

On pages 117-124 is a set of detailed maps of 16 of the most popular parks and monuments. An explanation of the unique scenic or historic attractions each of these parks offers accompanies each map.

U.S. National Parks Charts

The U.S. National Park Service administers more than 300 separate parks, monuments, recreation areas, battlefields and historic sites in every part of the United States. Many of these offer unequalled opportunities for engrossing travel. These sites range from such new areas as the Chattahoochee River National Recreation Area in Georgia to New York harbor's familiar Statue of Liberty. On pages 125-126 are easy-to-read National Park Service charts that list selected sites along with information concerning entrance fees, recreation facilities available, food services and accommodations.

U.S. Accommodations Listings

Unique to your Road Atlas are eight pages of U.S. accommodations listings beginning on page 127, arranged in alphabetical order by state and then by city or town. This section first lists hotel and motel chains with their addresses and toll-free reservation numbers where available. The following pages allow you at a glance to locate the affiliates of these major hotel and motel chains in the area you may be visiting. Referring to these listings makes it easy for you to plan your next night's lodging.

U.S. Campground Listings

Camping is an adventurous, unfettered way to sightsee and at the same time get close to nature. Sleeping under the night sky and cooking breakfast out-of-doors is one of the great travel experiences. Because millions of people now enjoy this relatively inexpensive way to travel, your Road Atlas contains campground listings for the United States. These pages, 135-140, list hundreds of selected campgrounds along with their addresses, phone numbers and number of sites. This information will be helpful in making advance reservations, a good idea any time of the year but especially during the peak tourist seasons.

Discount Coupons

To help you get the most for your travel dollar, your Road Atlas includes eight pages of discount coupons containing over 100 money-saving value vouchers. These coupons, on pages 141-148, provide special cash discounts at outstanding sightseeing attractions located throughout the United States. The participating attractions have been selected to give you and your family a variety of places to fit your individual interests and travel plans.

Please be certain to read the specific terms and conditions spelled out on the back of each coupon before presenting it at the participating attraction.

In-Car Entertainment

Travel provides excellent opportunities for the family to share exciting and horizon-broadening experiences together. However, part of any trip is the time spent traveling between destinations. Verbal car games are always a good way to help pass this tedious travel time. The **Fun Along the Way** section of your Road Atlas, on pages 149-150, includes games that both children and adults will enjoy playing. Included are such favorites as Buzz, City Scramble and Tall Tales.

Pages 151-152 present interesting **U.S. Facts and Figures** about the United States that can be used to initiate discussions.

The maps of the United States, Canada and Mexico in this Road Atlas contain a wealth of useful information. If you are to make full use of the information however, an understanding of the symbols used on the maps is important. These symbols and a brief explanation about them are shown below. Take a few minutes to study this legend; it will help you plan your trip. Then consult it frequently when traveling along your planned route.

As you will see when you use this Road Atlas, each map contains complete Interstate Highway coverage. Although nearly 95% of the planned U.S. Interstate Highway System is finished, a number of sections, especially in or near urban areas, are under construction or still in the planning stages. This Atlas indicates areas under construction at the time these maps were compiled for publication. If you are riding along an Interstate

and find no barricades and detour signs where the map shows a section to be under construction, you can proceed on the newly completed section without hesitation, confident that intersecting routes and exits will be clearly posted.

U.S. state maps begin on page 14. Maps of major urban areas may be found starting on page 74; they provide greater detail and include secondary routes not always shown on the state maps. Both state and city maps also spot recreational facilities and places of interest. Selected maps have legends which explain special symbols used.

When using the road maps of Canada and Mexico, be certain to refer to the special legends for these maps found on pages 69 and 72 respectively, as well as the places of interest marked on the maps.

MAP LEGEND

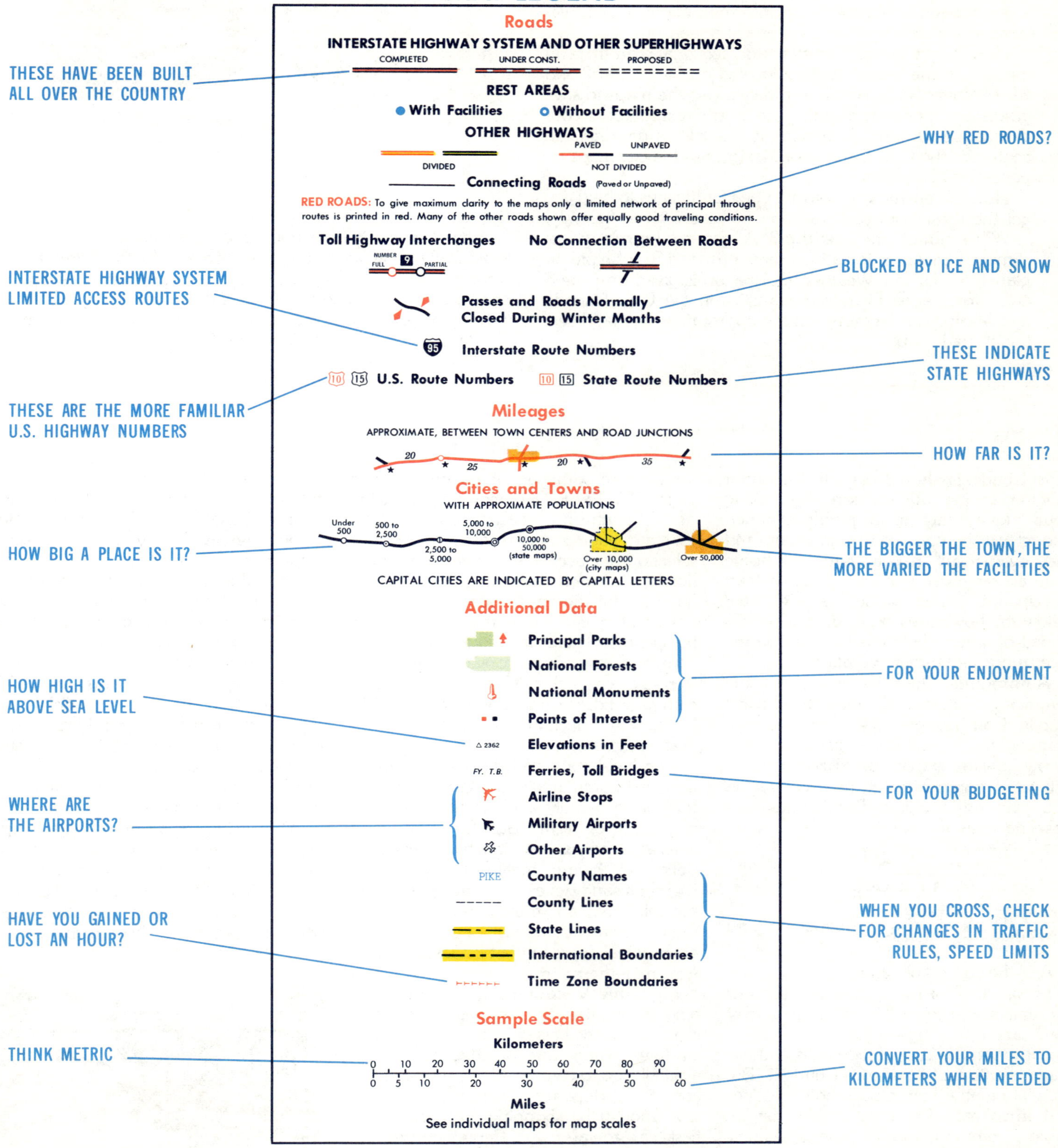

Roads

INTERSTATE HIGHWAY SYSTEM AND OTHER SUPERHIGHWAYS
COMPLETED — UNDER CONST. — PROPOSED

REST AREAS — With Facilities — Without Facilities

OTHER HIGHWAYS — PAVED — UNPAVED — DIVIDED — NOT DIVIDED

Connecting Roads (Paved or Unpaved)

RED ROADS: To give maximum clarity to the maps only a limited network of principal through routes is printed in red. Many of the other roads shown offer equally good traveling conditions.

Toll Highway Interchanges — NUMBER FULL 9 PARTIAL

No Connection Between Roads

Passes and Roads Normally Closed During Winter Months

95 **Interstate Route Numbers**

10 15 **U.S. Route Numbers**   10 15 **State Route Numbers**

Mileages
APPROXIMATE, BETWEEN TOWN CENTERS AND ROAD JUNCTIONS
20   25   20   35

Cities and Towns
WITH APPROXIMATE POPULATIONS
Under 500 — 500 to 2,500 — 2,500 to 5,000 — 5,000 to 10,000 — 10,000 to 50,000 (state maps) — Over 10,000 (city maps) — Over 50,000
CAPITAL CITIES ARE INDICATED BY CAPITAL LETTERS

Additional Data
Principal Parks
National Forests
National Monuments
Points of Interest
△ 2362 Elevations in Feet
FY. T.B. Ferries, Toll Bridges
Airline Stops
Military Airports
Other Airports
PIKE County Names
County Lines
State Lines
International Boundaries
Time Zone Boundaries

Sample Scale
Kilometers
0 10 20 30 40 50 60 70 80 90
0 5 10 20 30 40 50 60
Miles
See individual maps for map scales

Annotations
THESE HAVE BEEN BUILT ALL OVER THE COUNTRY
INTERSTATE HIGHWAY SYSTEM LIMITED ACCESS ROUTES
THESE ARE THE MORE FAMILIAR U.S. HIGHWAY NUMBERS
HOW BIG A PLACE IS IT?
HOW HIGH IS IT ABOVE SEA LEVEL
WHERE ARE THE AIRPORTS?
HAVE YOU GAINED OR LOST AN HOUR?
THINK METRIC
WHY RED ROADS?
BLOCKED BY ICE AND SNOW
THESE INDICATE STATE HIGHWAYS
HOW FAR IS IT?
THE BIGGER THE TOWN, THE MORE VARIED THE FACILITIES
FOR YOUR ENJOYMENT
FOR YOUR BUDGETING
WHEN YOU CROSS, CHECK FOR CHANGES IN TRAFFIC RULES, SPEED LIMITS
CONVERT YOUR MILES TO KILOMETERS WHEN NEEDED

Part of the excitement of traveling begins in the planning stages. Everyone can participate in poring over maps and selecting places to see. Sharing the preparations for travel can heighten the interest for everyone, adults and children alike, and can be a rewarding family experience.

Where to go? Choose from a wide variety of vacation areas by referring to the U.S. Places of Interest section found on pages 53-61, and the Canada and Mexico Places of Interest found along with their respective maps on pages 63-71.

Perhaps the national parks of the U.S. interest you. Detailed maps and descriptions of 16 favorite National Park Service areas can be found on pages 117-124; an extensive chart of National Park areas is on pages 125-126.

Write to one or more of the organizations listed on page 7 for free information. Reviewing brochures and pamphlets is a good way to become better informed about a particular state or vacation spot that especially appeals to you.

How to get there? The state and sectional road maps of the U.S., Canada and Mexico, located on pages 14-52 and 62-73, cover North America in detail. Use the maps to save gasoline and time by planning the most direct routes to your destinations. If your destination is one of the metropolitan centers of the U.S., Canada or Mexico, use the city maps on pages 74-116.

How expensive will your trip be? Guidelines to help you get the most from your vacation dollar are given below.

What about the weather? A general knowledge of climatic conditions in various parts of the U.S. is helpful in planning your trip. Weather information is given on page 8 and climographs for numerous cities in the U.S., Canada and Mexico can be found on the appropriate state and sectional road maps.

How far is it? Plan your day-to-day driving schedule realistically, to avoid getting overtired and to allow time for sightseeing along the way. Point-to-point distances are shown on the maps in this Atlas and can be quickly added up to give you daily mileage totals. An easy-to-use Mileage Guide can also be found on page 9.

Expenses

Travel budgets should play an important part in planning your vacation. Mishandling your money due to a poorly planned budget will hamper your enjoyment, and may even result in your returning home from an otherwise perfect trip earlier than anticipated. Some guidelines are listed below on how to estimate your expenses ahead of time to help ensure you a worry-free, more pleasurable vacation.

Automobile: The rising cost of gasoline is a major concern to all motorists. The first step in determining auto expenses for your planned trip is to look at gasoline costs. Know how many miles per gallon your car will get and approximately how many miles you plan on traveling. Then figure your gasoline costs this way:

$$\frac{\text{Planned Mileage}}{\text{Average Miles Per Gallon}} = \frac{\text{Gallons}}{\text{Needed}}$$

$$\frac{\text{Gallons}}{\text{Needed}} \times \frac{\text{Price Per}}{\text{Gallon}} = \text{Expected Cost}$$

Add the cost of tolls and any possible ferry fares, allow for the unexpected such as car repairs, and you will have an estimate of your car's operating expenses.

Lodging: The cost of lodging is another major portion of your travel budget. But there are many chain motels, hotels, lodges and inns that offer comfortable, dependable accommodations at reasonable rates. If you're traveling with children, look into the chains that allow children to stay in the same room with you at no charge. See pages 127-134 for a comprehensive listing of U.S. chain accommodations. To estimate your lodging costs, make your selection of places to stay, then contact the hotels via their toll-free "800" phone numbers for the latest room rates.

Camping can be a means to a low-budget trip. There are thousands of inexpensive campgrounds conveniently located throughout the U.S. Check the campgrounds section on pages 135-140 for aid in selecting your en route and destination stops.

Food: Estimating food costs is an essential step in preparing your travel budget. Be realistic, especially if planning a trip to a major city or resort. But this is one area where you can usually stretch your vacation dollar. A hearty breakfast, light lunch (perhaps sandwiches, fruit and beverage at a picnic area), and then your main meal at the end of a day's ride is not only the most economical approach but it also helps prevent mid-afternoon drowsiness.

Fly-Drive Vacations: If you are planning a fly-drive type of vacation, shop carefully for the best bargains and discount air fares. Then, with an idea of what is available, go to your travel agent to arrange convenient departure times. You can also save money by shopping around for the best car rental rates. Rental companies usually charge on a per-day plus mileage basis. If planning an extended trip, this can be expensive. Contact the national car rental companies (some have toll-free "800" numbers; check the Yellow Pages) to determine the availability of money-saving, unlimited mileage plans.

Miscellaneous: Traveling off-season or when schools are in session is frequently an economic plus. You may also find an improvement in service as well as in prices for lodgings and attractions.

Carrying large amounts of cash is unwise. Use traveler's checks as a safeguard against loss and theft. They are available in small and large denominations and can be cashed almost anywhere. Purchase them at banks for approximately $1 per $100 of checks.

When using credit cards for gas, lodgings and food, keep a close record of your purchases. This will provide you with an accurate account of your total travel expenses so you will not be unpleasantly surprised when the statements come in after you return home.

Traveling With Children

Millions of families know the joy of traveling on a vacation with children. Here are some tips that will help your family join those millions and to make your next vacation trip the most enjoyable possible, for children and adults alike.

Perhaps the first step towards an enjoyable vacation is care in packing for the children. This is nearly an art, acquired only through experience. Allowing children under 12 to pack their own gear is not advisable; left to themselves, the children will often pack nothing but toys and forget the essentials like socks. Let children select what they want but let an adult include what is practical.

When traveling with children from car bed age to about five or six, it is important to begin to instill good travel habits. Children this age are often difficult to travel with because they have short attention spans. Car rules are often necessary to prevent the driver from being distracted by squabbles. Have the children (as well as the adults) use seat belts and instruct them not to interfere with the driver.

Another helpful hint for traveling with children is to provide a comfortable place for them in the car. Give small children familiar playthings or a blanket and pillow. Older children may want to bring some reading material, crossword puzzles and games. Verbal car games are always a good way to pass traveling time. There are many of these games such as "20 Questions" and spotting license plates; see pages 149-150 for a sampling. There are also numerous books on the subject, but none can surpass parental imagination.

While on the road, select restaurants based on your knowledge of your children's appetites and interests. Adolescents especially may enjoy a night out for pizza or tacos. This can be a nutritious and relatively inexpensive way to satisfy your children's wants and needs.

It is also a good idea to plan stops at roadside rests and parks where children can run around and release accumulated energy. This is a perfect opportunity to toss around a Frisbee or football. Perhaps you might want to combine picnic lunches with these rest and recreation stops.

The children's enjoyment of a trip will be greatly influenced by the attitude of the accompanying adults. If you are interested in the areas and sites visited and have taken the time to learn something about them in advance, the children will be infected with the same spirit. As your trip progresses, it is an excellent idea to have the children, especially teenagers, participate by reading about and selecting attractions they would particularly like to visit.

One of the nicest things about traveling with children is the opportunity you have to introduce them to new and fascinating things. Clear, simple explanations of historic and natural sights will add to their enjoyment. In return, your children will often give you a fresh and interesting new outlook on traveling and sightseeing.

Where To Write For Information

UNITED STATES

ALABAMA. Bureau of Publicity & Info., 532 S. Perry St., Montgomery 36130.
ALASKA. Div. of Tourism, Dept. of Commerce & Economic Devel., Pouch E, Juneau 99811.
ARIZONA. Office of Tourism, 112 N. Central Ave., Phoenix 85004.
ARKANSAS. Dept. of Parks & Tourism, One Capitol Mall, Little Rock 72201.
CALIFORNIA. Chamber of Commerce, Box 1736, Sacramento 95808.
COLORADO. Office of Tourism, Div. of Commerce & Devel., 1313 Sherman St., Rm. 500, Denver 80203.
CONNECTICUT. Communications Services Div., Dept. of Commerce, 210 Washington St., Hartford 06106.
DELAWARE. State Travel Service, Div. of Economic Devel., 630 State College Rd., Dover 19901.
DISTRICT OF COLUMBIA. Dept. of Tourism, Washington Convention & Visitors Assn., 1575 Eye St., N.W., Washington, D.C. 20005.
FLORIDA. Florida News Bureau, Dept. of Commerce, Div. of Tourism, 107 W. Gaines St., Tallahassee 32304.
GEORGIA. Dept. of Industry & Trade, 1400 N. Omni International, Box 1776, Atlanta 30301.
HAWAII. Hawaii Visitors Bureau, Suite 801, Waikiki Business Plaza, 2270 Kalakaua Ave., Honolulu 96815.
IDAHO. Div. of Economic & Community Affairs, Capitol Building, Boise 83720.
ILLINOIS. Office of Tourism, Dept. of Commerce & Community Affairs, 205 W. Wacker Dr., Chicago 60606.
INDIANA. Tourism Devel. Div., Dept. of Commerce, 440 N. Meridian St., Indianapolis 46204.
IOWA. Iowa Devel. Comm., 250 Jewett Bldg., Des Moines 50309.
KANSAS. Dept. of Economic Devel., 6th Floor, 503 Kansas Ave., Topeka 66603.
KENTUCKY. Advertising and Travel Promotion Div., Capitol Annex Bldg., Frankfort 40601.
LOUISIANA. Office of Tourism, Box 44291, Capitol Station, Baton Rouge 70804.

MAINE. Publicity Bureau, 87 Winthrop St., Hallowell 04347.
MARYLAND. Dept. of Economic & Community Devel., Office of Tourist Devel., 1748 Forest Dr., Annapolis 21401.
MASSACHUSETTS. Dept. of Commerce and Development, Saltonstall Office Bldg., 100 Cambridge St., Boston 02202.
MICHIGAN. Dept. of Commerce, Box 30226, Law Bldg., Lansing 48909.
MINNESOTA. Tourism Div., Dept. of Economic Devel., 480 Cedar St., St. Paul 55101.
MISSISSIPPI. Board of Economic Devel., 1200 Walter Stillers Bldg., P.O. Box 849, Jackson 39205.
MISSOURI. Dept. of Consumer Affairs, Regulation & Licensing, Div. of Tourism, Box 1055, 308 E. High St., Jefferson City 65102.
MONTANA. Chamber of Commerce, Box 1730, Helena 59624.
NEBRASKA. Div. of Travel & Tourism, Dept. of Economic Devel., Box 94666, 301 Centennial Mall S., Lincoln 68509.
NEVADA. Dept. of Economic Devel., Capitol Complex, Carson City 89701.
NEW HAMPSHIRE. Office of Vacation Travel, Box 856, State House Annex, Concord 03301.
NEW JERSEY. Dept. of Labor & Industry, Div. of Travel & Tourism, Box 400, Trenton 08625.
NEW MEXICO. Tourism & Travel Div., Commerce & Industry Dept., Bataan Memorial Bldg., Santa Fe 87503.
NEW YORK. Bureau of State Information, Dept. of Commerce, 99 Washington Ave., Albany 12245.
NEW YORK CITY, N.Y. New York Convention & Visitors Bureau, 2 Columbus Circle, New York 10019.
NORTH CAROLINA. Div. of Travel & Tourism, Dept. of Commerce, 430 N. Salisbury St., Raleigh 27611.
NORTH DAKOTA. Tourism Promotion Div., Highway Dept., Capitol Grounds, Bismarck 58505.
OHIO. Office of Travel & Tourism, Box 1001, Columbus 43216.
OKLAHOMA. Tourism & Recreation Dept., 500 Will Rogers Memorial Bldg., Oklahoma City 73105.

OREGON. Travel Info. Section, Oreg. State Highway Div., 101 State Highway Bldg., Salem 97310.
PENNSYLVANIA. Dept. of Commerce, Bureau of Travel Devel., 402 S. Office Bldg., Harrisburg 17120.
RHODE ISLAND. Dept. of Economic Devel., Tourist Promotion Div., Gilbane Bldg., 7 Jackson Walkway, Providence 02903.
SOUTH CAROLINA. Dept. of Parks, Recreation & Tourism, Suite 113, Edgar A. Brown Bldg., 1205 Pendleton St., Columbia 29201.
SOUTH DAKOTA. Dept. of Economic & Tourism Devel., Joe Foss Bldg., Pierre 57501.
TENNESSEE. P.O. Box 23170, Tourist Devel., Nashville 37210.
TEXAS. Travel & Info. Div., Texas Highway Dept., 11th and Brazos, Austin 78701.
UTAH. Utah Travel Council, Council Hall, Capitol Hill, Salt Lake City 84114.
VERMONT. Agency of Devel. & Community Affairs, Travel Div., 61 Elm St., Montpelier 05602.
VIRGINIA. State Chamber of Commerce, Travel Devel. Dept., 611 E. Franklin St., Richmond 23219.
WASHINGTON. Dept. of Commerce & Economic Devel., Travel Devel., 312 1st Ave. N., Seattle 98109.
WEST VIRGINIA. Communications Div., Governor's Office of Economic & Community Devel., Bldg. 6, Rm. B-504, Charleston.25305.
WISCONSIN. Dept. of Business Devel., Div. of Tourism, P.O. Box 7970, Madison 53707.
WYOMING. Wyoming Travel Commission, I-25 at Etchepare Circle, Cheyenne 82002.

CANADA

CANADIAN GOVERNMENT OFFICE OF TOURISM, 235 Queen St., 4th Floor, Ottawa K1A 0H6.
ALBERTA. Travel Alberta, Capital Square, 12th Floor, 10065 Jasper Ave., Edmonton T5J 0H4.
BRITISH COLUMBIA. Tourism British Columbia, 1117 Wharf St., Victoria V8W 2Z2.

MANITOBA. Travel Manitoba, Dept. 1045, Legislative Bldg., Winnipeg R3C 0V8.
NEW BRUNSWICK. Tourism New Brunswick, Box 12345, Fredericton E3B 5C3.
NEWFOUNDLAND-LABRADOR. Newfoundland and Labrador Dept. of Development, Tourism Development Div., Box 2016, St. John's A1C 5R8.
NORTHWEST TERRITORIES. TravelArctic, Government of the N.W.T., Yellowknife X1A 2L9.
NOVA SCOTIA. Dept. of Tourism, Travel Information Div., Box 130, Halifax B3J 2M7.
ONTARIO. Ontario Travel, Queen's Park, Toronto M7A 2E5.
PRINCE EDWARD ISLAND. Tourism Services Div., Box 940, Charlottetown C1A 7M5.
QUÉBEC. Québec Tourisme, CP 20,000, Québec G1K 7X2.
SASKATCHEWAN. SaskTravel, Dept. of Tourism and Renewable Resources, 3211 Albert St., Regina S4S 5W6.
YUKON. Tourism Yukon, Box 2703, Whitehorse Y1A 2C6.

MEXICO

MEXICAN GOVERNMENT TOURIST OFFICES: Suite 1201, Cain Tower, Peachtree Center, Atlanta, Ga. 30303; Suite 1201, 9701 Wilshire Blvd., Beverly Hills, Calif. 90212; Suite 3612, John Hancock Center, Chicago, Ill. 60611; Suite 1230, Two Turtle Creek Village, Dallas, Tex. 75219; Suite 2010, First Denver Plaza Bldg., 633 17th St., Denver, Colo. 80202; Suite 1370, Lummus Tower, Houston, Tex. 77056; Suite 2804, 100 Biscayne Blvd., Miami, Fla. 33132; One Shell Square Bldg., Concourse, New Orleans, La. 70139; Suite 1002, 405 Park Ave., New York, N.Y. 10022; GPM South Tower, 800 N.W. Loop 410, San Antonio, Tex. 78216; Suite 1220, 600 B St., San Diego, Calif. 92101; Suite 2465, 50 California St., San Francisco, Calif. 94111; Suite 1535, 5151 E. Broadway, Tucson, Ariz. 85711; Suite 329 1156 15th St., N.W., Washington, D.C. 20005.

WEATHER—WHAT TO EXPECT

The maps below are designed to help you check seasonal weather conditions throughout the conterminous United States. The accompanying chart shows average summer and winter temperatures for selected U.S. cities plus the average amount of precipitation for these cities. With this information, you can plan your wardrobe as well as take necessary precautions for driving.

	Altitude in Feet	TEMPERATURES IN DEGREES FAHRENHEIT				PRECIPITATION IN INCHES ❋Less than ½ inch	
		Average July Temp.		Average Jan. Temp.			
		Max.	Min.	Max.	Min.	Aver. July Rain	Aver. Jan. Precip.
Albany, N.Y.	60	84	60	30	13	3	2
Albuquerque, N. Mex.	4950	92	65	47	24	1	❋
Amarillo, Tex.	3675	91	66	49	23	3	1
Atlanta Ga.	1040	87	69	51	33	5	4
Baltimore, Md.	60	87	67	42	25	4	3
Baton Rouge, La.	50	91	73	62	41	7	4
Birmingham, Ala.	600	90	70	54	34	5	5
Bismarck, N. Dak.	1650	84	57	19	-3	2	❋
Boise, Idaho	2740	91	59	37	21	❋	1
Boston, Mass.	15	81	65	36	23	3	4
Brownsville, Tex.	15	93	76	70	51	1	1
Buffalo, N.Y.	610	80	61	30	18	3	3
Burlington, Vt.	220	81	59	26	8	4	2
Charleston, S.C.	40	89	71	60	37	8	3
Charleston, W. Va.	600	86	64	44	25	5	3
Cheyenne, Wyo.	6060	84	55	38	15	2	❋
Chicago, Ill.	590	84	65	32	17	4	2
Cleveland, Ohio	660	82	61	33	20	3	3
Columbia, S.C.	215	92	70	57	34	6	3
Columbus, Ohio	770	85	62	36	20	4	3
Dallas, Tex.	510	96	74	56	34	2	2
Denver, Colo.	5280	87	59	44	16	2	1
Des Moines, Iowa	800	85	65	28	11	3	1
Detroit, Mich.	600	83	63	32	19	3	2
Duluth, Mich.	615	76	55	18	1	4	1
El Paso, Tex.	3760	95	70	57	30	2	❋
Fresno, Calif.	295	98	63	55	36	❋	2
Great Falls, Mont.	3310	84	55	29	12	1	1
Houston, Tex.	55	94	73	63	42	4	4
Indianapolis, Ind.	715	85	65	36	20	4	3
Jackson, Miss.	285	93	71	58	36	4	5
Jacksonville, Fla.	20	90	72	65	45	7	3
Kansas City, Mo.	750	88	70	36	18	4	1
Knoxville, Tenn.	920	88	68	49	32	5	5
Las Vegas, Nev.	2030	104	73	56	33	❋	❋
Little Rock, Ark.	290	93	70	50	30	3	4
Los Angeles, Calif.	150	83	64	67	47	❋	3
Louisville, Ky.	460	87	66	42	25	4	4
Medford, Oreg.	1380	90	54	44	29	❋	4
Memphis, Tenn.	275	92	72	49	32	4	5
Miami, Fla.	10	89	76	76	59	7	2
Milwaukee, Wisc.	600	80	59	27	11	3	2
Minneapolis, Minn.	825	82	61	21	3	4	1
Nashville, Tenn.	475	90	69	48	29	4	5
New Orleans, La.	10	90	73	62	44	7	5
New York, N.Y.	10	85	68	39	26	4	3
Norfolk, Va.	20	87	70	49	32	6	3
Oklahoma City, Okla.	1280	93	72	48	26	3	1
Omaha, Nebr.	980	90	66	33	12	4	1
Philadelphia, Pa.	25	87	67	40	24	4	3
Phoenix, Ariz.	1100	105	78	65	38	1	1
Pittsburgh, Pa.	780	84	65	37	24	4	3
Portland, Me.	75	79	57	31	12	3	3
Portland, Oreg.	20	79	55	44	33	❋	6
Presque Isle, Me.	450	76	54	20	1	4	2
Raleigh, N.C.	345	88	67	51	30	5	3
Rapid City, S. Dak.	3160	86	59	34	10	2	❋
Reno, Nev.	4405	91	47	45	18	❋	1
Richmond, Va.	160	88	68	47	28	6	3
St. Louis, Mo.	455	88	69	40	23	4	2
Salt Lake City, Utah	4220	93	61	37	19	1	1
San Antonio, Tex.	700	96	74	62	40	2	2
San Diego, Calif.	10	75	64	65	46	❋	2
San Francisco, Calif.	60	64	53	56	46	❋	4
Seattle, Wash.	25	76	56	45	35	1	6
Spokane, Wash.	1890	84	55	31	20	❋	2
Tallahassee, Fla.	55	91	72	64	41	9	4
Tampa, Fla.	20	90	74	71	50	8	2
Washington, D.C.	50	88	69	44	28	4	3
Wichita, Kans.	1320	92	70	41	21	4	1

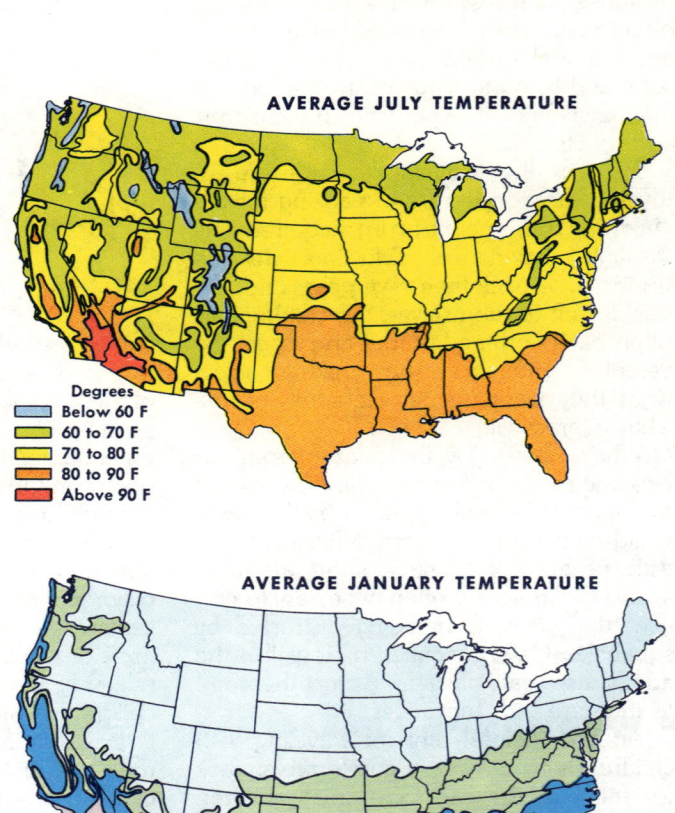

AVERAGE JULY TEMPERATURE

Degrees
- Below 60 F
- 60 to 70 F
- 70 to 80 F
- 80 to 90 F
- Above 90 F

AVERAGE JANUARY TEMPERATURE

Degrees
- Below 30 F
- 30 to 40 F
- 40 to 50 F
- 50 to 60 F
- Above 60 F

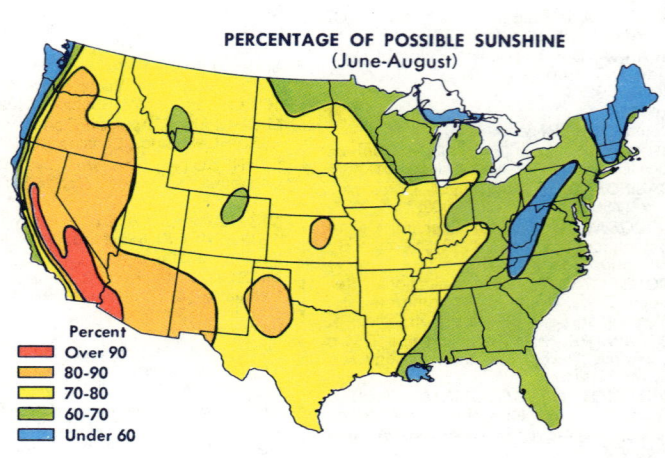

PERCENTAGE OF POSSIBLE SUNSHINE
(June-August)

Percent
- Over 90
- 80-90
- 70-80
- 60-70
- Under 60

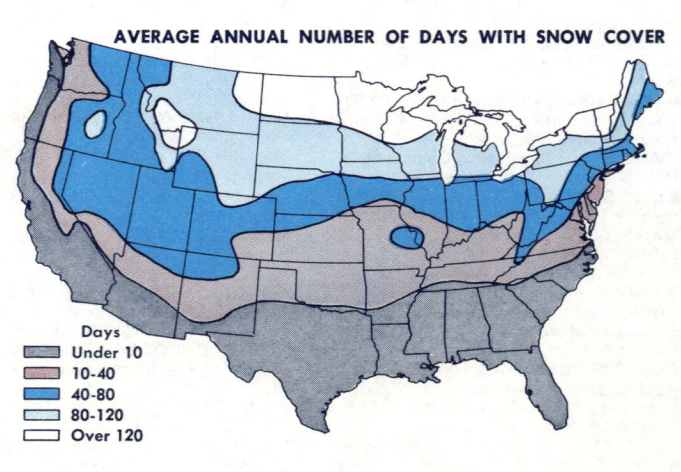

AVERAGE ANNUAL NUMBER OF DAYS WITH SNOW COVER

Days
- Under 10
- 10-40
- 40-80
- 80-120
- Over 120

MILEAGE GUIDE

APPROXIMATE MILEAGES	Albuquerque, N. Mex.	Atlanta, Ga.	Birmingham, Ala.	Boston, Mass.	Chicago, Ill.	Cleveland, Ohio	Dallas, Tex.	Denver, Colo.	Detroit, Mich.	El Paso, Tex.	Houston, Tex.	Indianapolis, Ind.	Kansas City, Mo.	Las Vegas, Nev.	Little Rock, Ark.	Los Angeles, Calif.	Louisville, Ky.	Memphis, Tenn.	Miami, Fla.	Minneapolis, Minn.	Montréal, Qué.	Nashville, Tenn.	New Orleans, La.	New York, N.Y.	Oklahoma City, Okla.	Omaha, Nebr.	Phoenix, Ariz.	Portland, Oreg.	Raleigh, N.C.	Richmond, Va.	St. Louis, Mo.	Salt Lake City, Utah	San Francisco, Calif.	Seattle, Wash.	Tampa, Fla.	Washington, D.C.
Acadia Nat. Pk., Me.	2420	1315	1450	270	1215	880	2020	2210	1050	2565	2085	1155	1635	3005	1700	3220	1210	1560	1805	1615	355	1355	1785	480	1900	1670	2850	3310	965	810	1400	2625	3375	3245	1640	700
Albany, N.Y.	2025	990	1075	165	810	465	1625	1795	540	2170	1735	765	1245	2615	1320	2825	810	1185	1440	1200	225	990	1425	145	1495	1255	2455	2890	620	465	1010	2205	2960	2830	1280	360
Albuquerque, N. Mex.		1390	1285	2180	1265	1565	650	430	1535	270	830	1260	775	590	885	830	1285	1010	1950	1175	2095	1220	1150	1970	550	845	425	1370	1730	1820	1020	600	1130	1450	1720	1825
Atlanta, Ga.	1390		150	1065	675	670	805	1385	715	1425	810	515	785	2025	515	2205	395	380	655	1070	1215	235	490	850	850	985	1815	2670	390	535	530	1900	2525	2725	455	630
Atlantic City, N.J.	1960	790	915	330	800	475	1510	1760	630	2105	1560	695	1175	2550	1180	2760	735	1040	1280	1200	485	835	1260	125	1430	1250	2390	2890	440	285	930	2205	2960	2830	1115	180
Baltimore, Md.	1865	670	795	395	675	345	1385	1650	515	2005	1465	570	1045	2445	1060	2660	615	920	1125	1090	545	710	1145	175	1355	1160	2290	2820	300	145	810	2100	2870	2840	960	40
Billings, Mont.	970	1825	1735	2175	1215	1550	1345	565	1480	1220	1580	1400	1030	985	1425	1265	1500	1485	2485	825	1915	1570	1830	2025	1150	830	1195	895	2000	1960	1270	555	1200	830	2265	1885
Birmingham, Ala.	1285	150		1185	635	700	665	1280	725	1275	660	490	695	1845	385	2055	370	245	755	1030	1260	190	345	970	705	895	1665	2535	540	685	490	1765	2390	2640	535	750
Bismarck, N. Dak.	1115	1510	1455	1785	820	1165	1160	700	1090	1315	1405	1025	780	1375	1160	1625	1130	1230	2155	425	1490	1260	1580	1610	950	580	1485	1290	1625	1595	970	935	1615	1220	1995	1515
Boston, Mass.	2180	1065	1185		965	630	1780	1960	700	2410	1845	915	1395	2770	1450	3030	945	1310	1515	1400	325	1100	1535	215	1665	1430	2605	3095	695	540	1160	2365	3115	3000	1355	435
Buffalo, N.Y.	1750	855	885	445	530	185	1360	1520	255	1900	1445	490	975	2340	1025	2555	515	890	1425	940	375	695	1225	365	1235	990	2340	2635	605	475	730	1925	2675	2545	1260	370
Butte, Mont.	1025	2060	1970	2410	1450	1785	1565	790	1715	1285	1805	1635	1270	850	1660	1130	1735	1725	2720	1060	2150	1805	2055	2260	1375	1065	1070	675	2235	2200	1510	420	1035	610	2500	2120
Calgary, Alta.	1465	2290	2220	2560	1575	1935	1905	1120	1865	1750	2140	1790	1575	1285	1955	1600	1900	2020	2950	1190	2280	2035	2410	2410	1735	1380	1520	820	2390	2350	1790	870	1360	740	2745	2270
Charleston, S.C.	1690	305	450	940	905	700	1080	1700	850	1705	1070	710	1085	2280	815	2475	615	680	590	1300	1090	535	725	725	1145	1285	2080	2930	245	400	840	2190	2810	2925	445	505
Charleston, W. Va.	1535	500	585	765	475	250	1055	1360	360	1675	1140	295	760	2125	720	2340	250	585	1020	870	780	390	905	560	1015	940	2020	2480	330	305	515	1795	2575	2510	855	350
Cheyenne, Wyo.	520	1455	1350	1910	955	1285	870	95	1215	755	1110	1075	645	870	1035	1150	1165	1100	2100	800	1780	1200	1365	1760	700	495	895	1195	1705	1685	900	455	1210	1265	1880	1620
Chicago, Ill.	1265	675	635	965		345	940	995	270	1425	1080	180	540	1755	645	2070	310	545	1335	395	825	440	925	790	800	475	1690	2130	805	775	295	1400	2150	2035	1135	695
Cleveland, Ohio	1565	670	700	630	345		1175	1340	170	1715	1260	305	790	2185	840	2370	330	705	1270	740	560	510	1040	460	1050	810	2035	2450	580	445	545	1745	2495	2455	1105	360
Columbia, S.C.	1595	205	355	905	800	630	1010	1570	740	1630	1015	610	965	2240	720	2460	505	580	640	1195	1055	420	690	690	1120	1170	2020	2820	210	365	725	2075	2750	2865	475	470
Columbus, Ohio	1425	570	565	725	310	135	1040	1230	185	1580	1125	175	645	2025	705	2215	185	560	1190	395	700	365	895	510	905	745	1860	2305	500	475	415	1700	2450	2335	1020	415
Dallas, Tex.	650	805	665	1780	940	1175		780	1180	625	240	885	500	1230	325	1400	840	460	1310	955	1705	680	510	1565	210	665	1010	2010	1190	1280	645	1245	1770	2095	1080	1345
Denver, Colo.	430	1385	1280	1960	995	1340	780		1265	690	1025	1085	600	760	945	1090	1110	1035	2045	835	1825	1150	1280	1785	610	530	800	1270	1660	1665	845	505	1215	1320	1805	1650
Des Moines, Iowa	985	890	800	1280	345	655	690	670	585	1130	935	465	195	1505	550	1715	565	625	1550	250	1145	635	1010	1130	555	130	1440	1785	1095	1060	375	1065	1815	1780	1340	985
Detroit, Mich.	1535	715	725	700	270	170	1180	1265		1690	1270	280	740	2110	840	2340	355	715	1370	685	555	535	1065	620	1025	735	1960	2380	695	610	520	1670	2420	2285	1170	520
El Paso, Tex.	270	1425	1275	2410	1425	1715	625	690	1690		745	1410	925	695	950	790	1460	1090	1950	1375	2240	1300	1095	2120	680	990	400	1630	1810	1900	1170	860	1170	1710	1725	1965
Fargo, N. Dak.	1240	1325	1260	1595	640	970	1075	865	905	1385	1315	825	625	1535	1020	1815	935	1075	1985	235	1295	1070	1435	1450	865	430	1655	1465	1425	1385	790	1120	1815	1400	1780	1305
Glacier Nat. Pk., Mont.	1280	2220	2135	2495	1535	1870	1745	945	1800	1540	1960	1720	1415	1085	1805	1370	1830	1870	2875	1130	2155	1985	2210	2345	1535	1210	1325	645	2345	2315	1660	675	1155	560	2020	2205
Grand Canyon, Ariz.	405	1850	1660	2535	1675	1970	1045	705	1895	575	1235	1615	1130	290	1290	535	1640	1415	2355	1615	2500	1625	1555	2325	955	1245	225	1275	2190	2195	1375	510	835	1360	2125	2285
Houston, Tex.	830	810	660	1845	1080	1260	240	1025	1270	745		990	725	1425	435	1540	930	555	1205	1180	1820	800	365	1630	455	900	1150	2200	1200	1345	785	1435	1945	2285	975	1410
Indianapolis, Ind.	1260	515	490	915	180	305	885	1085	280	1410	990		480	1850	560	2065	115	435	1170	585	835	290	815	705	745	565	1685	2195	625	595	245	1500	2245	2305	970	565
Jackson, Miss.	1055	390	240	1425	740	890	415	1180	915	1035	420	630	635	1645	255	1815	560	205	895	1025	1450	380	190	1210	570	835	1425	2430	780	925	485	1660	2190	2510	665	990
Jacksonville, Fla.	1570	310	420	1170	1000	925	1005	1705	1025	1645	900	825	1120	2245	815	2405	710	670	335	1395	1320	545	560	955	1150	1295	2015	3020	475	630	850	2190	2780	2910	190	735
Kansas City, Mo.	775	785	695	1395	540	790	500	600	740	925	725	480		1365	380	1580	510	455	1440	455	1320	550	800	1185	340	200	1200	1870	1060	1065	245	1105	1815	1940	1220	1045
Knoxville, Tenn.	1395	190	260	925	550	510	860	1325	525	1475	920	350	725	1985	525	2200	245	385	850	935	1040	170	610	710	845	925	1820	2595	335	425	480	1830	2540	2575	650	490
Las Vegas, Nev.	590	2025	1845	2770	1755	2185	1230	760	2110	695	1425	1850	1365		1475	285	1875	1600	2630	1660	2650	1810	1745	2560	1145	1375	295	1005	2320	2430	1610	440	580	1175	2420	2415
Little Rock, Ark.	885	515	385	1450	645	840	325	945	840	950	435	560	380	1475		1720	515	135	1135	800	1400	345	450	1235	345	580	1310	2195	860	945	360	1425	2010	2325	910	1015
Los Angeles, Calif.	830	2205	2055	3030	2070	2370	1400	1090	2340	790	1540	2065	1580	285	1720		2095	1815	2745	1885	2895	2025	1890	2820	1385	1655	385	990	2535	2645	1820	705	390	1160	2510	2630
Louisville, Ky.	1285	395	370	945	310	350	840	1110	355	1460	930	115	510	1875	515	2095		370	1055	705	890	175	710	735	770	710	1710	2315	550	555	270	1620	2375	2420	855	580
Madison, Wis.	1260	825	780	1100	140	475	965	960	405	1405	1150	325	475	1785	715	2065	440	620	1490	265	965	575	1010	950	825	420	1695	1960	925	890	360	1370	2125	1895	1280	810
Memphis, Tenn.	1010	380	245	1310	545	705	460	1035	715	1090	555	435	455	1600	135	1815	370		1000	810	1265	210	395	1095	460	650	1435	2290	720	810	285	1520	2145	2400	770	875
Mexico City, Mex.	1510	1800	1650	2835	2100	2325	1180	1895	2305	1245	990	2030	1670	1930	1475	1955	1980	1615	2195	2120	2865	1825	1355	2620	1380	1825	1635	2930	2190	2405	1810	2105	2415	2935	1965	2475
Miami, Fla.	1950	655	755	1515	1335	1270	1310	2045	1370	1950	1205	1170	1440	2630	1135	2745	1055	1000		1730	1665	895	870	1300	1460	1645	2355	3315	820	975	1200	2555	3140	3400	255	1080
Milwaukee, Wis.	1345	760	715	1050	90	425	1020	1000	350	1505	1150	275	540	1835	720	2150	390	625	1415	310	915	525	1000	875	875	520	1805	2050	885	865	360	1455	2190	1935	1215	785
Minneapolis, Minn.	1175	1070	1030	1400	395	740	955	835	685	1375	1180	585	455	1660	800	1885	705	810	1730		1155	835	1210	1185	790	355	1630	1695	1200	1170	550	1220	1970	1625	1530	1090
Montréal, Qué.	2095	1215	1260	325	825	560	1705	1825	555	2240	1820	835	1320	2650	1400	2895	890	1265	1665	1155		1070	1600	370	1570	1285	2525	2760	845	690	1075	2235	2990	2700	1505	585
Nashville, Tenn.	1220	235	190	1100	440	510	680	1150	535	1300	800	290	550	1810	345	2025	175	210	895	835	1070		530	885	670	745	1645	2420	510	600	295	1660	2355	2490	695	665
New Orleans, La.	1150	490	345	1535	925	1040	510	1280	1065	1095	365	815	800	1745	450	1890	710	395	870	1210	1600	530		1320	680	1000	1500	2520	880	1025	665	1755	2295	2605	640	1100
New York, N.Y.	1970	850	970	215	790	460	1565	1785	620	2120	1630	705	1185	2560	1235	2820	735	1095	1300	1185	370	885	1320		1455	1255	2395	2795	480	330	945	2190	2940	2825	1140	230
Oklahoma City, Okla.	550	850	705	1665	800	1050	210	610	1025	680	455	745	340	1145	345	1385	770	460	1460	790	1570	670	680	1455		455	975	1860	1180	1270	500	1090	1690	1940	1230	1335
Omaha, Nebr.	845	985	895	1430	475	810	665	530	735	990	900	565	200	1375	580	1655	710	650	1645	355	1285	745	1000	1255	455		1265	1650	1260	1170	445	935	1680	1740	1470	1130
Orlando, Fla.	1710	440	530	1325	1130	1085	1075	1815	1140	1695	980	940	1230	2300	910	2470	830	775	230	1525	1475	690	640	1125	1240	1430	2075	3070	625	780	990	2315	2835	3095	85	890
Ottawa, Ont.	2005	1130	1205	435	740	500	1615	1735	470	2150	1740	750	1220	2560	1305	2805	840	1180	1635	1035	115	1015	1540	430	1485	1195	2440	2635	810	655	980	2145	2900	2575	1460	545
Philadelphia, Pa.	1895	770	890	305	745	410	1485	1720	570	2045	1550	635	1110	2485	1155	2700	660	1015	1220	1140	455	805	1240	90	1380	1210	2320	2855	400	245	875	2145	2895	2880	1060	140
Phoenix, Ariz.	425	1815	1665	2605	1690	2035	1010	800	1960	400	1150	1685	1200	295	1310	385	1710	1435	2355	1630	2525	1645	1500	2395	975	1265		1295	2220	2255	1445	635	795	1465	2125	2250
Pittsburgh, Pa.	1620	710	750	565	465	130	1220	1435	290	1760	1305	350	830	2200	900	2415	375	765	1200	860	590	550	1080	365	1085	930	2035	2565	440	320	590	1865	2615	2570	1040	230
Portland, Me.	2265	1170	1290	110	1045	700	1885	2040	775	1365	1950	1005	1480	2800	1545	3115	1030	1415	1620	1440	270	1205	1640	320	1750	1510	2690	3050	800	645	1245	2445	3195	3080	1460	540
Portland, Oreg.	1370	2670	2535	3095	2130	2450	2010	1270	2380	1630	2200	2195	1870	1005	2195	990	2315	2290	3315	1695	2760	2420	2520	2795	1860	1650	1295		3005	2780	2115	770	655	170	3140	2820
Québec, Qué.	2245	1165	1425	365	990	725	1855	1975	720	2390	1985	1000	1480	2800	1565	3045	1055	1430	1815	1305	165	1235	1765	520	1720	1435	2675	2910	995	840	1240	2385	3140	2850	1620	735
Raleigh, N.C.	1730	390	540	695	805	580	1190	1660	695	1810	1200	625	1060	2320	860	2535	550	720	820	1200	845	510	880	480	1180	1260	2220	3005		155	815	2170	2875	3020	655	260
Rapid City, S. Dak.	805	1505	1415	1855	885	1230	1080	400	1160	1050	1320	1080	705	1085	1085	1380	1180	1150	2165	565	1710	1250	1535	1705	900	510	1235	1225	1680	1645	955	675	1420	1155	1945	1565
Reno, Nev.	1035	2415	2290	2895	1930	2270	1675	1015	2205	1135	1865	2060	1615	445	1920	475	2150	2060	2970	1760	2765	2185	2185	2750	1585	1460	735	560	2690	2670	1860	525	220	735	2745	2605
Richmond, Va.	1820	535	685	540	775	445	1280	1665	610	1900	1345	595	1065	2430	945	2645	555	810	975	1170	690	600	1025	330	1270	1170	2255	2780	155		820	2175	2880	2810	815	105
St. Louis, Mo.	1020	530	490	1160	295	545	645	845	520	1170	785	245	245	1610	360	1820	270	285	1200	550	1075	295	665	945	500	445	1445	2115	815	820		1355	2060	2185	1025	805
Salt Lake City, Utah	600	1900	1765	2365	1400	1745	1245	505	1670	860	1435	1500	1105	440	1425	705	1620	1520	2555	1220	2235	1660	1755	2190	1090	935	635	770	2170	2175	1355		750	830	2290	2095
San Antonio, Tex.	725	1010	860	2045	1210	1445	270	905	1455	570	200	1155	770	1240	600	1375	1115	740	1405	1225	1955	950	560	1835	470	935	975	2095	1400	1545	915	1325	1745	2290	1175	1610
San Diego, Calif.	780	2160	2010	2965	2045	2350	1355	1225	2325	735	1480	2045	1555	335	1640	120	2150	1820	2685	1990	2875	2030	1830	2755	1335	1625	360	1100	2540	2630	1805	775	520	1275	2455	2610
San Francisco, Calif.	1130	2525	2390	3115	2150	2495	1770	1215	2420	1170	1945	2245	1815	580	2010	390	2375	2145	3140	1970	2990	2355	2295	2940	1690	1680	795	655	2875	2880	2060	750		825	2855	2865
Seattle, Wash.	1450	2725	2640	3000	2035	2455	2095	1320	2285	1710	2285	2305	1940	1175	2325	1160	2420	2400	3400	1625	2700	2490	2605	2825	1940	1740	1465	170	3020	2810	2185	830	825		3165	2840
Sioux Falls, S. Dak.	995	1165	1075	1505	550	880	825	635	810	1145	1065	735	380	1415	775	1695	840	835	1825	235	1375	910	1190	1355	615	185	1420	1565	1335	1295	610	1000	1755	1500	1615	1215
Spokane, Wash.	1325	2370	2280	2705	1815	2080	1875	1095	2015	1620	2115	1935	1585	1070	1965	1235	2045	2030	3035	1345	2415	2120	2385	2560	1705	1390	1370	360	2535	2495	1830	720	900	275	2815	2415
Tampa, Fla.	1720	455	535	1355	1135	1105	1080	1805	1170	1725	975	970	1220	2420	910	2510	855	770	255	1530	1505	695	640	1140	1230	1470	2125	3140	655	815	1025	2290	2855	3165		920
Toronto, Ont.	1755	945	940	550	495	285	1365	1485	225	1905	1530	515	1005	2310	1085	2560	580	950	1515	890	335	770	1300	455	1235	945	2190	2575	705	555	760	1900	2655	2515	1330	465
Vancouver, B.C.	1605	2785	2695	3120	2230	2495	2245	1455	2425	1860	2415	2345	2000	1295	2380	1300	2460	2445	3445	1760	2935	2530	2755	2970	2085	1805	1610	315	2945	2905	2245	970	970	140	3225	2830
Washington, D.C.	1825	630	750	445	695	360	1345	1650	520	1965	1410	565	1045	2415	1015	2630	580	875	1080	1090	585	665	1100	230	1335	1130	2250	2820	260	105	805	2095	2865	2840	920	
Winnipeg, Man.	1445	1520	1455	1765	830	1165	1310	1075	1080	1690	1550	1020	855	1710	1235	2045	1130	1300	2180	455	1450	1265	1685	1645	1120	660	1910	1490	1620	1580	1015	1350	1990	1415	1975	1500
Yellowstone N. P., Wyo.	945	1890	1800	2365	1345	1690	1310	530	1615	1205	1565	1470	1105	780	1475	1050	1615	1555	2560	1000	2190	1655	1820	2135	1140	905	980	825	2150	2120	1350	330	980	760	2325	2040
Yosemite N. P., Calif.	945	2305	2200	2750	1915	2260	1585	990	2185	1050	1780	2005	1520	355	1830	270	2005	1955	2955	1750	2775	2165	2145	2670	1500	1450	650	795	2580	2585	1765	540	200	965	2730	2595

ONE CENTIMETER EQUALS ABOUT 220 KILOMETERS

ONE INCH EQUALS ABOUT 347 MILES

© GENERAL DRAFTING CO., INC.

GREENLAND (Den.)

CANADA

UNITED STATES

U.S. (ALASKA)

MEXICO

GUATEMALA
BELIZE
HONDURAS
EL SALVADOR
NICARAGUA
COSTA RICA
PANAMA

CUBA
JAMAICA
THE BAHAMAS
HAITI
DOM. REP.
PUERTO RICO

BARBADOS
TRINIDAD & TOBAGO
GRENADA
ST. VINCENT
ST. LUCIA
DOMINICA

ATLANTIC OCEAN

CANADIAN SHIELD

NEWFOUNDLAND

Gulf of St. Lawrence

L. Mattagami
L. Mistassini

INTERIOR LOWLANDS

GREAT LAKES

L. Superior
L. Michigan
L. Huron
L. Erie
L. Ontario

Winnipeg
Lake of the Woods
Thunder Bay
Duluth
Minneapolis
Des Moines
St. Louis
Kansas City
Omaha
Wichita
Little Rock
Oklahoma City
Dallas
Shreveport

GREAT PLAINS

COLORADO PLATEAU
GREAT BASIN
SIERRA NEVADA

Mt. Whitney

Death Valley

Los Angeles
San Diego

BAJA CALIFORNIA

TROPIC OF CANCER

SA. MADRE OCCIDENTAL
SA. MADRE ORIENTAL

LLANO ESTACADO

El Paso
Chihuahua
Torreón
Monterrey
Tampico

GULF OF MEXICO

Bahía de Campeche

ISTMO DE TEHUANTEPEC
YUCATAN PEN.
Mérida

BAHAMAS
Nassau

Straits of Florida
CUBA
Habana
Los Pinos

CAYMAN BASIN

Yucatán Channel

Gulf of Honduras
Belize

MIDDLE AMERICA TRENCH

GUATEMALA TRENCH

ISTMO DE PANAMA

PACIFIC OCEAN

HISPANIOLA
GREATER ANTILLES
Jamaica
Kingston
PUERTO RICO TRENCH
Santo Domingo
Puerto Rico
San Juan

Leeward Islands
Windward Is.
LESSER ANTILLES

VENEZUELAN BASIN

CARIBBEAN SEA

Barranquilla

TROPIC OF CANCER

APPALACHIAN

COASTAL PLAIN

St. John's
C. Race
Sydney
C. Breton I.
Halifax
C. Sable
Prince Edward I.
Sable I.
Fredericton
Portland
C. Cod
Boston
Hartford
New York
Philadelphia
Washington
Norfolk
C. Hatteras
Charleston
Savannah
Jacksonville
Tampa
L. Okeechobee
Miami
New Orleans

Bermuda

Mt. Washington

Quebec
Montreal
Ottawa
Burlington
Syracuse
Buffalo
Toronto
Rochester
Pittsburgh
Cleveland
Detroit
Cincinnati
Indianapolis
Charleston
Louisville
Nashville
Memphis
Birmingham
Atlanta
Charlotte
Jackson

Mt. Mitchell

Sault Ste. Marie
Timmins
Thunder Bay
L. Nipigon
Chicago
Milwaukee

Mississippi
Missouri
Platte
Arkansas
Red
Canadian
Brazos
Colorado
Rio Grande
Pecos
Gila

Ohio
Tennessee

Denver
Pikes Peak
Cheyenne
Rapid City
BLACK HILLS

Amarillo
Albuquerque
Santa Fe
Phoenix
Tucson

Salt Lake City
Great Salt Lake

ROCKY MOUNTAINS

Colorado R.
Grand Canyon

Gila R.

San Antonio
Houston
Acapulco
Guadalajara
Mazatlán
Culiacán
La Paz

Golfo de California

SA. MADRE DEL SUR

Managua
Tegucigalpa
San Salvador
Guatemala
Puerto Cabezas
Co. Chirripó

LAND HEIGHTS

meters	feet
4000	13123
2000	6562
1000	3281
200	656
land below sea level	
200	656
3000	9843
6000	19685

OCEAN DEPTHS

11

U.S. ROAD MAP

United States
EXCEPT ALASKA AND HAWAII

Interstate Highway System and Other Superhighways
— COMPLETED
===== PROPOSED OR UNDER CONST.

Other Principal Highways

95 **Interstate Route Numbers**

11 6 **U.S. Route Numbers**

15 17 **State Route Numbers**

Approximate populations of cities and towns
○ Under 1,000
◉ 25,000 – 100,000
⊕ 1,000 – 25,000
◉ Over 100,000

CAPITAL CITIES ARE INDICATED BY CAPITAL LETTERS

ONE CENTIMETER EQUALS ABOUT 112 KILOMETERS
0 100 200 300 400 500 Km.
0 50 100 150 200 250 300 Mi.
ONE INCH EQUALS ABOUT 177 MILES
© GENERAL DRAFTING CO., INC.

MASSACHUSETTS—NEW HAMPSHIRE—
CONNECTICUT—RHODE ISLAND—VERMONT

FOR LEGEND SEE PAGE 5

ONE CENTIMETER EQUALS ABOUT 18.2 KILOMETERS

ONE INCH EQUALS ABOUT 28.7 MILES

© GENERAL DRAFTING CO., INC.

14

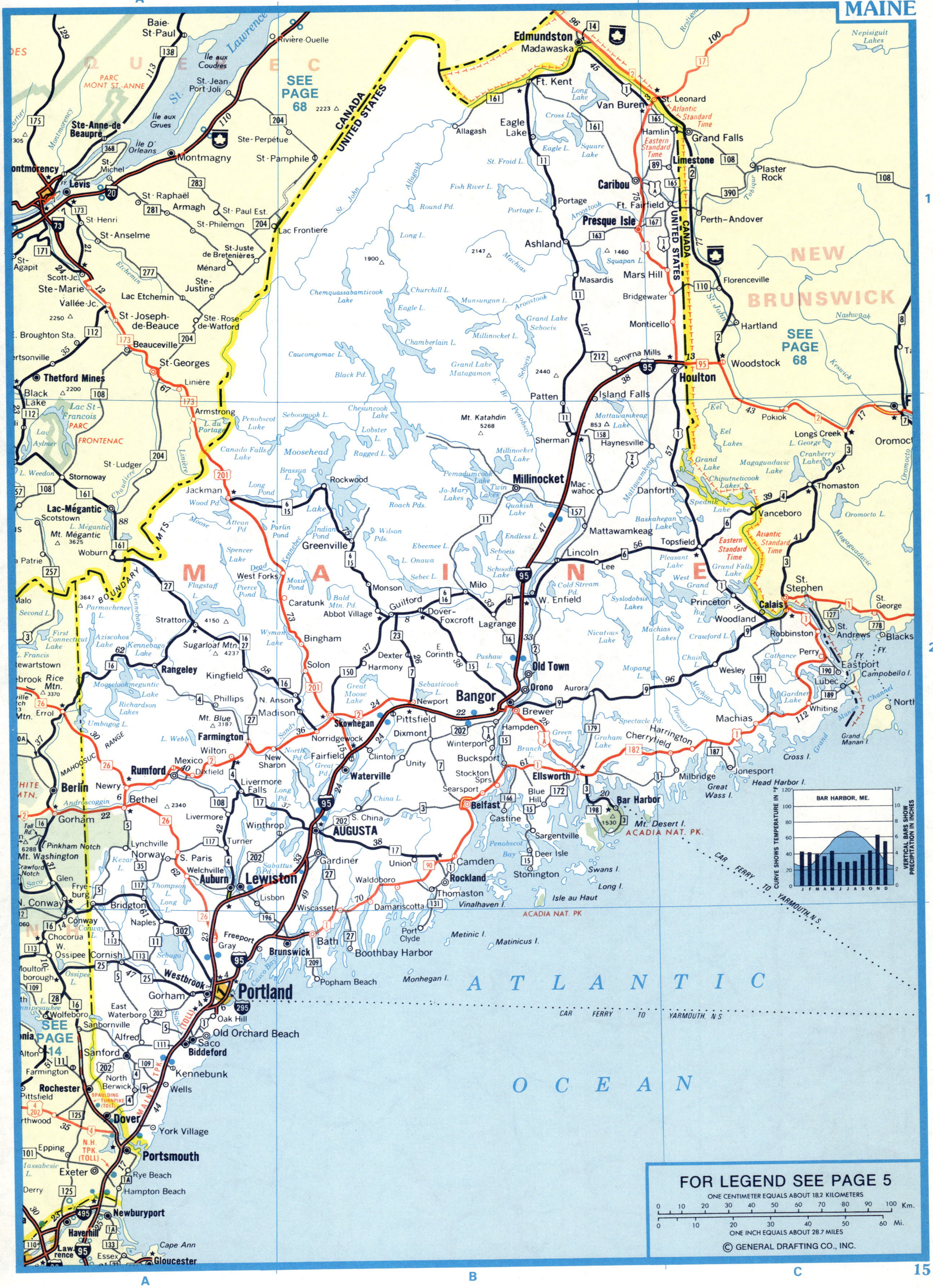

QUEBEC

NEW BRUNSWICK

MAINE

ATLANTIC OCEAN

SEE PAGE 68

SEE PAGE 14

Major cities and towns:
Edmundston, Madawaska, Ft. Kent, Van Buren, St. Leonard, Grand Falls, Caribou, Ft. Fairfield, Limestone, Plaster Rock, Presque Isle, Perth-Andover, Ashland, Mars Hill, Florenceville, Bridgewater, Monticello, Hartland, Woodstock, Houlton, Smyrna Mills, Island Falls, Patten, Haynesville, Sherman, Millinocket, Macwahoc, Danforth, Vanceboro, Mattawamkeag, Lincoln, Lee, Topsfield, Princeton, Woodland, Calais, St. Stephen, Robbinston, St. Andrews, Greenville, Monson, Milo, Dover-Foxcroft, Lagrange, W. Enfield, Old Town, Orono, Bangor, Brewer, Machias, Eastport, Lubec, Cherryfield, Harrington, Milbridge, Ellsworth, Bar Harbor, Acadia Nat. Pk., Belfast, Camden, Rockland, Augusta, Waterville, Skowhegan, Farmington, Rumford, Bethel, Lewiston, Auburn, Brunswick, Bath, Boothbay Harbor, Portland, Westbrook, Gorham, Saco, Biddeford, Old Orchard Beach, Kennebunk, Wells, Sanford, York Village, Portsmouth, Dover, Rochester, Newburyport, Haverhill, Gloucester

Ste-Anne-de-Beaupré, Montmorency, Lévis, Montmagny, Thetford Mines, St-Georges, Beauceville, Lac-Mégantic, Jackman, Rockwood, Stratton, Rangeley, Kingfield, Phillips, Bingham, Solon, Madison, N. Anson, Norridgewock, Pittsfield, Newport, Dexter, Harmony, E. Corinth, Guilford, Abbot Village, Caratunk, West Forks, Berlin, Gorham, Mt. Washington, N. Conway, Conway, Fryeburg, Bridgton, Naples, Sanford

Mt. Katahdin 5268

Moosehead Lake, Chesuncook Lake, Chamberlain L., Grand Lake, Eagle Lake, Long L., Square Lake, Cross L., Churchill L., Flagstaff L., Sebago L., Penobscot Bay, Casco Bay, Mt. Desert I.

Atlantic Standard Time / Eastern Standard Time

CANADA / UNITED STATES

BAR HARBOR, ME.
CURVE SHOWS TEMPERATURE IN °F
VERTICAL BARS SHOW PRECIPITATION IN INCHES
J F M A M J J A S O N D

CAR FERRY TO YARMOUTH, N.S.

FOR LEGEND SEE PAGE 5

ONE CENTIMETER EQUALS ABOUT 18.2 KILOMETERS
0 10 20 30 40 50 60 70 80 90 100 Km.
0 10 20 30 40 50 60 Mi.
ONE INCH EQUALS ABOUT 28.7 MILES

© GENERAL DRAFTING CO., INC.

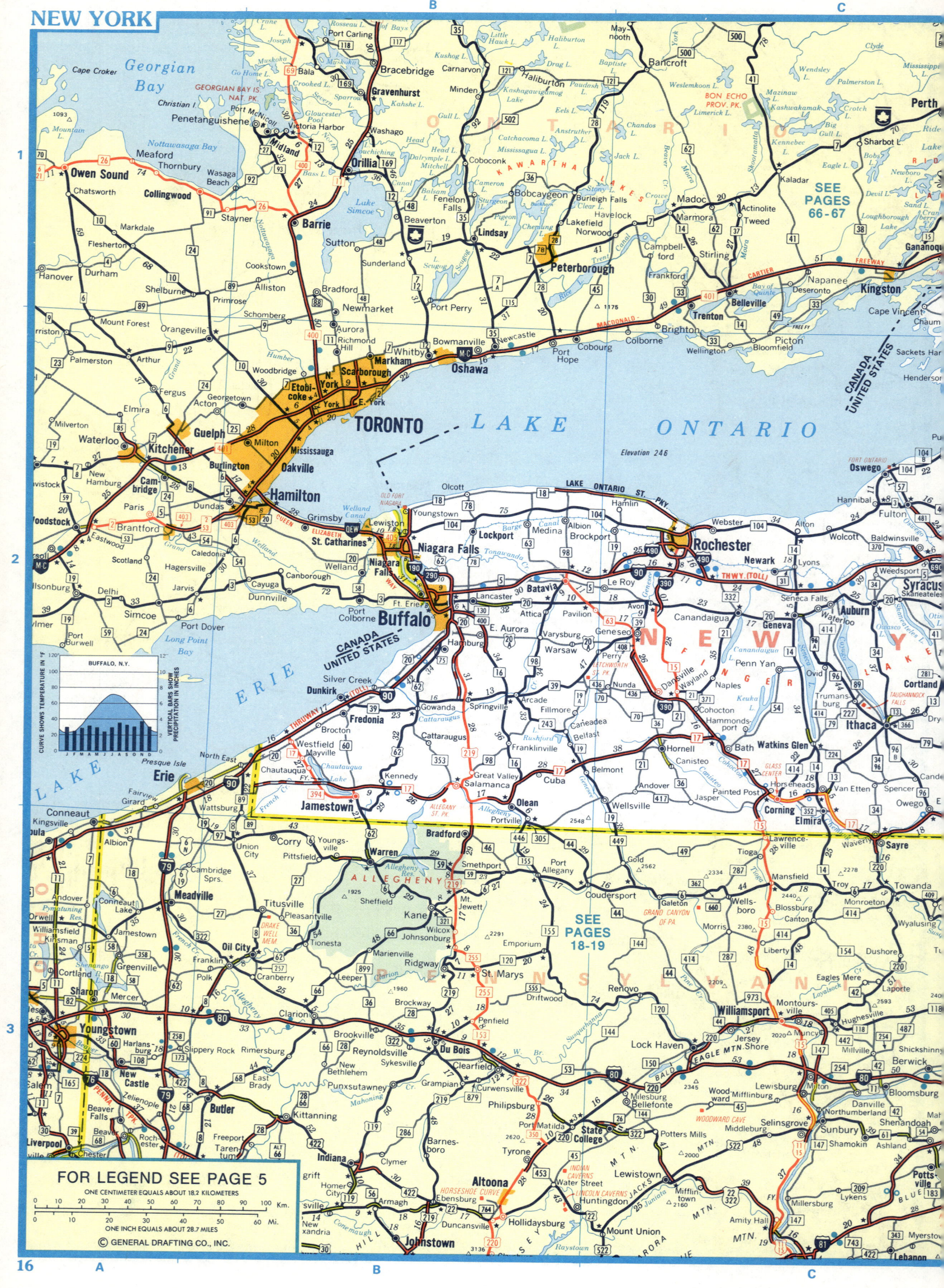

FOR LEGEND SEE PAGE 5

ONE CENTIMETER EQUALS ABOUT 18.2 KILOMETERS

ONE INCH EQUALS ABOUT 28.7 MILES

© GENERAL DRAFTING CO., INC.

SEE PAGE 67

CANADA
UNITED STATES

QUEBEC

VERMONT

NEW HAMPSHIRE

MASSACHUSETTS

CONNECTICUT

RHODE ISLAND

NEW YORK

ADIRONDACK PARK

ADIRONDACK MTS.

CATSKILL MTS.

CATSKILL PARK

GREEN MTS.

WHITE MTN.

NEW JERSEY

PENNSYLVANIA

ATLANTIC OCEAN

Long Island Sound

LONG ISLAND

FIRE ISLAND NAT. SEASHORE

MONTPELIER
CONCORD
ALBANY
HARTFORD
PROVIDENCE
New York
Burlington
Watertown
Utica
Rome
Schenectady
Troy
Saratoga Sprs.
Glens Falls
Plattsburgh
Malone
Massena
Ogdensburg
Potsdam
Canton
Saranac Lake
Lake Placid
Tupper Lake
Rutland
Pittsfield
Worcester
Springfield
Manchester
Nashua
Keene
Bennington
Brattleboro
Poughkeepsie
Kingston
Newburgh
Middletown
Port Jervis
Scranton
Allentown
Reading
Bethlehem
Easton
New Brunswick
Newark
Jersey City
Elizabeth
Yonkers
White Plains
Bridgeport
New Haven
Danbury
Waterbury
Meriden
New Britain
Middletown
New London
Norwich
Pawtucket
Woonsocket
Bristol
Stamford
Peekskill
Ossining
Hempstead
Freeport
Bay Shore
Patchogue
Riverhead
Southampton
East Hampton
Montauk
Montauk Pt.

SEE PAGE 14
SEE PAGE 19
SEE PAGE 14

LAKE PLACID, N.Y.
CURVE SHOWS TEMPERATURE IN °F
VERTICAL BARS SHOW PRECIPITATION IN INCHES
J F M A M J J A S O N D

NEW YORK, N.Y.
CURVE SHOWS TEMPERATURE IN °F
VERTICAL BARS SHOW PRECIPITATION IN INCHES
J F M A M J J A S O N D

17

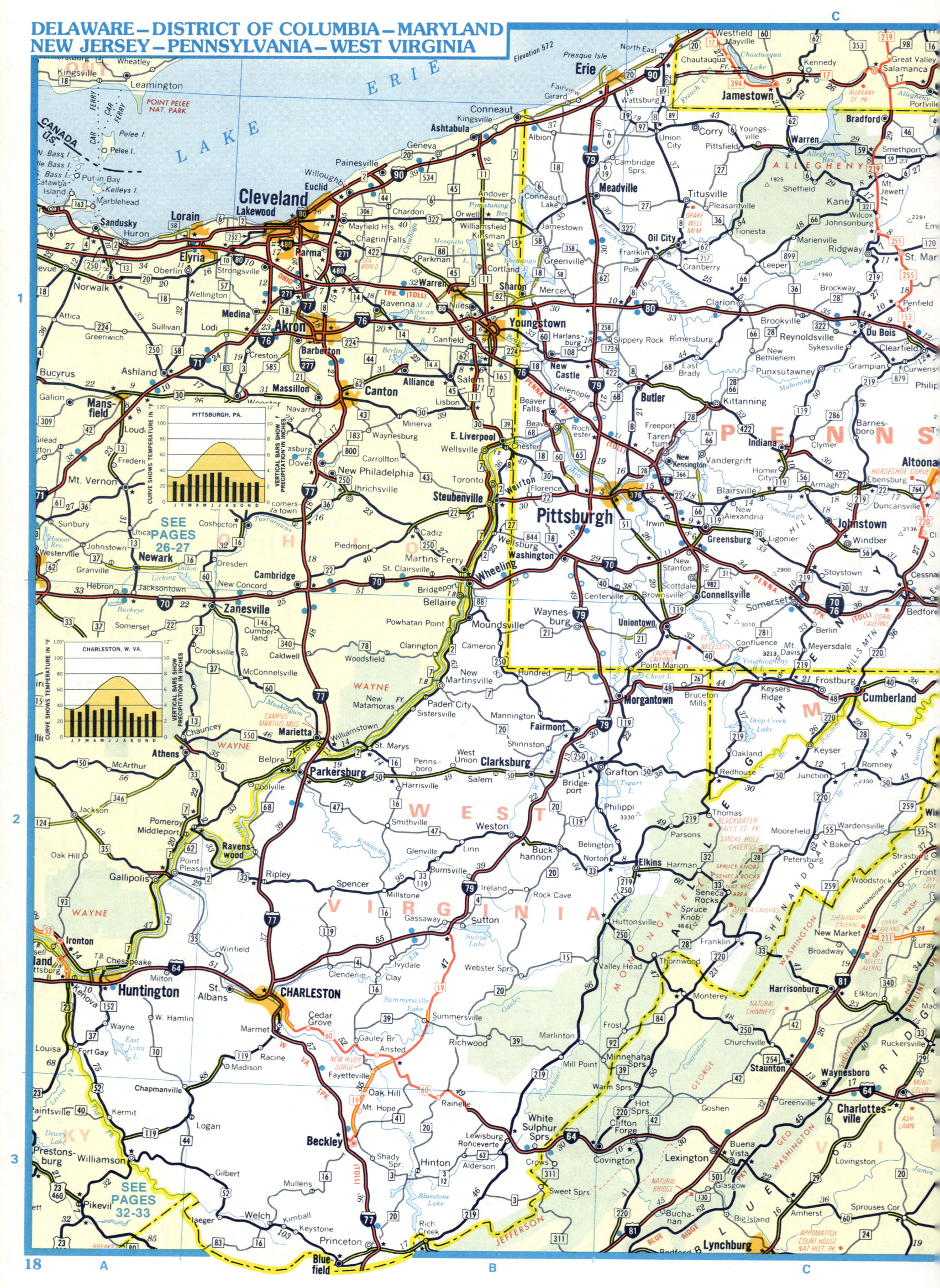

LAKE ERIE

CANADA
U.S.

Erie

Cleveland
Lakewood

Pittsburgh

PITTSBURGH, PA.

SEE
PAGES
26-27

CHARLESTON, W. VA.

Wheeling

Parkersburg

W E S T

V I R G I N I A

Huntington

CHARLESTON

Beckley

SEE
PAGES
32-33

Blue-
field

18   A   B   C

FOR LEGEND SEE PAGE 5

ONE CENTIMETER EQUALS ABOUT 18.2 KILOMETERS

ONE INCH EQUALS ABOUT 28.7 MILES

© GENERAL DRAFTING CO., INC.

19

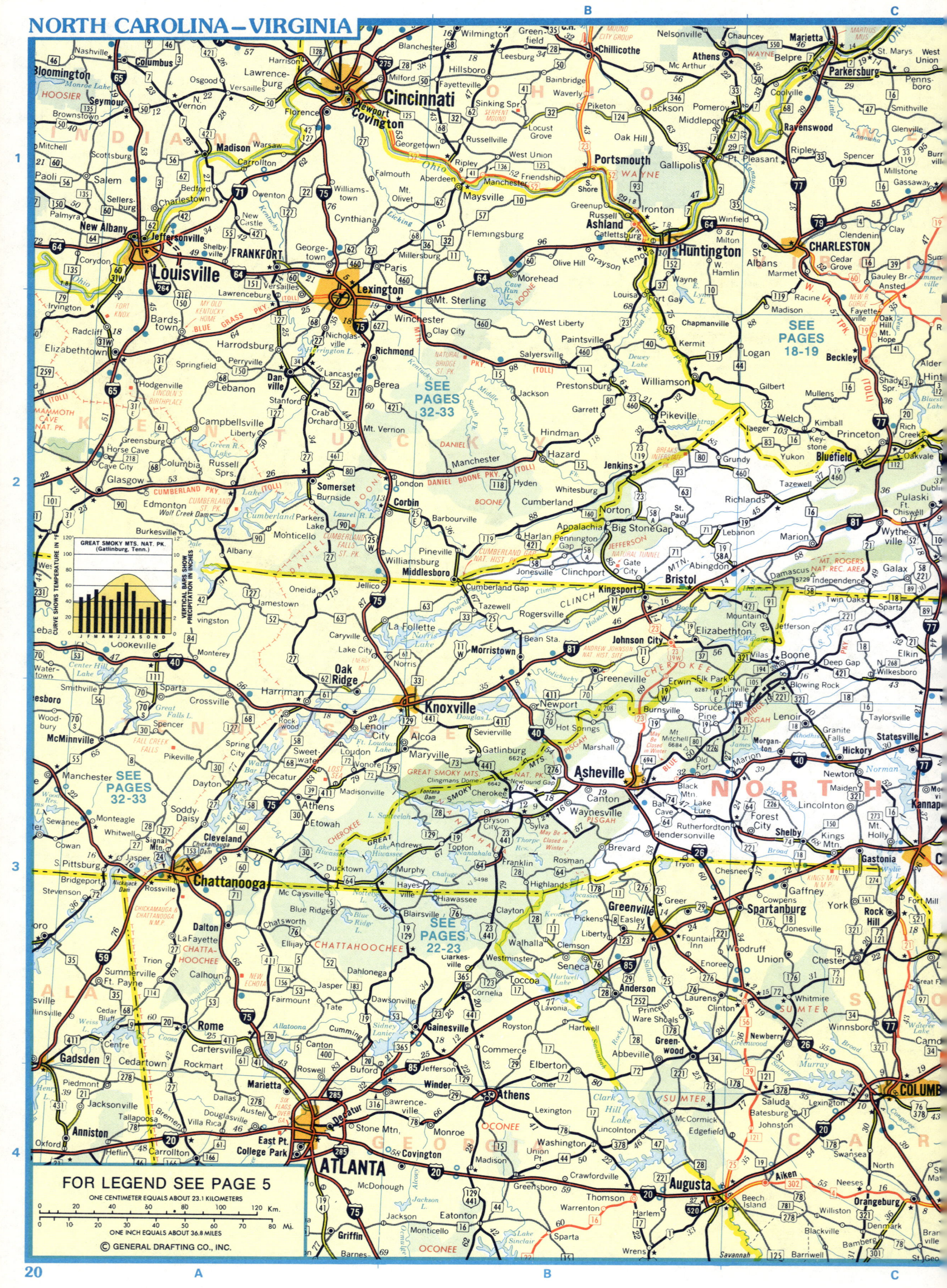

GREAT SMOKY MTS. NAT. PK.
(Gatlinburg, Tenn.)

SEE PAGES 32-33

SEE PAGES 18-19

SEE PAGES 32-33

SEE PAGES 22-23

FOR LEGEND SEE PAGE 5

ONE CENTIMETER EQUALS ABOUT 23.1 KILOMETERS
ONE INCH EQUALS ABOUT 36.8 MILES
© GENERAL DRAFTING CO., INC.

Major cities and features (partial):

Baltimore, WASHINGTON, ANNAPOLIS, DOVER, Frederick, Winchester, Harrisonburg, Staunton, Charlottesville, Richmond, Lynchburg, Roanoke, Danville, Petersburg, Williamsburg, Newport News, Hampton, Norfolk, Portsmouth, Virginia Beach, Chesapeake, Suffolk

RALEIGH, Durham, Greensboro, High Point, Winston-Salem, Burlington, Chapel Hill, Fayetteville, Goldsboro, Kinston, New Bern, Morehead City, Jacksonville, Wilmington, Rocky Mount, Wilson, Greenville, Washington, Kitty Hawk, Nags Head, Manteo, Cape Hatteras, Ocracoke

SEE PAGES 18-19

SEE PAGE 23

VIRGINIA, NORTH CAROLINA, SOUTH CAROLINA, MARYLAND, DEL., W. VA.

ATLANTIC OCEAN, Chesapeake Bay, Albemarle Sound, Pamlico Sound

Climate charts:

NORFOLK, VA.
Curve shows temperature in °F — Vertical bars show precipitation in inches
J F M A M J J A S O N D

WILMINGTON, N.C.
Curve shows temperature in °F — Vertical bars show precipitation in inches
J F M A M J J A S O N D

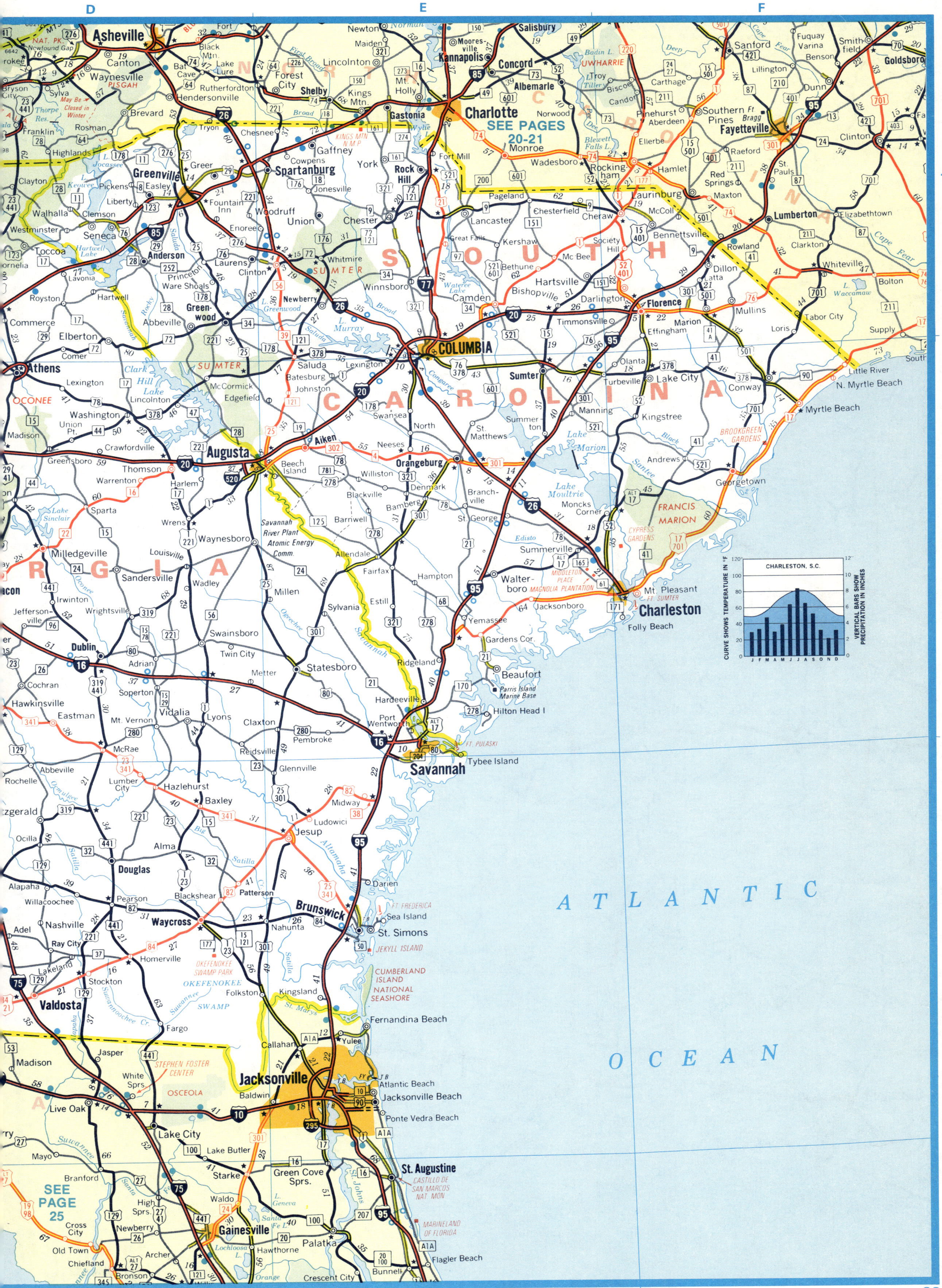

SOUTH CAROLINA

NORTH CAROLINA

GEORGIA

Asheville, Canton, Waynesville, PISGAH, Brevard, Rosman, Highlands, Franklin, Bryson City, Sylva, Cherokee, NAT. PK., Newfound Gap, Thorpe Res., May Be Closed in Winter, Black Mtn., Bat Cave, Lake Lure, Forest City, Rutherfordton, Hendersonville, Tryon, Chesnee, Gaffney, Cowpens, Greer, Spartanburg, Jonesville, Shelby, Kings Mtn., Gastonia, Mt. Holly, Lincolnton, Maiden, Newton, Salisbury, Kannapolis, Concord, Albemarle, Charlotte

SEE PAGES 20-21

Monroe, Fort Mill, Rock Hill, York, Chester, Great Falls, Lancaster, Pageland, Chesterfield, Cheraw, Wadesboro, Rockingham, Hamlet, Laurinburg, McColl, Bennettsville, Rowland, Dillon, Latta, Marion, Mullins, Loris, Conway, Little River, N. Myrtle Beach, Myrtle Beach

Salisbury, Lexington, Troy, Carthage, Pinehurst, Aberdeen, Southern Pines, Ft. Bragg, Fayetteville, Raeford, Red Springs, Maxton, Lumberton, Elizabethtown, Whiteville, Bolton, Supply, Tabor City, Cape Fear

Sanford, Lillington, Dunn, Clinton, Goldsboro, Smithfield, Benson, Fuquay Varina

COLUMBIA, Saluda, Batesburg, Lexington, Edgefield, Johnston, Aiken, Beech Island, Williston, Blackville, Denmark, Bamberg, Barnwell, Allendale, Fairfax, Estill, Hampton, Walterboro, Yemassee, Ridgeland, Hardeeville, Beaufort, Parris Island Marine Base, Hilton Head I.

Winnsboro, Camden, Bishopville, Hartsville, Darlington, Florence, Timmonsville, Effingham, Olanta, Turbeville, Lake City, Manning, Kingstree, Andrews, Georgetown, Sumter, St. Matthews, Orangeburg, Branchville, St. George, Summerville, Moncks Corner, Lake Marion, Lake Moultrie, FRANCIS MARION, Mt. Pleasant, Charleston, Folly Beach

BROOKGREEN GARDENS, CYPRESS GARDENS, MIDDLETON PLACE, MAGNOLIA PLANTATION

Athens, Lexington, Washington, Union Pt., Crawfordville, Sparta, Milledgeville, Sandersville, Wrightsville, Dublin, Swainsboro, Metter, Statesboro, Sylvania, Millen, Waynesboro, Wrens, Thomson, Harlem, Augusta, Warrenton, Greensboro, Macon, Jeffersonville, Irwinton, Eastman, McRae, Vidalia, Lyons, Claxton, Glennville, Reidsville, Hazlehurst, Baxley, Jesup, Ludowici, Midway, Savannah, Tybee Island, FT. PULASKI, Port Wentworth, Pembroke

Hawkinsville, Cochran, Adrian, Soperton, Mt. Vernon, Lumber City, Rochelle, Abbeville, Fitzgerald, Ocilla, Douglas, Alma, Blackshear, Waycross, Patterson, Brunswick, Nahunta, St. Simons, Sea Island, FT. FREDERICA, JEKYLL ISLAND, Darien, Sapelo Island

Alapaha, Willacoochee, Nashville, Adel, Pearson, Homerville, OKEFENOKEE SWAMP PARK, OKEFENOKEE SWAMP, STEPHEN FOSTER CENTER, Fargo, Valdosta, Ray City, Lakeland, Stockton, Jasper, Madison, Mayo, Branford, White Sprs., Live Oak, Lake City, Lake Butler, Lake Geneva, OSCEOLA

CUMBERLAND ISLAND NATIONAL SEASHORE, Kingsland, Folkston, Callahan, Fernandina Beach, Yulee, Jacksonville, Baldwin, Atlantic Beach, Jacksonville Beach, Ponte Vedra Beach, St. Augustine, CASTILLO DE SAN MARCOS NAT. MON., Green Cove Sprs., Starke, Waldo, Gainesville, Newberry, High Sprs., Lake Butler, Palatka, Hawthorne, Crescent City, MARINELAND OF FLORIDA, Flagler Beach, Bunnell

SEE PAGE 25

ATLANTIC OCEAN

CHARLESTON, S.C.
CURVE SHOWS TEMPERATURE IN °F
VERTICAL BARS SHOW PRECIPITATION IN INCHES
J F M A M J J A S O N D

FOR LEGEND SEE PAGE 5

ONE CENTIMETER EQUALS ABOUT 30.7 KILOMETERS

ONE INCH EQUALS ABOUT 48.7 MILES

© GENERAL DRAFTING CO., INC.

LITTLE ROCK, ARK.

JACKSON, MISS.

NEW ORLEANS, LA.

CURVE SHOWS TEMPERATURE IN °F — VERTICAL BARS SHOW PRECIPITATION IN INCHES

SEE PAGE 37

SEE PAGES 32-33

SEE PAGES 38-39

SEE PAGE 22

SEE PAGES 40-41

MISSOURI

TENNESSEE

ARKANSAS

MISSISSIPPI

LOUISIANA

ALABAMA

TEXAS

OZARK MOUNTAINS

MARK TWAIN NAT. FOR.

OUACHITA MTS.

DELTA

KISATCHIE

GULF OF MEXICO

LITTLE ROCK · Hot Springs · North Little Rock · Memphis · Fort Smith · Pine Bluff · Texarkana · El Dorado · Camden · Shreveport · Monroe · Alexandria · Natchez · Vicksburg · JACKSON · Meridian · Hattiesburg · Laurel · Baton Rouge · New Orleans · Lafayette · Lake Charles · Beaumont · Port Arthur · Houma · Biloxi · Gulfport · Mobile · Greenville · Greenwood · Tupelo · Columbus · Oxford · Clarksdale · Jonesboro · Fayetteville · Harrison · Mountain Home · Conway · Searcy

24

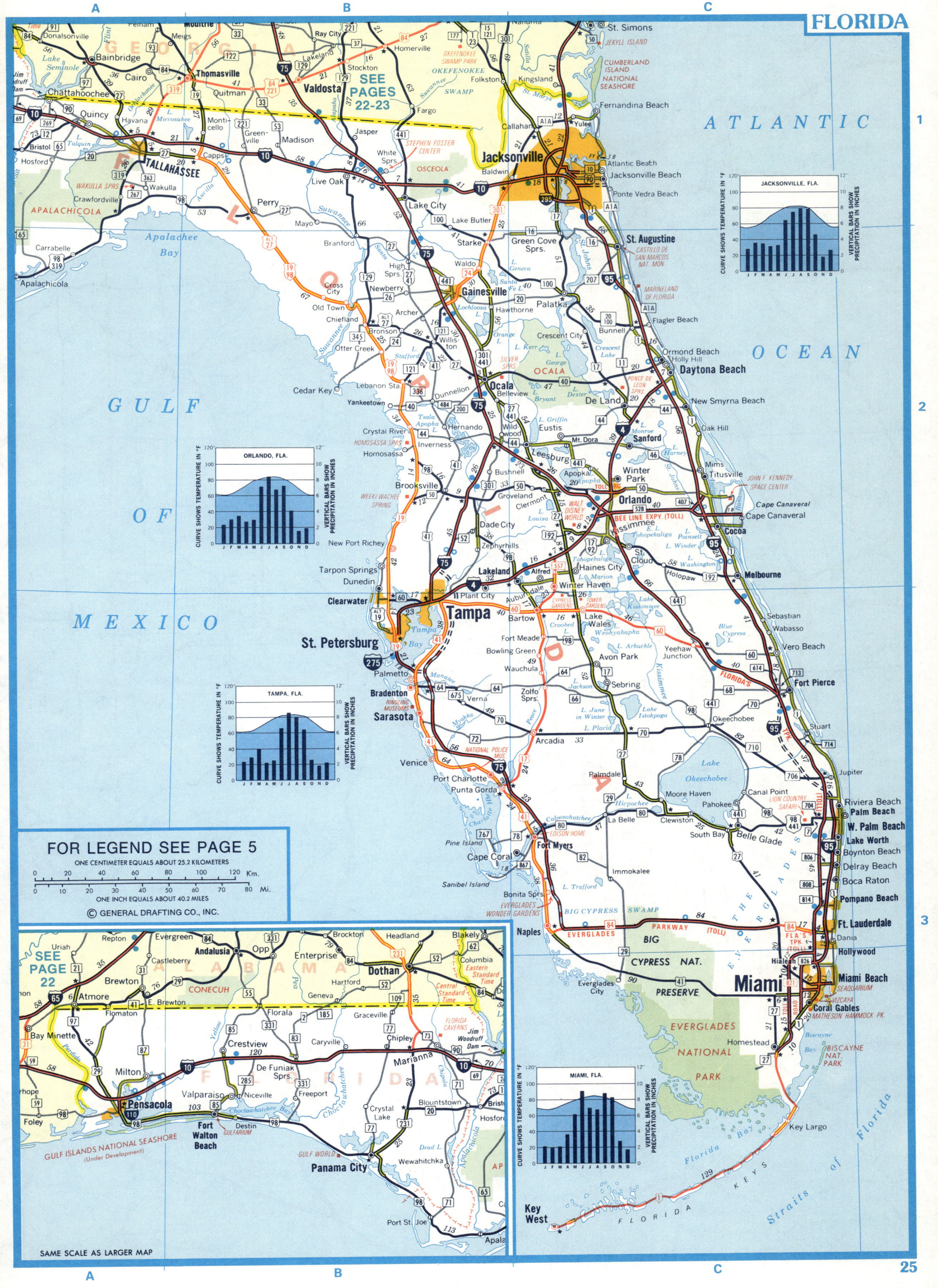

FOR LEGEND SEE PAGE 5

ONE CENTIMETER EQUALS ABOUT 18.3 KILOMETERS

0 10 20 30 40 50 60 70 80 90 100 Km.

ONE INCH EQUALS ABOUT 28.7 MILES

0 10 20 30 40 50 60 Mi.

© GENERAL DRAFTING CO., INC.

MICHIGAN

ONE CENTIMETER EQUALS ABOUT 30.7 KILOMETERS
ONE INCH EQUALS ABOUT 48.7 MILES

LAKE SUPERIOR

CANADA
UNITED STATES

Isle Royale Nat. Park
Grand Portage
Marais
SUMMER PASS. FY.
SUMMER PASS FERRY
Central Standard Time
Eastern Standard Time
Apostle Islands
Apostle Is. Nat. Lakeshore (Devel.)
Copper Harbor
Calumet
Laurium
Hancock
Houghton
Ontonagon
Baraga
Mass
L'Anse
Michigamme
Marquette
Negaunee
Ishpeming
Munising
Paradise
Newberry
Sault
HIAWATHA
Michipicoten Island
Elevation 602
Grand I.
PICTURED ROCKS NAT. LAKESHORE (Under Devel.)
Wakefield
Bessemer
Hurley
Ironwood
Montreal
OTTAWA
Watersmeet
Bergland
Bruce Crossing
Covington
Republic
Amasa
Crystal Falls
Sagola
Trenary
Rapid River
Indian L.
St. Ignace
Manistique
Straits of Mackinac
Park Falls
Iron River
Iron Mtn.
Norway
Gladstone
Escanaba
Powers
Rhinelander
Monico
Florence
Pembine
Niagara
Cedar River
Gills Rock
Sister Bay
Menominee
Chambers
Washington I.
N. Manitou I.
S. Fox I.
N. Fox I.
Beaver I.
Petoskey
Charlevoix
Northport
Eastern Standard Time
Central Standard Time
Merrill
Antigo
Mountain
Pound
Stephenson
Wausaukee

SEE PAGE 30

SEE PAGE 31

Seney
McMillan
Newberry
Eckerman
Germfask
Blaney Park
Naubinway
Trout Lake
Manistique
Epoufette
Brevort
HIAWATHA
Straits of Mackinac
Garden I.
Hog I.
High I.
St. James
Beaver I.
Cross Village
Harbor Sprs.
Charlevoix
Petoskey
Boyne City
East Jordan
Alanson
Walloon L.
N. Manitou I.
S. Manitou I.
Northport
Leland
Suttons Bay
Elk Rapids
Eastport
Mancelona
Empire
Glen L.
Kalkaska
Grayling
Frankfort
Interlochen
Traverse City
Elberta
Benzonia
Manton
Bear Lake
Mesick
Onekama
Manistee
Cadillac
Lake City
McBain
Marion
Harrison
Scottville
Baldwin
Reed City
Ludington
Pentwater
Hart
Shelby
Big Rapids
Mt. Pleasant
Hesperia
Remus
White Cloud
Fremont
Six Lakes
Whitehall
Newaygo
Howard City
Stanton
Muskegon
Kent City
Cedar Springs
Greenville
Coopersville
Sparta
Grand Haven
Rockford
Ionia
Portland
Grand Rapids
Wyoming
Holland
Middleville
Hastings
Charlotte
LANSING
Saugatuck
Wayland
Nashville
Allegan
Plainwell
Battle Creek
South Haven
Bangor
Kalamazoo
Marshall
Hartford
Paw Paw
Benton Harbor
Watervliet
Decatur
St. Joseph
Dowagiac
Mendon
Union City
Coldwater
Berrien Sprs.
Cassopolis
Three Rivers
Niles
Mottville
Sturgis
Elkhart
INDIANA

LAKE MICHIGAN

Stevens Pt.
New London
Waupaca
Black Creek
Green Bay
De Pere
Algoma
Kewaunee
Plainfield
Denmark
Kaukauna
Appleton
Menasha
Neenah
Brillion
Wautoma
Oshkosh
Chilton
Two Rivers
Manitowoc
Berlin
Green Lake
Ripon
Kiel
Princeton
Fond du Lac
Sheboygan
Waupun
Plymouth
Sheboygan Falls
WISCONSIN
Portage
Beaver Dam
Theresa
Mayville
West Bend
Columbus
Watertown
Oconomowoc
Port Washington
Sun Prairie
Jefferson
Ft. Atkinson
Shorewood
Waukesha
MILWAUKEE
Cudahy
S. Milwaukee
Edgerton
Whitewater
Janesville
Elkhorn
Union Grove
Racine
Beloit
Lake Geneva
Burlington
Kenosha
Harvard
Zion
Rockford
Belvidere
Woodstock
Mundelein
Waukegan
Lake Forest
Highland Park
Oregon
Rochelle
Marengo
Winnetka
Evanston
Dixon
Sycamore
Elgin
St. Charles
CHICAGO
Aurora
Sandwich
East Chicago
Michigan City
Gary

ILLINOIS

SEE PAGE 30
SEE PAGE 31

28

Marie
Sault Ste. Marie
Echo Bay · Dunns Valley · 638 · 554 · White · 639 · Elliot Lake · Whiskey L. · Vermilion L. · Copper Cliff · Coniston · Hagar · Warren · St. Charles · Verner · Sturgeon Falls
561 · 50 · Desbarats · Bruce Mines · 129 · 546 · Dunlop L. · Agnew L. · Whitefish · 17 · 69 · N.D. du Lac · Estaire · 64 · 50 · Cache Bay · North Bay
548 · Hilton Beach · Thessalon · 17 · Cobden · 108 · McKerrow · Nairn Centre · Espanola · Bear L. · 637 · Alban · 607 · 64 · Noelville · Nipissing
548 · St. Joseph I. · Tenby Bay · Blind River · 94 · Spragge · Massey · Spanish · 68 · Hannah L. · KILLARNEY PROV. PK. · 522 · GRUNDY LAKE PROV. PK. · Port Loring · 522 · Restoule L. · 524
129 · Cedarville · Drummond · 134 · Johnswood · Cockburn Island · Barrie I. · Gore Bay · Kagawong · Little Current · Killarney · Sheguiandah · Britt · Kawigamog Lake · Magnetawan

North Channel

De Tour Village · Drummond Island · Meldrum Bay · 540 · Kagawong · 540 · 30,000 · Islands · 124 · McKellar
Mackinac I. · Bois Blanc I. · MANITOULIN · Mindemoya L. · 542 · 68 · Manitowaning · Pointe au Baril · Parry Sound · MUSKOKA
Cheboygan · Great Duck I. · Britainville · Providence Bay · 68 · Tehkummah · Manitoulin · 141 · Parry I.
33 · Black L. · 23 · Rogers City · ISLAND · S. Baymouth · Georgian Bay Is. Nat. Pk. · Otter L. · Bala
68 · 93 · Onaway · Grand L. · Cove I. · Tobermory · Cape Hurd · Fitzwilliam I. · GEORGIAN BAY IS. NAT. PK. · St. Joseph · 69

LAKE HURON

SAULT STE. MARIE, MICH.

Alpena · 32 · Thunder Bay · Fletcher Pd. · Cape Croker · 6 · Christian I. · Port McNicoll · Penetanguishene · Midland
Fairview · Curran · 72 · Harrisville · 61 · Wiarton · 1093 · Mountain L. · Nottawasaga Bay · 27 · 93
Mio · 33 · HURON · Au Sable · Oscoda · Southampton · 70 · Meaford · Thornbury · Wasaga Beach · 92
18 · Rose City · 23 · Port Elgin · 6 · 11 · Owen Sound · 26 · Collingwood · Stayner
West Branch · 55 · Tawas City · E. Tawas · Underwood · SEE PAGES 66-67 · Chatsworth · 10 · Markdale · 91 · 24 · Cookstown
56 · Alger · Kincardine · Flesherton · 68 · 10 · 89 · Alliston · Schomberg
30 · 61 · Standish · Port Austin · Caseville · 25 · Amberley · 21 · 9 · Walkerton · Hanover · Durham · Shelburne · Primrose · Woodbridge
Baldwin · 23 · Pinconning · Pigeon · 128 · 53 · Harbor Beach · 86 · Lucknow · 88 · Mount Forest · Orangeville · 10 · 22
75 · 46 · Elkton · 142 · Bad Axe · 142 · 34 · Wingham · Harriston · Arthur · Fergus · Georgetown · Acton
Bay City · Sebewaing · Unionville · Goderich · 12 · Clinton · Seaforth · Listowel · Palmerston · 37 · Elmira · Guelph · Milton
Midland · 47 · Saginaw · 15 · Richville · Sandusky · Port Sanilac · 8 · Mitchell · Milverton · Waterloo · Kitchener · 13 · Burlington
Merrill · 46 · 40 · Vassar · 69 · Marlette · Port Sanilac · Exeter · 83 · Stratford · New Hamburg · Cambridge · Dundas
St. Charles · 57 · 52 · 31 · 53 · 19 · 33 · Grand Bend · St. Marys · Tavistock · Paris · 403 · Brantford
Owosso · Flint · Davison · Lapeer · Imlay City · 72 · PINERY PROV. PK. · 21 · Forest · Woodstock · Scotland · Hagersville
21 · 56 · Emmett · 19 · Port Huron · BLUE WATER BRIDGE · Strathroy · London · Ingersoll · Delhi · 24 · Jarvis
69 · Perry · Fenton · Oxford · Romeo · Sarnia · 402 · Lambeth · Tillsonburg · Simcoe · Port Dover
43 · Williamston · 25 · Clarkston · Lake Orion · Petrolia · 80 · Glencoe · Talbotville Royal · Aylmer · 24 · Long Point Bay
496 · Perry · Fowlerville · 59 · Pontiac · Sterling Hts. · Wardsville · St. Thomas · Port Burwell · Port Rowan
Mason · Howell · Brighton · Royal Oak · Warren · Dresden · Thamesville · Wallacetown · Port Stanley · LONG POINT PROV. PK.
127 · Leslie · Stockbridge · Southfield · St. Clair Shores · Algonac · Ridgetown · DETROIT, MICH. · Long Point
Chelsea · Detroit · Lake St. Clair · Chatham · 40 · Morpeth · RONDEAU PROVINCIAL PARK
Jackson · Ann Arbor · Dearborn · Windsor · Tilbury · 401 · Blenheim · Elevation 572 · Presque Isle · Erie · 90
Saline · Taylor · Essex · Wheatley · North East
Milan · Ypsilanti · Amherstburg · Kingsville · Leamington · POINT PELEE NAT. PARK · Fairview · Girard · Wattsburg
Tecumseh · 275 · 75 · Monroe · Pelee I. · Conneaut
Adrian · Dundee · 25 · Middle Bass I. · FOR LEGEND SEE PAGE 5
Hudson · Sylvania · Toledo · N. Bass I. · SEE PAGES 26-27 · Painesville · ONE CENTIMETER EQUALS ABOUT 18.2 KILOMETERS
OHIO · Willoughby · Euclid · ONE INCH EQUALS ABOUT 28.7 MILES · © GENERAL DRAFTING CO., INC.

UNITED STATES · CANADA

Georgian Bay

WISCONSIN

FOR LEGEND SEE PAGE 5

ONE CENTIMETER EQUALS ABOUT 30.7 KILOMETERS
0 20 40 60 80 100 120 140 160 Km.
0 20 40 60 80 100 Mi.
ONE INCH EQUALS ABOUT 48.7 MILES

© GENERAL DRAFTING CO., INC.

SEE PAGE 35
SEE PAGE 37
SEE PAGES 28-29
SEE PAGE 31

LAKE SUPERIOR
Elevation 602

ONTARIO
QUETICO PROVINCIAL PARK
VOYAGEURS NAT. PK. (under devel.)

Fort Frances
International Falls
Atikokan
Thunder Bay
SIBLEY PROV. PK.
Grand Portage
ISLE ROYALE NAT. PARK
Copper Harbor
Marquette
Munising
PICTURED ROCKS NAT. LAKESHORE
Manistique
Escanaba
Iron Mtn. Norway
Menominee
Marinette

Duluth
Superior
Ashland
APOSTLE IS. NAT. LAKESHORE
Apostle Islands
Bayfield
Washburn
Hurley
Ironwood
Bessemer
Hibbing
Virginia
Chisholm
Grand Rapids
Two Harbors
Silver Bay
Grand Marais
Ely

MINNESOTA
MICHIGAN
WISCONSIN
ILLINOIS
IOWA

St. Paul
Minneapolis
Rochester
Winona
La Crosse
Eau Claire
Chippewa Falls
Menomonie
Marshfield
Wausau
Stevens Pt.
Wisconsin Rapids
Green Bay
Appleton
Oshkosh
Fond du Lac
Sheboygan
Madison
Janesville
Beloit
Milwaukee
Racine
Kenosha
Rockford
Chicago
Dubuque
Des Moines
Davenport
Rock Island
Peoria

LAKE MICHIGAN
Sturgeon Bay
Manitowoc
Two Rivers
Port Washington
Holland
Muskegon
Grand Haven
South Haven
Benton Harbor

Central Standard Time   Eastern Standard Time

Rhinelander
Eagle River
Minocqua
Antigo
Shawano
Tomahawk
Merrill
Rice Lake
Spooner
Hayward
Park Falls
Ladysmith
Medford
Neillsville
Black River Falls
Sparta
Tomah
Mauston
Portage
Baraboo
Wisconsin Dells
Waupun
Ripon
Waupaca
New London
Clintonville
De Pere
Neenah
Menasha
Chilton
Kiel
West Bend
Watertown
Waukesha
Jefferson
Ft. Atkinson
Edgerton
Monroe
Platteville
Dodgeville
Prairie du Chien
Richland Center
Reedsburg
Viroqua

GREAT AMERICA
Great Lakes Naval Training Cen.
Waukegan
Evanston
Gary
Joliet
Aurora
Elgin

30

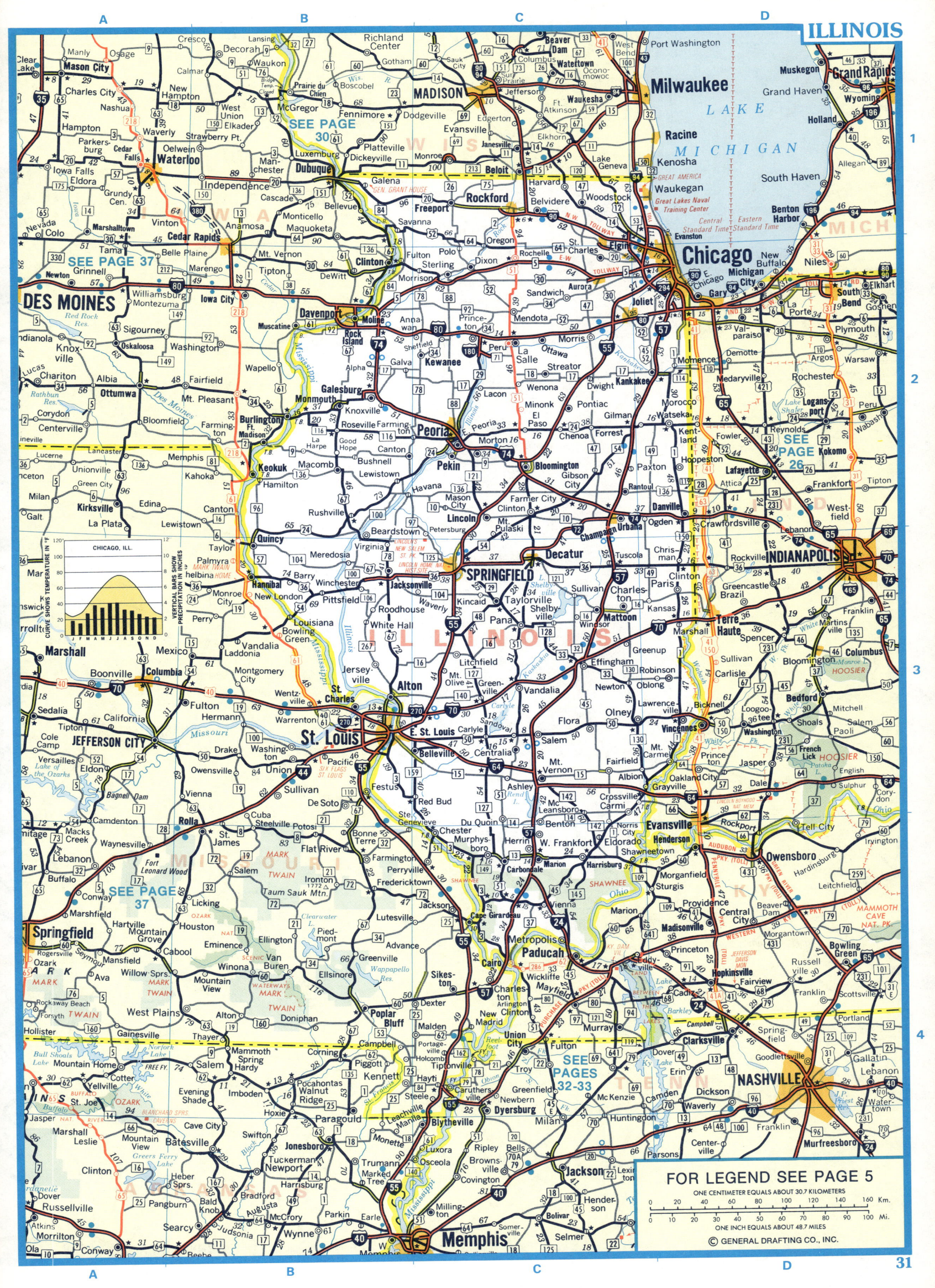

FOR LEGEND SEE PAGE 5

ONE CENTIMETER EQUALS ABOUT 30.7 KILOMETERS
ONE INCH EQUALS ABOUT 48.7 MILES
© GENERAL DRAFTING CO., INC.

FOR LEGEND SEE PAGE 5

ONE CENTIMETER EQUALS ABOUT 23.1 KILOMETERS

ONE INCH EQUALS ABOUT 36.8 MILES

© GENERAL DRAFTING CO., INC.

LOUISVILLE, KY.

MEMPHIS, TENN.

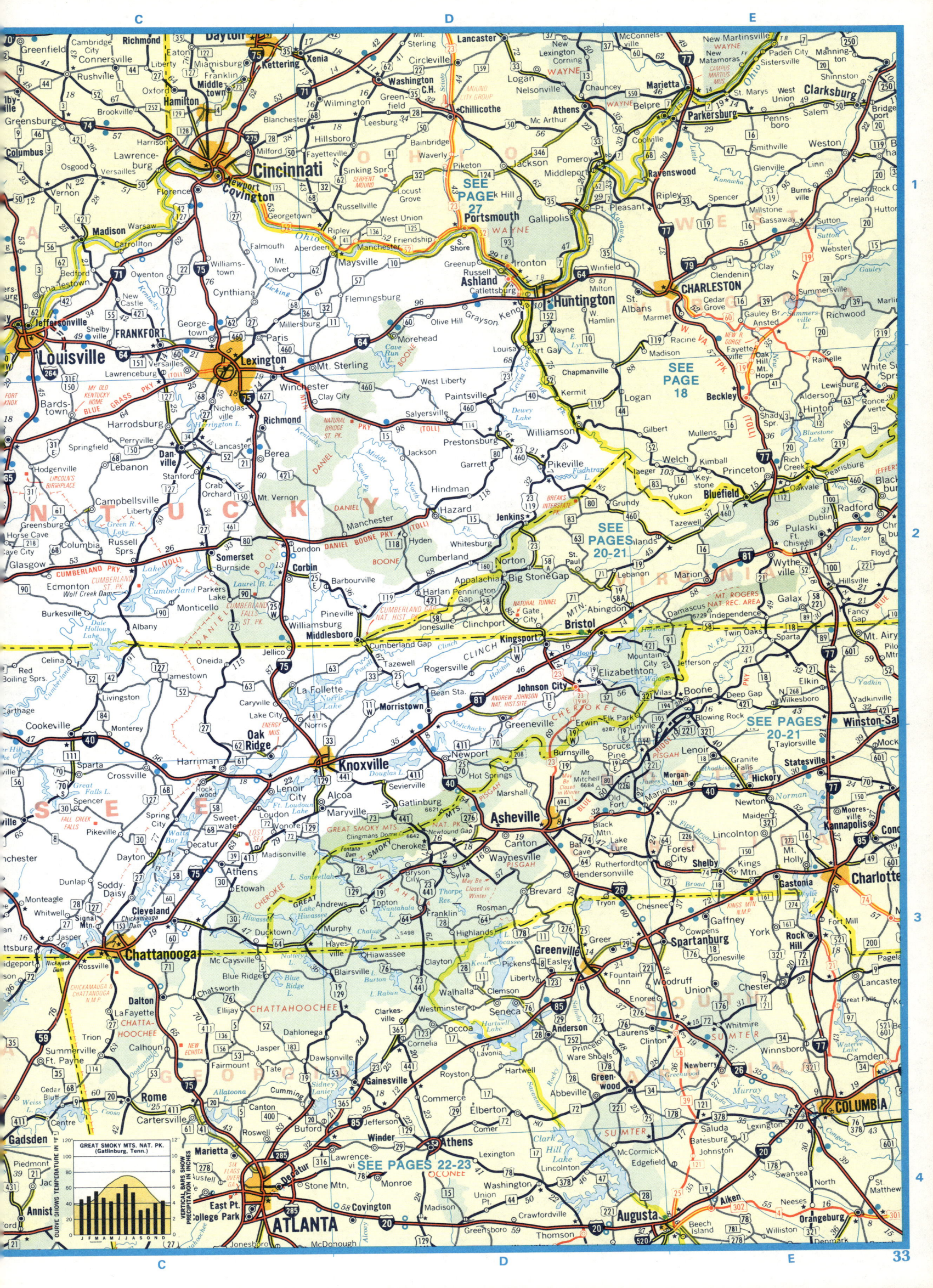

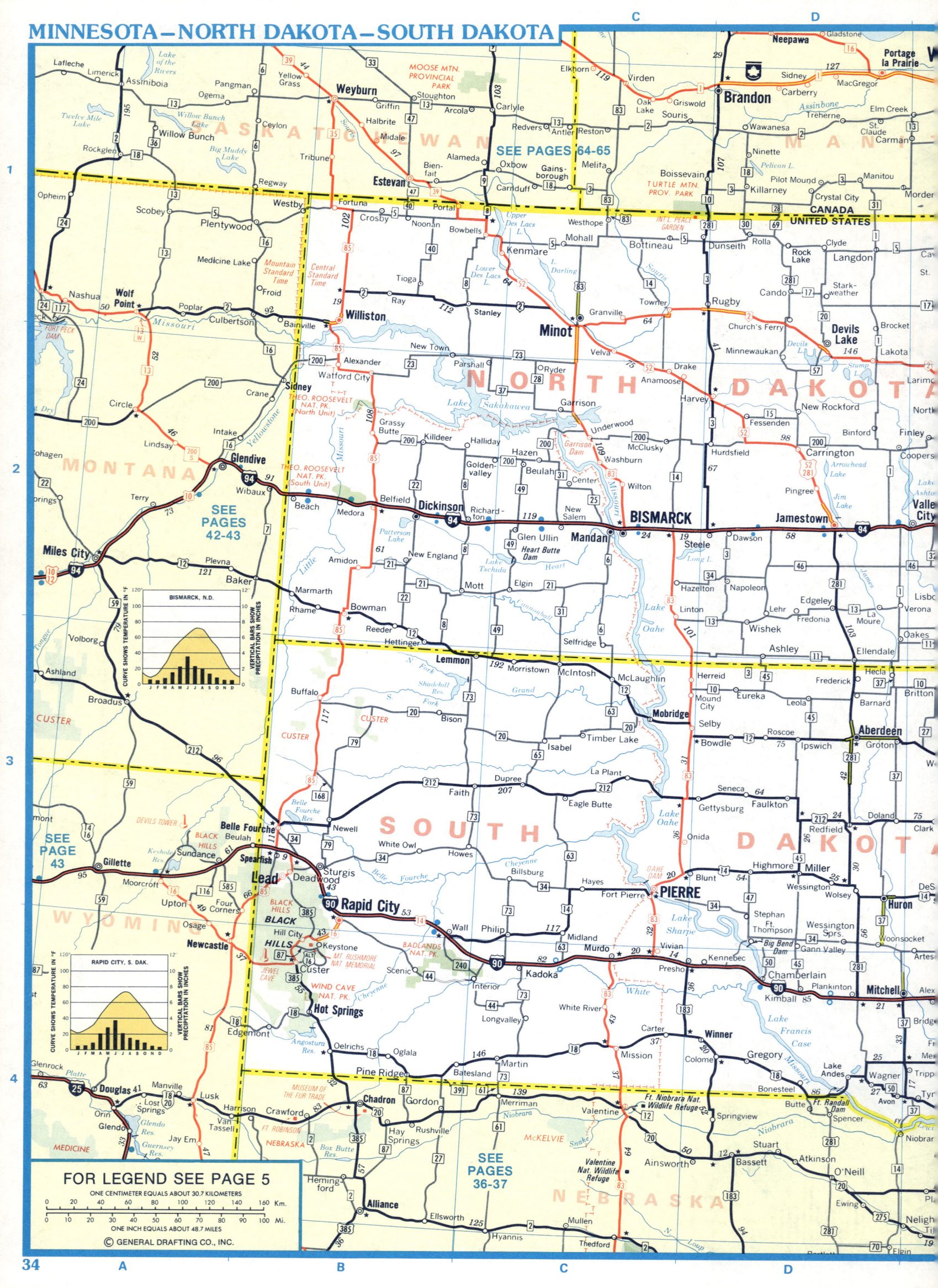

MINNESOTA—NORTH DAKOTA—SOUTH DAKOTA

SEE PAGES 64-65

CANADA
UNITED STATES

NORTH DAKOTA

SOUTH DAKOTA

MONTANA

WYOMING

NEBRASKA

SEE PAGES 42-43

SEE PAGE 43

SEE PAGES 36-37

BISMARCK, N.D.
CURVE SHOWS TEMPERATURE IN °F
VERTICAL BARS SHOW PRECIPITATION IN INCHES
J F M A M J J A S O N D

RAPID CITY, S. DAK.
CURVE SHOWS TEMPERATURE IN °F
VERTICAL BARS SHOW PRECIPITATION IN INCHES
J F M A M J J A S O N D

FOR LEGEND SEE PAGE 5

ONE CENTIMETER EQUALS ABOUT 30.7 KILOMETERS
0 20 40 60 80 100 120 140 160 Km.
0 10 20 30 40 50 60 70 80 90 100 Mi.
ONE INCH EQUALS ABOUT 48.7 MILES
© GENERAL DRAFTING CO., INC.

34

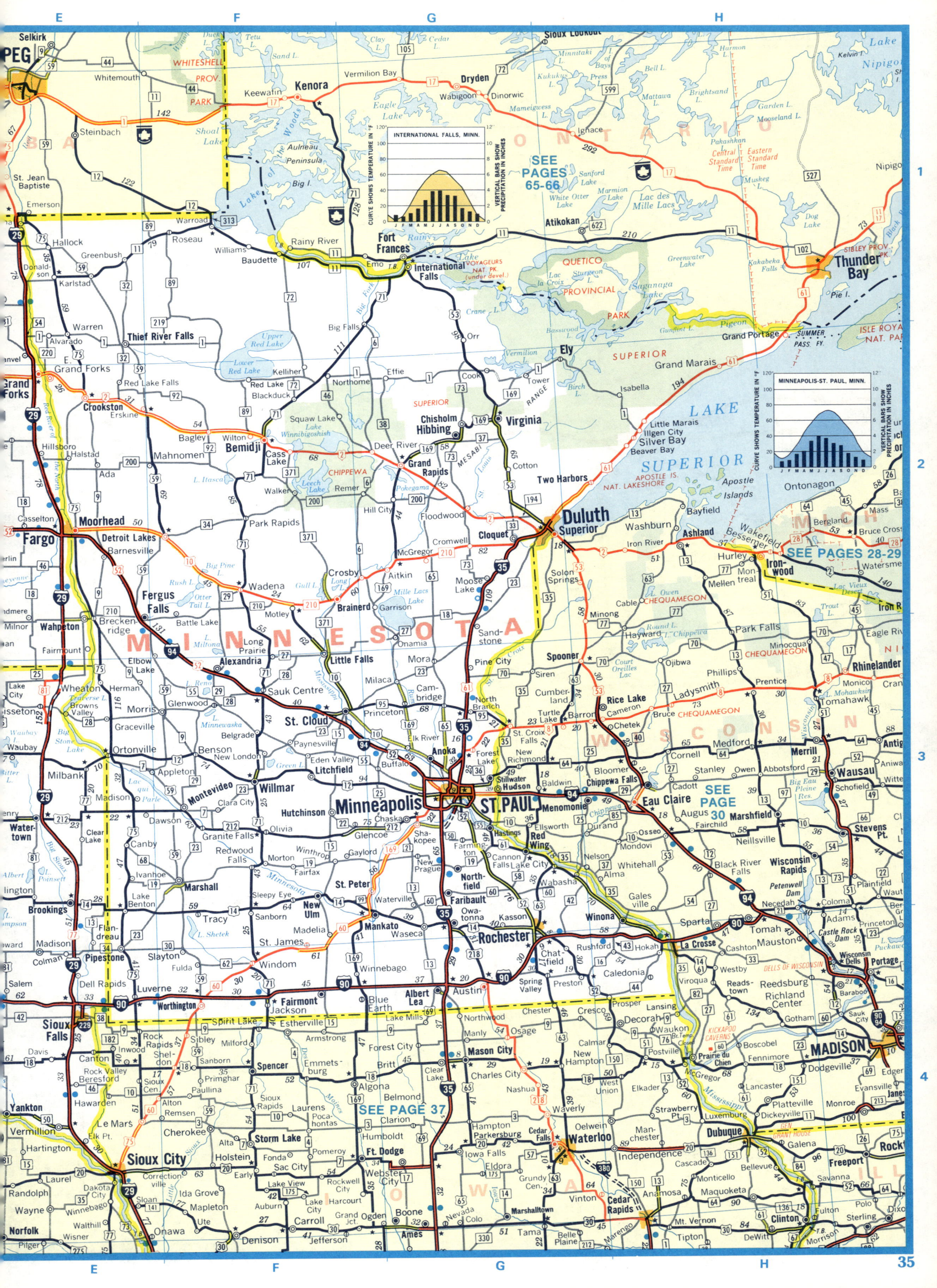

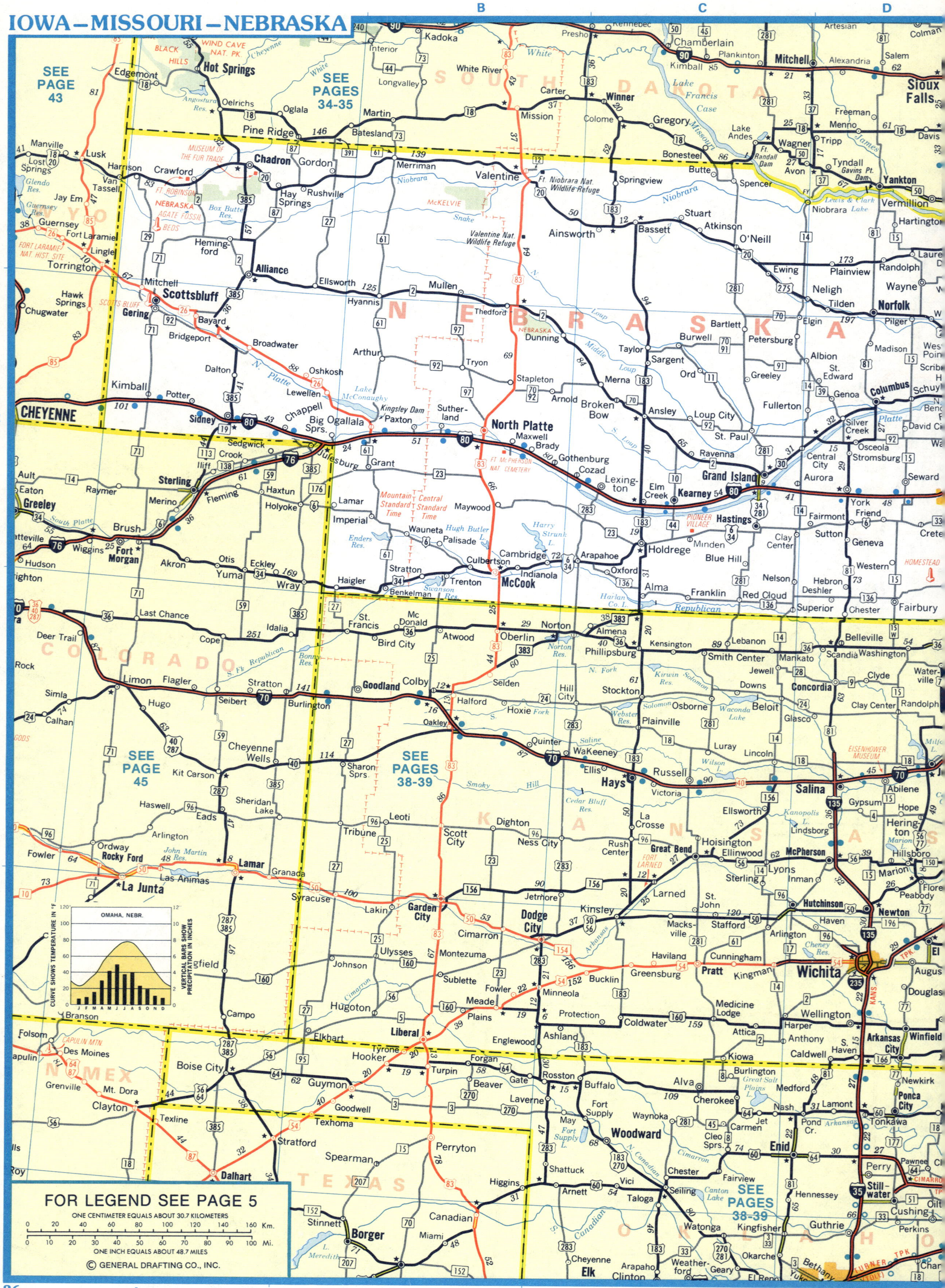

IOWA – MISSOURI – NEBRASKA

SEE PAGE 43

SEE PAGES 34-35

SEE PAGE 45

SEE PAGES 38-39

SEE PAGES 38-39

SOUTH DAKOTA

NEBRASKA

COLORADO

KANSAS

OKLAHOMA

TEXAS

WYO.

N. MEX

OMAHA, NEBR.

CURVE SHOWS TEMPERATURE IN °F.
VERTICAL BARS SHOW PRECIPITATION IN INCHES

J F M A M J J A S O N D

FOR LEGEND SEE PAGE 5

ONE CENTIMETER EQUALS ABOUT 30.7 KILOMETERS
0 20 40 60 80 100 120 140 160 Km.
0 10 20 30 40 50 60 70 80 90 100 Mi.
ONE INCH EQUALS ABOUT 48.7 MILES

© GENERAL DRAFTING CO., INC.

36

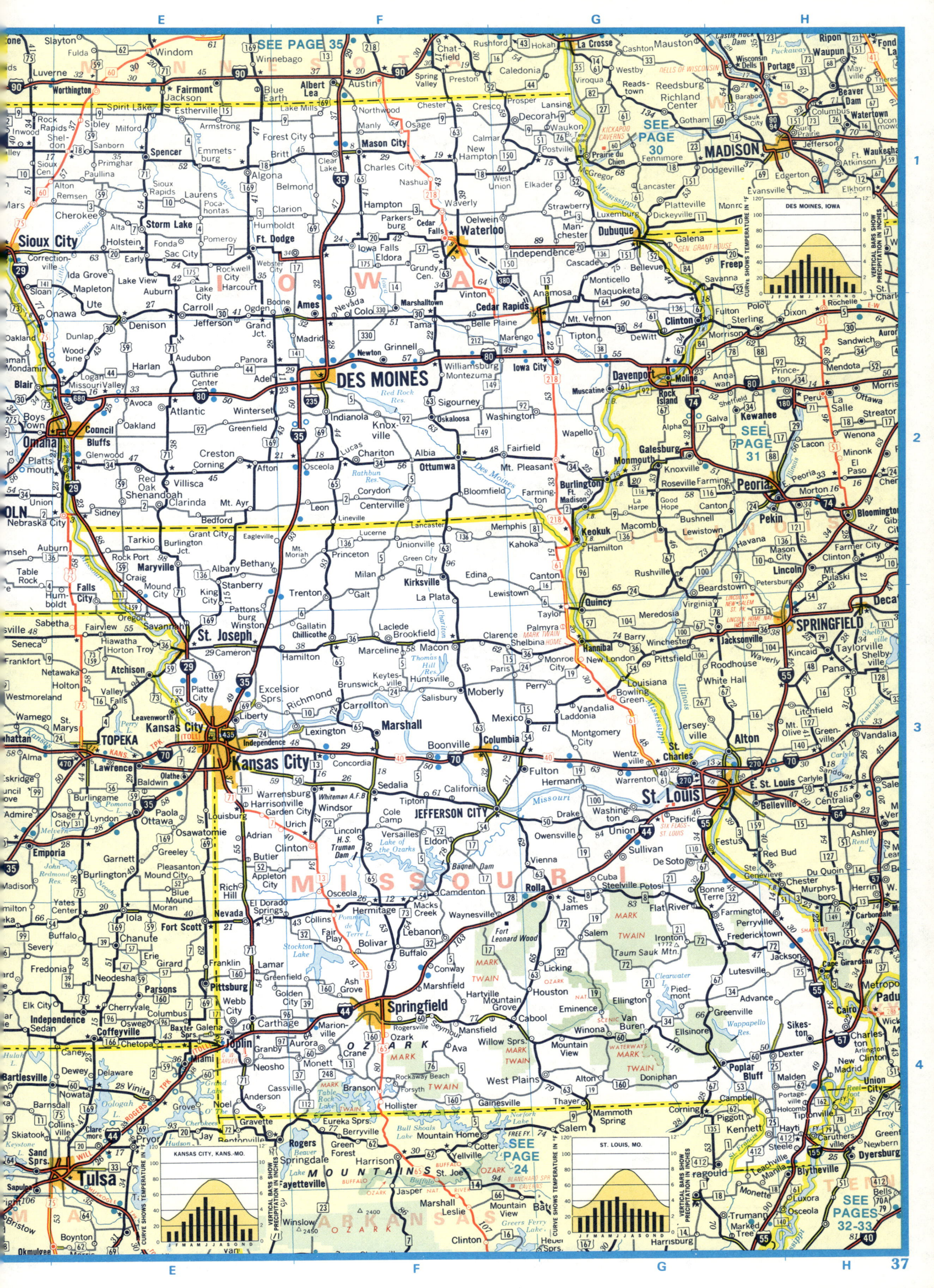

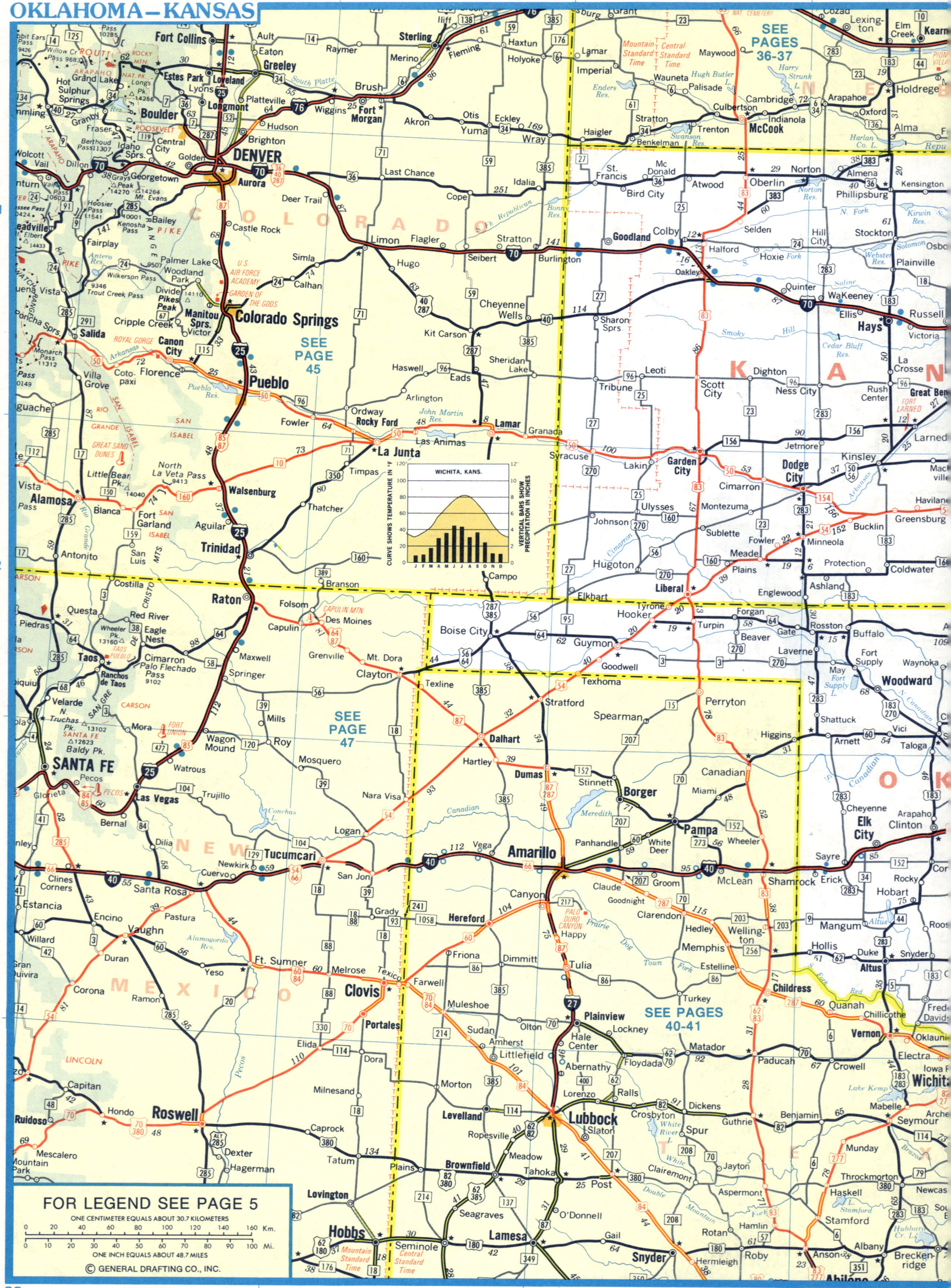

STATES: COLORADO · KANSAS · NEW MEXICO · OKLAHOMA · TEXAS

Grid references (top): B · C · D
Grid references (left): 1 · 2 · 3 · 4

SEE PAGES 36-37
SEE PAGE 45
SEE PAGE 47
SEE PAGES 40-41

Chart (inset):
WICHITA, KANS.
CURVE SHOWS TEMPERATURE IN °F
VERTICAL BARS SHOW PRECIPITATION IN INCHES
J F M A M J J A S O N D

Selected place names (Colorado): Fort Collins, Ault, Eaton, Greeley, Estes Park, Loveland, Longmont, Boulder, Platteville, Brighton, Denver, Aurora, Golden, Central City, Idaho Springs, Georgetown, Hot Sulphur Springs, Grand Lake, Dillon, Vail, Leadville, Fairplay, Bailey, Woodland Park, Manitou Springs, Colorado Springs, Victor, Cripple Creek, Canon City, Florence, Pueblo, Salida, Villa Grove, Alamosa, Blanca, Fort Garland, Aguilar, Walsenburg, Trinidad, Raton, Deer Trail, Limon, Flagler, Stratton, Burlington, Last Chance, Cope, Idalia, Simla, Hugo, Calhan, Haswell, Eads, Arlington, Ordway, Rocky Ford, Fowler, Las Animas, Lamar, Granada, La Junta, Timpas, Thatcher, Branson

Kansas: Sterling, Holyoke, Haxtun, Goodland, Colby, Oberlin, Norton, Phillipsburg, Stockton, Plainville, Russell, Hays, Victoria, La Crosse, Great Bend, Larned, Kinsley, Dodge City, Garden City, Leoti, Tribune, Scott City, Ness City, Rush Center, Dighton, Syracuse, Lakin, Ulysses, Johnson, Hugoton, Liberal, Meade, Sublette, Montezuma, Cimarron, Bucklin, Minneola, Greensburg, Coldwater, Ashland, Englewood, Plains, Protection

New Mexico: Questa, Red River, Eagle Nest, Taos, Cimarron, Springer, Maxwell, Raton, Capulin, Des Moines, Clayton, Grenville, Mt. Dora, Santa Fe, Las Vegas, Mora, Wagon Mound, Roy, Mills, Mosquero, Logan, Tucumcari, Santa Rosa, Vaughn, Encino, Corona, Duran, Fort Sumner, Melrose, Clovis, Portales, Roswell, Ruidoso, Hobbs, Lovington, Tatum

Texas/Oklahoma: Dalhart, Texline, Stratford, Dumas, Borger, Pampa, Amarillo, Canyon, Hereford, Friona, Dimmitt, Tulia, Plainview, Lubbock, Levelland, Brownfield, Lamesa, Childress, Vernon, Wichita Falls, Woodward, Elk City, Clinton, Shamrock, McLean, Wellington, Altus, Mangum, Hobart, Snyder

Bottom grid references: A · B · C · D

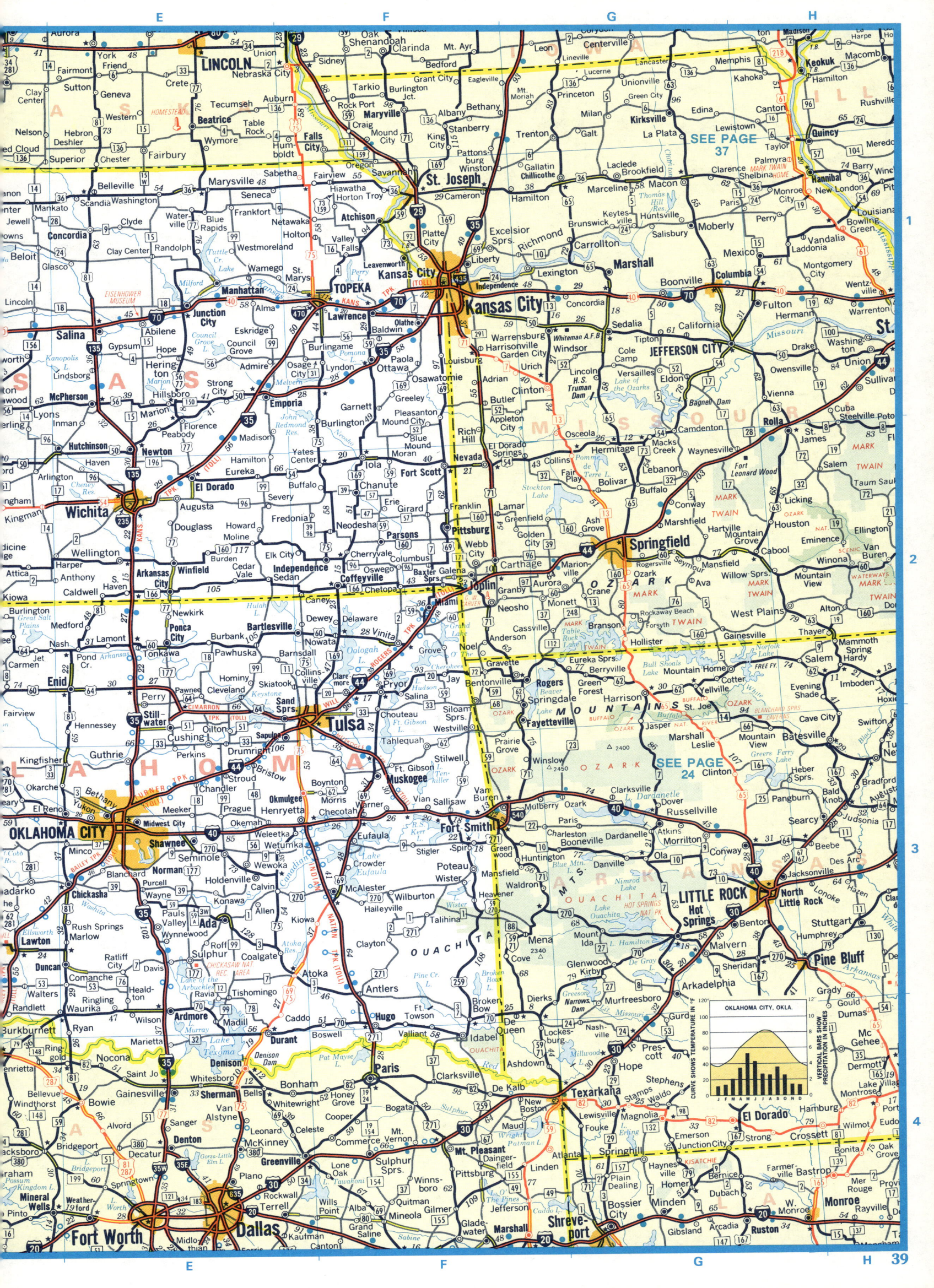

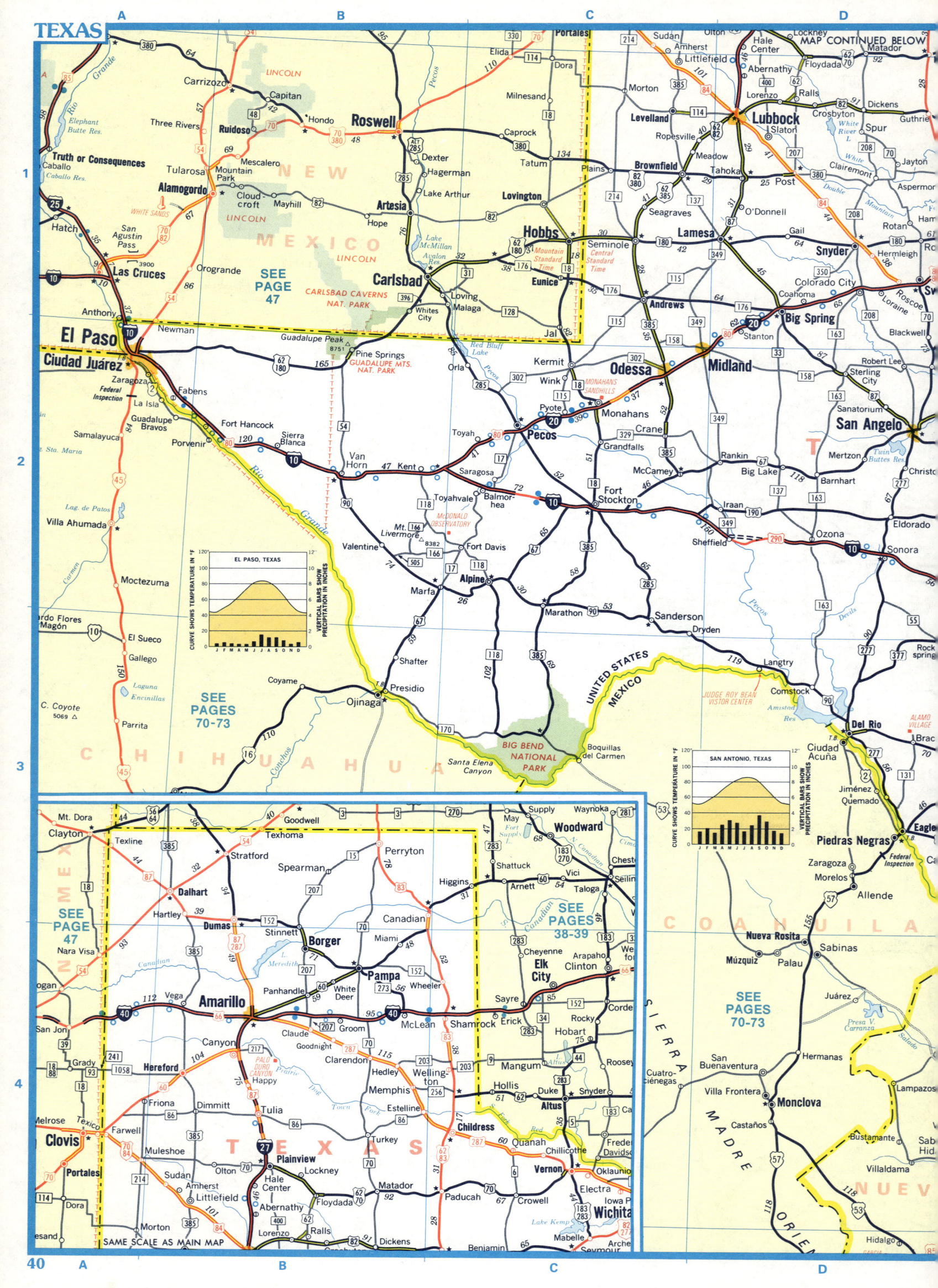

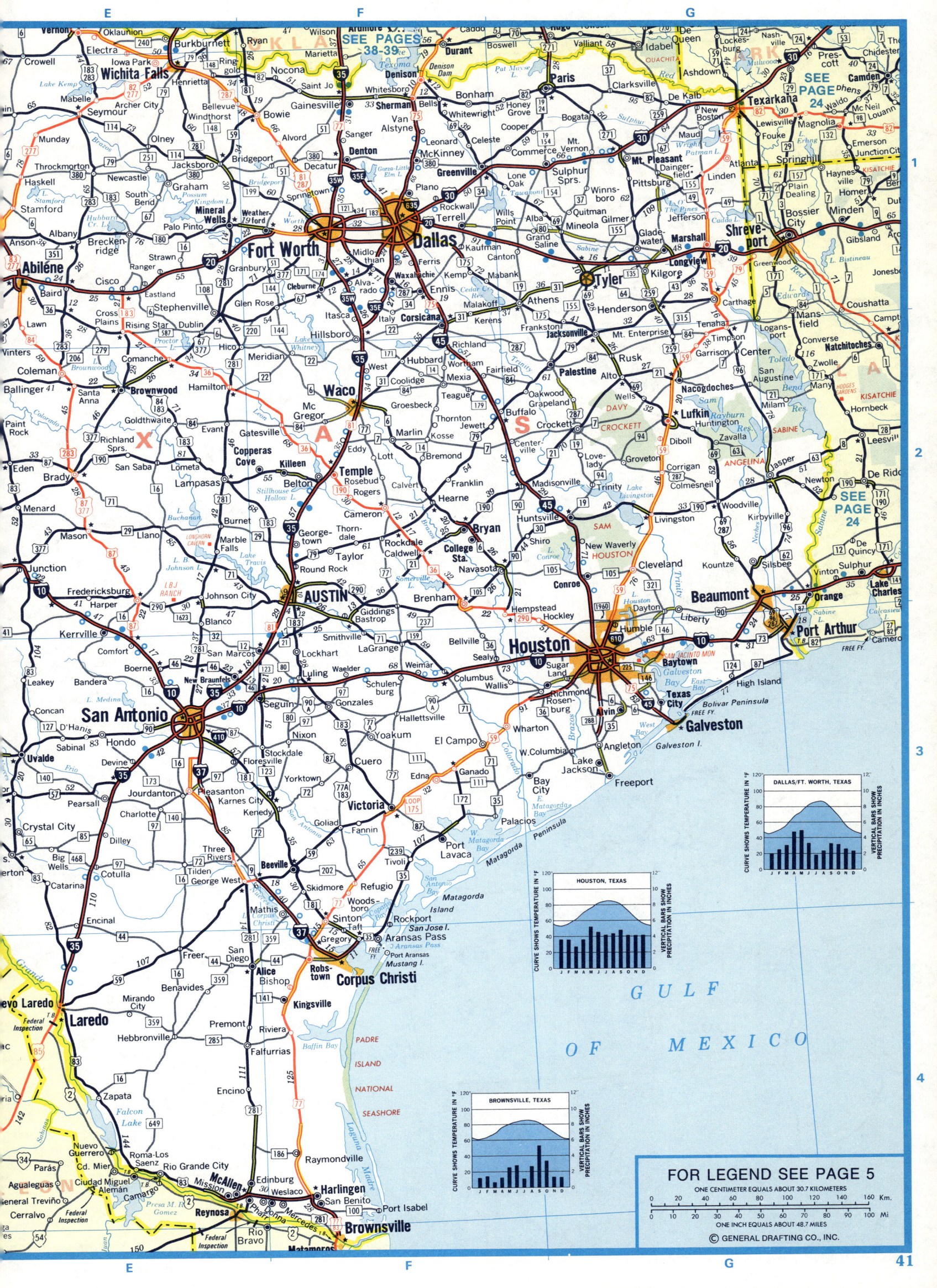

FOR LEGEND SEE PAGE 5

ONE CENTIMETER EQUALS ABOUT 30.7 KILOMETERS

ONE INCH EQUALS ABOUT 48.7 MILES

© GENERAL DRAFTING CO., INC.

IDAHO–MONTANA–WYOMING

GLACIER NAT. PK., MONT.

CURVE SHOWS TEMPERATURE IN °F
VERTICAL BARS SHOW PRECIPITATION IN INCHES

BOISE, IDAHO

CURVE SHOWS TEMPERATURE IN °F
VERTICAL BARS SHOW PRECIPITATION IN INCHES

Pacific Standard Time
Mountain Standard Time

SEE PAGE 48

SEE PAGE 48

SEE PAGE 49

SEE PAGE 44

FOR LEGEND SEE PAGE 5

ONE CENTIMETER EQUALS ABOUT 30.7 KILOMETERS
0 10 20 30 40 50 60 70 80 90 100 Km.
0 10 20 30 40 50 60 Mi.
ONE INCH EQUALS ABOUT 48.7 MILES

© GENERAL DRAFTING CO., INC.

42

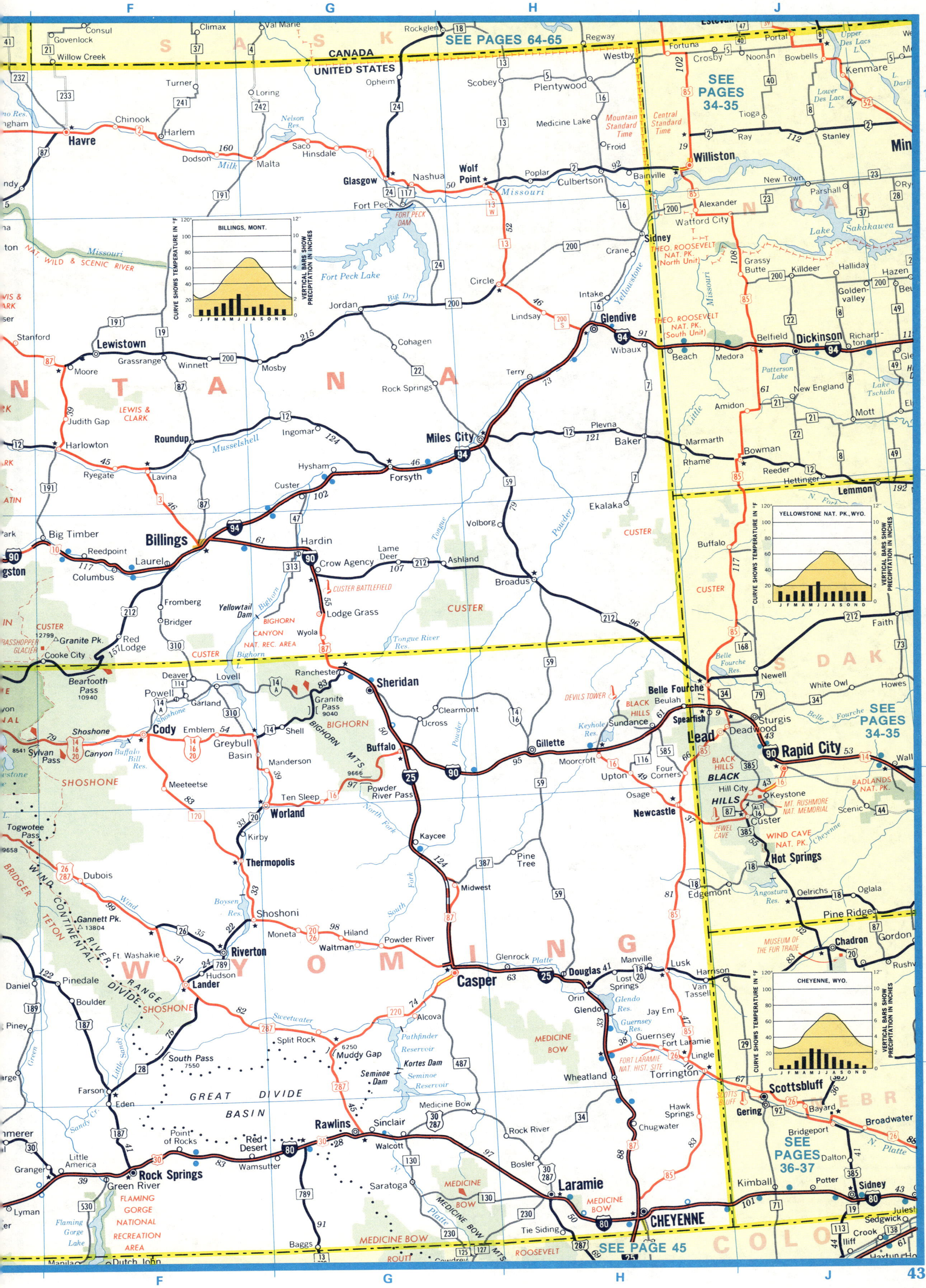

SEE PAGES 64-65
SEE PAGES 34-35
SEE PAGES 34-35
SEE PAGES 36-37
SEE PAGE 45

CANADA
UNITED STATES

SASK

Consul   Climax   Val Marie   Rockglen   Regway   Estevan   Portal
Govenlock   Turner   Loring   Opheim   Scobey   Plentywood   Westby   Fortuna   Crosby   Noonan   Bowbells   Kenmare
Willow Creek   Chinook   Harlem   Medicine Lake   Froid   Williston   Ray   Stanley   Min
Havre   Dodson   Malta   Saco   Hinsdale   Glasgow   Nashua   Wolf Point   Poplar   Culbertson   Bainville   New Town   Parshall
Fort Peck   Alexander   Watford City

MONTANA   WYOMING   N DAK   S DAK   COLO   NEBR

Stanford   Lewistown   Grassrange   Winnett   Mosby   Jordan   Cohagen   Circle   Intake   Lindsay   Glendive   Wibaux   Beach   Medora   Belfield   Dickinson   Richardton
Moore   Judith Gap   Harlowton   Roundup   Ingomar   Rock Springs   Miles City   Terry   Baker   Plevna   Marmarth   Bowman   New England   Mott
Ryegate   Lavina   Hysham   Forsyth   Volborg   Ekalaka   Rhame   Reeder   Hettinger   Lemmon
Big Timber   Reedpoint   Billings   Custer   Hardin   Lame Deer   Ashland   Broadus   Buffalo   Faith
Laurel   Columbus   Crow Agency   Custer Battlefield   Newell   White Owl   Howes
Fromberg   Bridger   Yellowtail Dam   Lodge Grass   Wyola   Belle Fourche   Beulah   Sturgis
Red Lodge   Cooke City   Sheridan   Clearmont   Gillette   Sundance   Spearfish   Lead   Deadwood   Rapid City   Wall
Beartooth Pass   Deaver   Lovell   Ranchester   Ucross   Moorcroft   Newcastle   Hill City   Keystone   Scenic   Badlands Nat. Pk.
Powell   Garland   Shell   Buffalo   Upton   Osage   Custer   Hot Springs
Cody   Emblem   Greybull   Basin   Manderson   Four Corners   Edgemont   Oelrichs   Pine Ridge
Meeteetse   Ten Sleep   Powder River Pass   Kaycee   Pine Tree   Chadron   Gordon
Worland   Kirby   Midwest   Lusk   Harrison
Thermopolis   Shoshoni   Moneta   Hiland   Waltman   Powder River   Glenrock   Douglas   Manville   Van Tassell
Dubois   Riverton   Hudson   Lander   Casper   Orin   Lost Springs   Jay Em   Guernsey   Fort Laramie   Lingle   Torrington   Scottsbluff   Gering
Pinedale   Boulder   Alcova   Glendo   Bridgeport   Bayard   Broadwater
Piney   Farson   Split Rock   Muddy Gap   Seminoe Dam   Wheatland   Hawk Springs   Chugwater
Daniel   Eden   South Pass   Kortes Dam   Medicine Bow   Dalton
Point of Rocks   Red Desert   Rawlins   Sinclair   Rock River   Bosler   Kimball   Potter
Rock Springs   Green River   Wamsutter   Walcott   Saratoga   Laramie   Sidney
Granger   Little America   Baggs   Tie Siding   Cheyenne   Crook   Iliff
Lyman   Dutch John

Havre   Milk   Missouri   Fort Peck Lake   Fort Peck Dam   Yellowstone   Lake Sakakawea   Missouri
Nat. Wild & Scenic River   Big Dry   Musselshell   Patterson Lake   Lake Tschida
Lewis & Clark   Tongue   Powder   Little Missouri
Bighorn Canyon Nat. Rec. Area   Bighorn   Tongue River Res.   Custer
Shoshone   Bighorn Mts.   Powder River   Black Hills   Belle Fourche Res.
Beartooth Pass 10940   Granite Pass 9040   Devils Tower   Keyhole Res.   Mt. Rushmore Nat. Memorial
Granite Pk. 12799   Cloud Pk. 13804   Jewel Cave   Wind Cave Nat. Pk.
Sylvan Pass 8541   Buffalo Bill Res.   North Fork   Angostura Res.
Shoshone   Boysen Res.   Museum of the Fur Trade
Togwotee Pass 9658   Wind River   Continental Divide   Shoshone   South Fork
Gannett Pk. 13804   Sweetwater   Pathfinder Reservoir   Glendo Res.   Guernsey Res.   Fort Laramie Nat. Hist. Site
Ft. Washakie   South Pass 7550   6250   Kortes Dam   Seminoe Reservoir   Medicine Bow   Scotts Bluff
Wind River Range   Great Divide Basin   Seminoe Dam
Flaming Gorge National Recreation Area   Little Sandy Cr.   Green   Platte
Flaming Gorge Lake   Sandy Cr.   Medicine Bow Mts.   Medicine Bow Route   Roosevelt   N. Platte

Central Standard Time   Mountain Standard Time

Theo. Roosevelt Nat. Pk. (North Unit)   Theo. Roosevelt Nat. Pk. (South Unit)
Grassy Butte   Killdeer   Halliday   Hazen   Golden Valley   Amidon

Chart: BILLINGS, MONT. — CURVE SHOWS TEMPERATURE IN °F / VERTICAL BARS SHOW PRECIPITATION IN INCHES — J F M A M J J A S O N D
Chart: YELLOWSTONE NAT. PK., WYO. — CURVE SHOWS TEMPERATURE IN °F / VERTICAL BARS SHOW PRECIPITATION IN INCHES — J F M A M J J A S O N D
Chart: CHEYENNE, WYO. — CURVE SHOWS TEMPERATURE IN °F / VERTICAL BARS SHOW PRECIPITATION IN INCHES — J F M A M J J A S O N D

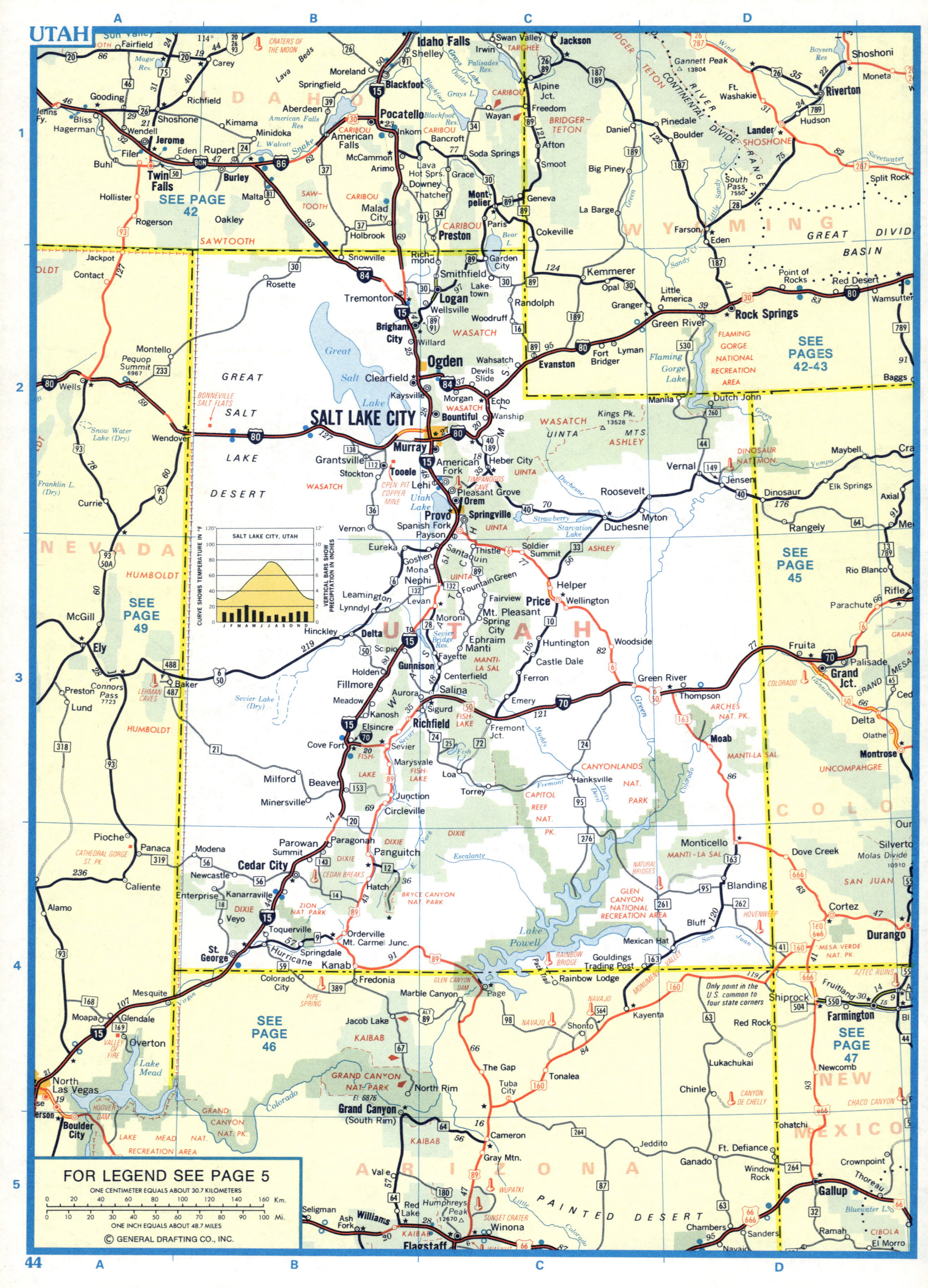

UTAH

SEE PAGE 42

SEE PAGES 42-43

SEE PAGE 45

SEE PAGE 49

SEE PAGE 46

SEE PAGE 47

SALT LAKE CITY, UTAH
Curve shows temperature in °F
Vertical bars show precipitation in inches
J F M A M J J A S O N D

FOR LEGEND SEE PAGE 5

ONE CENTIMETER EQUALS ABOUT 30.7 KILOMETERS
0 20 40 60 80 100 120 140 160 Km.
0 10 20 30 40 50 60 70 80 90 100 Mi.
ONE INCH EQUALS ABOUT 48.7 MILES

© GENERAL DRAFTING CO., INC.

44

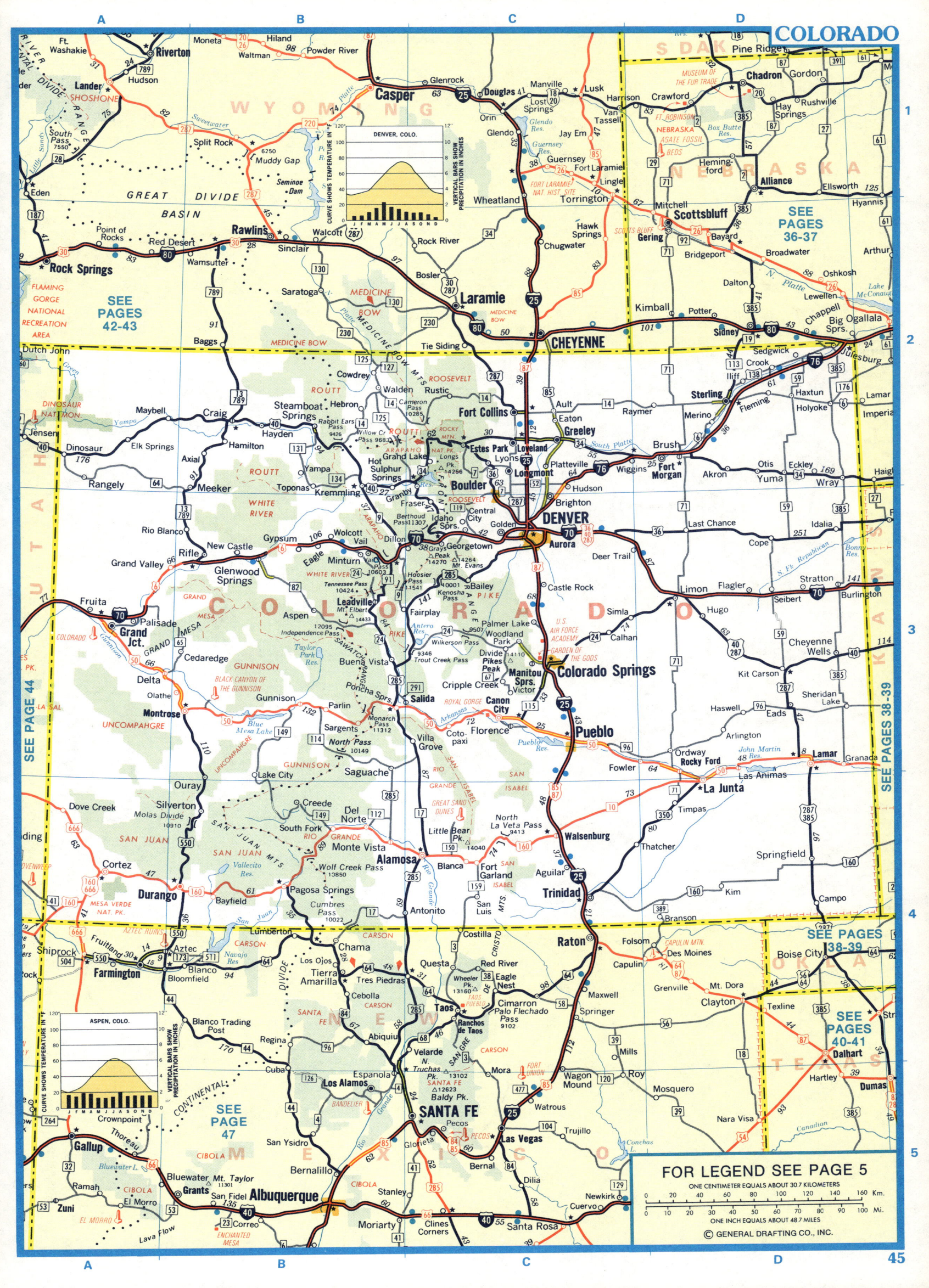

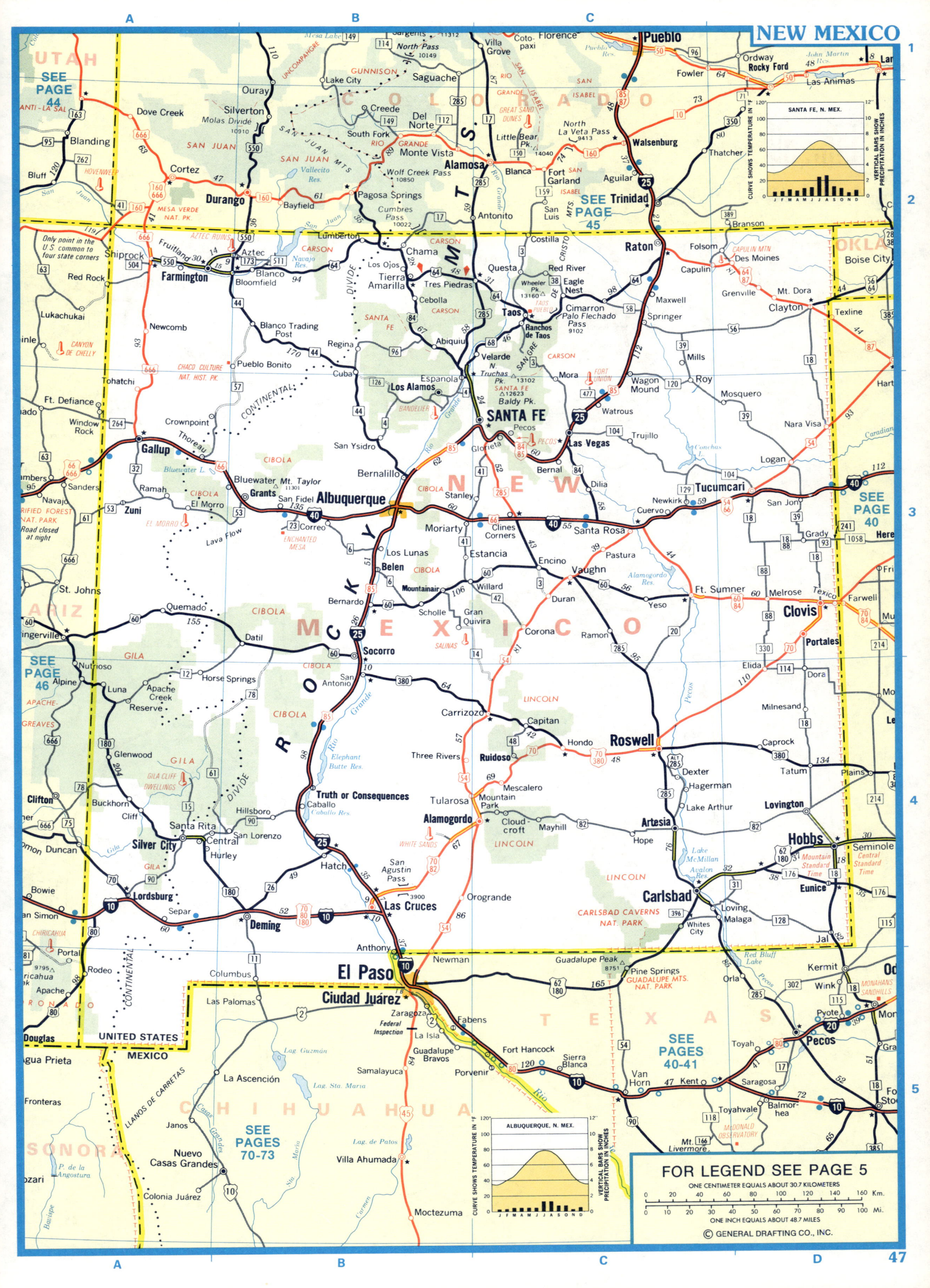

NEW MEXICO

FOR LEGEND SEE PAGE 5

ONE CENTIMETER EQUALS ABOUT 30.7 KILOMETERS
ONE INCH EQUALS ABOUT 48.7 MILES

© GENERAL DRAFTING CO., INC.

47

OREGON — WASHINGTON

FOR LEGEND SEE PAGE 5

ONE CENTIMETER EQUALS ABOUT 30.7 KILOMETERS

ONE INCH EQUALS ABOUT 48.7 MILES

© GENERAL DRAFTING CO., INC.

SEE PAGE 48

SEE PAGE 42

SEE PAGE 44

SEE PAGE 46

SEE PAGES 50-51

RENO, NEV.

LAS VEGAS, NEV.

FOR LEGEND SEE PAGE 5

ONE CENTIMETER EQUALS ABOUT 30.7 KILOMETERS

ONE INCH EQUALS ABOUT 48.7 MILES

© GENERAL DRAFTING CO., INC.

ALASKA/HAWAII

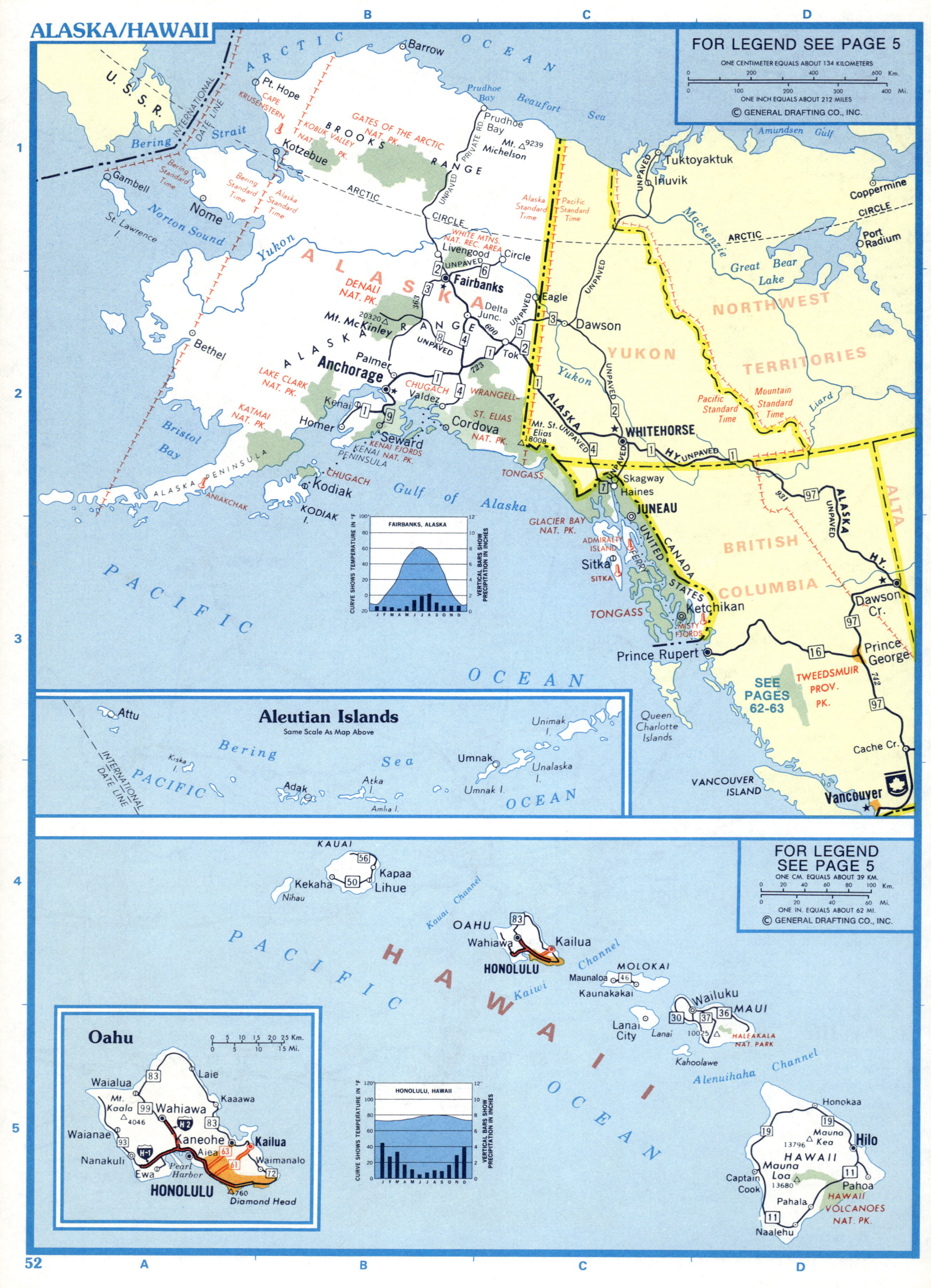

FOR LEGEND SEE PAGE 5

ONE CENTIMETER EQUALS ABOUT 134 KILOMETERS
ONE INCH EQUALS ABOUT 212 MILES
© GENERAL DRAFTING CO., INC.

Aleutian Islands
Same Scale As Map Above

FAIRBANKS, ALASKA
CURVE SHOWS TEMPERATURE IN °F
VERTICAL BARS SHOW PRECIPITATION IN INCHES

SEE PAGES 62-63

FOR LEGEND SEE PAGE 5

ONE CM. EQUALS ABOUT 39 KM.
ONE IN. EQUALS ABOUT 62 MI.
© GENERAL DRAFTING CO., INC.

Oahu

0 5 10 15 20 25 Km.
0   5   10   15 Mi.

HONOLULU, HAWAII
CURVE SHOWS TEMPERATURE IN °F
VERTICAL BARS SHOW PRECIPITATION IN INCHES

America's bountiful scenic wonders, its diversified array of historic and cultural highlights and varied recreational facilities can provide refreshing stopovers as well as destinations on your vacation trip. Many of these outstanding places of interest are listed on the following nine pages. A brief description provides the highlights of each attraction. Under each state name is a reference to the page on which the map of that state can be found. Index references (B-1) locate the attractions and towns on that map. Shown separately on pages 74 to 112 are detailed downtown and vicinity maps of more than 150 of the major cities and metropolitan areas throughout the United States showing these places of interest, plus others shown on the maps.

Please note that the facts concerning these attractions were up-to-date at the time this Atlas was published but are subject to change.

ALABAMA

See map on page 22

AVE MARIA GROTTO (B-1). Miniature reproductions of more than 150 religious and other famous buildings from many countries. Home contains antique furnishings.
BELLINGRATH GARDENS AND HOME (A-3). Moss-draped trees, year-round displays of flowers and shrubs. Home contains antique furnishings.
BIRMINGHAM (B-2). A huge, 60-ton statue of **Vulcan,** Roman god of fire, overlooks city. **Arlington,** an antebellum home, was built in 1822. **Birmingham Zoo** is one of the largest in the Southeast.
BRIDGEPORT (C-1). Nearby **Russell Cave Nat. Monument** preserves cavern which sheltered prehistoric Indians for about 8,000 years.
HUNTSVILLE (B-1). **Alabama Space and Rocket Center** has one of world's largest space-age exhibits. Bus tours of **Marshall Space Flight Center.**
MOBILE (A-3). Historic seaport with Old World atmosphere. **USS Alabama,** World War II battleship, is a war memorial. **Azalea Trail Festival** Feb.-Mar.
MONTGOMERY (B-2). Alabama's **State Capitol** building was also the first capitol of the Confederacy. Adjacent is the **First White House of the Confederacy,** a museum.

ALASKA

See map on page 52

ANCHORAGE (B-2). This modern city is Alaska's largest. Skiing and other winter sports at **Mt. Alyeska.** Takeoff point for plane trips to the Aleutian Islands and to the Arctic.
DENALI NAT. PARK (B-2). Highest mountain in North America. A subarctic wilderness of glaciers and tundra. Wilderness home for grizzly bears, moose and caribou. Camping and hiking. Summer temperatures range from 30° to 80° F.
FAIRBANKS (B-2). **Alaskaland,** built for the 1967 Centennial, is an exposition of Alaska history. The **University of Alaska** is world's northernmost college.
HAINES (C-2). Home of the famed Chilkat Indian dancers; frequent performances in summer. One of two terminals of the Marine Highway, car and passenger ferry service from Seattle, Wash., and Prince Rupert, B.C., through the Inside Passage.
JUNEAU (C-2). **Alaska State Museum** has a collection of Eskimo, Indian and gold-rush exhibits. Nearby is **Mendenhall Glacier,** a river of blue ice.

ARIZONA

See map on page 46

APACHE TRAIL (C-4). Ariz. 88 from Apache Junction to Globe traverses rugged, spectacular mountain scenery. Route passes by **Canyon, Apache** and **Roosevelt** lakes, through **Fish Creek Canyon,** to **Tonto Nat. Monument** with prehistoric cliff dwellings.
CASA GRANDE RUINS NAT. MONUMENT (C-4). Prehistoric ruins are dominated by this structure, once four stories high with walls four feet thick at base.
GLEN CANYON DAM (C-2). Seen from **Glen Canyon Bridge** 700 feet above river. **Visitor Center** on rim of canyon. Self-guided tours of dam.
GRAND CANYON NAT. PARK (B-2). Endless pageant of colors in canyon of the Colorado River is one of nation's magnificent sights. (See page 123.)
LAKE MEAD (A-2). Formed by **Hoover Dam,** 726 feet high, with 550 miles of shoreline. Huge recreation area. Tours of dam. Boat cruises from marina.
METEOR CRATER (C-3). Unique pit about a mile wide and 600 feet deep was created by a meteorite.
MONTEZUMA CASTLE NAT. MONUMENT (C-3). Five-story, 20-room prehistoric cliff dwellings.
OAK CREEK CANYON (B-3). U.S. 89A twists through vistas of broad gorges, red sandstone formations and verdant foliage.
PETRIFIED FOREST NAT. PARK (D-3). Largest known concentration of petrified wood in the world seen from park road. Views of **Painted Desert.**
SAGUARO NAT. MONUMENT (C-4). Giant cacti and many other desert plants. Scenic loop auto drive.

TOMBSTONE (D-5). Famous early mining camp. **Bird Cage Theater** and **Boot Hill Graveyard** among historic landmarks.

ARKANSAS

See map on page 24

BLANCHARD SPRINGS CAVERNS (A-2). Underground trails enter two huge rooms and pass outstanding stone formations. Guided tours.
HOT SPRINGS NAT. PARK (A-2). World-famous springs in scenic **Ouachita Mountains.** Mineral baths regulated by U.S. Government. (See page 119.)
LITTLE ROCK (A-2). **Territorial Restoration** recalls pre-statehood era in its restored structures. Arkansas' **First State Capitol** is a museum of state history.
MOUNTAIN VIEW (A-2). Center of folk music, arts, crafts. Demonstrations at **Ozark Folk Center.**
MURFREESBORO (A-2). Visitors may keep any diamonds they find at **Crater of Diamonds.**
OUACHITA MOUNTAINS (A-2). Fishing, swimming, hiking and camping in pine-covered mountain area. Recreation areas and 55-mile-long **Talimena Scenic Drive** in **Ouachita Nat. Forest. Petit Jean State Park,** near Morrilton (A-2), noted for scenic beauty of cascades and canyons.
OZARK MOUNTAINS (A-1). Scenic region with many resorts and recreational facilities. **Ozark Nat. Forest** includes **Magazine Mountain,** highest peak in the state.

CALIFORNIA

See maps on pages 50 and 51

ANAHEIM (F-4). **Disneyland,** internationally known family amusement center where America's past, present and future are depicted in rides and exhibits.
DEATH VALLEY NAT. MONUMENT (F-2). Expanse of desert wasteland with fantastic geological formations, canyons, cliffs and sand dunes.
FORT BRAGG (A-2). Forty-mile trip on **California Western Railroad** through beautiful redwood region.
HEARST SAN SIMEON STATE HISTORICAL MONUMENT (B-4). Mountaintop estate, former home of William Randolph Hearst. Collection of art treasures.
LAKE TAHOE (B-2). Mile-high lake in Sierra Nevada area is center of a popular resort.
LASSEN VOLCANIC NAT. PARK (B-1). Lassen Peak, which last erupted in 1921, has exhibits of volcanic activity around it.
LOS ANGELES (E-4). Exposition Park contains **California Museum of Science and Industry** with exhibits on health and contemporary resources, and **Los Angeles County Natural History Museum** housing historic and natural science displays. **Marineland,** Palos Verdes Estates, features shows with trained whales, porpoises and sea lions. The former luxury liner, **Queen Mary,** is berthed in Long Beach and offers tours, shops, restaurants and a hotel on board. Visitors are taken on tours of sound stages and movie sets at **Universal Studios,** Universal City. **Griffith Park,** Glendale, one of the world's largest city parks, houses **Griffith Observatory and Planetarium.**
MONTEREY (A-3). Picturesque first capital of California. **Old Custom House,** fishing docks and canneries; **17-Mile Drive.** Charming old **Carmel Mission** is at nearby Carmel.
MOTHER LODE REGION. Calif. 49 (B-2) traverses old camps and towns of the '49er gold rush. The state has preserved **Columbia** as an early mining town, and **Marshall** gold discovery site near Coloma.
MUIR WOODS NAT. MONUMENT (see San Francisco city map on page 104). Grove of redwoods on Mt. Tamalpais.

Bright Angel Point, Grand Canyon National Park, Arizona

DAVID MUENCH

PALOMAR OBSERVATORY (F-4). 200-inch reflecting telescope.

REDWOOD NAT. PARK (A-1). Under development, preserves coastal areas with magnificent groves of redwoods.

SAN DIEGO (F-4). Balboa Park, in center of city, covers 1,400 acres with botanical gardens, museums, galleries, the **Reuben H. Fleet Space Theater** and the renowned **San Diego Zoo,** displaying the world's largest animal population. **Sea World** in Mission Bay Park presents shows with trained whales, dolphins and seals. **San Diego Wild Animal Park** preserves wildlife in natural settings, seen from monorail.

SAN FRANCISCO (A-3). Golden Gate Park is one of the world's most beautiful parks. Contains **M.H. de Young Memorial Museum** and the **California Academy of Sciences. Fisherman's Wharf** is home of the fishing fleet and many fine restaurants. Nearby are the **Historic Ships; Balclutha,** an old-time square-rigger; and two shopping and entertainment centers: **The Cannery** and **Ghirardelli Square. Alcatraz Island,** the infamous former Federal prison, can be seen on guided tours leaving Pier 43. Reservations required. The **49-Mile Drive,** marked auto route, passes by major places of interest. **Marine World/Africa U.S.A.,** Redwood City, 58-acre amusement park, features shows and exhibits of marine life.

SAN JUAN CAPISTRANO (F-4). Old mission church noted for charm of its arcades and gardens.

SANTA BARBARA (E-3). Well-known resort. **Santa Barbara Mission** is one of California's loveliest.

SANTA CLARA (see San Francisco city map on page 104). **Great America** is a five-theme family amusement complex

SEQUOIA AND KINGS CANYON NAT. PARKS (C-3). Giant sequoias include General Sherman Tree, the largest known, 272 feet high, 101 feet around at base. Wilderness region of deep canyons and lakes.

WINERY TOURS. Wineries in **Napa, St. Helena, Calistoga** (A-2), welcome visitors.

YOSEMITE NAT. PARK (C-3). Famed for Yosemite Valley with its waterfalls, groves of giant sequoias, **El Capitan,** huge granite monolith, and **Half Dome,** rising 4,850 feet. (See page 122.)

COLORADO
See map on page 45

ASPEN (B-3). **Aspen Musical Festival and School**

presents concerts and recitals in summer. **Maroon Lake** and **Maroon Bells** are glacial areas nearby.

BLACK CANYON OF THE GUNNISON NAT. MONUMENT (B-3). Ten-mile area of the most spectacular section of canyon of the Gunnison River. The maximum depth of sheer-walled canyon is 2,425 feet.

CENTRAL CITY (C-3). Preserved old mining town holds theater festival during the summer in **Opera House.** Western memorabilia in **Teller House.**

COLORADO NAT. MONUMENT (A-3). Fantastically shaped rock formations and sheer-walled canyons carved by erosion. Scenic rim drive.

COLORADO SPRINGS (C-3). **Garden of the Gods** has acres of oddly eroded sandstone. **Will Rogers Shrine of the Sun** reached by Cheyenne Mountain Highway. **U.S. Air Force Academy.** Auto road from Cascade and cog railway from Manitou Springs to summit of Pikes Peak.

DENVER (C-3). **City Park** has a zoo which displays nearly 400 species in natural habitats. **Forney Transportation Museum** contains more than 250 antique vehicles, including trains, cars and planes. **U.S. Mint** produced its first coin here in 1906. All denominations minted here today. Outstanding collection of art objects from around the world in the **Art Museum.**

DINOSAUR NAT. MONUMENT (A-2). Vast primitive area of 330 square miles with spectacular canyons. Dinosaur fossils. Also see Utah text.

GREAT SAND DUNES NAT. MONUMENT (C-4). Great expanse of colorful, shifting dunes.

MESA VERDE NAT. PARK (A-4). Largest and best-preserved cliff-dweller ruins in America. Most important is Cliff Palace with 200 rooms and 23 ceremonial chambers. Guided trips and campfire talks.

ROCKY MOUNTAIN NAT. PARK (C-2). Huge area of mountains, forests, alpine lakes and rugged gorges. Longs Peak (14,256 feet) is highest of 65 peaks over 10,000 feet. (See page 120.)

ROYAL GORGE (C-3). Scenic gorge of Arkansas River spanned by one of world's highest suspension bridges, 1,053 feet above stream. **Incline Railway** to gorge bottom; **Aerial Tramway** across it.

CONNECTICUT
See map on page 14

GILLETTE CASTLE (A-3). Unique former home of late actor William Gillette.

HARTFORD (A-3). Colt Collection of Firearms in **Museum of Connecticut History, State Library.**

Nook Farm includes Mark Twain Home and Harriet Beecher Stowe House. **Old State House,** nation's oldest state house, has original Gilbert Stuart portrait of Washington and collection of political memorabilia.

MYSTIC (A-3). Renowned maritime museum, **Mystic Seaport,** has famous ships and 60 buildings with vast collection of seafaring memorabilia.

NEW HAVEN (A-3). Home of **Yale University. Peabody Museum** has natural history collection.

DELAWARE
See map on page 19

DOVER (E-2). **State House,** built more than 150 years ago, is still in use. **State Museum.**

NEW CASTLE, near Wilmington (E-2). Quaint community of old homes and buildings retains colonial atmosphere. Of interest is **Amstel House Museum.**

WILMINGTON (E-2). **Hagley Museum** exhibits tell story of Brandywine Valley industry, features historic du Pont black powder mills. **Winterthur,** former Henry du Pont home, is a magnificent museum devoted to American arts from 1684 to 1840.

DISTRICT OF COLUMBIA
See map on page 19

WASHINGTON (D-2). **Air and Space Museum,** a Smithsonian Institution unit, is a spectacular showcase of nation's air and space achievements. Other Smithsonian museums include the **National Museum of American History, National Museum of Natural History,** and the **National Gallery of Art** and its **East Building.** Tours of **Capitol** include visit to Senate or House. Portion of **White House** open to visitors on tours. Other highlights are **Washington Monument, Lincoln Memorial, Arlington National Cemetery.**

FLORIDA
See map on page 25

BISCAYNE NAT. PARK (C-3). Keys, coral reefs and marine life comprise this water-oriented park.

CAPE CANAVERAL (C-2). **John F. Kennedy Space Center** can be toured by auto Sun. only. Bus tours.

EVERGLADES NAT. PARK (C-3). Wilderness of swamps and hammocks. Wildlife and tropical growth seen from trails; **tram tour** of Shark Valley area.

Vernal Falls, Yosemite National Park, California

DAVID MUENCH

Washington, D.C.

CHUCK O'REAR

Cypress Gardens, Florida

FRED SIEB

GARDENS. Among Florida's famous gardens are **Cypress Gardens** (C-3), **Maclay State Gardens** near Tallahassee (A-1) and **Ravine State Gardens,** Palatka (C-2).

LAKE WALES (C-3). **Bok Tower Gardens,** in garden setting of Mountain Lake Sanctuary, features carillon concerts. Nearby **Passion Play** is presented outdoors. **Masterpiece Gardens** features a mosaic reproduction of da Vinci's "The Last Supper."

LION COUNTRY SAFARI (C-3). Visitors drive their cars through park where lions, giraffes, elephants, other wild animals roam uncaged.

MARINELAND OF FLORIDA (C-2). Many varieties of deep-sea life in an ocean floor setting in outdoor aquaria. Performances by trained porpoises.

MIAMI (C-3). Performances by trained whales, porpoises and sea lions, live sharks in shark channel and other marine specimens on display in ocean floor settings at **Miami Seaquarium.** Hundreds of cobras and other reptiles on display, some "milked" for venom, at **Miami Serpentarium.** Magnificent Italian mansion with acres of gardens, **Vizcaya** is the former James B. Deering estate; now Dade County Art Museum.

ORLANDO (C-2). **Circus World,** south of city, presents live circus acts, magic show, animal rides, musical show on circus theme. **Sea World of Florida** offers shows by trained whales, porpoises and seals, and a Shark Encounter exhibit. Water slides and other water-oriented fun at **Wet 'n' Wild.**

ST. AUGUSTINE (C-2). Quaint houses, narrow streets preserve charm of nation's oldest city, founded 1565. Extensive restoration in **San Agustin Antiguo** perpetuates Hispanic origin. **Castillo de San Marcos Nat. Monument,** Spanish fortress (1672).

ST. PETERSBURG (B-3). Attractions include **Sunken Gardens** with thousands of exotic plants.

SARASOTA (B-3). **Ringling Museum of Art, Ringling Residence, Museum of the Circus. Cars and Music of Yesterday. Sarasota Jungle Gardens.**

SPRINGS. Florida's famed springs include **Weeki Wachee Spring** (B-2) where swimmers perform underwater; **Silver Springs** (B-2) and **Wakulla Springs** (A-1) with glass-bottom and jungle boat rides; **Homosassa Springs** (B-2) where fresh- and saltwater fish live together in natural aquarium.

TAMPA (B-3). Hundreds of roaming animals in African veldt area in **Dark Continent,** theme amusement park, seen from monorail, railway, sky ride; trained animal and bird shows; brewery tours.

WALT DISNEY WORLD (C-2). Family vacation and entertainment complex includes six themed "lands" which depict the history, fantasy and future of America: Main Street, U.S.A., Frontierland, Fantasyland, Adventureland, Liberty Square and Tomorrowland with its popular Space Mountain. Live shows, parades, shops, restaurants, hotels, campground and recreational facilities.

GEORGIA
See map on pages 22 and 23

ATLANTA (C-2). **Six Flags Over Georgia,** a 276-acre family amusement park, depicts the history of Georgia through rides and shows. **Stone Mountain Park,** 16 miles east, has varied recreational facilities, auto museum, game ranch, train ride, water sports and memorial to Confederate heroes carved on side of mountain. **Atlanta Memorial Arts Center** is home of Atlanta Symphony, other performing arts, High Museum of Art. **Underground Atlanta** comprises buildings left underground when viaducts were constructed in the 1920's and which have been restored to an 1890's atmosphere with shops, restaurants and entertainment.

FORT PULASKI NAT. MONUMENT (E-2). Well-preserved fortress built 1829-1847.

LUMPKIN (C-2). Nearby **Westville** is functioning village of original structures demonstrating crafts and way of life of Georgia in 1850's.

MACON (D-2). Archaeological record of prehistoric Indians preserved in seven temple mounds at **Ocmulgee Nat. Monument.**

OKEFENOKEE SWAMP PARK (D-3). Vast jungle-like wilderness is sanctuary inhabited by rare birds, alligators and myriad fish. Boat tours and boardwalks.

WARM SPRINGS (C-2). **Little White House** was the Georgia home of Franklin D. Roosevelt, now a national shrine. Museum adjacent.

Chicago, Illinois

HAWAII
See map on page 52

HAWAII (D-5). On the "Big Island" is **Hawaii Volcanoes Nat. Park,** an area of active volcanoes. Crater rim drive and views of extensive lava fields. In the resort area of **Kona** on the west coast there are great fern forests, acres of wild orchids, beaches, deep-sea fishing boats and **Hulihee Palace,** summer home of Hawaiian royalty, now a museum.

KAUAI (B-4). Hawaii's "Garden Isle" where South Pacific was filmed and where, it is said, the rainbow was born. It is the wettest, greenest and oldest geologically of the islands.

MAUI (D-4). An island of lush jungles, beaches and waterfalls. **Haleakala Nat. Park** preserves a dormant volcano with trails into the huge crater.

OAHU (A-5). Honolulu is the capital and home for more than a third of Hawaii's population. The $30 million **State Capitol** is the focal point for the civic center. Guided tours of the **University of Hawaii's** landscaped grounds and East-West Center. International surfing championships held usually in Dec. **Cruises** include trips to Pearl Harbor. World-famous **Waikiki Beach,** with huge shopping centers, hotels and apartments, is a center for tourist activities.

IDAHO
See map on page 42

BOISE (B-4). State capital and gateway to hunting and fishing in **Boise Nat. Forest** area. **Pioneer Village, Boise Art Gallery** and **Idaho Historical Museum** in Julia Davis Park.

CHALLIS (C-3). **Salmon River Gorge** cuts through a wild area of forests and mountains.

COEUR D'ALENE (B-2). **Coeur d'Alene Lake** is one of America's most beautiful. **Heyburn State Park.**

CRATERS OF THE MOON NAT. MONUMENT (D-4). Over 80 square miles of lava flows, terraces, caverns and bridges resulting from volcanic activity. Loop drive in spring and summer. (See page 122.)

HELLS CANYON (B-3). Deepest gorge in the United States. Scenic areas offer beautiful views.

SUN VALLEY (C-4). Skiing is the featured winter attraction at this year-round resort. Chair lifts to summits. Swimming pool, other sports facilities.

TWIN FALLS (C-4). Nearby is **Shoshone Falls,** 210-foot-high cataract. **Perrine Memorial Bridge** spans Snake River 486 feet below.

ILLINOIS
See map on page 31

CHICAGO (D-2). **Field Museum of Natural History** houses famous collection of primitive art, Stone Age dioramas, halls of plants, fossils, jades and gems, habitat groups of birds and animals. **Museum of Science and Industry** contains an operating coal mine and a German submarine which may be explored. Applied science, engineering and industry exhibits include many that can be hand operated. Panoramas of city and environs from observatories of **Sears Tower,** world's tallest building, and the **John Hancock Center.** About 550 species of fish are represented in **Shedd Aquarium.**

GALENA (B-1). Built on five levels, old town reflects its lead-mining period when it thrived between 1840 and 1880. **Ulysses S. Grant Home** and **Old Market** are state memorials.

GREAT AMERICA (D-1). Family-oriented amusement complex with more than 125 thrill rides, entertainment, shops, boutiques, restaurants.

LINCOLN VILLAGE (C-3). Reproduction in Lincoln's **New Salem State Park** of village where Lincoln lived 1831-1837.

SPRINGFIELD (C-3). The state capital, home of Abraham Lincoln for many years, his final resting place. **Lincoln Home** contains period furnishings. **Lincoln Tomb** in Oak Ridge Cemetery.

INDIANA
See map on page 26

INDIANAPOLIS (A-2). **State Capitol; Benjamin Harrison Home,** where 23rd President lived; **Indianapolis Motor Speedway,** scene of famous 500-mile race. **Conner Prairie Pioneer Settlement** north of city depicts Indiana's early days.

LINCOLN BOYHOOD NAT. MEMORIAL (A-2). Abraham Lincoln lived in area 1816-1830. Grave of Lincoln's mother, Nancy Hanks Lincoln. **Living Historical Farm.**

MICHIGAN CITY (A-1). West of this resort city is the Indiana sand dune area preserved in **Indiana Dunes State Park** and **Indiana Dunes Nat. Lakeshore.**

NASHVILLE (A-2). Quaint village in area of rolling hills is an artist colony. Some galleries and shops of artisans welcome visitors.

SANTA CLAUS (A-2). **Santa Claus Land** is a family entertainment park with rides and shows.

SPRING MILL STATE PARK (A-2). Re-creation of pioneer village of 1850's.

IOWA
See map on page 37

DECORAH (G-1). Large springs, ice caves and bluffs in this scenic region. **Norwegian-American Museum.**

DES MOINES (F-2). **Living History Farms** re-create 130 years of agriculture in period farms. **Des Moines Art Center** shows 19th- and 20th-century art.

HERBERT HOOVER NAT. HISTORIC SITE (G-2). Birthplace and home of 31st President. **Hoover Presidential Library.**

MAQUOKETA (G-1). Thirteen caves, some illuminated, hiking trails around rim of deep ravine, in **Maquoketa Caves State Park.**

McGREGOR (G-1). Outstanding scenery in vicinity, including **Pikes Peak State Park** high above the Mississippi River. **Effigy Mounds Nat. Monument** contains examples of prehistoric burial mounds.

KANSAS
See map on pages 38 and 39

ABILENE (E-1). **Eisenhower Center** includes Dwight D. Eisenhower Home, Museum, Library and the Place of Meditation, his final resting place.

DODGE CITY (C-2). Former "Cowboy Capital." **Boot Hill,** famous cowboy burial ground, **Beeson Museum.**

FORT LARNED NAT. HISTORIC SITE (D-1). Military post built to protect Santa Fe Trail trade.

KANSAS CITY (F-1). **Agricultural Hall of Fame and National Center** pays tribute to agriculture in America. Exhibits include horse-drawn vehicles, antique cars, forging display. **Huron Indian Cemetery,** Huron Park. Burial ground of Wyandots. Indian and pioneer items in **Wyandotte County Museum.**

TOPEKA (F-1). **State House** and **State Historical Society Museum. Reinisch Rose Gardens** in Gage Park.

WICHITA (E-2). **Historical Museum** has Indian relic displays. **Cowtown** preserves flavor of city's era as a cattle center. Aircraft production center.

KENTUCKY
See map on pages 32 and 33

BARDSTOWN (C-2). **My Old Kentucky Home,** plantation manor immortalized in song by Stephen Foster. **"Stephen Foster Story"** outdoor musical in summer.

CUMBERLAND FALLS STATE RESORT PARK (C-2). Park contains one of the largest waterfalls in eastern U.S.

FRANKFORT (C-1). **State Capitol** resembles U.S. Capitol. **Old State Capitol** contains unique double stairway.

HARRODSBURG (C-2). Reproduction of historic Fort Harrod in **Old Fort Harrod State Park.**

LAND BETWEEN THE LAKES (A-2). Vast recreation area of the Tennessee Valley Authority between Kentucky Lake and Lake Barkley.

LEXINGTON (C-1). Heart of the **Blue Grass Region,** breeding ground of thoroughbred horses. Some horse farms welcome visitors. **Kentucky Horse Park** is dedicated to the thoroughbred with exhibits and special events. **Ashland,** home of Henry Clay.

LINCOLN BIRTHPLACE NAT. HISTORIC SITE (C-2). Traditional Lincoln birthplace cabin is preserved in marble and granite memorial.

LOUISVILLE (C-1). **Churchill Downs** is scene of annual Kentucky Derby. **Kentucky Derby Museum** on grounds. **Zoological Garden** exhibits 500 animals.

MAMMOTH CAVE NAT. PARK (B-2). Enormous limestone caverns, underground river and lake, odd formations. Guided tours.

LOUISIANA
See map on page 24

BATON ROUGE (B-4). Observation tower of the **State Capitol** offers views of the Mississippi. **Old Arsenal Museum** tells story of Louisiana history.

HODGES GARDENS (A-4). Year-round blooms in gardens set against natural backgrounds.

JUNGLE GARDENS (A-5). Extensive collection of subtropical plants. Home of thousands of water birds.

NATCHITOCHES (A-4). Many architectural reminders of French and Spanish founders. Antebellum homes and plantations. Tours of some homes.

NEW ORLEANS (C-4). Fascinating city of Old World charm and atmosphere with renowned restaurants, quaint patios, antique shops centered in the **Vieux Carré,** the old French Quarter. **French Market,** restored 150-year-old landmark, has series of shops. Facing Jackson Square are **Cathedral of St. Louis,** one of nation's fabled churches, flanked by **Cabildo** and **Presbytère,** both of which have State Museum exhibits. **Bourbon Street** is mecca of night life and **Royal Street** contains antique shops.

ST. MARTINVILLE (B-4). The grave of Emmeline Labiche, the Evangeline of Longfellow's poem. **Acadian House Museum** in Longfellow-Evangeline State Park.

MAINE
See map on page 15

ACADIA NAT. PARK (B-2). Spectacular ocean and mountain scenery. Paved roads along ocean and to top of Cadillac Mountain. Camping, hiking, fishing, museums, naturalist service. (See page 118.)

BATH (B-3). Historic shipbuilding town. Four sites of **Maine Maritime Museum** include shipyard where wooden sailing vessels were built.

KENNEBUNK (A-3). **Seashore Trolley Museum** exhibits old-time trolley cars and offers rides to visitors.

PORTLAND (A-3). Old seaport and gateway to Maine's vacationland of beaches, lakes and forests. **Wadsworth-Longfellow House.**

SEARSPORT (B-2). **Penobscot Marine Museum** is housed in six buildings.

MARYLAND
See map on pages 18 and 19

ANNAPOLIS (D-2). Founded 1649. Old section is an **Historic District. State House,** oldest in use. **U.S. Naval Academy** and museum.

ANTIETAM NAT. BATTLEFIELD (D-2). Scene of battle in 1862. Museum at cemetery entrance.

BALTIMORE (D-2). Nearby **Fort McHenry,** birthplace of "The Star Spangled Banner," contains restored barracks with War of 1812 exhibits. **Constellation,** U.S. Navy's first ship. **B & O Transportation Museum** of railroad equipment.

OCEAN CITY (E-2). Oceanfront resort. Bathing in Atlantic Ocean. Marlin fishing and other amusements are available.

Jenney Grist Mill, Plymouth, Massachusetts

FRED SIEB

MASSACHUSETTS
See map on page 14

BOSTON (B-2). The **Freedom Trail,** a two-mile walking path through downtown Boston, covers 16 places of outstanding historic interest. Starts at **Boston Common** and ends at the **USS Constitution,** "Old Ironsides," at Boston Naval Shipyard. Includes **Faneuil Hall,** the original "Cradle of Liberty" and the setting for pre-Revolutionary meetings, and Faneuil Hall **Marketplace,** Dock Square. Next door is restored **Quincy Market,** with restaurants, delicatessans and a variety of shops. **Old North Church, Old State House, Paul Revere House** are historic landmarks.

CAPE COD (B-3). **Cape Cod Nat. Seashore** preserves one of last remaining natural expanses of beach and dunes on Atlantic.

DEERFIELD (A-2). Old village with many 17th- and 18th-century structures. Museum houses.

EDAVILLE RAILROAD, near Wareham (B-3). Narrow-gauge, five-mile-long line runs through cranberry bogs.

FALL RIVER (B-3). **USS Massachusetts,** World War II battleship, submarine **Lionfish** and destroyer **Kennedy** in Battleship Cove.

GLOUCESTER (B-2). Old fishing port and resort center. Bronze statue of **Gloucester Fisherman** on waterfront. **Hammond Museum** contains collection of rare objects.

LEXINGTON AND CONCORD (B-2). Sites of Revolutionary battles preserved in **Minute Man Nat. Historical Park: North Bridge, Battle Green, Battle Road.** Visitor Centers near North Bridge, Concord and alongside Battle Road.

NANTUCKET (B-3). Formerly great whaling center is still picturesque. Museum.

NEW BEDFORD (B-3). Relics of whaling industry in **Whaling Museum.**

PLYMOUTH (B-3). Traditional landing place of Pilgrims is canopied under a granite portico. **Plimoth Plantation** includes reproduction of ship "Mayflower II" and Pilgrim Village. **Jenney Grist Mill** is a reconstructed 1636 water-powered corn-grinding mill.

SALEM (B-2). **House of Seven Gables** made famous by Hawthorne. **Salem Maritime Nat. Historic Site** contains Old Custom House and Derby Wharf.

SAUGUS, near Boston (B-2). **Saugus Iron Works Nat. Historic Site,** "birthplace" of nation's iron industry, has reconstructed furnace and mill.

SPRINGFIELD (A-2). **Springfield Armory,** nation's oldest arsenal, has extensive small arms collection. **Basketball Hall of Fame.**

STURBRIDGE (B-2). **Old Sturbridge Village** is a living re-creation of an early 19th-century rural New England village demonstrating Yankee arts and crafts.

MICHIGAN
See map on pages 28 and 29

BATTLE CREEK (C-2). Cereal manufacturing center. Kellogg Co. offers tours.

DEARBORN (D-2). **Greenfield Village,** a reconstructed town with historic buildings and **Henry Ford Museum** of Americana.

DETROIT (D-2). Several automobile plants are open to visitors. **Belle Isle** has museum and aquarium.

HOLLAND (B-2). **Tulip Time Festival** preserves many Dutch customs. **Baker Furniture Museum. Windmill Island** has miniature Little Netherlands, gardens, old windmill.

ISLE ROYALE NAT. PARK (A-1). Scenic wilderness recreation area.

MACKINAC ISLAND (C-1). No autos allowed on this historic island, now a pleasant summer resort.

SAULT STE. MARIE (C-1). Famous **Soo Locks** raise ships 21 feet between Lakes Huron and Superior.

MINNESOTA
See map on page 35

BEMIDJI (F-2). Hub of extensive lake and forest recreation area. **Historical and Wildlife Museum.** Giant statues of Paul Bunyan and Babe the Blue Ox.

DULUTH (H-2). One of the world's largest inland ports, the westernmost port on the St. Lawrence Seaway. **Harbor tours** provide views of ore docks, grain elevators. **Aerial Lift Bridge** and **Marine Museum** at harbor entrance. **Skyline** and **North Shore drives** are also of interest.

HIBBING (G-2). Largest city in the 110-mile-long Mesabi Iron Range, part of state's "iron ore country." Observatory at huge open-pit iron mine.

MINNEAPOLIS (G-3). **Walker Art Center, University of Minnesota, Minnehaha Falls, Minneapolis Institute of Arts, Nicollet Mall.**

ST. PAUL (G-3). **State Capitol** resembles St. Peter's in Rome. **City Hall and Court House, St. Paul Arts and Science Center.**

TOWER (G-2). **Tower-Soudan State Park** features state's first and deepest underground iron mine.

VOYAGEURS NAT. PARK (G-1). Beautiful northern lakes were once route of French-Canadian voyageurs.

MISSISSIPPI

See map on page 24

ANTEBELLUM HOMES. Tradition of Old South preserved in these lovely homes. Many open during Spring Pilgrimages in **Natchez** (B-4), **Columbus** (C-3) and **Holly Springs** (C-2).

GREENWOOD (B-3). **Florewood River Plantation** is re-creation of working cotton plantation of 1850's.

GULF COAST. Old World atmosphere, colorful fishing fleets and many shrimp packing plants at **Biloxi** (C-4). Near Biloxi is **Beauvoir,** last home of Jefferson Davis, President of the Confederacy.

VICKSBURG (B-3). In **Vicksburg Nat. Military Park** are extensive remains of trenches and earthworks. Markers tell the story of the siege in 1863. Tours of **U.S. Waterways Experiment Station.**

MISSOURI

See map on page 37

BRANSON (F-4). Ozark Mountain craft demonstrations in **Silver Dollar City** west of town. Country music shows, water recreation, amusement rides.

HANNIBAL (G-3). Boyhood home of Mark Twain, the author and humorist, and the **Mark Twain Museum.** Nearby in Monroe City is the **Mark Twain Birthplace Memorial Shrine and State Park.**

INDEPENDENCE (E-3). **Harry S. Truman Library** is a research center and museum displaying Truman memorabilia.

KANSAS CITY (E-3). More than 70 rides, shows and other attractions at **Worlds of Fun,** 48-acre theme park near city. **Nelson Gallery of Art and Atkins Museum** houses one of the nation's finest collections, including an extensive display of Oriental art.

ST. JOSEPH (E-3). Of interest is **Pony Express Stables Museum,** tracing the history of the West and Pony Express.

ST. LOUIS (G-3). **Gateway Arch** of **Jefferson Nat. Expansion Memorial** offers rides to top. Included in national memorial are **Old Courthouse,** with exhibits on westward expansion, and **Old Cathedral. Six Flags-St. Louis** at nearby Eureka is a 200-acre family amusement center. **Missouri Botanical Garden** features geodesic dome with tropical plantings. In Forest Park are conservatory, municipal opera, **St. Louis Art Museum, St. Louis Zoological Park** and **Mc Donnell Planetarium.** Near city is **National Museum of Transport** with motor vehicle, railroad, aircraft, communication displays.

MONTANA

See map on pages 42 and 43

BUTTE (D-2). Mining center often called "Richest Hill on Earth" has produced much of U.S. copper and other ores for 100 years.

CUSTER BATTLEFIELD NAT. MONUMENT (G-3). Site of Battle of Little Big Horn where Lt. Col. George Custer and his troops were slain by the Sioux, June 25, 1876.

FORT PECK DAM (G-2). Hydraulically filled earth dam harnesses Missouri River.

GLACIER NAT. PARK (D-1). Rugged mountain area with over 50 glaciers, 200 lakes. **The Going-to-the-Sun Road** crosses the Continental Divide at 6,664 feet. (See page 123.)

GREAT FALLS (E-2). **Charles M. Russell Museum** and studio.

LEWIS AND CLARK CAVERN (E-3). Limestone cavern is one of Northwest's most beautiful.

NEBRASKA

See map on pages 36 and 37

CRAWFORD (A-1). **Fort Robinson,** old frontier

Gateway Arch, St. Louis, Missouri

JACK ZEHRT

military post established in 1874, is preserved. Museum portrays story of the Indian wars.

LINCOLN (D-2). **State Historical Society Museum** has large collection of Western Americana. **University of Nebraska State Museum** has science and wildlife displays and planetarium. **State Capitol** houses the only unicameral legislature in U.S.

MINDEN (C-2). **Pioneer Village** houses vast collection of Americana of the plains.

NORTH PLATTE (B-2). **Scouts Rest Ranch** was home of "Buffalo Bill" Cody.

OMAHA (D-2). One of the world's largest livestock markets and meat-packing centers. **Joslyn Art Museum, Henry Doorly Zoo, Strategic Air Command Museum** at Offutt AFB with outdoor displays of aircraft and missiles.

SCOTTS BLUFF NAT. MONUMENT (A-2). Immense sandstone promontory towering above Old Oregon Trail. Trail history depicted in museum. Excellent view from top reached by auto road.

NEVADA

See map on page 49

CATHEDRAL GORGE STATE PARK (C-3). Picturesque formations of arches, spires and peculiar forms carved by erosion from stone.

LAKE MEAD (C-3). Formed by **Hoover Dam** (726 feet high) with 550 miles of shoreline. Tours of dam.

LAS VEGAS (C-3). Famous entertainment center featuring gambling and big-name shows. West of city is scenic **Red Rock Canyon.**

LEHMAN CAVES NAT. MONUMENT (C-2). Limestone caverns contain many formations of varied colors.

PYRAMID LAKE (A-2). Sparkling blue waters set in a barren but colorful setting.

RENO (A-2). Seat of **University of Nevada** with its **Mackay School of Mines, Mining Museum** and **Atmospherium-Planetarium.** Antique and vintage cars in **Harrah's Auto Collection.**

VALLEY OF FIRE STATE PARK (C-3). Red sandstone eroded by the elements in unusual formations.

VIRGINIA CITY (A-2). Historic mining town in Comstock Lode. Some old landmarks remain, including **Piper's Opera House, Old Court House.**

NEW HAMPSHIRE

See map on page 14

LAKE WINNIPESAUKEE (B-2). State's largest lake. Cruises spring to fall.

KANCAMAGUS HIGHWAY (B-2). N.H. 112 between Conway and North Woodstock, a 34-mile scenic route through **White Mountain Nat. Forest,** provides unspoiled views.

PORTSMOUTH (B-2). **Strawbery Banke,** ten-acre site containing over 30 buildings of the 17th-19th centuries, is being restored. Crafts, exhibits.

WHITE MOUNTAINS (B-1). Summit of 6,288-foot Mt. Washington reached by cog railway, toll road, chauffeured van. In **Franconia Notch** are the **Flume,** a chasm traversed by boardwalks, and **The Old Man of the Mountains,** famous natural profile. **Aerial Tramway** carries passengers up Cannon Mt. and **Skimobile** runs up Cranmore Mt.

NEW JERSEY

See map on page 19

ATLANTIC CITY (E-2). A world-famous seaside resort. Casino gambling, top entertainers.

CAPE MAY (E-2). Victorian atmosphere retained at seaside resort.

EDISON NAT. HISTORIC SITE, West Orange, west of Newark (E-1). Laboratories, workshops, home and office of Thomas A. Edison.

FLEMINGTON (E-1). Workers demonstrate 18th-century crafts in **Liberty Village.**

MILLVILLE (E-2). **Wheaton Village** preserves the old-time way of making glass in reconstruction of glass-making town. Glass blowing, other crafts.

MORRISTOWN (E-1). **Morristown Nat. Historical Park** includes Washington's headquarters in 1779-1780. Historical museum. **Wick House** nearby.

NETCONG (E-1). Grist mill, store, church, homes comprise nearby **Waterloo Village,** an original colonial community.

PRINCETON (E-1). **Nassau Hall** at **Princeton University** was meeting place of Continental Congress in 1783.

SIX FLAGS GREAT ADVENTURE (E-1). Drive-through wild animal park and extensive family amusement park.

SMITHVILLE (E-2). **Old Village** contains restored buildings of 1700's. Crafts.

TRENTON (E-1). **State House, Old Barracks** built in 1758. **Cultural Center** includes State Museum and a planetarium.

NEW MEXICO

See map on page 47

ALBUQUERQUE (B-3). State's largest city, founded

American Falls, Niagara Falls, New York

FRED SIEB

1706. **Old Town Plaza** preserves original buildings. Aerial tram, longest on continent, climbs west side of Sandia Peak. **Turquoise Trail,** 28-mile paved auto road, ascends other side of peak.

AZTEC RUINS NAT. MONUMENT (A-2). Pueblo ruin of 500 rooms, noted for its **Great Kiva,** a ceremonial chamber.

BANDELIER NAT. MONUMENT (B-3). Cliff dwellings, ruins of communal houses, home of prehistoric Indians 1200-1500 A.D. Museum.

CAPULIN MOUNTAIN NAT. MONUMENT (D-2). Cone of extinct volcano rises 1,000 feet above the plain to 8,182 feet. Road to top; trail down into crater.

CARLSBAD CAVERNS NAT. PARK (C-4). Largest limestone caverns yet discovered. Guided tours conducted through miles of lighted corridors.

CHACO CULTURE NAT. HISTORICAL PARK (A-2). One of the outstanding archaeological areas in the U.S. More than a dozen large ruins, occupied in 12th century. Site includes 800-room **Pueblo Bonito.**

ENCHANTED MESA (B-3). Castle-like sandstone butte 430 feet high. To the west is **Acoma Pueblo,** an Indian village and historic mission church.

FORT UNION NAT. MONUMENT (C-2). Ruins of largest military post on southwestern frontier.

GALLUP (A-3). As many as 20 tribes participate in annual **Intertribal Indian Ceremonial** in mid-Aug.

SANTA FE (C-3). Romantic old state capital retains charm of bygone days. **Palace of the Governors,** 370 years old and nation's oldest public building, has art and historic museum. **San Miguel Church.**

TAOS (C-2). Renowned artist colony. North of town is **Taos Pueblo,** famous Indian community.

WHITE SANDS NAT. MONUMENT (B-4). Winds have piled the deposits of almost pure gypsum into great dunes covering a huge area. Auto route access.

NEW YORK

See map on pages 16 and 17

ADIRONDACK MOUNTAINS (D-2). Outdoor sports area. Toll road up 4,867-foot **Whiteface Mountain** (E-1). **High Falls Gorge** encloses lovely rapids and falls of the Ausable River.

ALBANY (E-2). **Empire State Plaza** houses state offices; **State Museum** is located in State Education Building.

AUSABLE CHASM (E-1). Vast canyon cut by river. Footpaths skirt curious rock formations, bridges span gorge and boats ply waters.

BUFFALO (B-2). **Albright-Knox Art Gallery** displays contemporary paintings as well as European and Oriental art. **Buffalo Museum of Science** houses dinosaur and other natural history exhibits.

CATSKILL GAME FARM, near Catskill (D-2). Tame deer, other animals.

COOPERSTOWN (D-2). **Baseball Hall of Fame** honors baseball immortals, commemorates birthplace of game. **Farmers' Museum** has antique farming implements, craft demonstrations, re-created village crossroads 1790-1860. **Fenimore House** has folk art, James Fenimore Cooper family mementos.

CORNING GLASS CENTER, Corning (C-2). Glass made by hand at Steuben Factory. **Hall of Science and Industy, Museum of Glass.**

FINGER LAKES (C-2). Resort center in central New York. Water sports, scenic drives, outstanding state parks. Tours of wineries in Hammondsport and Naples.

FORT TICONDEROGA, near Ticonderoga (E-2). Built in 1755 by French, captured by British (1759) and by Ethan Allen (1775). Barracks and museum.

HOWE CAVERNS (D-2). Great chambers cut by underground stream. Guided tours with boat ride.

HYDE PARK (E-3). Home of Franklin D. Roosevelt, library-museum and grave. Nearby is the Renaissance-style **Frederick W. Vanderbilt Mansion** with elaborate furnishings.

LAKE GEORGE (E-2). Resort center with "million-dollar" beach features a variety of family attractions.

LAKE PLACID (D-1). Internationally famous resort for summer and winter sports. Summer ice skating in **Olympic Arena.**

LETCHWORTH STATE PARK (C-2). 600-foot deep, 17-mile-long Genesee River gorge has three waterfalls. Museum, picnicking.

NEW YORK CITY (E-3). **American Museum of Natural History** contains Halls of Dinosaurs, Minerals and Gems, Earth History and Ocean Life. Sky shows in adjacent **Hayden Planetarium.** One of the world's premier collections from all periods in the **Metropolitan Museum of Art. Lincoln Center** offers entertainment in the arts; includes **Avery Fisher Hall** and the **Metropolitan Opera House.** Statue of Liberty houses **Museum of Immigration.** Boats leave from Battery Park. Panoramas of city from the **World Trade Center** and **Empire State Building. Bronx Zoo** is one of the nation's largest.

NIAGARA FALLS (B-2). One of the natural wonders of the world is actually three falls: the American and Bridal Veil Falls in the U.S. and the Horseshoe or Canadian Falls. **Goat Island** separates American and Canadian Falls. Miniature train makes the trip between Goat Island and Prospect Point. **Maid of the Mist** sightseeing boat cruises pass the American and Canadian falls. **Cave of the Winds** guided tour goes to bottom of Bridal Veil Falls. Vista of American Falls from **Prospect Point Tower.**

ROCHESTER (C-2). Home of **Eastman Kodak Company.** Tours of Elmgrove Plant (cameras, projectors); Kodak Park Division (films); Hawkeye Plant (optical goods). **International Museum of Photography.**

SARATOGA SPRINGS (D-2). State-owned spa noted for its mineral waters. **National Museum of Racing. Saratoga Nat. Historical Park** nearby.

THOUSAND ISLANDS (C-1). Sightseeing boats from Alexandria Bay and Clayton tour the many islands and channels of this vacationland.

WATKINS GLEN (C-2). Gorge more than 200 feet deep and nearly two miles long has waterfalls and grottoes. **Grand Prix** race in Oct.

NORTH CAROLINA

See map on pages 20 and 21

ASHEVILLE (B-3). Lavishly furnished **Biltmore House,** former George W. Vanderbilt home. **Thomas Wolfe Memorial,** boyhood home of author.

BLUE RIDGE PARKWAY (B-3, C-2). Scenic highway along summit of Blue Ridge Mountains. Breathtaking panoramas.

CAPE HATTERAS NAT. SEASHORE (F-3). Maintains the natural state of the Outer Banks, a long strip of barrier islands which separate the bay and ocean. Dunes, unusual flora.

CHAPEL HILL (D-3). **Morehead Planetarium** on campus of University of North Carolina presents nightly and weekend performances.

CHARLOTTE (C-3). **Carowinds,** south of city off Int. 77, is a seven-theme entertainment park.

DURHAM (D-3). Home of **Duke University.** Carillon recitals from **University Chapel. Sarah P. Duke Gardens** with plantings of irises and cherry trees.

FORT RALEIGH (F-3). Site of Sir Walter Raleigh's "Lost Colony" and birthplace of Virginia Dare, first English child born in America. Fort and museum. **"Lost Colony"** pageant in summer.

GREAT SMOKY MOUNTAINS NAT. PARK (B-3). See descriptive listing under Tennessee.

RALEIGH (D-3). On or near Capitol Square are

Museum of Art, Museum of History with exhibits on state's development, and Natural History Museum.

WILMINGTON (E-3). USS North Carolina, World War II battleship, is moored as a memorial. Near city are famous gardens: Greenfield, Orton Plantation and Airlie. Chandler's Wharf is a 19th-century waterfront restoration.

WINSTON-SALEM (C-2). Old Salem comprises restored buildings from early Moravian settlement.

WRIGHT BROTHERS NAT. MEMORIAL (F-2). Memorial on Kill Devil Hill marks the scene of the first successful airplane flight on Dec. 17, 1903.

NORTH DAKOTA
See map on pages 34 and 35

BADLANDS. Scenic region along Little Missouri River. De Mores State Historic Park near Medora (B-2) is a mansion where a French nobleman attempted to establish a cattle empire in the 1880's.

BISMARCK (C-2). The State Capitol, 18 stories high, has an observation deck. On grounds is State Historical Museum.

GARRISON DAM (C-2). One of the largest rolled-earth dams is a key unit in control of Missouri River.

INTERNATIONAL PEACE GARDEN (C-1). This formal garden commemorates amicable relations between the United States and Canada.

THEODORE ROOSEVELT NAT. PARK (B-2). Badlands region where Roosevelt ranched in 1880's.

TURTLE MOUNTAINS. Region of rolling hills and sparkling lakes near Dunseith (D-1). Lake Metigoshe State Park is in the area.

OHIO
See map on pages 26 and 27

CINCINNATI (B-2). Kings Island, north of city, is a family amusement park complex featuring rides, live shows and Lion Country Safari drive. Findlay Market sells fresh meats and produce in one of the oldest open-air markets in the U.S. Three states may be seen on a clear day from observatory on 42nd floor of Carew Tower. Cincinnati Zoo is one of the country's oldest and largest.

CLEVELAND (D-1). Ohio's largest city has many museums, including Museum of Art and Western Reserve Historical Society Museum.

COSHOCTON (C-1). Era of 1800's when Ohio-Erie Canal prospered re-created in Roscoe Village. Restored lock and buildings. Boat rides in summer.

DAYTON (B-2). Air Force Museum traces development of military aviation from the Wright Brothers. Also Museum of Natural History with planetarium and Dayton Art Institute; Carillon Park with museums, exhibits and Deeds Carillon.

MARIETTA (D-2). Site of earliest permanent white settlement in Ohio. Campus Martius Museum houses Indian and pioneer relics. Sternwheeler on exhibit in River Museum.

SANDUSKY (C-1). Amusement park at Cedar Point offers thrill rides and shows by trained marine animals.

SCHOENBRUNN VILLAGE, New Philadelphia (D-1). Authentic restoration of the state's first town, a Moravian village founded 1772.

SEA WORLD (D-1). Marine-life park has live shows with dolphins, sea lions and a killer whale; water-skiing shows, marine exhibits.

SERPENT MOUND (C-2). 1,330-foot serpentine structure is one of most remarkable Indian effigy mounds in America.

TOLEDO (C-1). Famous glass collection in Museum of Modern Art. Zoological Park is outstanding.

OKLAHOMA
See map on pages 38 and 39

ALABASTER CAVERNS STATE PARK, west of Alva (D-2). One of largest known gypsum caves extends one-half mile underground. Guided tours.

ANADARKO (D-3). Authentic early Plains Indian villages and their way of life re-created in Indian City, U.S.A.

CHICKASAW NAT. RECREATION AREA (E-3). Wooded area traversed by picturesque streams and known for springs and waterfalls.

CLAREMORE (F-2). Will Rogers Memorial houses humorist's mementos. J.M. Davis Gun Museum has collection of over 25,000 weapons.

OKLAHOMA CITY (D-3). Oklahoma Historical Society Building has one of nation's most complete displays of Indian relics. National Cowboy Hall of Fame is dedicated to pioneers who won the West. Frontier City, U.S.A. is a replica of 1800's town with rides and shops. Oklahoma City Zoo.

TAHLEQUAH (F-3). Capital of Cherokee Indian Nation. "Tsa-La-Gi," Cherokee village, and "Trail of Tears," outdoor drama presented in summer.

TULSA (F-2). Called the "Oil Capital of the World." Gilcrease Institute exhibits Western art and archaeology. Philbrook Art Center features European and American Indian art. World Museum/Art Center combines unique architecture with collection of international artifacts.

OREGON
See map on page 48

BEND (B-3). Awesome volcanic region to the south includes Lava Butte, 500-foot-high cinder cone with road to top, and Lava River Caves State Park, Lava Lands Visitor Center.

BONNEVILLE DAM (A-3). Massive project for hydroelectric power and navigation on Columbia River. Navigation lock, fish ladders and underwater fish viewing windows are of interest.

CRATER LAKE NAT. PARK (A-4). Indigo-blue lake, deepest in the U.S., in crater of extinct volcano six miles across. Road around rim. (See page 124.)

ENTERPRISE (D-3). Gateway to beautiful Wallowa Lake State Park, one of Northwest's grandest scenic areas. Gondola lift ride up Mt. Howard. To the east is the Hells Canyon Nat. Recreation Area of the Snake River. Overlook at Hat Point.

MOUNT HOOD (B-3). Perpetually snowcapped peak, over two miles high. Fishing, swimming, hiking, camping in surrounding area. Winter sports at Timberline Lodge. Mt. Hood Loop Highway (Int. 84, Oreg. 35, U.S. 26).

OREGON CAVES NAT. MONUMENT (A-4). Weird and fantastic limestone formations. Tours.

OREGON COAST. Most of Oregon's 400-mile shore is publicly owned. The coastal terrain is characterized by smooth beaches, rocky cliffs, offshore stone buttes. Much of sand dunes region is preserved in a national recreation area.

PORTLAND (A-3). Leading seaport, noted also for beautiful gardens. Rose Festival in mid-June.

Hood River Valley and Mt. Hood, Oregon

DAVID MUENCH

Portland Art Museum, Oregon Museum of Science and Industry, Washington Park Zoo, Western Forestry Center.

WINSTON (A-4). Nearly 70 species of animals roam free in Wildlife Safari, drive-through park.

PENNSYLVANIA
See map on pages 18 and 19

GETTYSBURG NAT. MILITARY PARK (D-2). Scene of Civil War battle in 1863. Miles of roadways wind past monuments. (See page 120.)

HERSHEY (D-1). Home of world's largest maker of chocolate and cocoa. Chocolate World transports visitors past animated exhibits telling story of chocolate from bean to candy. Hershey Gardens, Hershey Museum, Hersheypark with amusement rides and ZooAmerica, an environmental zoo.

LANCASTER (D-1). Amish, Mennonites and other religious sects centered in Lancaster. Wheatland, home of President James Buchanan. Nearby is the Pennsylvania Farm Museum of Landis Valley.

LONGWOOD GARDENS, near West Chester (E-2). Extensive outdoor gardens, four acres under glass.

PENNSBURY MANOR, near Morrisville, southwest of Trenton (E-1). Re-creation of William Penn's home includes the manor house, outbuildings and gardens.

PHILADELPHIA (E-1). Independence Nat. Historic Park includes Independence Hall, where Declaration of Independence was adopted, Congress Hall, Franklin Court, Liberty Bell Pavilion and Visitor Center which has exhibits. (See page 118.) Betsy Ross House is a restored colonial dwelling traditionally regarded as place where American flag was made. Franklin Institute has exhibits of applied science and mechanics.

PITTSBURGH (B-1). In the Golden Triangle where the Allegheny and Monongahela rivers merge to form the Ohio is Point State Park and Gateway Center of five skyscrapers. Carnegie Institute has millions of specimens in the Museum of Natural History. Museum of Art has a permanent collection of paintings, prints and decorative arts. Panoramas of downtown Pittsburgh from Mt. Washington, reached by Monongahela Incline and Duquesne Incline.

POCONOS (E-1). Scenic mountain resort region. Summer and winter sports. The Quiet Valley Living Historical Farm is near Stroudsburg.

ROADSIDE AMERICA (D-1). Large indoor display of miniature farms, villages and cities.

TITUSVILLE (C-1). Drake Well Park is on the site of original well drilled in 1859, marking the beginning of the oil industry. Museum.

VALLEY FORGE (E-1). Site of winter encampment of the Continental Army, 1777-1778. Here are Washington's Headquarters, the Memorial Chapel, replicas of soldiers' huts and other landmarks.

RHODE ISLAND
See map on page 14

NEWPORT (B-3). Marine views and estates along the Ocean Drive and Cliff Walk. The Breakers, magnificent Vanderbilt mansion built 1895, is foremost of several manor houses open to visitors. Among historic landmarks are Touro Synagogue, nation's oldest, and International Tennis Hall of Fame.

PAWTUCKET (B-3). Slater Mill, Wilkinson Mill and Sylvanus Brown House constitute museum of early crafts.

PROVIDENCE (B-3). State House is built of white Georgia marble. Brown University. Roger Williams Park noted for gardens.

SOUTH CAROLINA
See map on page 23

BEAUFORT (E-2). Many handsome antebellum homes. St. Helena's Church (1724) is fine example of colonial architecture.

BROOKGREEN GARDENS (F-2). Outstanding sculpture collection in a setting of boxwood, flowers and moss-hung oaks.

CHARLESTON (F-2). City of fine old churches, winding streets, lovely gardens and iron lace gateways. Outstanding points of interest: Charles Towne Landing, 667-acre site of first permanent English settlement in state, Dock Street Theatre, Fort Sumter, reached by boat. Nearby Cypress Gardens, Middleton Place and Magnolia Plantation.

COLUMBIA (E-1). **Riverbanks Zoo** features animal and bird exhibits in unrestricted areas.

EDISTO GARDENS, at Orangeburg (E-2). Banks of azaleas, rose bushes, other plants.

FLORENCE (D-1). Jet aircraft and missiles at **Air and Missile Museum.**

KINGS MOUNTAIN NAT. MILITARY PARK (E-1). Site of British surrender to frontiersmen in 1780.

MYRTLE BEACH (F-2). State park at shore resort. One of the finest beaches on Atlantic coast.

SOUTH DAKOTA
See map on pages 34 and 35

BADLANDS NAT. PARK (B-4). Multicolored, odd-shaped formations eroded from soft rock and clay contain fossils 40 million years old.

BLACK HILLS (B-3). Vacation playground with lovely lakes, picturesque streams. At Spearfish (A-3), **Black Hills Passion Play** presented. **Spearfish Canyon** nearby. Cave tours at **Jewel Cave Nat. Monument** in summer.

DEADWOOD-LEAD (B-3). "Twin cities" are historic gold mining centers. Tours of **Homestake Mine.**

MADISON (E-3). **Prairie Village** has 30 original pioneer structures furnished in 1890's period.

MT. RUSHMORE NAT. MEMORIAL (B-4). Gigantic heads of Washington, Jefferson, Lincoln and Theodore Roosevelt carved from side of Mt. Rushmore.

RAPID CITY (B-3). **Dinosaur Park** has life-size models. Fossils and minerals at **Museum of Geology. Sioux Indian Museum and Crafts Center.**

WIND CAVE NAT. PARK (B-4). Strong currents blow alternately in and out of entrance. Bison herd.

TENNESSEE
See map on pages 32 and 33

ANDREW JOHNSON NAT. HISTORIC SITE (D-2). Home, tailor shop and grave of 17th President.

CHATTANOOGA (C-3). Monuments mark sites of Civil War battles. **Incline Railway** carries passengers to top of Lookout Mountain. **Chattanooga Choo-Choo,** a restored Victorian railroad depot, has shops and restaurants. **Chickamauga and Chattanooga Nat. Military Park, Ruby Falls** and **Rock City** with unusual stone formations are also here.

COLUMBIA (B-3). The ancestral home of James K. Polk, 11th President.

FALL CREEK FALLS STATE RESORT PARK (C-3). Water cascade plunges 256 feet, is highest falls east of Rockies.

GREAT SMOKY MOUNTAINS NAT. PARK (D-3). Roads and trails wind through this mountain area named for the smoke-like haze hovering over it. Many peaks tower over 6,000 feet. There is an abundance of wildlife, large virgin hardwood, numerous waterfalls, a wealth of wildflowers and many trout streams. Water sports at **Fontana Lake.** Camping at designated places. **"Unto These Hills"** pageant and **Indian Village** at Cherokee. (See pages 118-119.)

MEMPHIS (A-3). Overton Park contains **Memphis Zoological Garden and Aquarium. Botanical Garden** in Audubon Park. **Graceland,** estate and mansion of the late Elvis Presley, contains his grave. **Beale Street** is where W.C. Handy, "Father of the Blues," wrote his famous "Memphis Blues," "St. Louis Blues" and other melodies.

NASHVILLE (B-2). **Opryland U.S.A.** family amusement park on American music theme. **Grand Ole Opry** on grounds. **Parthenon** in Centennial Park is a full-sized reproduction of the famed Parthenon in Athens. **Country Music Hall of Fame and Museum** honors famed performers. Near city is **The Hermitage,** the home of President Andrew Jackson.

OAK RIDGE (C-2). **American Museum of Science and Energy** explains uses of energy in exhibits.

SHILOH NAT. MILITARY PARK (A-3). Scene of the Battle of Shiloh in 1862, first major battle in the western campaigns of the Civil War.

SWEETWATER (C-3). Boats carry visitors over **Lost Sea,** a mysterious 4½-acre underground lake.

TENNESSEE WALKING HORSES. Shelbyville (B-3) is center of walking horse area.

TEXAS
See map on pages 40 and 41

AUSTIN (F-2). **LBJ Library** preserves papers of Lyndon B. Johnson; exhibits on career and term in office. The **State Capitol** open on tours. To the west at Johnson City is **L. B. Johnson Nat. Historic Site,** including boyhood home and birthplace of 36th President. Bus tours of **LBJ Ranch.**

BIG BEND NAT. PARK (C-3). Area of striking scenery in Chisos Mountains and Santa Elena and Boquillas canyons, geological phenomena, unusual plants and animals.

CORPUS CHRISTI (F-4). Popular seaside area and gateway to **Padre Island.**

DALLAS (F-1). **Fair Park** attractions include Texas Hall of State, aquarium, garden center, amusement area, Cotton Bowl, music hall and museums of Fine Arts, Health and Science and Natural History. Site of the Texas State Fair held each Oct. **John F. Kennedy Museum** offers a sound and light presentation, "The Incredible Hours."

FORT WORTH (F-1). **Six Flags Over Texas,** in Arlington between Fort Worth and Dallas, has more than 100 rides including the ShockWave double-loop roller coaster and Log Flume. Live shows and entertainment. **Amon Carter Museum of Western Art** houses an outstanding collection of Western paintings, sculpture, books and objects of art, principally works of Frederic Remington, Georgia O'Keeffe and Charles M. Russell. **Botanic Garden** in Trinity Park.

GUADALUPE MOUNTAINS NAT. PARK (B-2). Area of scenic highlands, desert, canyons and cliffs.

HOUSTON (G-3). **Lyndon B. Johnson Space Center,** southeast of city center, is the NASA facility for training astronauts, space vehicle development and manned space flight control. Visitor Orientation Center contains spacecraft and space flight articles. **Astrodome** was world's first air-conditioned, domed, all-purpose Stadium. Tours available. **Astroworld,** family theme and entertainment park, features more than 100 rides and attractions including the Texas Cyclone roller coaster. **Hermann Park** is a wooded area with miles of scenic drives, picnic areas, Zoological Gardens and Museum of Natural Science.

HUNTSVILLE (G-2). The "Mount Vernon of Texas." Last home and burial place of Sam Houston, first President of the Republic of Texas. **Sam Houston Memorial Museum.**

LAKE TEXOMA (F-1). One of the largest man-made lakes in U.S. Nearby is **Eisenhower Birthplace** where 34th President was born.

LUBBOCK (D-1). **Texas Tech University Museum** has science, environment, art displays. Outdoor exhibits of adjacent **Ranching Heritage Center** tell story of ranching in the Southwest.

MIDLAND (D-2). The **Permian Basin Petroleum Museum** depicts development of industry.

PALO DURO CANYON (B-4). Deeply eroded, colorful canyon in plains region. Scenic auto drives. **"Texas"** outdoor drama in summer.

SAN ANTONIO (E-3). **The Alamo,** premier shrine of Texas liberty, is where 188 Texans defended the mission/fort to the death against 2,500 Mexican troops of General Santa Anna in 1836. **HemisFair Plaza,** site of the 1968 HemisFair, has Tower of the Americas, and museums on Texan and Mexican cultures. **Paseo del Rio** is a visitors' delight with tree-lined walks along the San Antonio River, bordered by shops, galleries and restaurants. Zoological Gardens, Aquarium and Oriental Garden in **Brackenridge Park.**

SAN JACINTO BATTLEGROUND (G-3). On these grounds a small Texas army under Sam Houston routed Mexican professionals in April, 1836, to win independence for Texas. Limestone shaft 570 feet high has observatory and museum in base. Nearby is the **USS Texas,** retired veteran of two world wars and the first battleship to become a state shrine.

UTAH
See map on page 44

ARCHES NAT. PARK (D-3). Arches and windows carved by erosion through solid stone; some of them are over 100 feet high.

BRYCE CANYON NAT. PARK (B-4). Countless eroded and brilliantly colored rock formations.

DINOSAUR NAT. MONUMENT (D-2). Vast primitive area of 330 square miles with spectacular canyons. Dinosaur fossils preserved in **Visitor Center.** Also see Colorado text.

MONUMENT VALLEY (C-4). Buttes carved by erosion from red sandstone rise like skyscrapers.

NATURAL BRIDGES NAT. MONUMENT (C-4). Three immense sandstone bridges, and remains of prehistoric cliff dwellings.

SALT LAKE CITY (B-2). Headquarters of Mormon Church centered in 10-acre **Temple Square.** Fine collection in **Pioneer Museum.**

ZION NAT. PARK (B-4). Zion Canyon, a magnificent

Monument Valley, Utah

gorge with precipitous walls as high as 2,000 feet. **Great White Throne** is imposing natural feature.

VERMONT

See map on page 14

BARRE (A-1). **Rock of Ages** granite quarry and **Craftsman Center** welcome visitors.
BELLOWS FALLS (A-2). **Steamtown U.S.A.** offers 26-mile round trip on old-fashioned steam train. Also here is an extensive collection of locomotives and museum with displays pertaining to railroading.
MARBLE EXHIBIT, near Rutland (A-2). **Vermont Marble Company** exhibits marble products, shows motion pictures of quarrying.
MT. MANSFIELD (A-1). Toll road and gondola to summit, highest point in Vermont.
SHELBURNE MUSEUM, near Burlington (A-1). Collection of Americana housed in re-created 19th-century New England village.
SKY LINE DRIVE. Six-mile toll road to summit of 3,816-foot Equinox Mountain (A-2).

VIRGINIA

See map on pages 20 and 21

ALEXANDRIA (E-1). Atmosphere of bygone days is reflected in restored homes on cobblestone streets, and such places as **Ramsay House Visitor Center, Stabler-Leadbeater Apothecary, Gadsby's Tavern, Carlyle House, Christ Church, George Washington Masonic Nat. Memorial.** Shops and galleries.
APPOMATTOX COURT HOUSE NAT. HISTORICAL PARK (D-2). Scene of Lee's surrender to Grant, April 9, 1865. **McLean House,** where terms were drafted, is restored.
BLUE RIDGE PARKWAY (C,D-2). Almost 500 miles along crest of mountains in Virginia and North Carolina between Shenandoah and Great Smoky Mountains national parks.
CHARLOTTESVILLE (D-2). Seat of **University of Virginia. Monticello,** home of Thomas Jefferson, and **Ash Lawn,** James Monroe's residence, are nearby.
FREDERICKSBURG (E-1). **Kenmore,** home of Washington's sister, Betty Washington Lewis. **Mary Washington Home,** the home of his mother, and **James Monroe Museum and Memorial Library. National Military Park** covers four Civil War battlefields in the area.

JAMESTOWN (E-2). Site of first permanent English settlement in America (1607). **Festival Park** nearby includes replicas of colonists' three ships.
KINGS DOMINION (E-2). Family amusement park, including **Lion Country Safari.**
LEXINGTON (D-2). Home of **Virginia Military Institute, Washington and Lee University.** Tomb of Robert E. Lee in Lee Memorial Chapel.
MARINERS MUSEUM, near Newport News (E-2). One of the world's largest collections of ship figureheads, models, prints, other nautical material.
MOUNT VERNON (E-1). Washington's beautiful estate overlooking the Potomac River provides a glimpse of the life and times of our first President.
NATURAL BRIDGE (D-2), near Lexington. Massive rock arch carrying U.S. 11 has 90-foot span 215 feet above creek in gorge below.
RICHMOND (E-2). Among the points of interest are the **Capitol,** the **Museum of the Confederacy, St. John's Church** where Patrick Henry made his "Liberty or Death" speech, **Richmond Nat. Battlefield Park.** East of Richmond along Va. 5 are the renowned James River plantations, **Berkeley** and **Shirley,** that depict colonial plantation days.
SHENANDOAH VALLEY. Historic, recreational and scenic area in northwestern Virginia. **Natural Chimneys** (D-1), towering monoliths of weathered stone, and many limestone caverns are open to public.
SKYLINE DRIVE. Spectacular scenic highway follows crest of Blue Ridge Mountains for over 100 miles in **Shenandoah Nat. Park** (D-1). Beautiful views of region from 66 overlooks. (See page 120.)
STRATFORD HALL (E-1). Birthplace of Robert E. Lee. Mansion and gardens of this old colonial plantation have been restored.
WILLIAMSBURG (E-2). Williamsburg was the colonial capital of Virginia from 1699 to 1780. Today this historic restoration has 88 original 18th- and 19th-century buildings, many gardens and greens. The Information Center is the usual starting point for all visits. **The Old Country,** Bush Gardens family amusement park, is five miles east on U.S. 60.
YORKTOWN (E-2). Lord Cornwallis surrendered to Washington at **Moore House** in 1781.

WASHINGTON

See map on page 48

GINKGO PETRIFIED FOREST STATE PARK (B-2). Preserves fossil forest once buried under lava. Many species of petrified trees.
GRAND COULEE DAM (C-1). One of the world's largest concrete structures. Self-guided tour.
LAKE CHELAN (B-1). Boat trips on lake that fills glacial gorge in the Cascades.
MT. RAINIER NAT. PARK (B-2). Glacial system on highest peak in Cascade Range is one of the largest in the U.S. (See page 124.)
OLYMPIC PENINSULA (A-2). Olympic Loop Highway (U.S. 101) circles the peninsula. **Olympic Nat. Park** is home of country's largest herd of Roosevelt elk. Rain forests. (See page 124.)
SAN JUAN ISLANDS (A-1). Group of 172 enchanting islands in northern Puget Sound. Ferries make trips from Anacortes and stop at four islands.
SEATTLE (A-2). **Pike Place Market** is a sprawling public market in business since 1907; offers farm produce, handcrafted items. Inexpensive restaurants with harbor views. Waterfront area also has **Seattle Aquarium,** harbor cruises, **Pier 70** with shops and restaurants. **Pioneer Square,** bounded by James St., 3rd Ave. and King St., just off the waterfront, is a restored area that now houses interesting shops and restaurants. **Seattle Center** was the site of the 1962 World's Fair and has the **Pacific Science Center, Space Needle** and varied exhibit areas.
WENATCHEE (B-2). Apple-growing center and home of **Apple Blossom Festival** in spring.

WEST VIRGINIA

See map on pages 18 and 19

BECKLEY (B-3). **Exhibition Mine** in New River Park. Displays trace history of coal mining.
CHARLESTON (B-2). Beautifully situated state capital is center for glass and chemicals.
HARPERS FERRY (D-2). **Harpers Ferry Nat. Historical Park** preserves historic town, scene of John Brown's raid in 1859 and Civil War battles.
MONONGAHELA NAT. FOREST. Vast preserve along eastern border of the state with headquarters at Elkins (C-2). Scenic drives. **Blackwater Falls, Spruce Knob-Seneca Rocks Nat. Recreation Area.**
NEW RIVER GORGE (B-2). Magnificent mountain and river scenery along deep gorge of the New River.

WISCONSIN

See map on page 30

BARABOO (C-4). **Circus World Museum** preserves colorful circus paraphernalia; live animals, circus acts.
GREEN BAY (D-3). State's oldest city. Historic structures in **Heritage Hill State Park. National Railroad Museum** has steam train rides.
MADISON (C-4). Capital and seat of the **University of Wisconsin. U.S. Forest Products Laboratory** is devoted to research on wood products. Tours.
MILWAUKEE (D-4). One of the nation's brewing centers offers tours of **Miller Brewing Co., Joseph Schlitz Brewing Co., Pabst Brewing Co. Experimental Aircraft Assn. Air Museum,** Franklin, displays all types of aircraft as well as engines, propellers, models, photographs. Outdoor displays. Conservatory in **Mitchell Park** is housed in three domes creating climates found in all parts of the world.
ST. CROIX FALLS (A-3). **Interstate Park** on the St. Croix River is noted for its geological oddities.
WISCONSIN DELLS (C-4). Cruise boats traverse the scenic, rocky gorge of the Wisconsin River. Tours of the Upper Dells and Lower Dells.

WYOMING

See map on pages 42 and 43

DEVILS TOWER NAT. MONUMENT (H-3). Rising abruptly 865 feet above ridge on which it stands, the tower resembles a gigantic tree stump. Geological museum and prairie dog colony.
FORT LARAMIE NAT. HISTORIC SITE (H-4). Built in 1834, the fort served as a trading and military post. Abandoned in 1890, it is now restored.
GRAND TETON NAT. PARK (E-4). Park includes most scenic part of Teton Range. More than 22 peaks over 10,000 feet high comprise one of the steepest escarpments in the world. (See page 122.)
YELLOWSTONE NAT. PARK (E-3). Country's first national park. More than 10,000 thermal features include **Old Faithful,** 400 other geysers, hot springs and pools. **Yellowstone Lake.** (See page 121.)

E.R. DEGGINGER

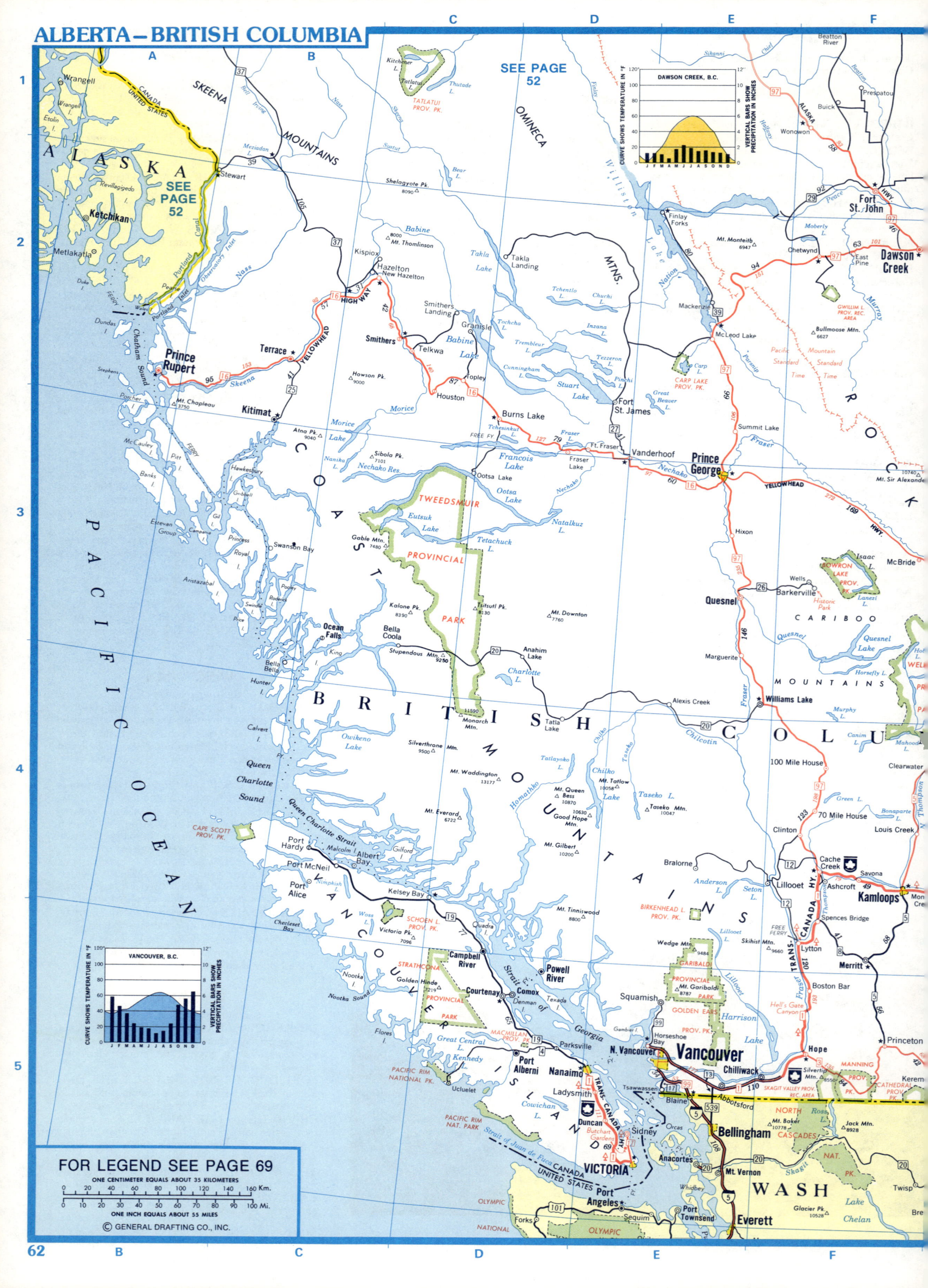

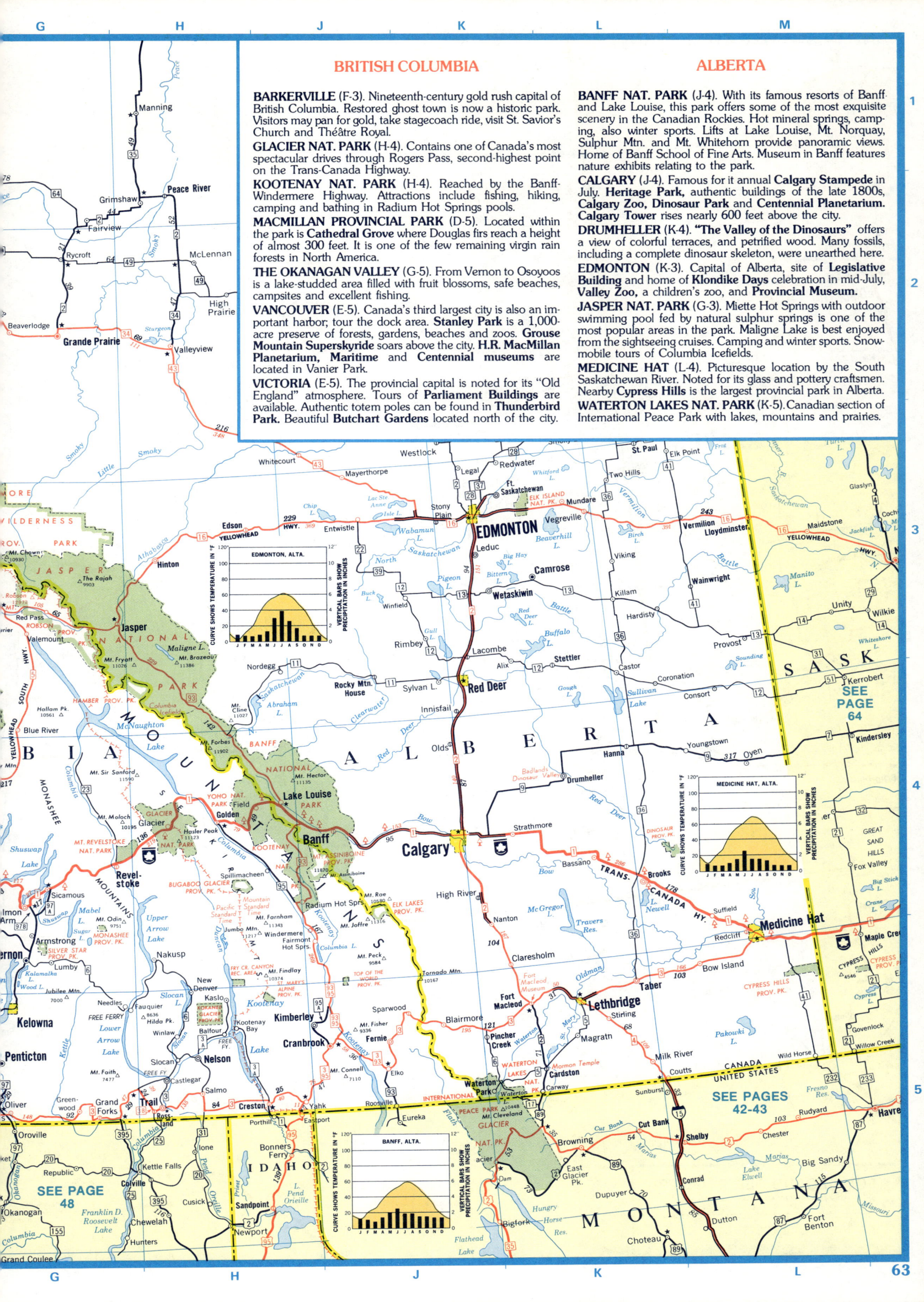

BRITISH COLUMBIA

BARKERVILLE (F-3). Nineteenth-century gold rush capital of British Columbia. Restored ghost town is now a historic park. Visitors may pan for gold, take stagecoach ride, visit St. Savior's Church and Théâtre Royal.

GLACIER NAT. PARK (H-4). Contains one of Canada's most spectacular drives through Rogers Pass, second-highest point on the Trans-Canada Highway.

KOOTENAY NAT. PARK (H-4). Reached by the Banff-Windermere Highway. Attractions include fishing, hiking, camping and bathing in Radium Hot Springs pools.

MACMILLAN PROVINCIAL PARK (D-5). Located within the park is **Cathedral Grove** where Douglas firs reach a height of almost 300 feet. It is one of the few remaining virgin rain forests in North America.

THE OKANAGAN VALLEY (G-5). From Vernon to Osoyoos is a lake-studded area filled with fruit blossoms, safe beaches, campsites and excellent fishing.

VANCOUVER (E-5). Canada's third largest city is also an important harbor; tour the dock area. **Stanley Park** is a 1,000-acre preserve of forests, gardens, beaches and zoos. **Grouse Mountain Superskyride** soars above the city. **H.R. MacMillan Planetarium, Maritime** and **Centennial museums** are located in Vanier Park.

VICTORIA (E-5). The provincial capital is noted for its "Old England" atmosphere. Tours of **Parliament Buildings** are available. Authentic totem poles can be found in **Thunderbird Park.** Beautiful **Butchart Gardens** located north of the city.

ALBERTA

BANFF NAT. PARK (J-4). With its famous resorts of Banff and Lake Louise, this park offers some of the most exquisite scenery in the Canadian Rockies. Hot mineral springs, camping, also winter sports. Lifts at Lake Louise, Mt. Norquay, Sulphur Mtn. and Mt. Whitehorn provide panoramic views. Home of Banff School of Fine Arts. Museum in Banff features nature exhibits relating to the park.

CALGARY (J-4). Famous for it annual **Calgary Stampede** in July. **Heritage Park,** authentic buildings of the late 1800s, **Calgary Zoo, Dinosaur Park** and **Centennial Planetarium. Calgary Tower** rises nearly 600 feet above the city.

DRUMHELLER (K-4). "The Valley of the Dinosaurs" offers a view of colorful terraces, and petrified wood. Many fossils, including a complete dinosaur skeleton, were unearthed here.

EDMONTON (K-3). Capital of Alberta, site of **Legislative Building** and home of **Klondike Days** celebration in mid-July, **Valley Zoo,** a children's zoo, and **Provincial Museum.**

JASPER NAT. PARK (G-3). Miette Hot Springs with outdoor swimming pool fed by natural sulphur springs is one of the most popular areas in the park. Maligne Lake is best enjoyed from the sightseeing cruises. Camping and winter sports. Snowmobile tours of Columbia Icefields.

MEDICINE HAT (L-4). Picturesque location by the South Saskatchewan River. Noted for its glass and pottery craftsmen. Nearby **Cypress Hills** is the largest provincial park in Alberta.

WATERTON LAKES NAT. PARK (K-5). Canadian section of International Peace Park with lakes, mountains and prairies.

SEE PAGE 64

SEE PAGES 42-43

SEE PAGE 48

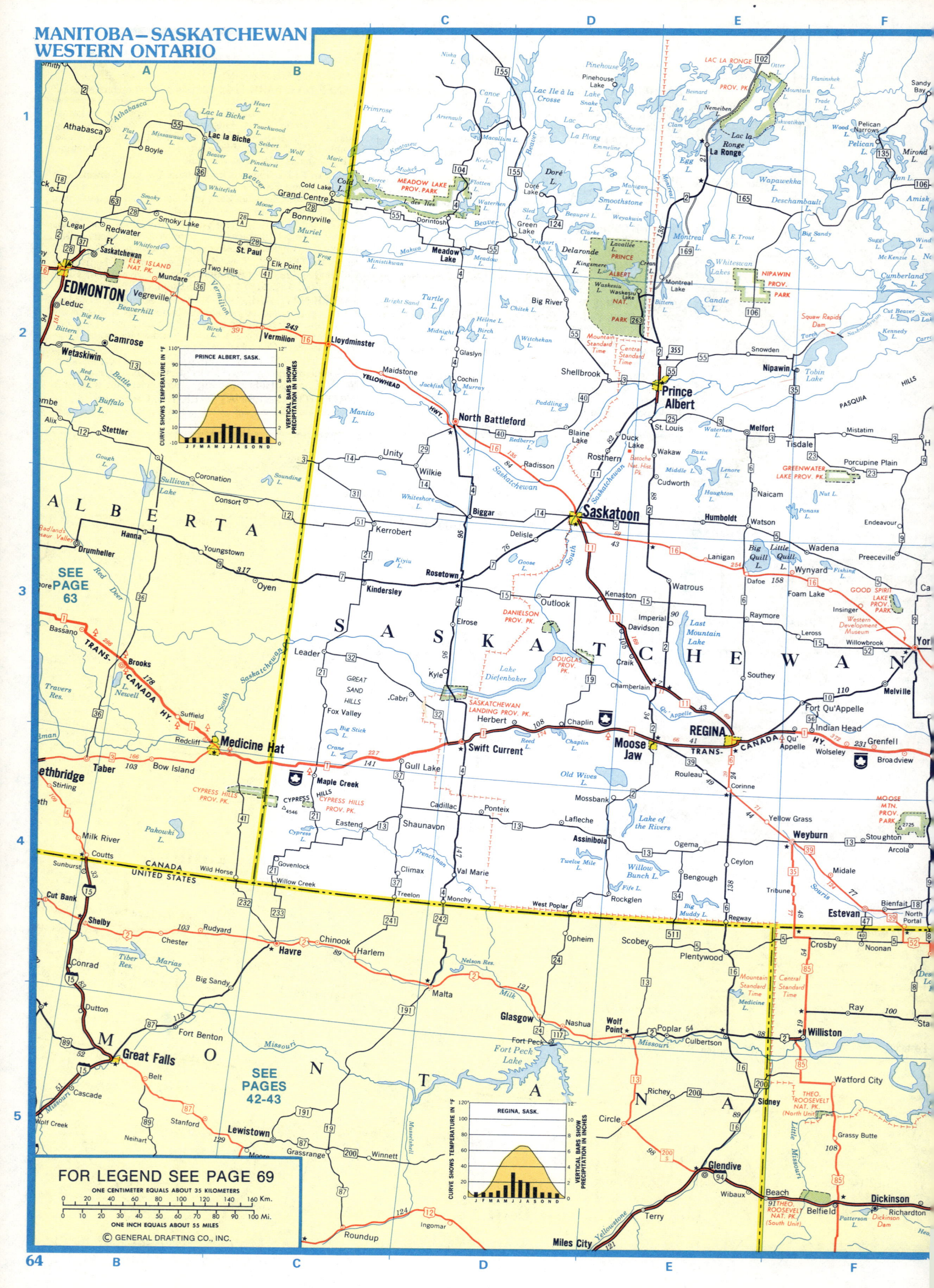

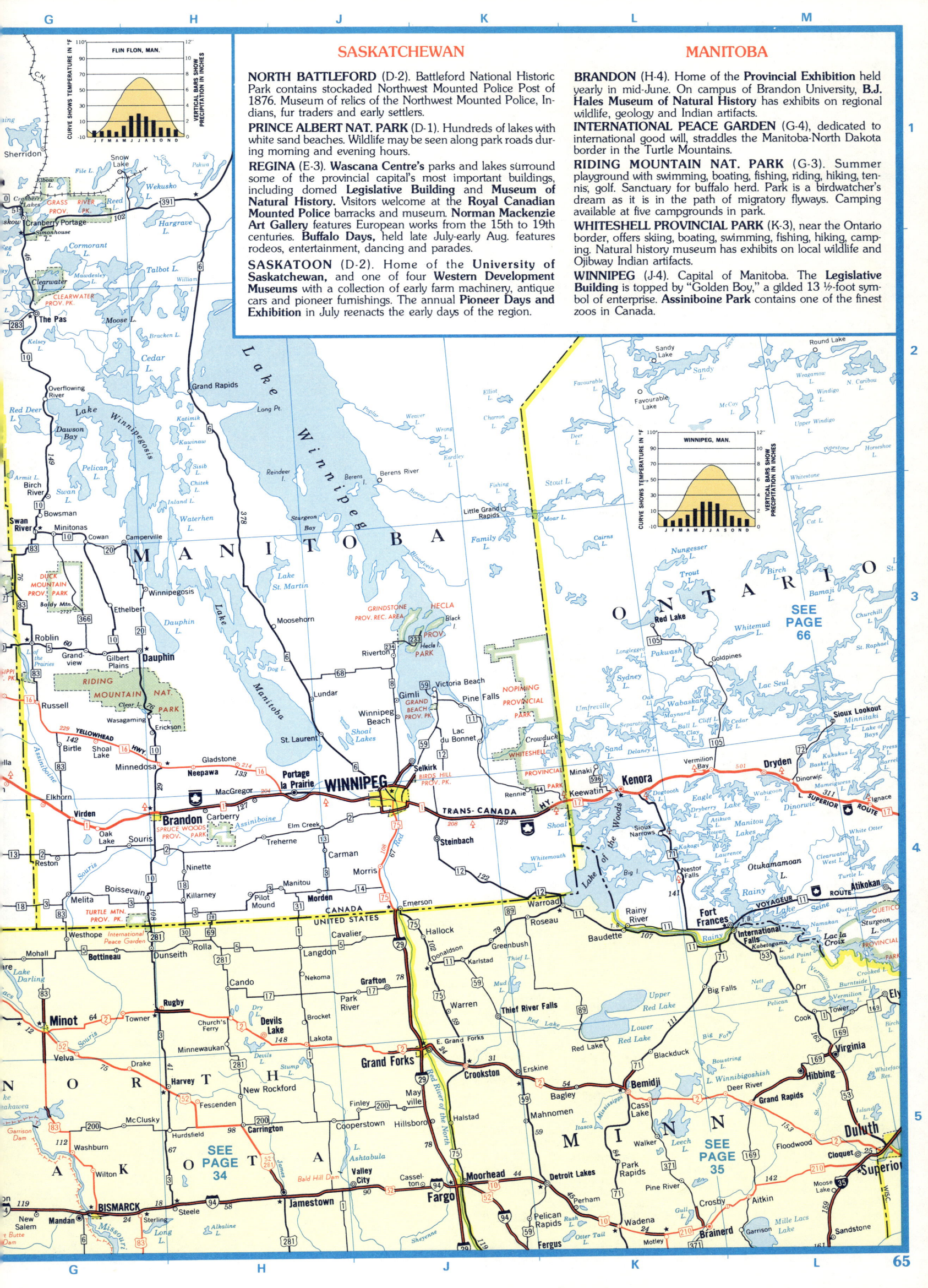

SASKATCHEWAN

NORTH BATTLEFORD (D-2). Battleford National Historic Park contains stockaded Northwest Mounted Police Post of 1876. Museum of relics of the Northwest Mounted Police, Indians, fur traders and early settlers.

PRINCE ALBERT NAT. PARK (D-1). Hundreds of lakes with white sand beaches. Wildlife may be seen along park roads during morning and evening hours.

REGINA (E-3). **Wascana Centre's** parks and lakes surround some of the provincial capital's most important buildings, including domed **Legislative Building** and **Museum of Natural History.** Visitors welcome at the **Royal Canadian Mounted Police** barracks and museum. **Norman Mackenzie Art Gallery** features European works from the 15th to 19th centuries. **Buffalo Days,** held late July-early Aug. features rodeos, entertainment, dancing and parades.

SASKATOON (D-2). Home of the **University of Saskatchewan,** and one of four **Western Development Museums** with a collection of early farm machinery, antique cars and pioneer furnishings. The annual **Pioneer Days and Exhibition** in July reenacts the early days of the region.

MANITOBA

BRANDON (H-4). Home of the **Provincial Exhibition** held yearly in mid-June. On campus of Brandon University, **B.J. Hales Museum of Natural History** has exhibits on regional wildlife, geology and Indian artifacts.

INTERNATIONAL PEACE GARDEN (G-4), dedicated to international good will, straddles the Manitoba-North Dakota border in the Turtle Mountains.

RIDING MOUNTAIN NAT. PARK (G-3). Summer playground with swimming, boating, fishing, riding, hiking, tennis, golf. Sanctuary for buffalo herd. Park is a birdwatcher's dream as it is in the path of migratory flyways. Camping available at five campgrounds in park.

WHITESHELL PROVINCIAL PARK (K-3), near the Ontario border, offers skiing, boating, swimming, fishing, hiking, camping. Natural history museum has exhibits on local wildlife and Ojibway Indian artifacts.

WINNIPEG (J-4). Capital of Manitoba. The **Legislative Building** is topped by "Golden Boy," a gilded 13½-foot symbol of enterprise. **Assiniboine Park** contains one of the finest zoos in Canada.

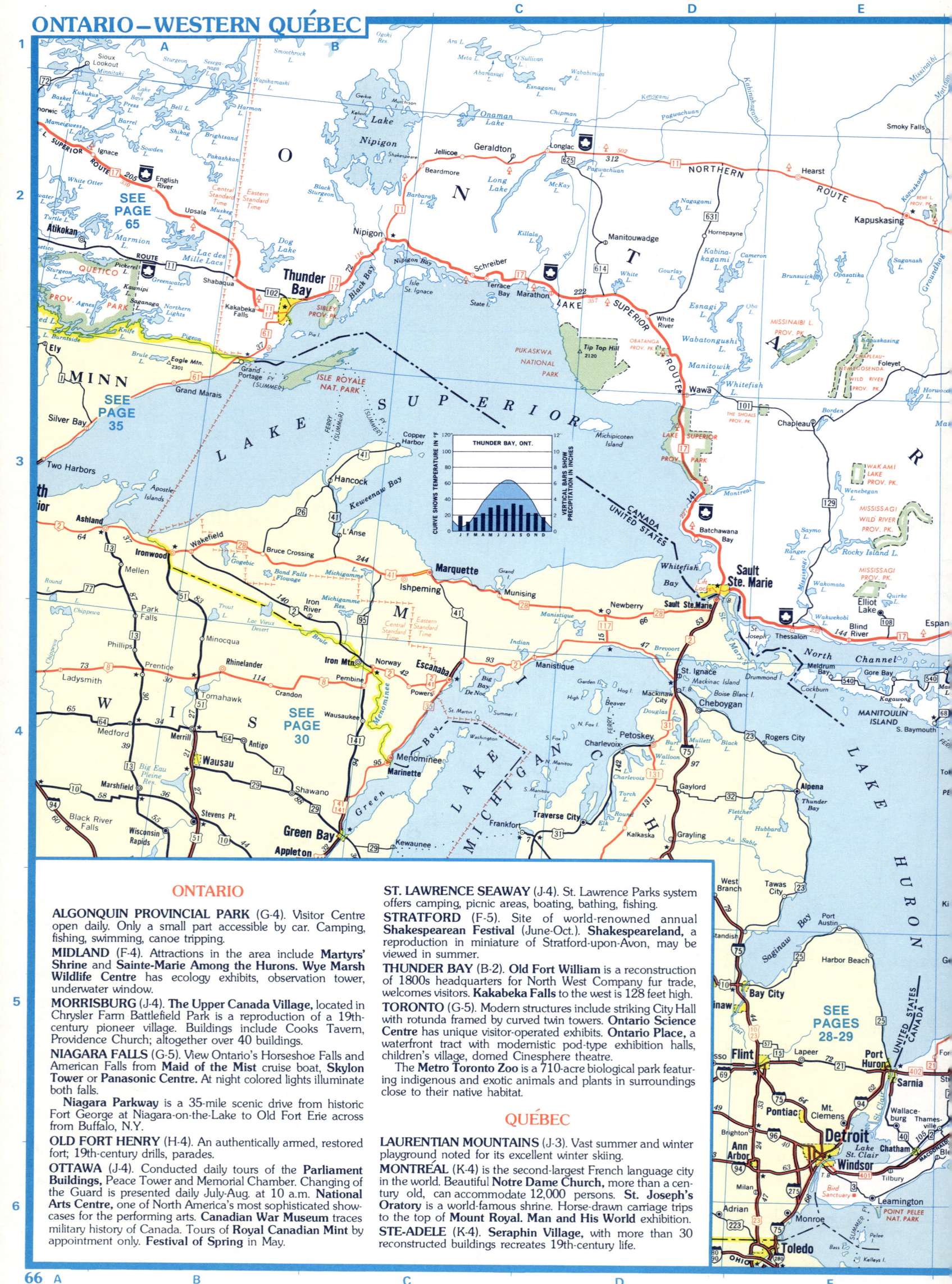

THUNDER BAY, ONT. (chart)

ONTARIO

ALGONQUIN PROVINCIAL PARK (G-4). Visitor Centre open daily. Only a small part accessible by car. Camping, fishing, swimming, canoe tripping.

MIDLAND (F-4). Attractions in the area include **Martyrs' Shrine** and **Sainte-Marie Among the Hurons. Wye Marsh Wildlife Centre** has ecology exhibits, observation tower, underwater window.

MORRISBURG (J-4). **The Upper Canada Village,** located in Chrysler Farm Battlefield Park is a reproduction of a 19th-century pioneer village. Buildings include Cooks Tavern, Providence Church; altogether over 40 buildings.

NIAGARA FALLS (G-5). View Ontario's Horseshoe Falls and American Falls from **Maid of the Mist** cruise boat, **Skylon Tower** or **Panasonic Centre.** At night colored lights illuminate both falls.

Niagara Parkway is a 35-mile scenic drive from historic Fort George at Niagara-on-the-Lake to Old Fort Erie across from Buffalo, N.Y.

OLD FORT HENRY (H-4). An authentically armed, restored fort; 19th-century drills, parades.

OTTAWA (J-4). Conducted daily tours of the **Parliament Buildings,** Peace Tower and Memorial Chamber. Changing of the Guard is presented daily July-Aug. at 10 a.m. **National Arts Centre,** one of North America's most sophisticated showcases for the performing arts. **Canadian War Museum** traces military history of Canada. Tours of **Royal Canadian Mint** by appointment only. **Festival of Spring** in May.

ST. LAWRENCE SEAWAY (J-4). St. Lawrence Parks system offers camping, picnic areas, boating, bathing, fishing.

STRATFORD (F-5). Site of world-renowned annual **Shakespearean Festival** (June-Oct.). **Shakespeareland,** a reproduction in miniature of Stratford-upon-Avon, may be viewed in summer.

THUNDER BAY (B-2). **Old Fort William** is a reconstruction of 1800s headquarters for North West Company fur trade, welcomes visitors. **Kakabeka Falls** to the west is 128 feet high.

TORONTO (G-5). Modern structures include striking City Hall with rotunda framed by curved twin towers. **Ontario Science Centre** has unique visitor-operated exhibits. **Ontario Place,** a waterfront tract with modernistic pod-type exhibition halls, children's village, domed Cinesphere theatre.

The **Metro Toronto Zoo** is a 710-acre biological park featuring indigenous and exotic animals and plants in surroundings close to their native habitat.

QUÉBEC

LAURENTIAN MOUNTAINS (J-3). Vast summer and winter playground noted for its excellent winter skiing.

MONTRÉAL (K-4) is the second-largest French language city in the world. Beautiful **Notre Dame Church,** more than a century old, can accommodate 12,000 persons. **St. Joseph's Oratory** is a world-famous shrine. Horse-drawn carriage trips to the top of **Mount Royal. Man and His World** exhibition.

STE-ADELE (K-4). **Seraphin Village,** with more than 30 reconstructed buildings recreates 19th-century life.

QUÉBEC

THE GASPÉ PENINSULA (E-2). All-paved, scenic drive follows rugged coastline; picturesque fishing villages. In the harbor at **Percé** is a giant, pierced rock. Excursion boats from Percé take visitors to **Bonaventure Island Park,** a bird sanctuary since 1963. **Forillion Nat. Park,** one of Canada's newest parks, was established in 1971. The park offers excellent views with cliffs towering over the Bay of Gaspé.

MATANE (D-2). The **Parc de Matane** contains some of the most beautiful scenery in Québec. Activities include skiing, swimming and hiking.

QUÉBEC CITY (B-3). Only walled city in North America, sits atop a rocky eminence. Winding, narrow streets are unique. **Battlefields Park,** or the Plains of Abraham, commemorates Generals Wolfe and Montcalm, who both died in battle here.

Montmorency Falls, more than 100 feet higher than Niagara, is nearby. **Shrine of Ste. Anne de Beaupre,** 22 miles east, attracts religious pilgrimages from around the world.

NEW BRUNSWICK

CARAQUET (E-3). The **Village Historique Acadian** consists of 40 restored buildings devoted to the Acadians. Demonstrations of Acadian lifestyle and crafts are presented by people in period costumes.

FREDERICTON (D-4). Parliament Building open to tourists. **Beaverbrook Art Gallery** houses early and contemporary British and Canadian art.

FUNDY NAT. PARK (E-4). Scenic recreational area on Bay of Fundy. Facilities include a heated saltwater pool and a 9-hole golf course. Camping.

KINGS LANDING HISTORICAL SETTLEMENT (D-3). Sixty buildings, staffed by costumed guides, present scenes of farm living of 1790 to 1870 period.

MONCTON (E-3). Acadian Historical Museum houses looms and 17th-century French pipe organ.

SAINT JOHN (E-4). Martello Tower, built 1814, houses firearms and antiquities. At **Reversing Falls** the water flows out to sea at low tide, then back up the river at high tide.

NOVA SCOTIA

CAPE BRETON HIGHLANDS NAT. PARK (G-3). Summer recreation includes bathing, boating, golf, tennis, deep-sea fishing, camping. The **Cabot Trail** offers spectacular ocean views. **Alexander Graham Bell National Historic Park** at Baddeck.

FORTRESS OF LOUISBOURG NAT. HISTORIC PARK (H-4). Great walled city and military station built 1717-1740 by the French is being reconstructed. Guided tours of completed structures available.

GRAND PRÉ NAT. HISTORIC PARK (E-4). Acadian village 1675-1755 contains chapel replica, statue of Evangeline and museum.

HALIFAX (F-4). Province House is outstanding example of Georgian architecture. **Citadel National Historic Park,** a fortress dating back to 1749, is site of Nova Scotia centennial art gallery and museum.

PORT ROYAL NAT. HISTORIC PARK (E-4). Reconstruction of first permanent European settlement in Canada (1605).

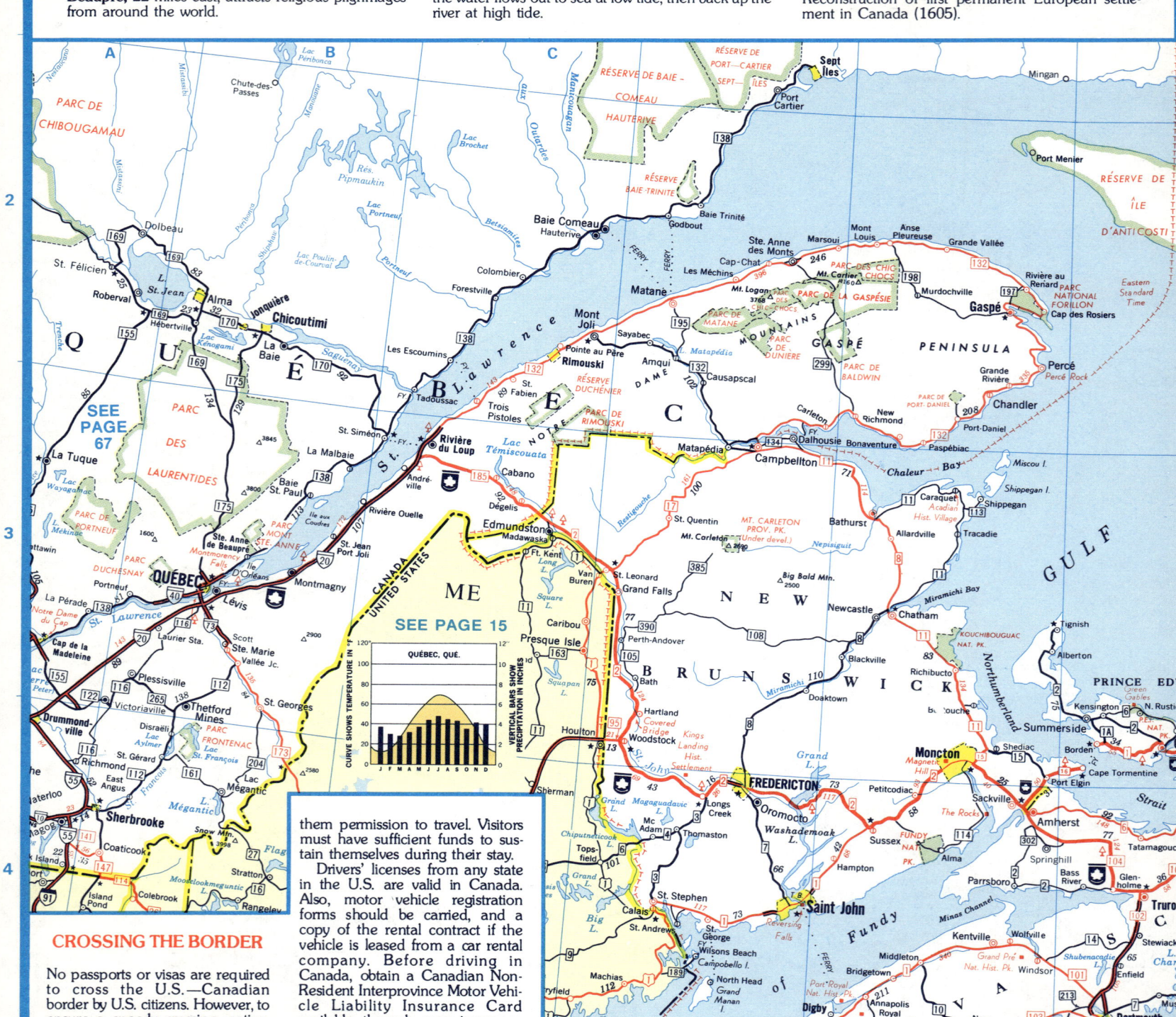

CROSSING THE BORDER

No passports or visas are required to cross the U.S.—Canadian border by U.S. citizens. However, to ensure a speedy crossing, native-born U.S. citizens should carry some identifying papers such as a birth certificate, voter's registration card or passport. Naturalized citizens should carry their naturalization papers or some other evidence of citizenship should they be asked for documentation. Persons under 18 who are not accompanied by an adult should bring a letter with them from a parent or guardian giving them permission to travel. Visitors must have sufficient funds to sustain themselves during their stay.

Drivers' licenses from any state in the U.S. are valid in Canada. Also, motor vehicle registration forms should be carried, and a copy of the rental contract if the vehicle is leased from a car rental company. Before driving in Canada, obtain a Canadian Non-Resident Interprovince Motor Vehicle Liability Insurance Card available through your insurance company. All provinces require visiting motorists to provide evidence of financial responsibility in case of accident. Proof is usually shown by this guarantee of liability insurance.

Motorists planning to operate their Citizen's Band radios must obtain a permit from a Regional Office of the Canadian Dept. of Communications. Applications must be made six months in advance.

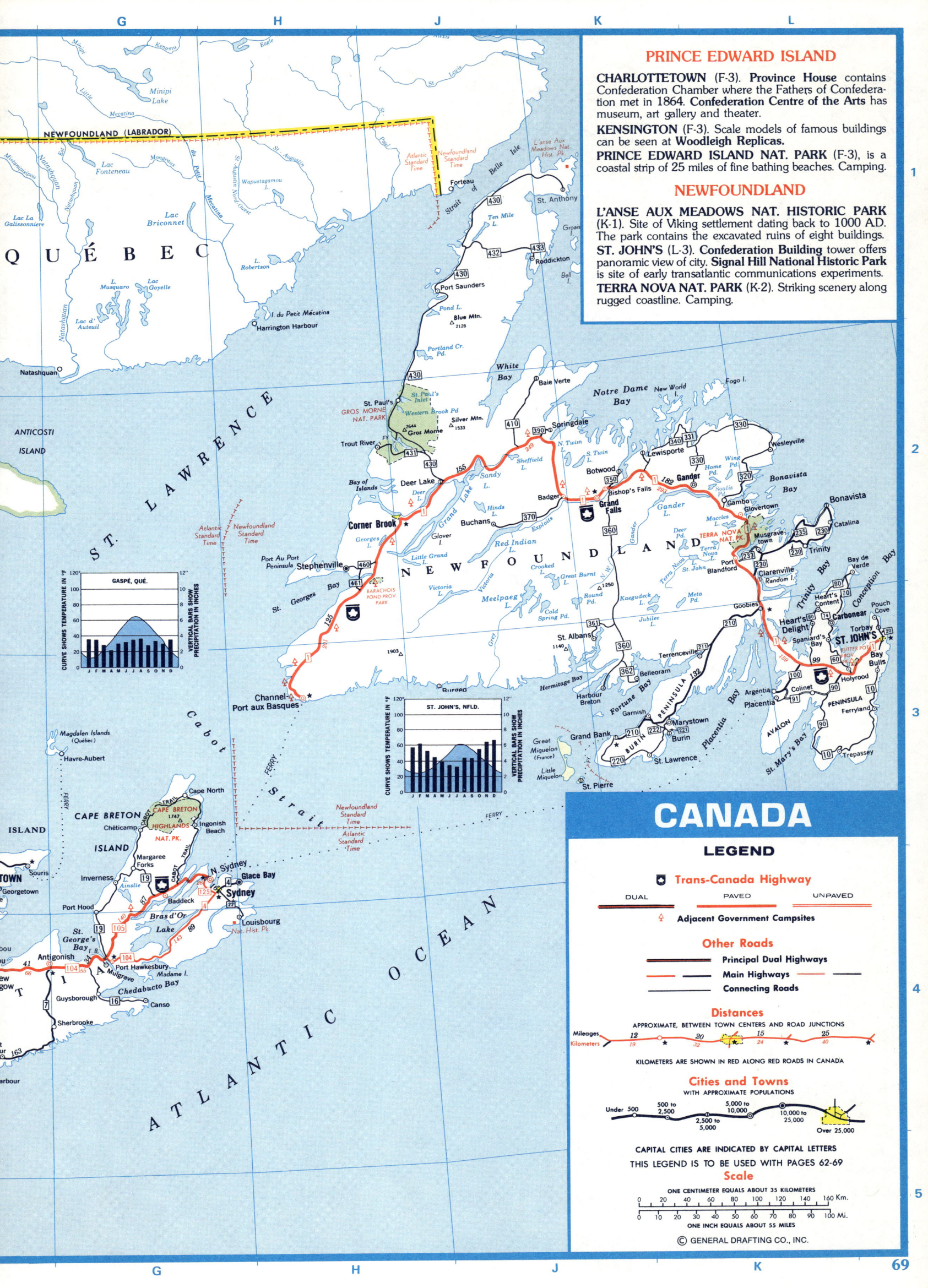

PRINCE EDWARD ISLAND

CHARLOTTETOWN (F-3). **Province House** contains Confederation Chamber where the Fathers of Confederation met in 1864. **Confederation Centre of the Arts** has museum, art gallery and theater.

KENSINGTON (F-3). Scale models of famous buildings can be seen at **Woodleigh Replicas.**

PRINCE EDWARD ISLAND NAT. PARK (F-3), is a coastal strip of 25 miles of fine bathing beaches. Camping.

NEWFOUNDLAND

L'ANSE AUX MEADOWS NAT. HISTORIC PARK (K-1). Site of Viking settlement dating back to 1000 A.D. The park contains the excavated ruins of eight buildings.

ST. JOHN'S (L-3). **Confederation Building** tower offers panoramic view of city. **Signal Hill National Historic Park** is site of early transatlantic communications experiments.

TERRA NOVA NAT. PARK (K-2). Striking scenery along rugged coastline. Camping.

CANADA

LEGEND

Trans-Canada Highway

DUAL   PAVED   UNPAVED

⚑ **Adjacent Government Campsites**

Other Roads

Principal Dual Highways
Main Highways
Connecting Roads

Distances

APPROXIMATE, BETWEEN TOWN CENTERS AND ROAD JUNCTIONS

Mileages   12   15   25
Kilometers   19   32   24   40

KILOMETERS ARE SHOWN IN RED ALONG RED ROADS IN CANADA

Cities and Towns

WITH APPROXIMATE POPULATIONS

Under 500   500 to 2,500   2,500 to 5,000   5,000 to 10,000   10,000 to 25,000   Over 25,000

CAPITAL CITIES ARE INDICATED BY CAPITAL LETTERS

THIS LEGEND IS TO BE USED WITH PAGES 62-69

Scale

ONE CENTIMETER EQUALS ABOUT 35 KILOMETERS
0 20 40 60 80 100 120 140 160 Km.
0 10 20 30 40 50 60 70 80 90 100 Mi.
ONE INCH EQUALS ABOUT 55 MILES

© GENERAL DRAFTING CO., INC.

MEXICO, NORTHERN

SEE PAGE 46

SEE PAGE 47

UNITED STATES

Baja California

Scale
0   100   200 Km.
0   50   100   Mi.

© GENERAL DRAFTING CO., INC.

70

MEXICO

ACAPULCO (C-9). One of the world's finest natural harbors. At **La Quebrada,** divers plunge into a rocky gorge. Miles of beaches for swimming, glorious scenery and luxurious hotels, night spots, boutiques.

CANCUN (H-6). Recently developed Caribbean island resort offers sand-and-sea recreation in natural setting. Island is a sanctuary for exotic wildlife and site of **El Rey archaeological zone.**

CHIHUAHUA (D-2). Here is the **museum-home** of **Pancho Villa,** bandit, revolutionary folk hero. Departure point for spectacular **train ride** to Copper Canyon.

CUERNAVACA (D-8). A semitropical resort since Aztec times. **Palace of Cortés** houses a museum.

DURANGO (E-4). The location for scores of Hollywood's westerns, the frontier town set outside the city may be visited. **State capitol** was mansion of silver baron. Opulent former home has window frames cast in solid silver.

GUADALAJARA (B-7). Mexico's second largest city boasts frescoes by Orozco adorning several public buildings.

GUANAJUATO (C-7). Picturesque city with narrow cobblestone streets lined with ornate mansions, it played a prominent role in independence struggle. Birthplace of famed muralist Diego Rivera is now a museum.

GUAYMAS (B-3). Long favored by sport fishermen; deep-sea fishing tournaments in summer.

LA PAZ (C-5). Inaccessible by land until 1973 completion of Highway 1. It now boasts a jetport, mainland ferry service, unspoiled beaches and superb fishing.

MAZATLAN (D-5). Important in the nation's shipping and fishing industries. It is also a favored resort.

MERIDA (G-6). Founded by the Spaniards in 1542, it is the most important city on the Yucatán Peninsula. Aside from its own colonial attractions, it is the gateway to **Maya archaeological zones** like Chichén Itzá, Uxmal and Kabah.

MEXICO CITY (D-8). Oldest metropolis in the Americas, the Mexican capital has 13 million inhabitants. Facing the Zócalo, downtown plaza, are the imposing **Cathedral** and the **National Palace** with Rivera murals. In Chapultepec Park are **National Museum of Anthropology, Museum of Modern Art** and **Museum of Natural History.**

MONTERREY (G-4). Second in industrial importance, it reflects little of its colonial past. The **Cathedral** contains fine murals. Visitors welcome at **Cuauhtemoc Brewery.**

MORELIA (C-8). A colonial jewel, its **Cathedral** is regarded as Mexico's most beautiful. Regional crafts—needlework, copper and lacquer ware, woodcarving, pottery—for sale in the **Palacio del Artesano.**

OAXACA (E-9). Famous for handicrafts sold in the market on Saturdays. **Archaeological zones** of Monte Alban, Mitla, Yagul, Zaachila are nearby.

PUEBLA (D-8). Noted for colonial architecture, Talavera tiles and pottery. Interior of 17th-century **Santo Domingo Church** is adorned with carved, painted and gilt ornamentation.

PUERTO VALLARTA (A-7). Picturebook Pacific coast resort town on 25-mile-long Bahía de Banderas. Natural beauty and superb climate have made "P.V." popular with foreigners for vacation and residence.

QUERETARO (C-7). Still a colonial city, modern industry is mushrooming on its perimeter. Notable structures include **Churches of Santa Clara** and **Santa Rosa** and a 200-year-old **aqueduct.**

SAN LUIS POTOSI (C-6). A mining, industrial and railroad center founded early in the 16th century, it retains a colonial flavor. Its silken rebozos (stoles) are one of area crafts featured in **Handicrafts Museum.**

SAN MIGUEL ALLENDE (C-7). Cobblestoned colonial town is a Mexican national historic monument. Fantastic spires of **La Parrochial Church** were designed by an illiterate architect inspired by drawings of Gothic cathedrals.

TAXCO (D-8). Red tile roofs on houses swathed in bougainvillaea mount one above another up on the hillsides in this old mining town, now a national monument. It is a city of talented silver craftsmen.

URUAPAN (C-8). "The flower garden of Mexico," famous for hand-painted lacquer ware. To the west is the volcano **Paricutín,** which erupted in a cornfield in 1943 and built a huge cone before its last eruption in 1952.

Pátzcuaro, east of Uruapan, has changed little since colonial days. Magnificent view of Lake Pátzcuaro from top of **Statue of Morelos** on an island in the lake.

VERACRUZ (E-8). Mexico's main seaport, founded 1519 by Hernando Cortés. City retains much of its 16th-century flavor, although it is a popular resort with facilities for swimming, deep-sea fishing and boating.

ZACATECAS (F-5). Silver mining town with colonial buildings and cobblestone streets. Facade of its **Cathedral** is intricately sculptured. Ruins of the fortified city **La Quemada,** (c. 900 A.D.) are 30 miles south.

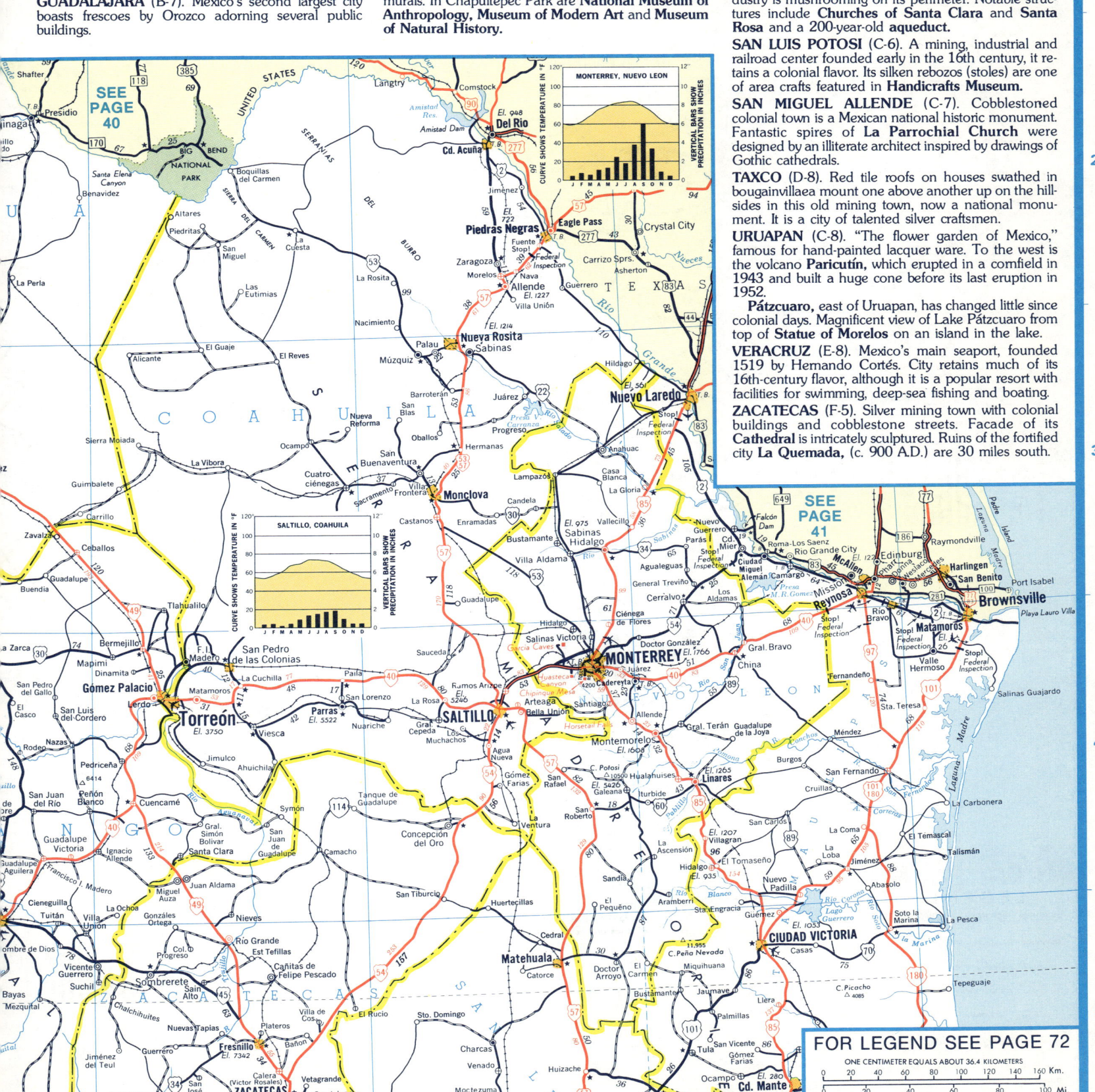

FOR LEGEND SEE PAGE 72

ONE CENTIMETER EQUALS ABOUT 36.4 KILOMETERS

ONE INCH EQUALS ABOUT 57.4 MILES

© GENERAL DRAFTING CO., INC.

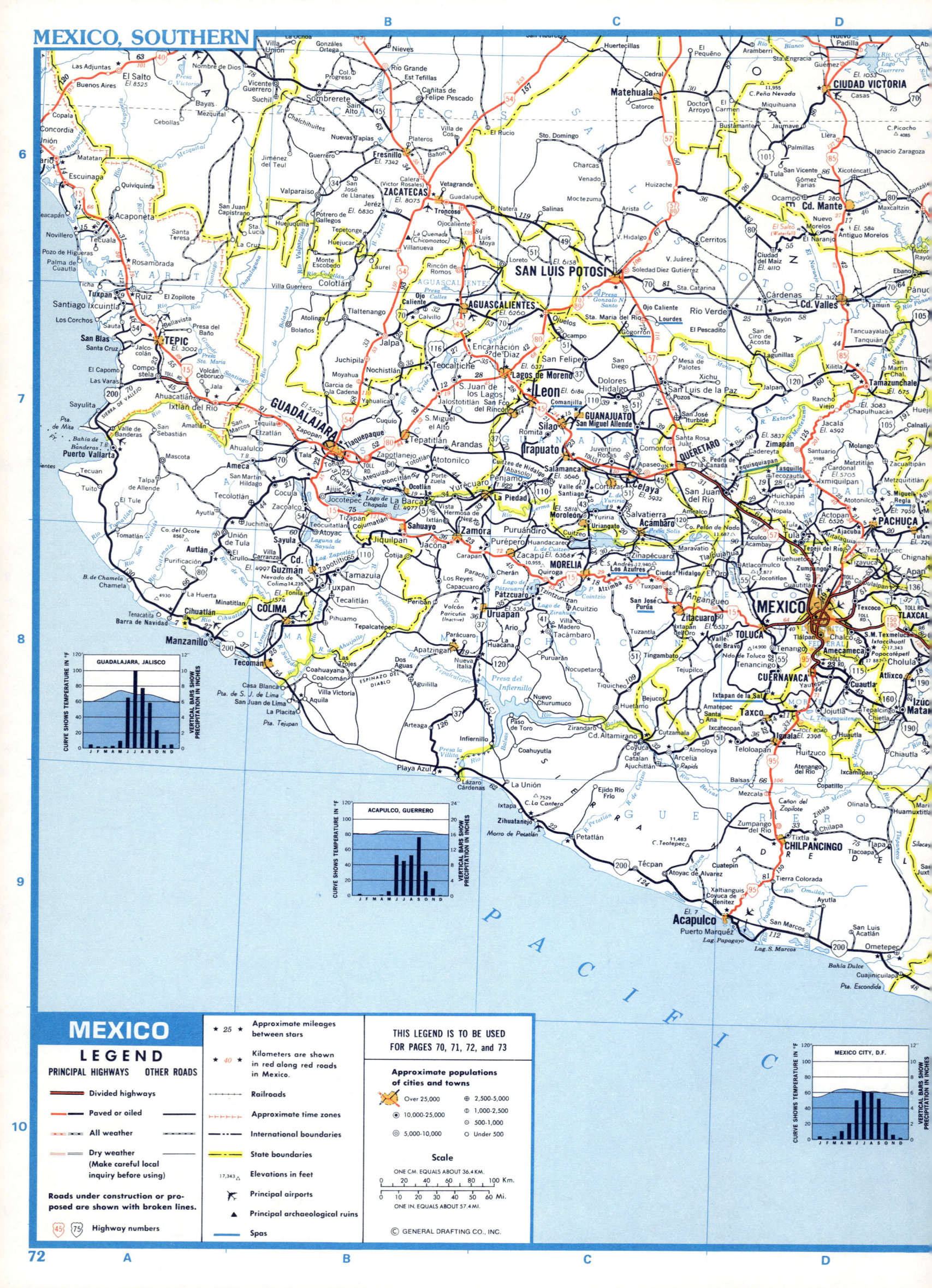

MEXICO, SOUTHERN

MEXICO
LEGEND
PRINCIPAL HIGHWAYS   OTHER ROADS

— Divided highways
— Paved or oiled
— All weather
— Dry weather (Make careful local inquiry before using)

Roads under construction or proposed are shown with broken lines.

45 75 Highway numbers

★ 25 Approximate mileages between stars
40 Kilometers are shown in red along red roads in Mexico.

╫ Railroads
╫ Approximate time zones
▬ International boundaries
━ State boundaries
17,343 △ Elevations in feet
✈ Principal airports
▲ Principal archaeological ruins
━ Spas

THIS LEGEND IS TO BE USED FOR PAGES 70, 71, 72, and 73

Approximate populations of cities and towns
✈ Over 25,000
◉ 10,000-25,000
◎ 5,000-10,000
⊕ 2,500-5,000
⊕ 1,000-2,500
⊙ 500-1,000
○ Under 500

Scale
ONE CM. EQUALS ABOUT 36.4 KM.
0 20 40 60 80 100 Km.
0 10 20 30 40 50 60 Mi.
ONE IN. EQUALS ABOUT 57.4 MI.

© GENERAL DRAFTING CO., INC.

GUADALAJARA, JALISCO
CURVE SHOWS TEMPERATURE IN °F
VERTICAL BARS SHOW PRECIPITATION IN INCHES

ACAPULCO, GUERRERO
CURVE SHOWS TEMPERATURE IN °F
VERTICAL BARS SHOW PRECIPITATION IN INCHES

MEXICO CITY, D.F.
CURVE SHOWS TEMPERATURE IN °F
VERTICAL BARS SHOW PRECIPITATION IN INCHES

72

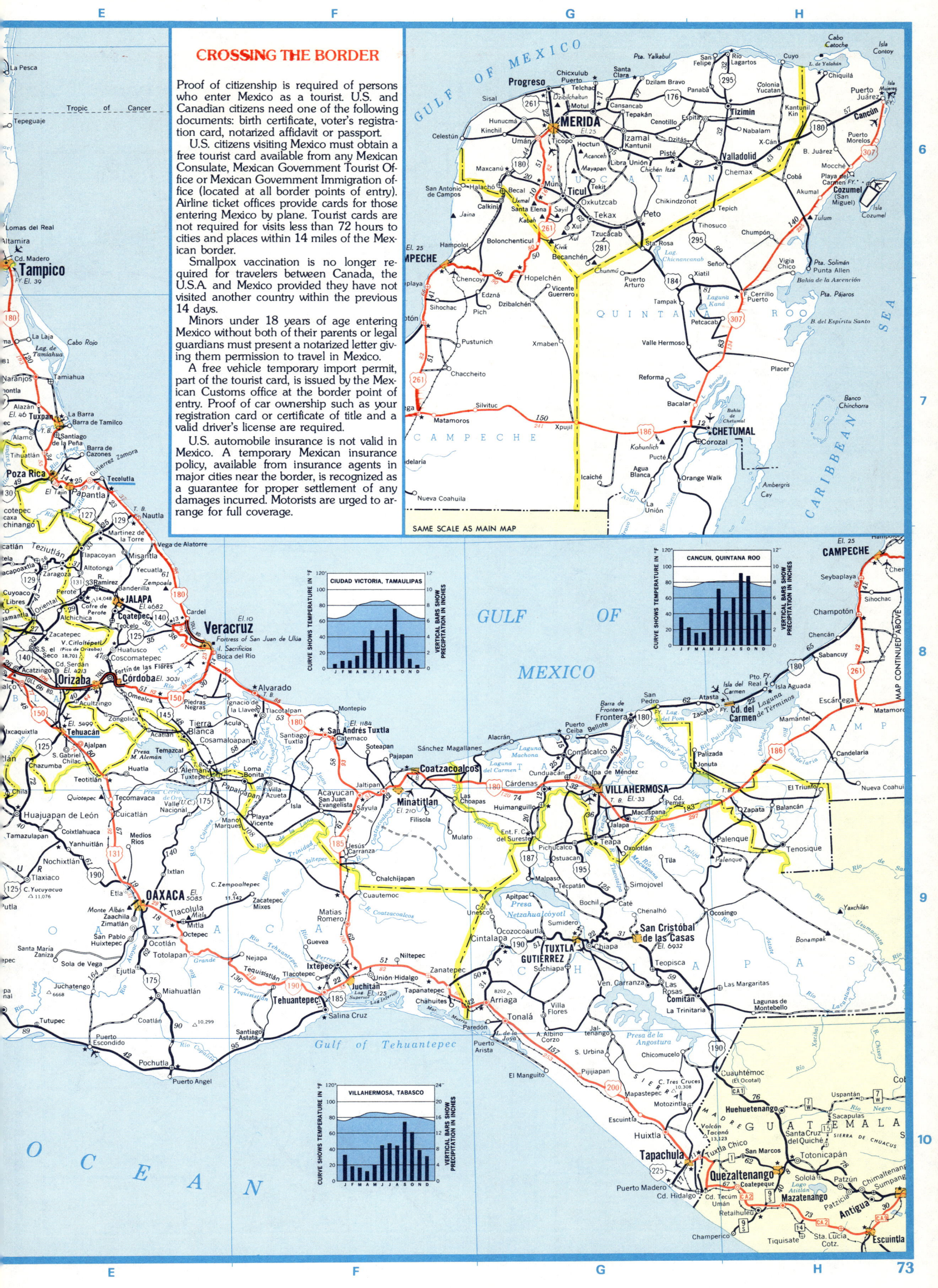

CROSSING THE BORDER

Proof of citizenship is required of persons who enter Mexico as a tourist. U.S. and Canadian citizens need one of the following documents: birth certificate, voter's registration card, notarized affidavit or passport.

U.S. citizens visiting Mexico must obtain a free tourist card available from any Mexican Consulate, Mexican Government Tourist Office or Mexican Government Immigration office (located at all border points of entry). Airline ticket offices provide cards for those entering Mexico by plane. Tourist cards are not required for visits less than 72 hours to cities and places within 14 miles of the Mexican border.

Smallpox vaccination is no longer required for travelers between Canada, the U.S.A. and Mexico provided they have not visited another country within the previous 14 days.

Minors under 18 years of age entering Mexico without both of their parents or legal guardians must present a notarized letter giving them permission to travel in Mexico.

A free vehicle temporary import permit, part of the tourist card, is issued by the Mexican Customs office at the border point of entry. Proof of car ownership such as your registration card or certificate of title and a valid driver's license are required.

U.S. automobile insurance is not valid in Mexico. A temporary Mexican insurance policy, available from insurance agents in major cities near the border, is recognized as a guarantee for proper settlement of any damages incurred. Motorists are urged to arrange for full coverage.

SAME SCALE AS MAIN MAP

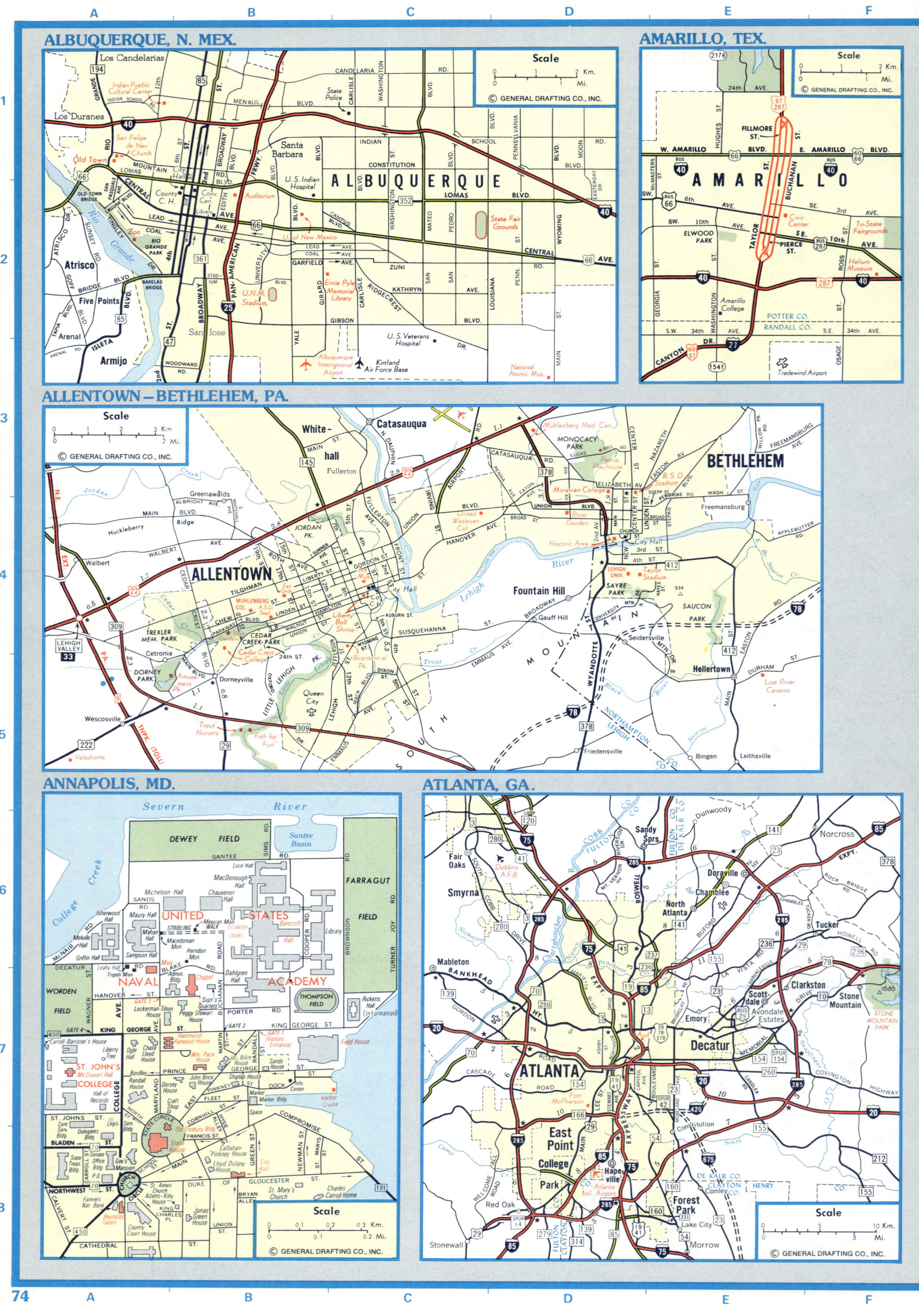

ALBUQUERQUE, N. MEX.

AMARILLO, TEX.

ALLENTOWN — BETHLEHEM, PA.

ANNAPOLIS, MD.

ATLANTA, GA.

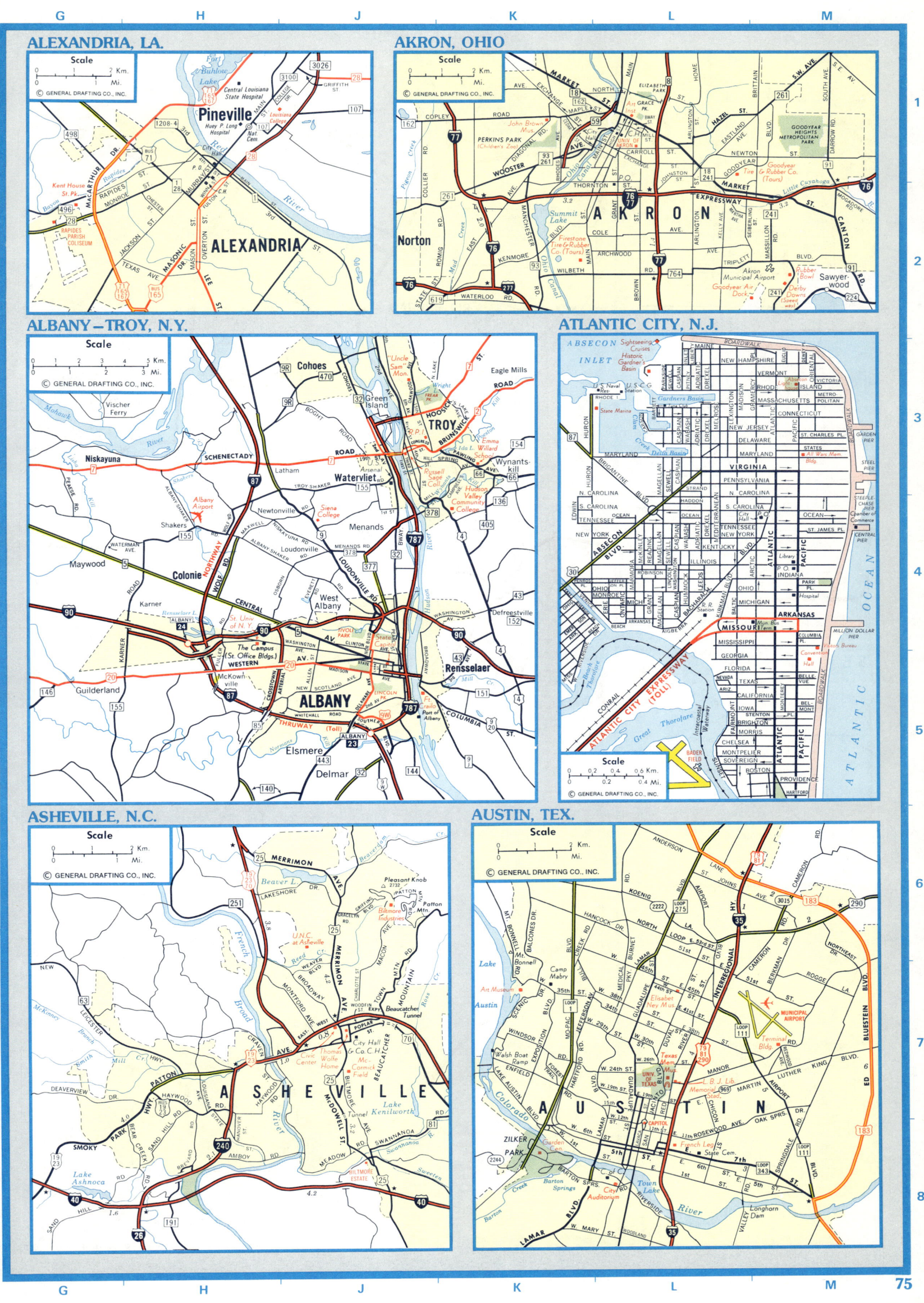

ALEXANDRIA, LA.

AKRON, OHIO

ALBANY—TROY, N.Y.

ATLANTIC CITY, N.J.

ASHEVILLE, N.C.

AUSTIN, TEX.

75

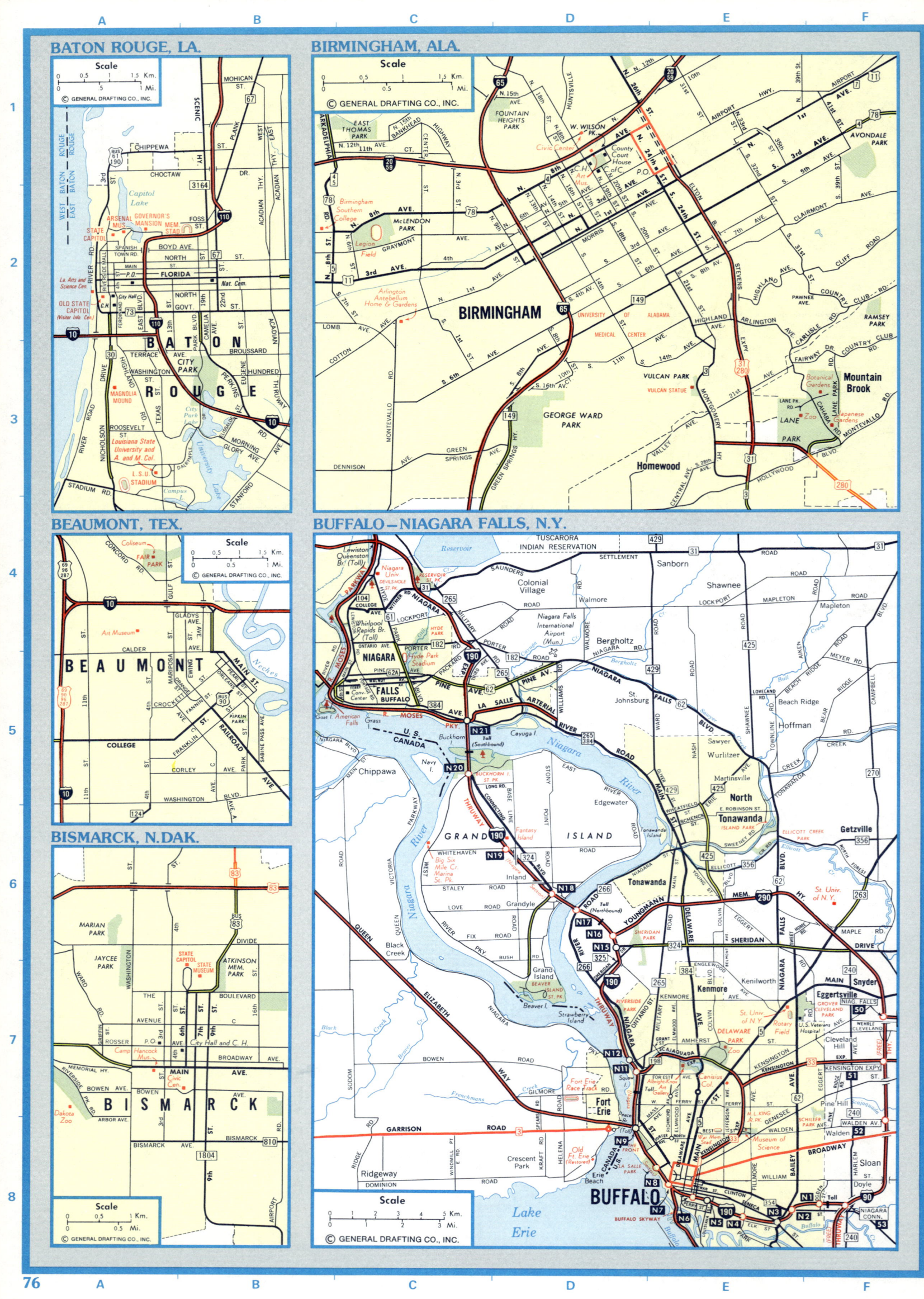

BATON ROUGE, LA.

BIRMINGHAM, ALA.

BEAUMONT, TEX.

BISMARCK, N.DAK.

BUFFALO — NIAGARA FALLS, N.Y.

BILLINGS, MONT.

CLEVELAND, OHIO

BOISE, IDAHO

BOSTON, MASS.

BILOXI—GULFPORT, MISS.

BUTTE, MONT.

77

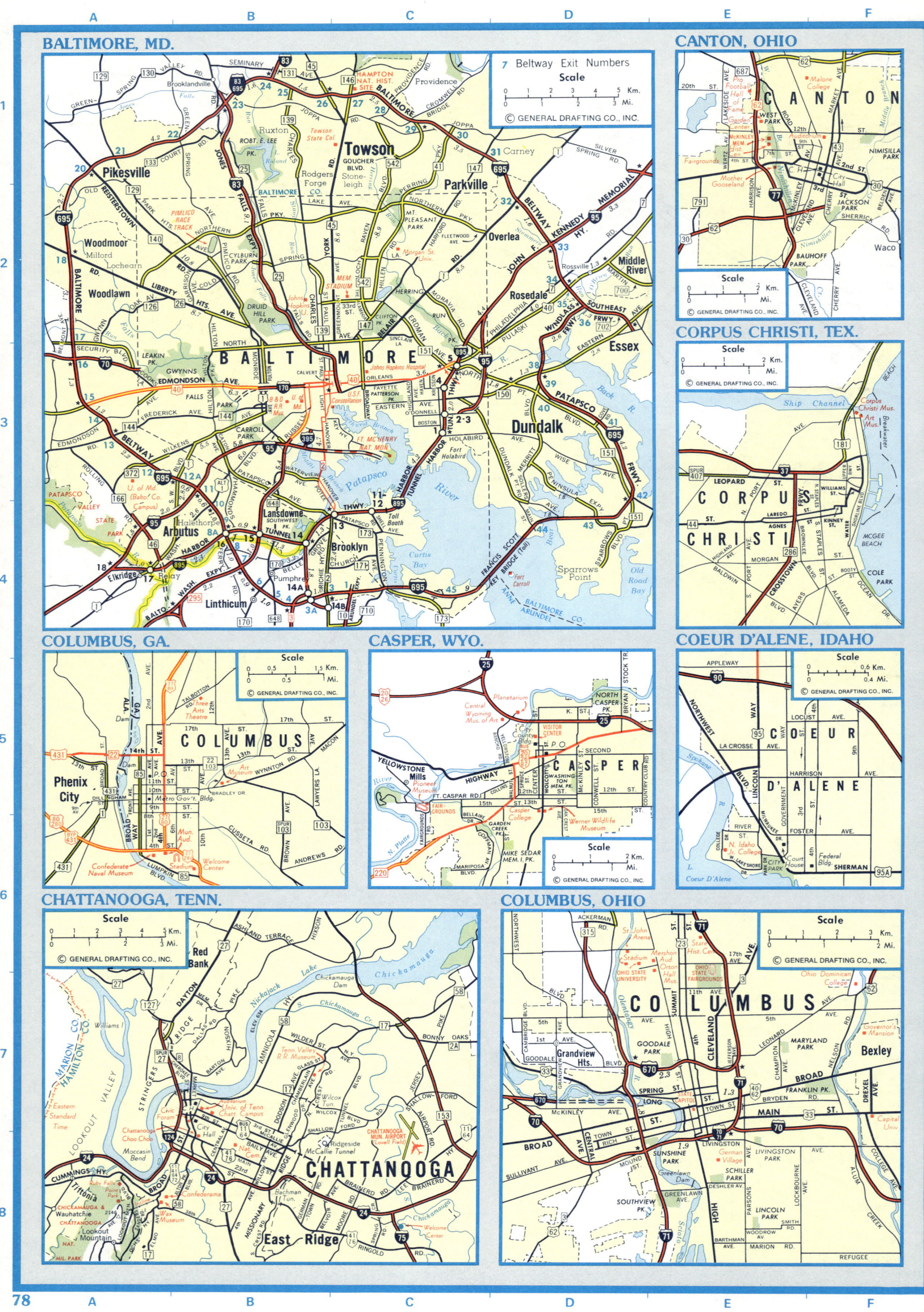

BALTIMORE, MD.

CANTON, OHIO

CORPUS CHRISTI, TEX.

COLUMBUS, GA.

CASPER, WYO.

COEUR D'ALENE, IDAHO

CHATTANOOGA, TENN.

COLUMBUS, OHIO

7 Beltway Exit Numbers
Scale
© GENERAL DRAFTING CO., INC.

CHEYENNE, WYO.

COLUMBIA, S.C.

CHICAGO, ILL

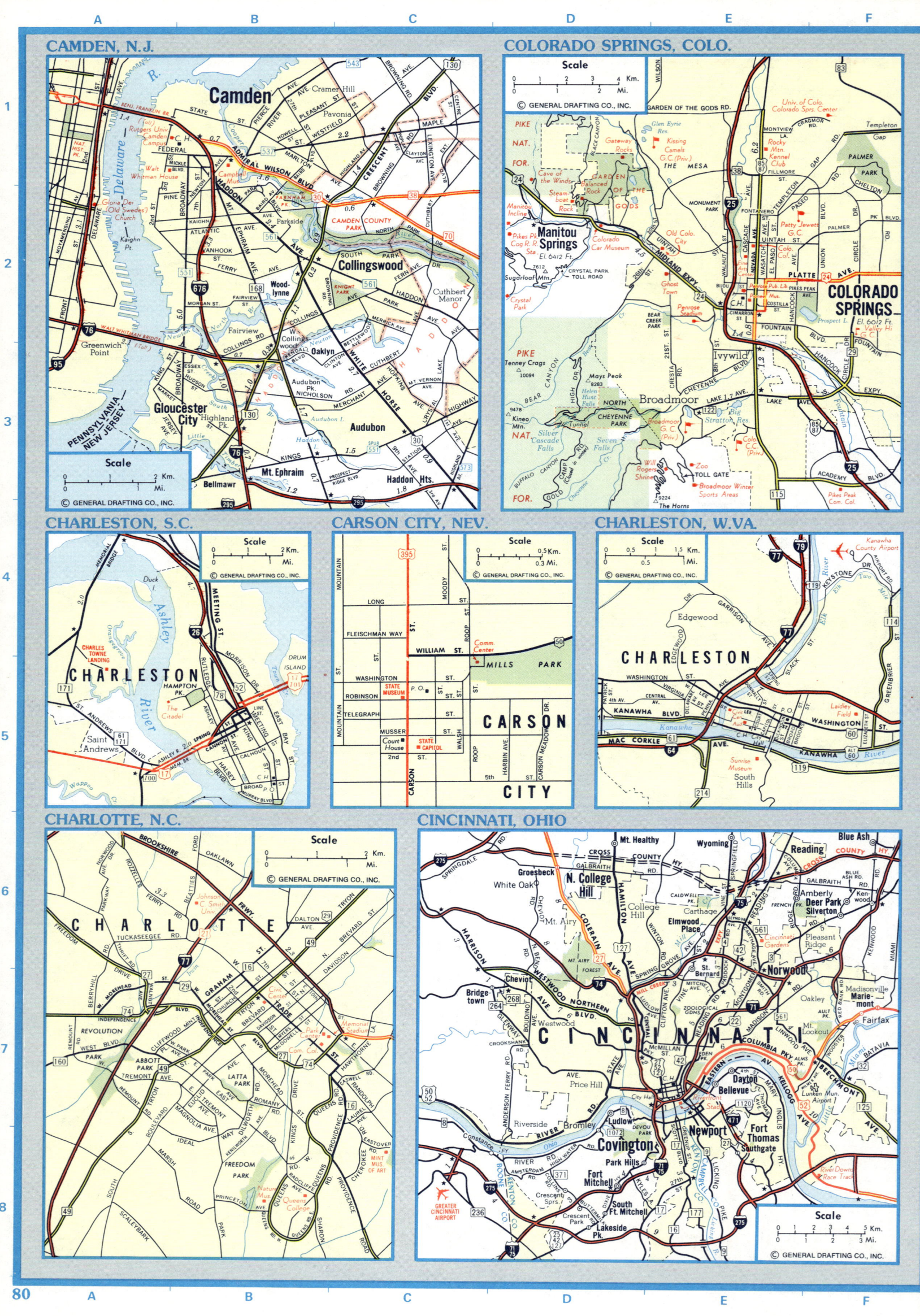

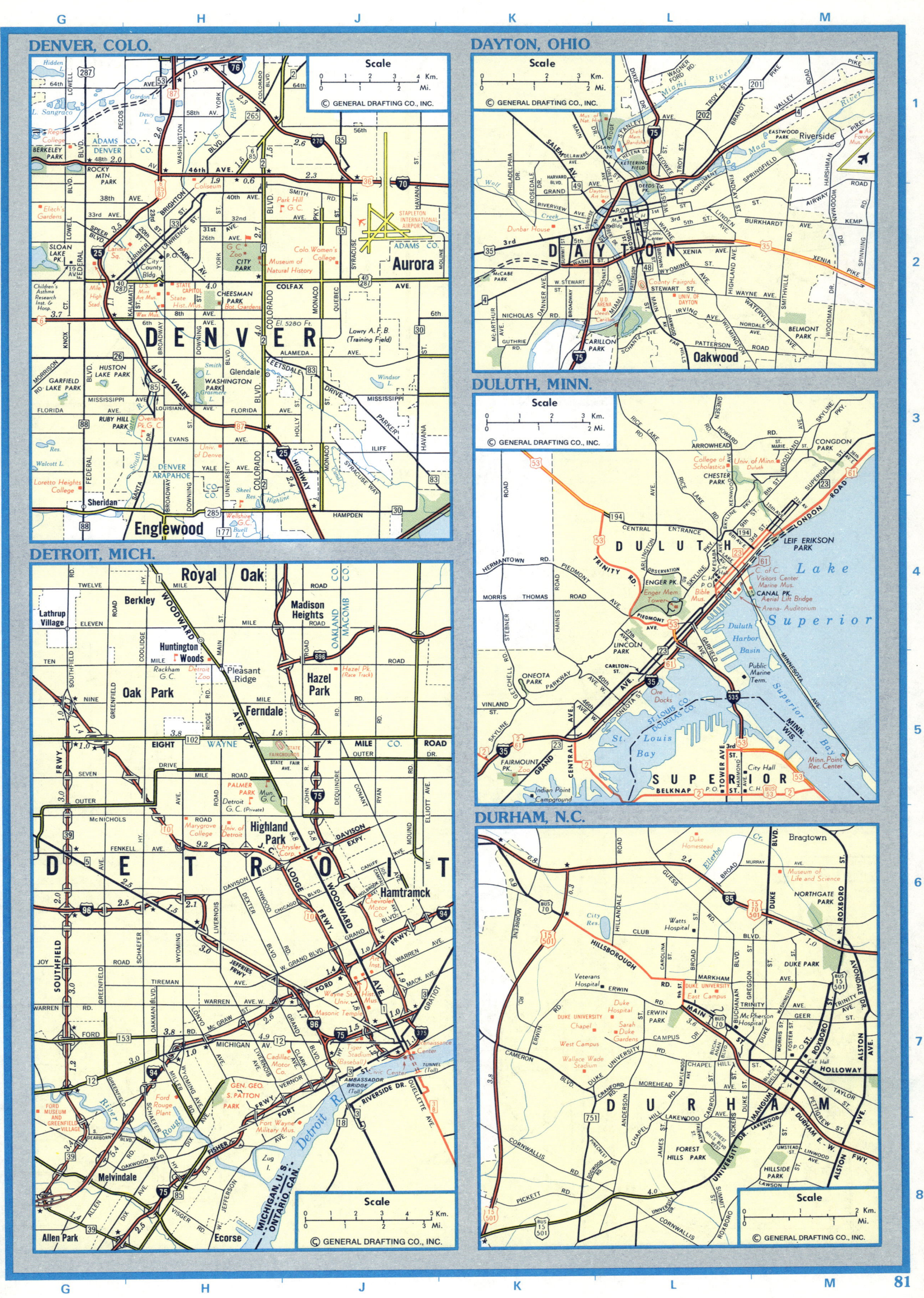

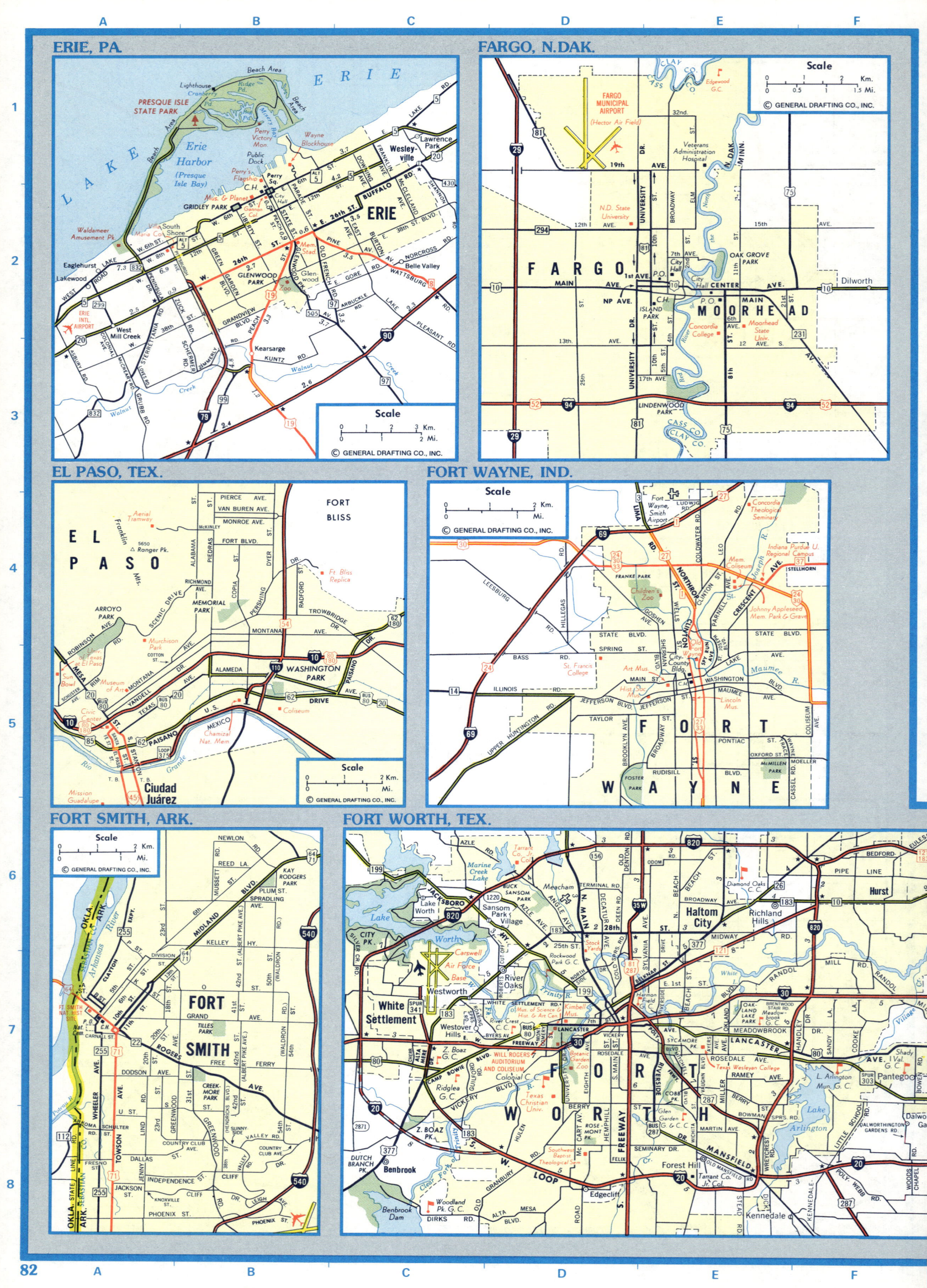

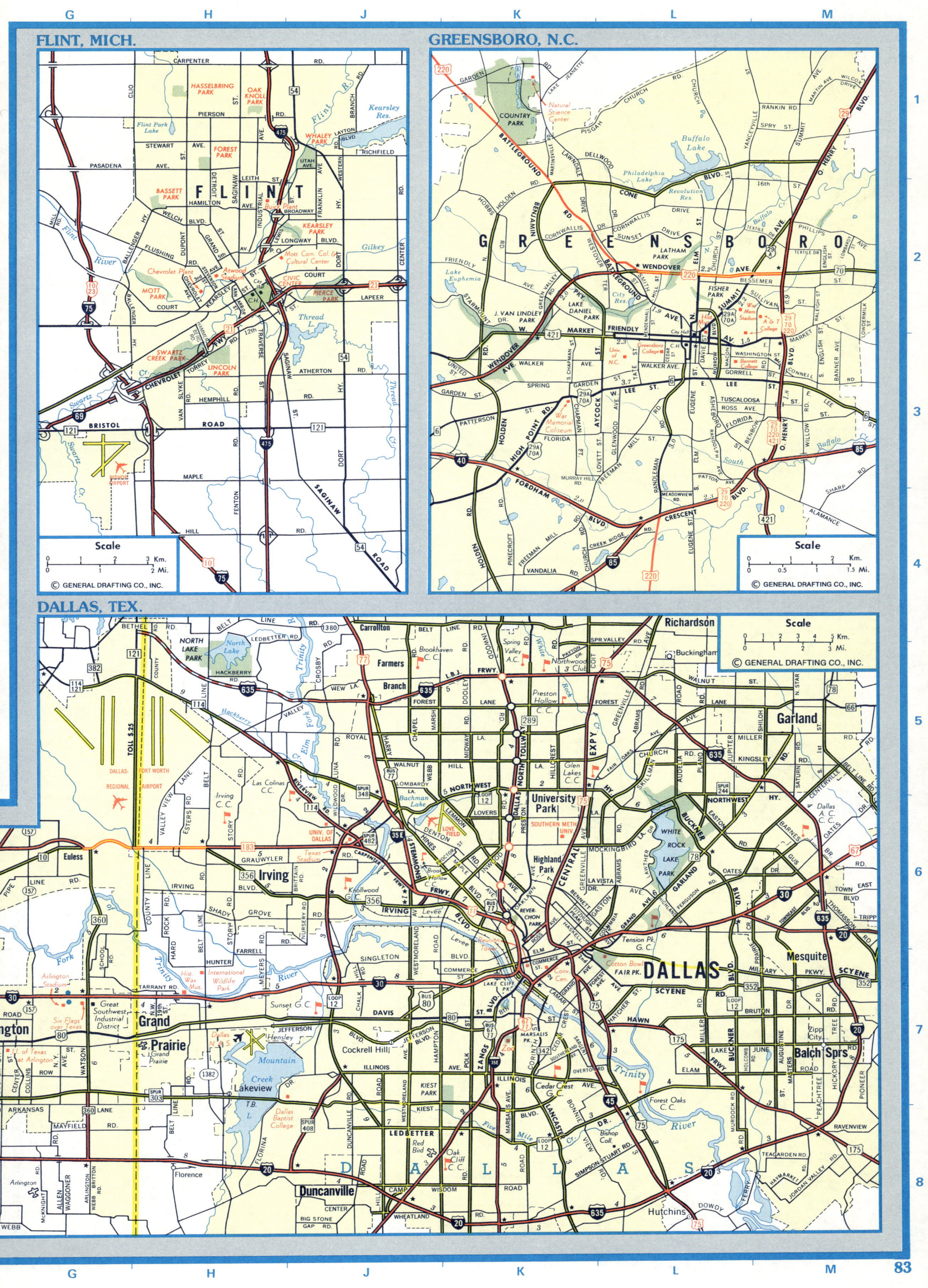

FLINT, MICH.

GREENSBORO, N.C.

DALLAS, TEX.

Scale

© GENERAL DRAFTING CO., INC.

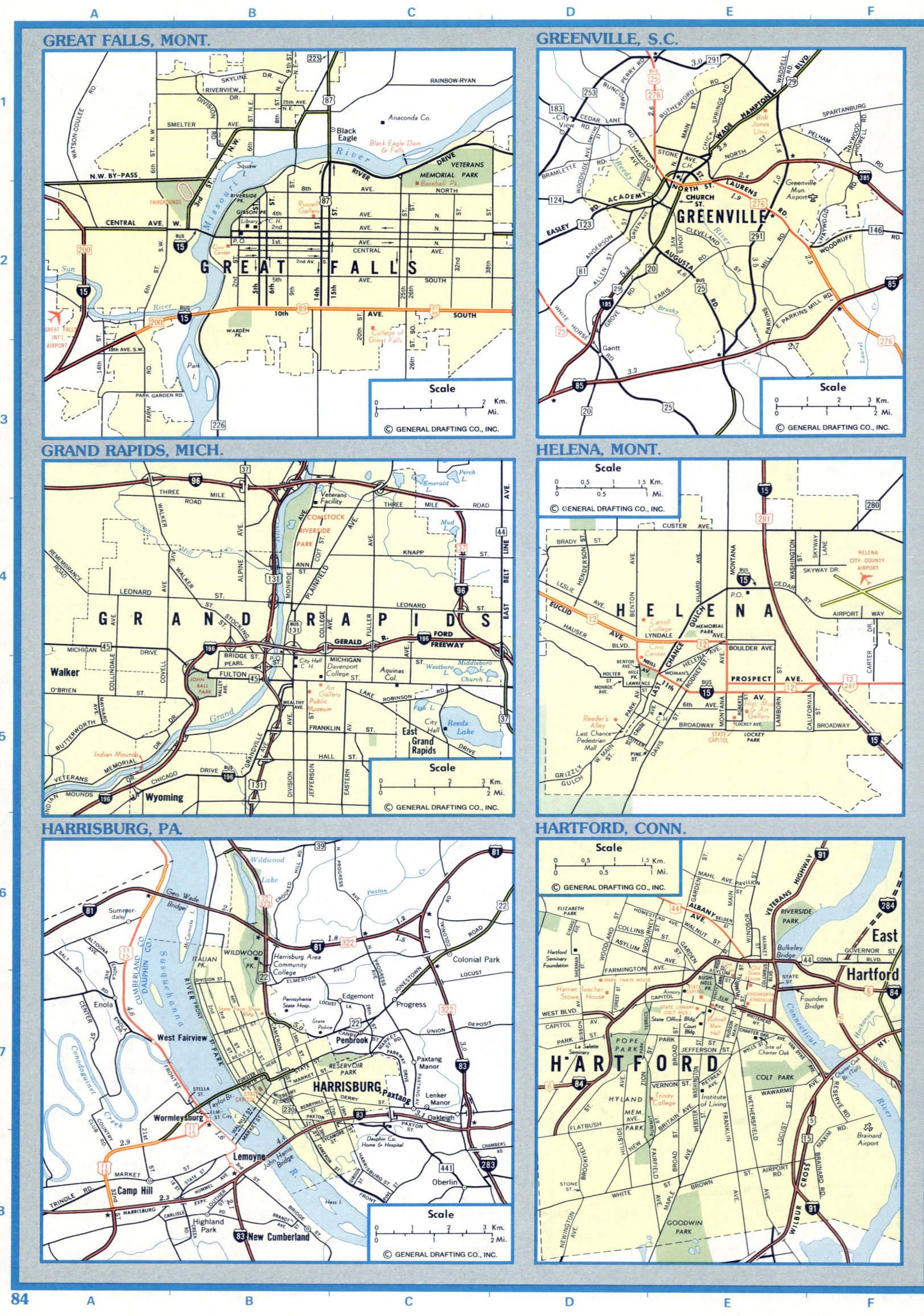

GREAT FALLS, MONT.

GREENVILLE, S.C.

GRAND RAPIDS, MICH.

HELENA, MONT.

HARRISBURG, PA.

HARTFORD, CONN.

HOUSTON, TEX.

Scale
0 1 2 3 4 5 6 Km.
0 1 2 3 4 Mi.
© GENERAL DRAFTING CO., INC.

GALVESTON, TEX.

Scale
0 1 2 3 4 5 Km.
0 1 2 3 Mi.
© GENERAL DRAFTING CO., INC.

HUNTINGTON, W.VA.

Scale
0 1 2 3 Km.
0 1 2 Mi.
© GENERAL DRAFTING CO., INC.

IDAHO FALLS, IDAHO

Scale
0 0.5 1 Km.
0 0.5 1 Mi.
© GENERAL DRAFTING CO., INC.

INDIANAPOLIS, IND.

JACKSON, MISS.

KALAMAZOO, MICH.

JACKSONVILLE, FLA.

KNOXVILLE, TENN.

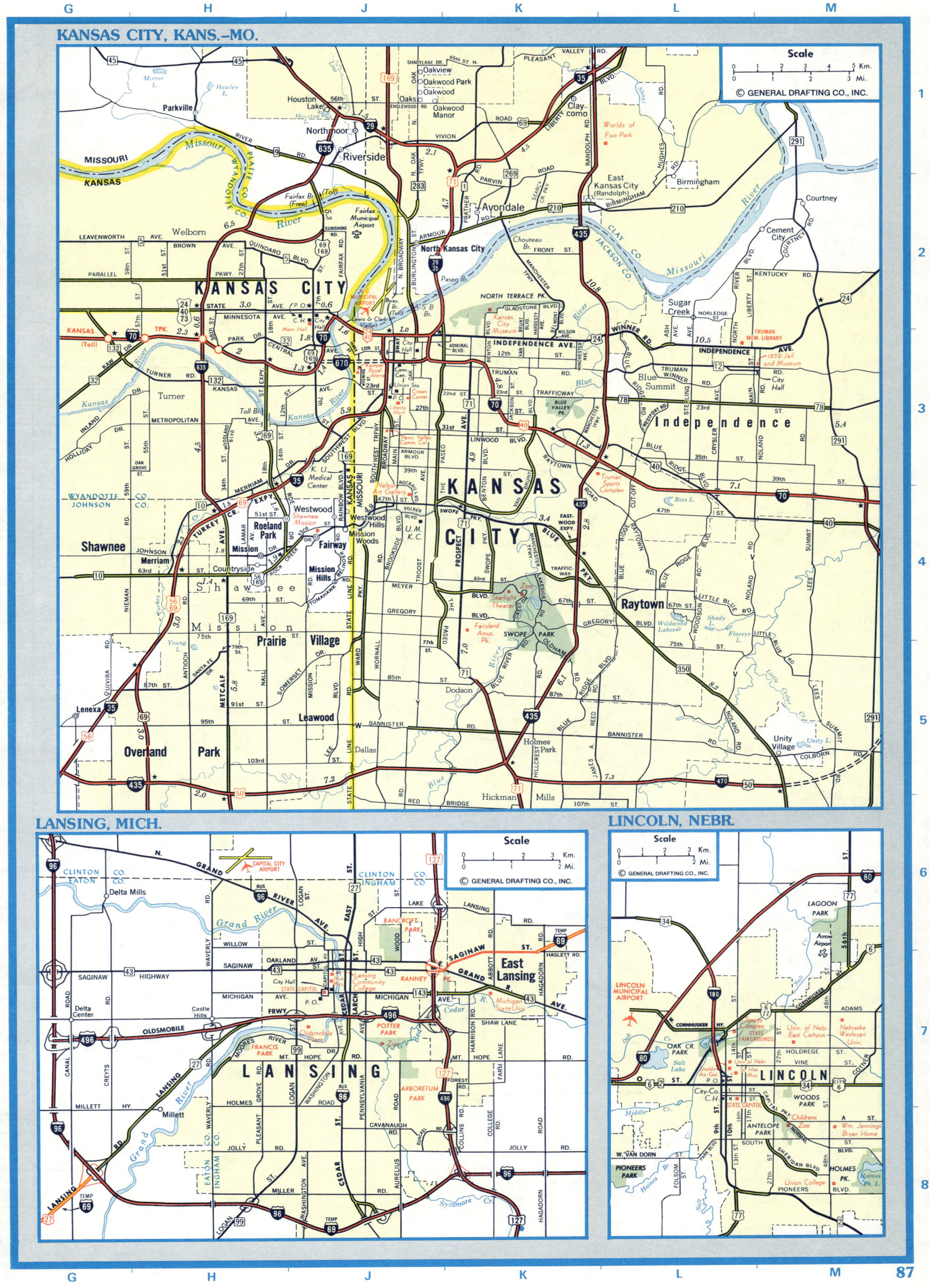

LANSING, MICH.

LINCOLN, NEBR.

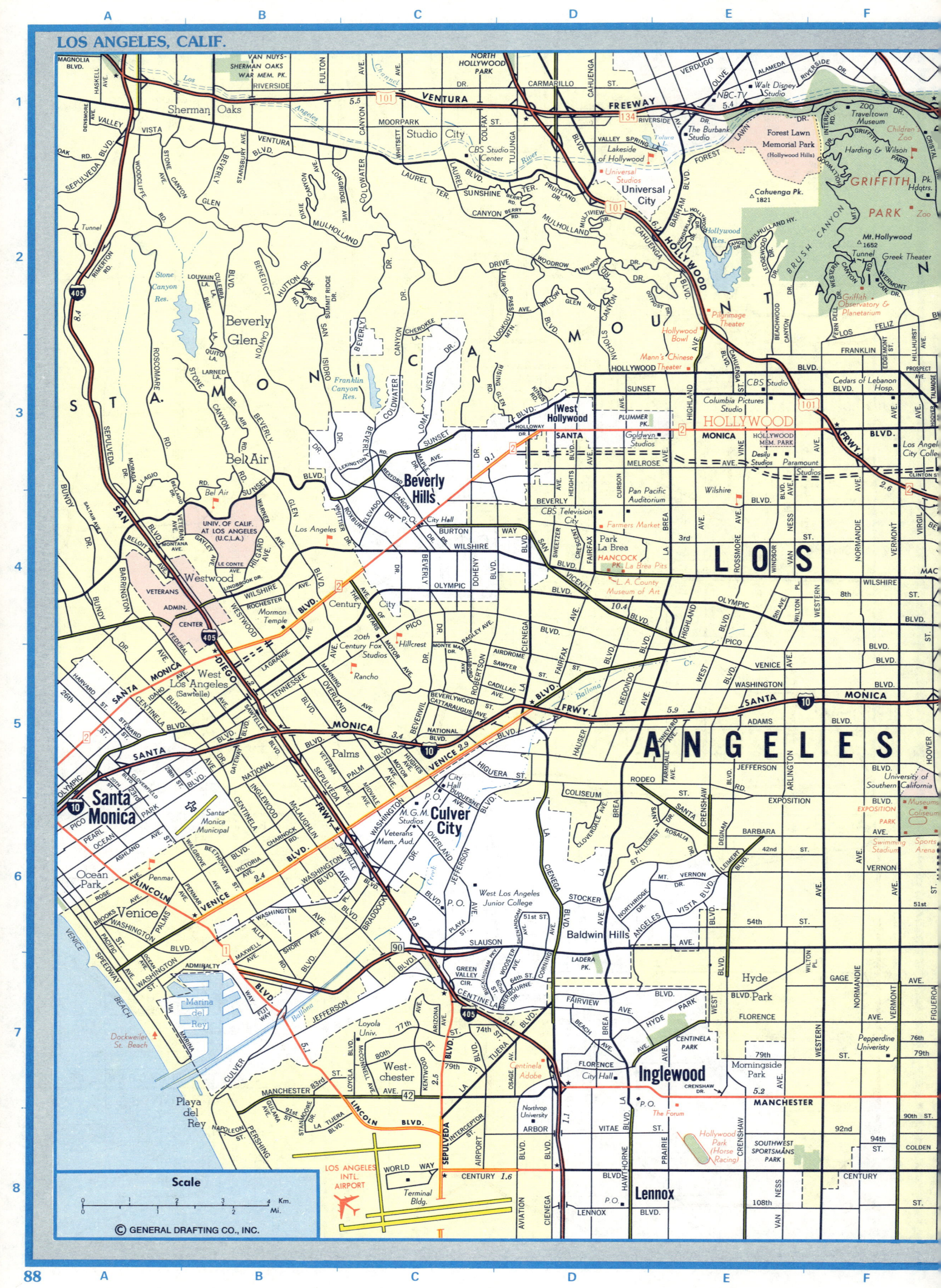

LOS ANGELES, CALIF.

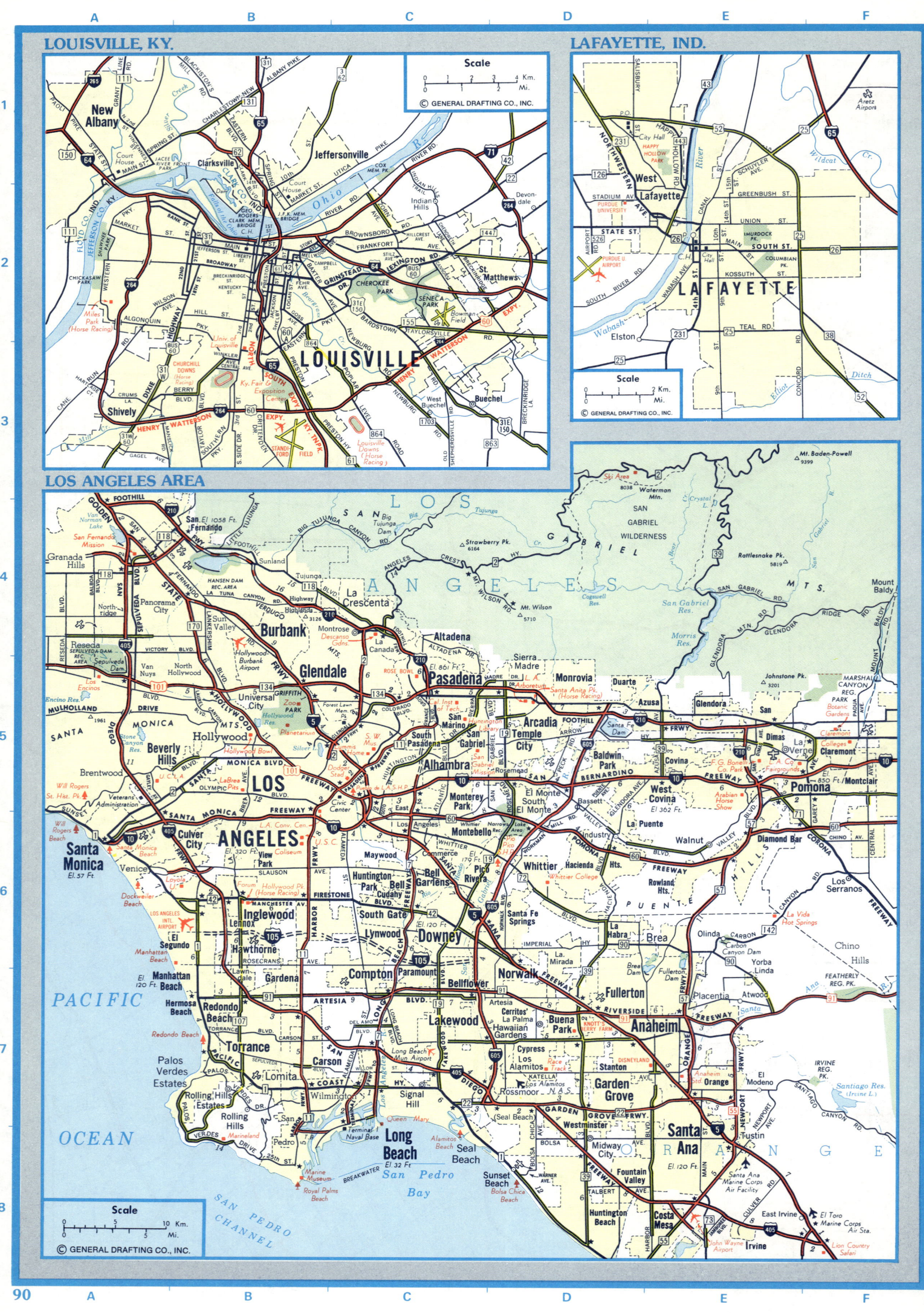

LITTLE ROCK, ARK.

LAS VEGAS, NEV.

LEXINGTON, KY.

MINNEAPOLIS—ST. PAUL, MINN.

MIAMI TO FORT LAUDERDALE, FLA.

MAP CONTINUED AT RIGHT

© General Drafting Co., Inc.

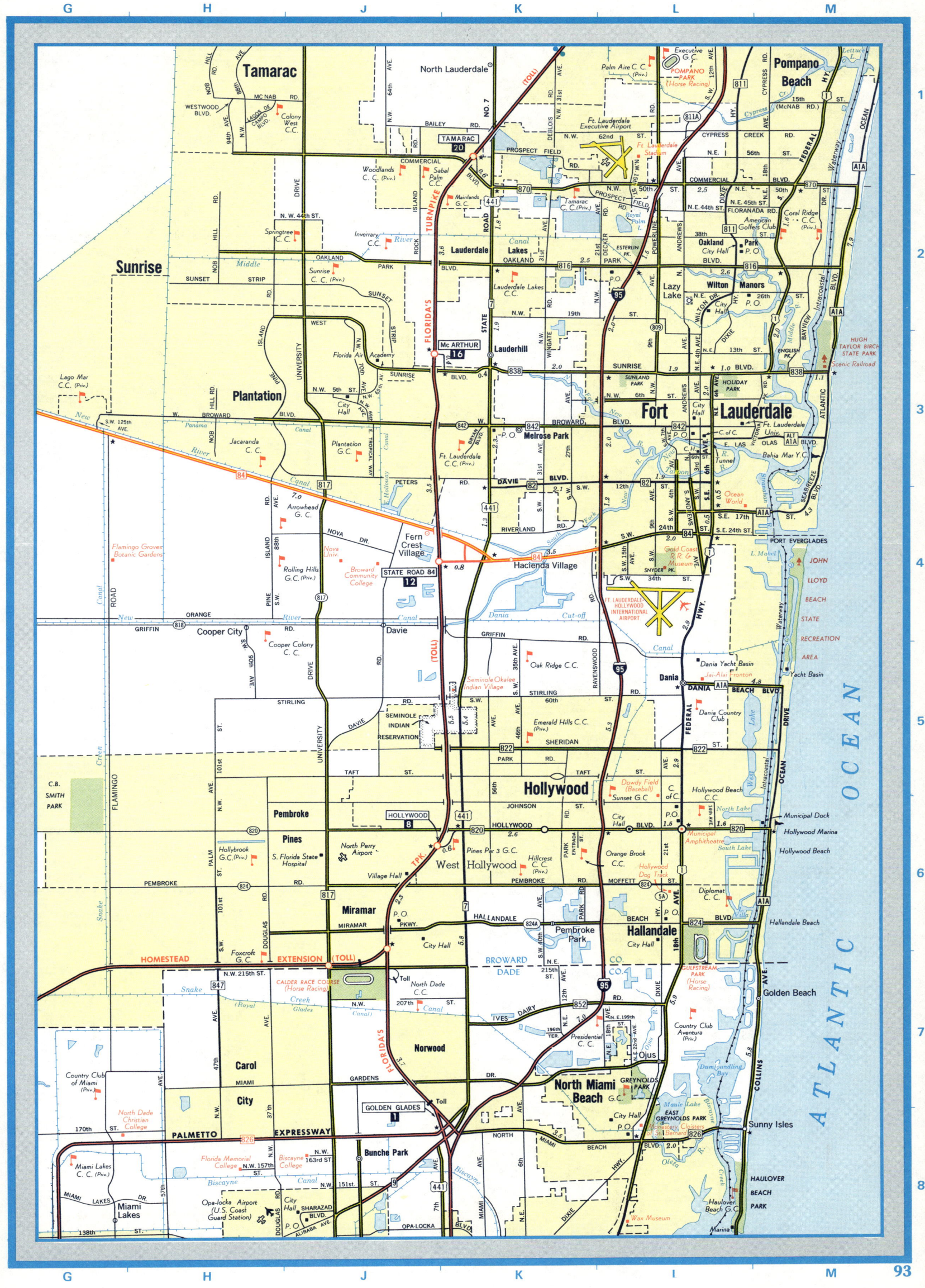

MEMPHIS, TENN.

MOBILE, ALA.

MONROE, LA.

MILWAUKEE, WIS.

MONTGOMERY, ALA.

Scale

© GENERAL DRAFTING CO., INC.

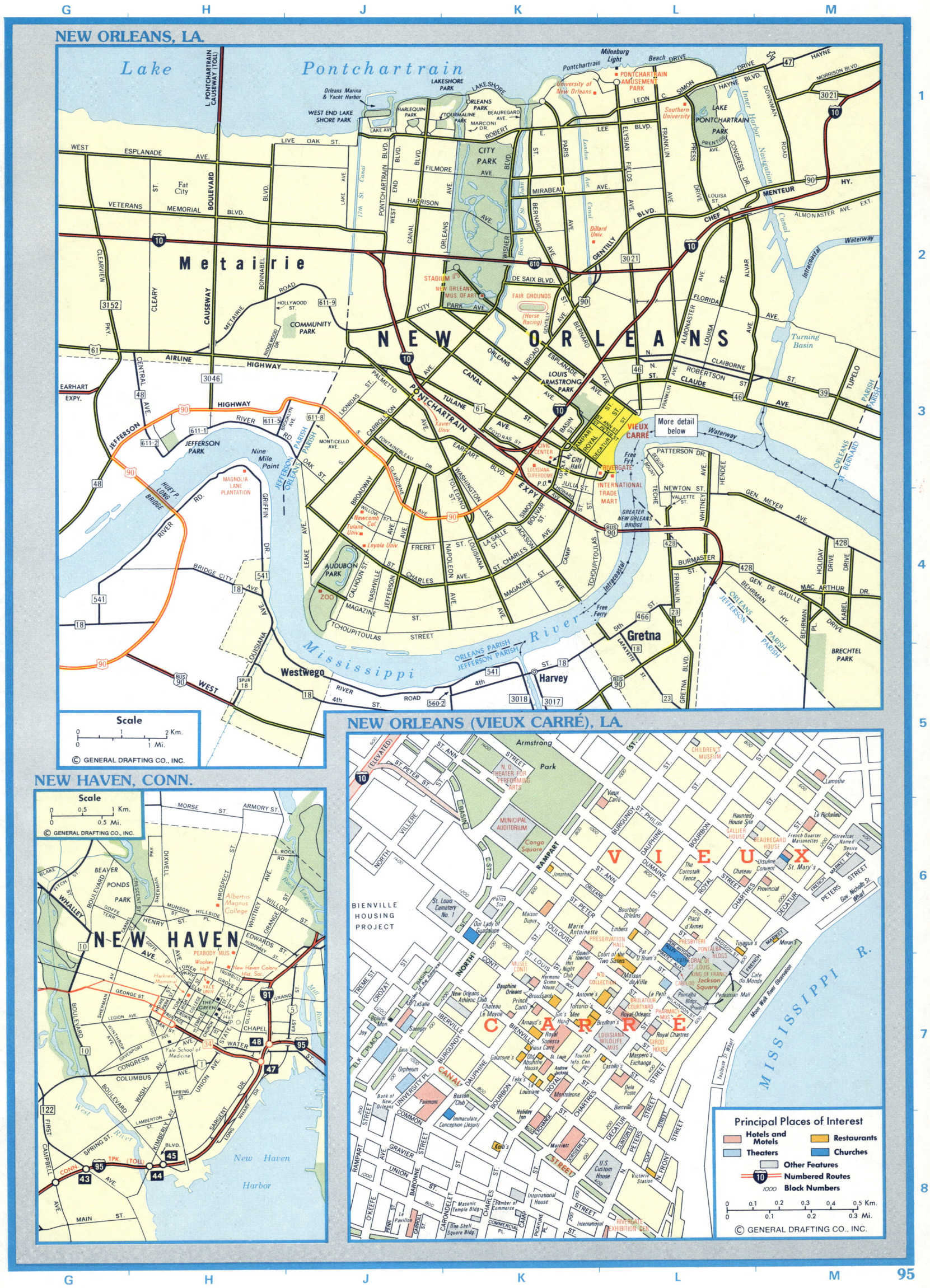

NEW ORLEANS, LA.

NEW ORLEANS (VIEUX CARRÉ), LA.

NEW HAVEN, CONN.

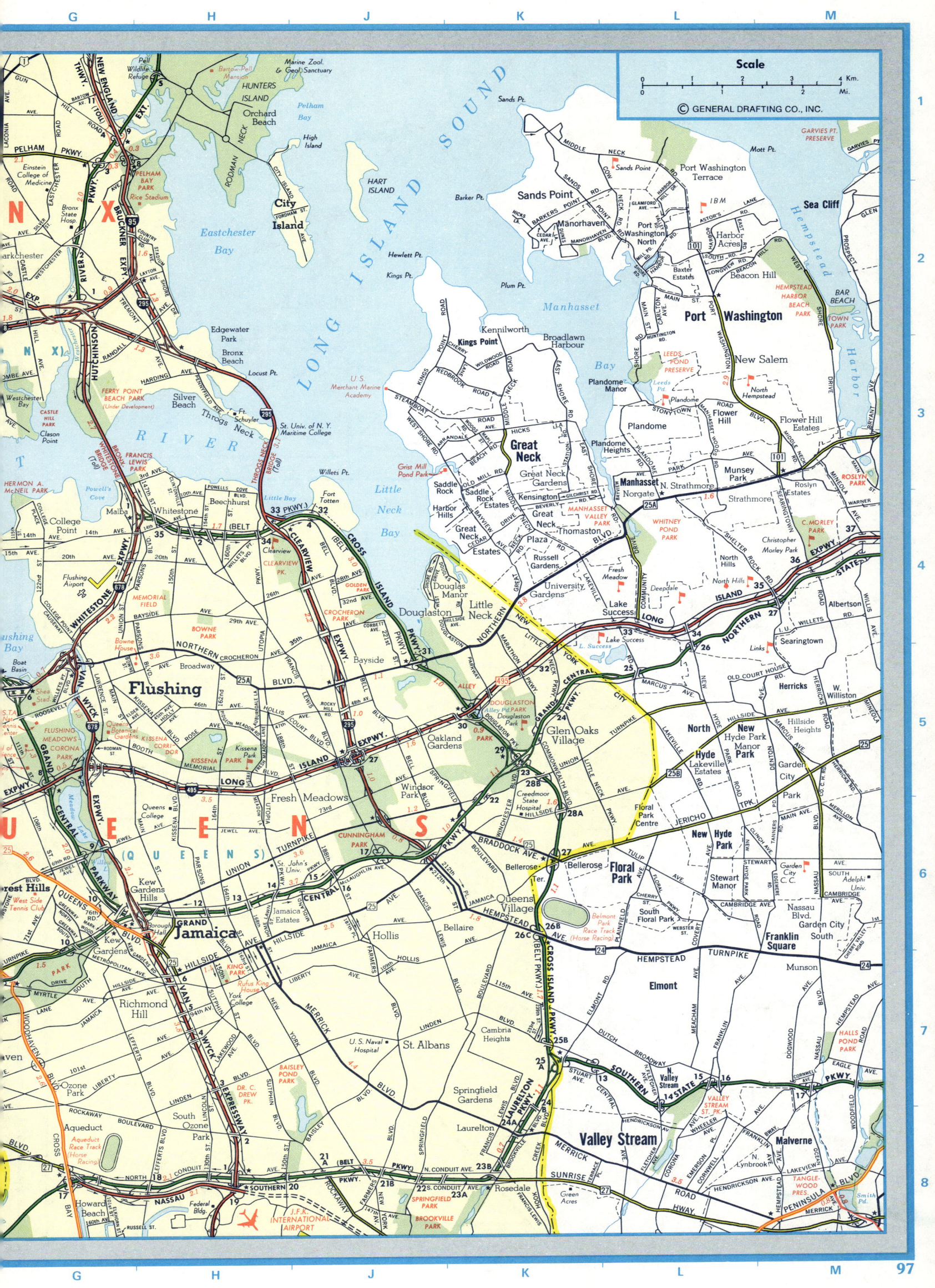

NASHVILLE, TENN.

NORFOLK, VA.

OMAHA, NEBR.

ORLANDO, FLA.

OKLAHOMA CITY, OKLA.

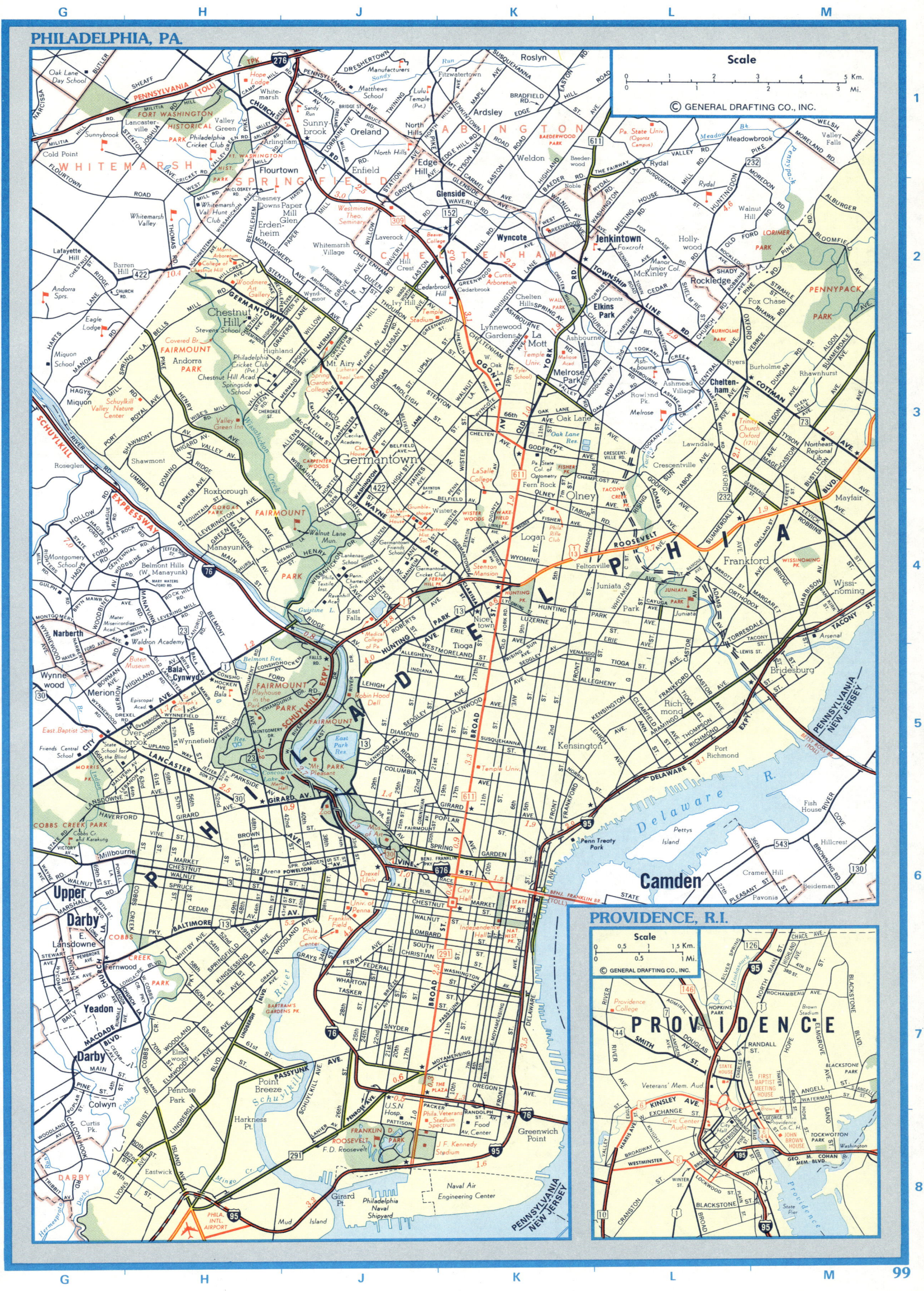

PHILADELPHIA, PA.

PROVIDENCE, R.I.

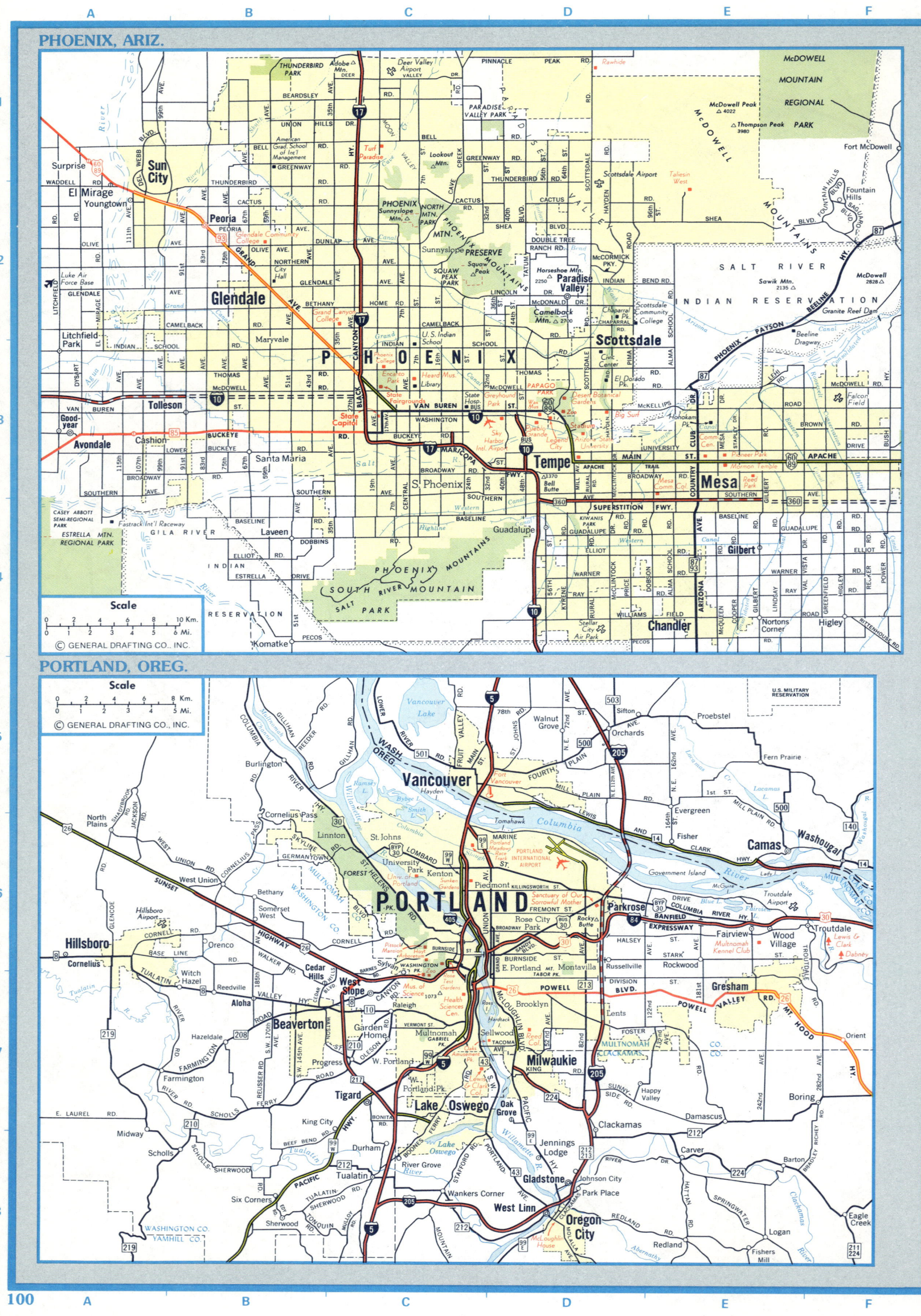

PHOENIX, ARIZ.

PORTLAND, OREG.

100

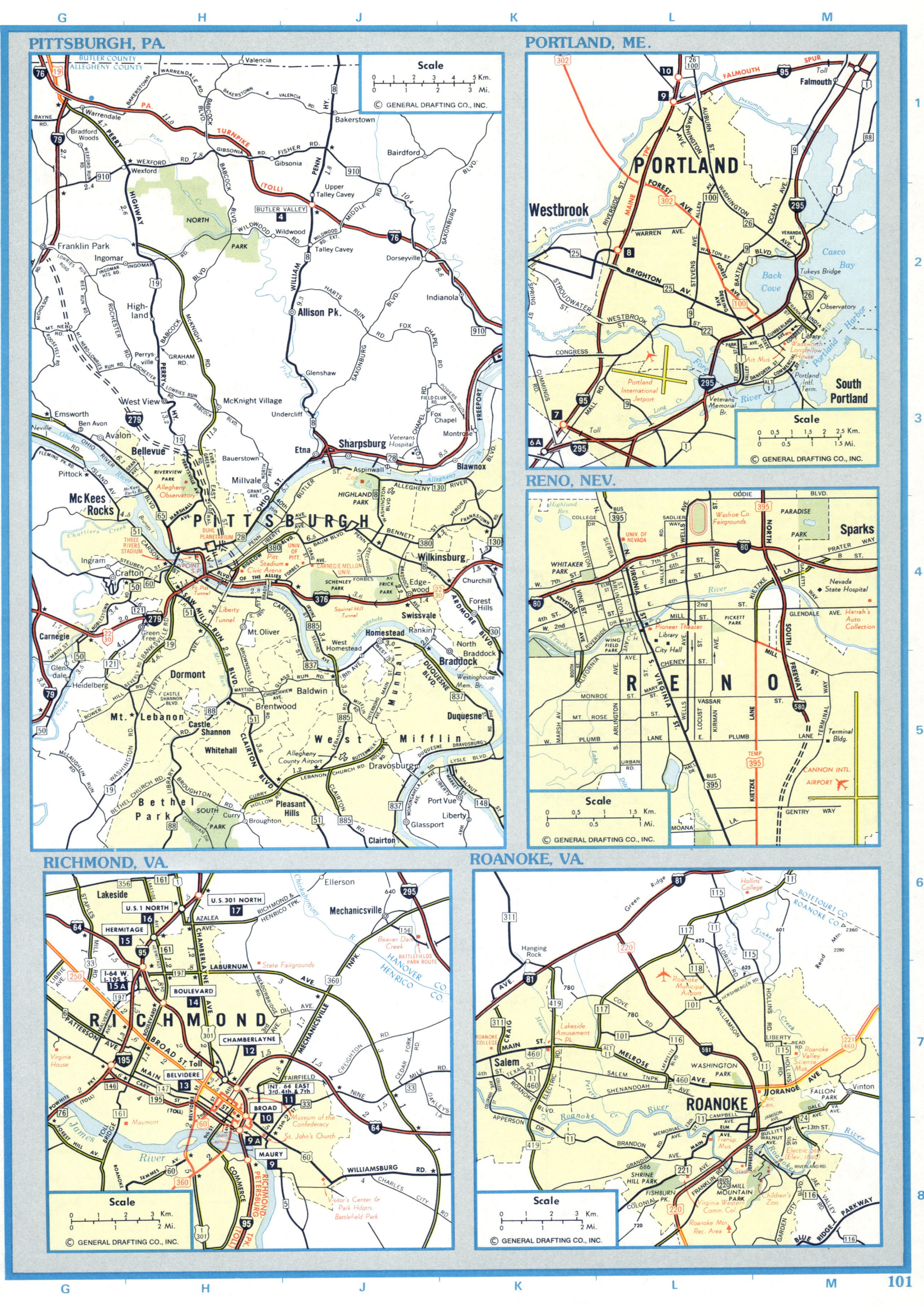

PITTSBURGH, PA.

PORTLAND, ME.

RENO, NEV.

RICHMOND, VA.

ROANOKE, VA.

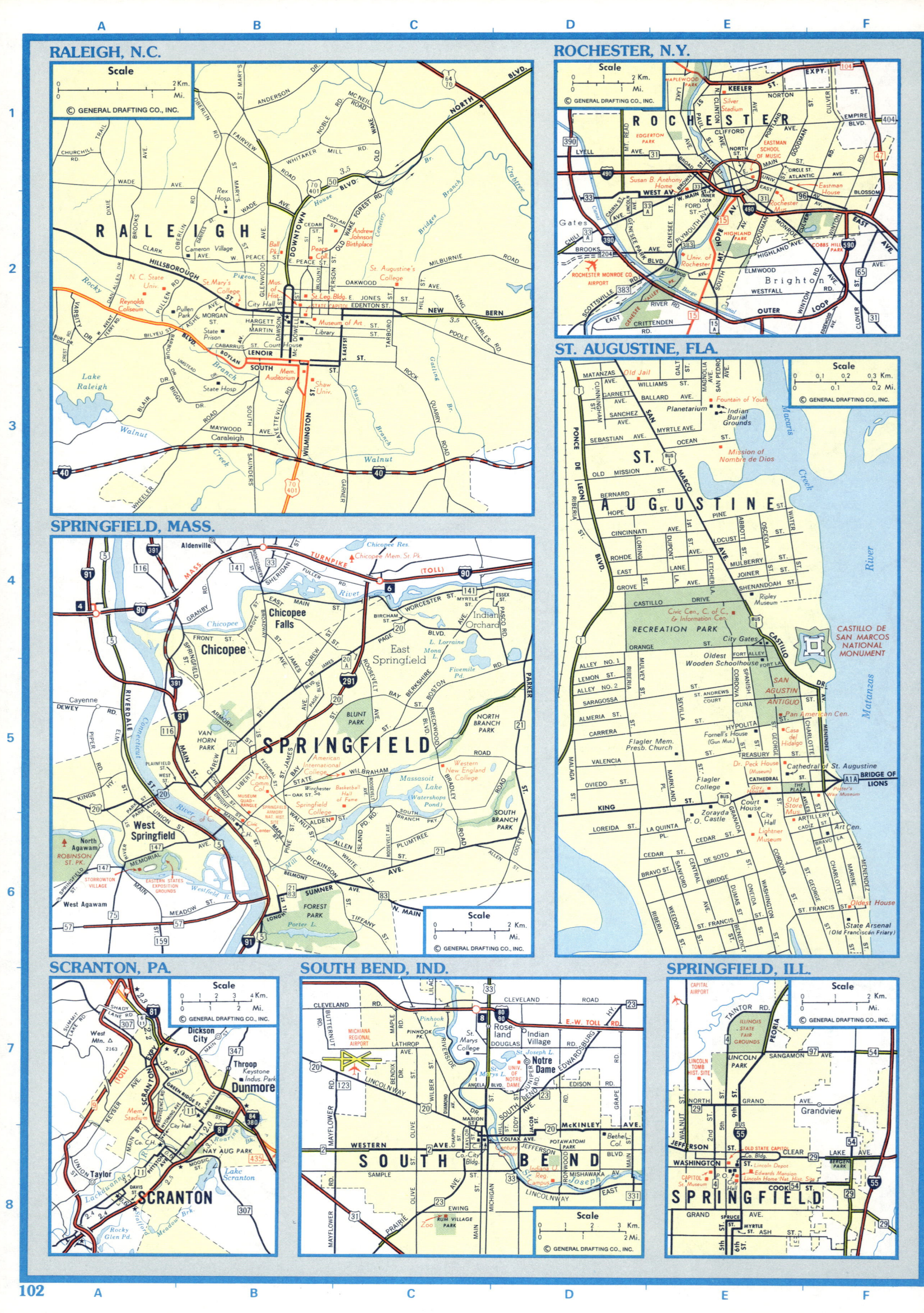

RALEIGH, N.C.

ROCHESTER, N.Y.

ST. AUGUSTINE, FLA.

SPRINGFIELD, MASS.

SCRANTON, PA.

SOUTH BEND, IND.

SPRINGFIELD, ILL.

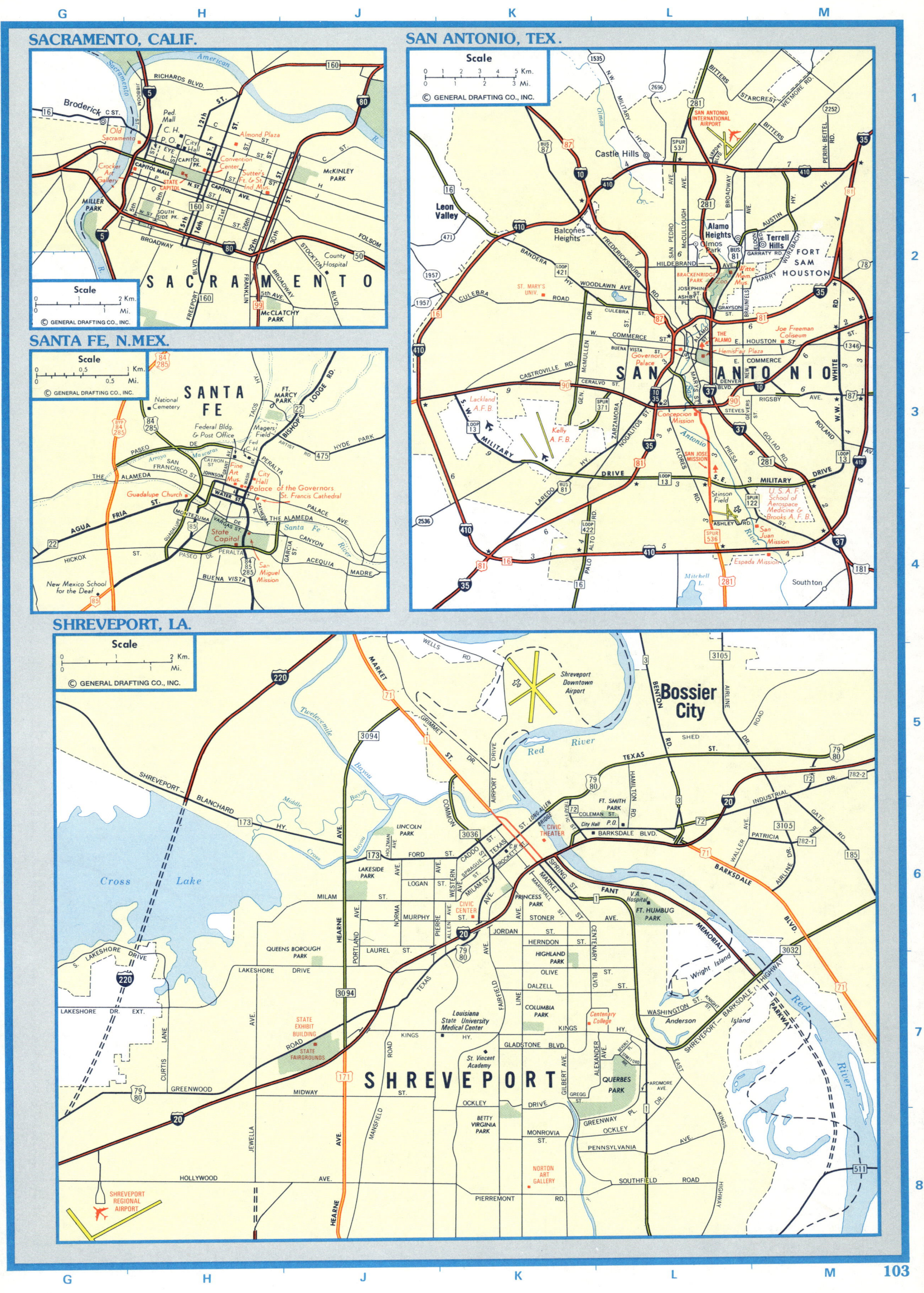

SACRAMENTO, CALIF.

SAN ANTONIO, TEX.

SANTA FE, N.MEX.

SHREVEPORT, LA.

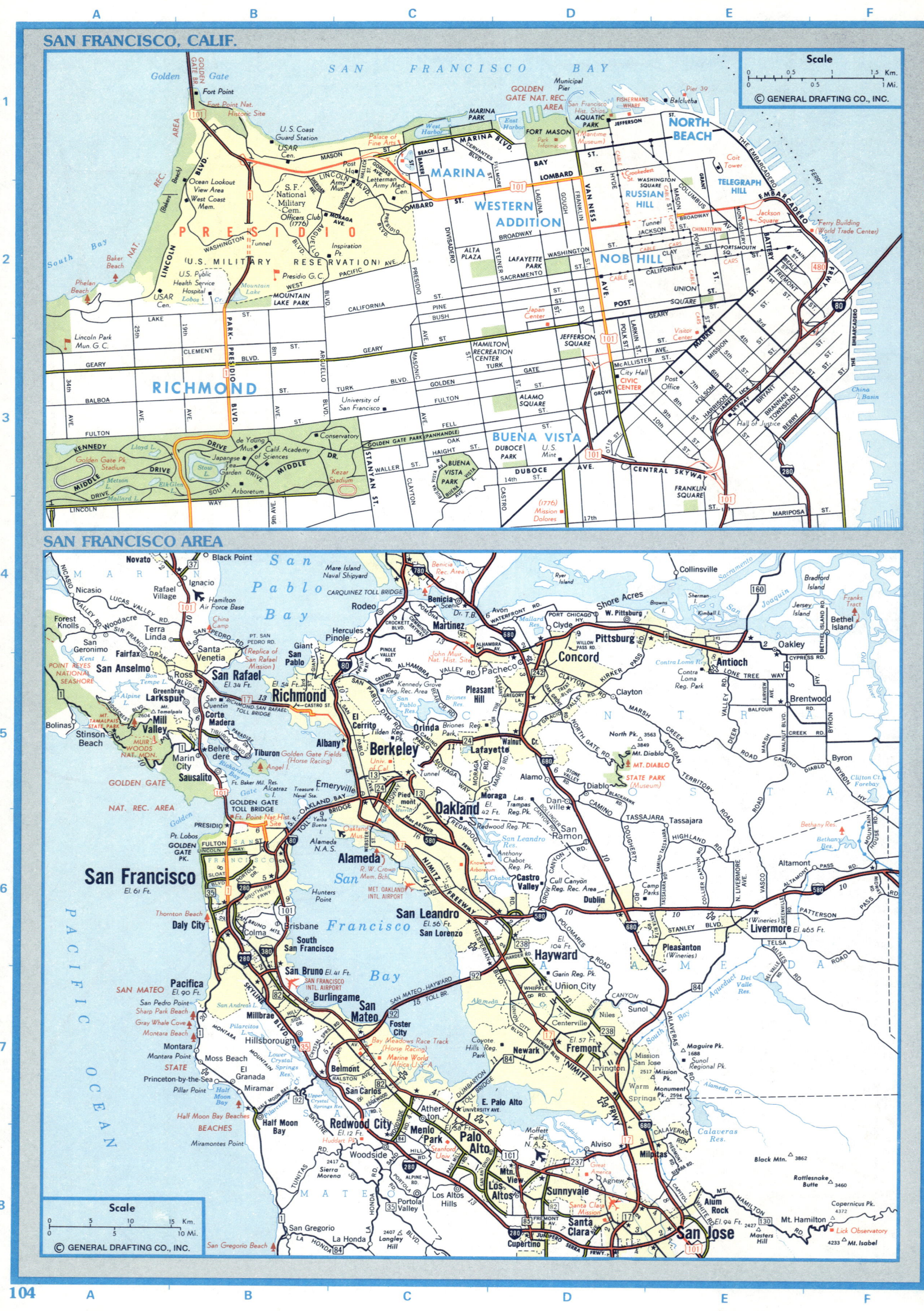

SPARTANBURG, S.C.

SYRACUSE, N.Y.

SPOKANE, WASH.

SAN DIEGO, CALIF.

SAVANNAH, GA.

Scale

© GENERAL DRAFTING CO., INC.

ST. LOUIS, MO.

ST. PETERSBURG, FLA.

SARASOTA, FLA.

TALLAHASSEE, FLA.

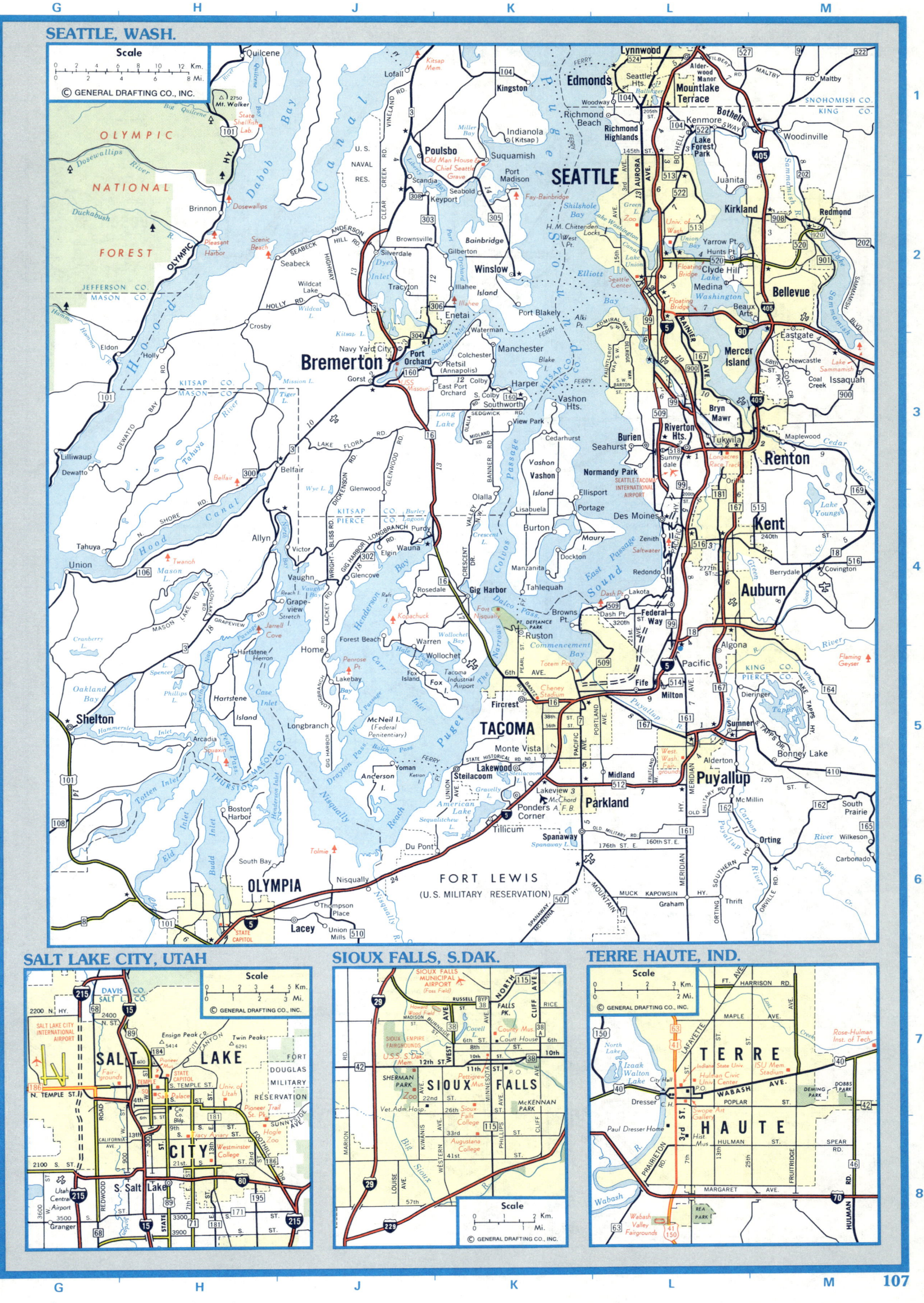

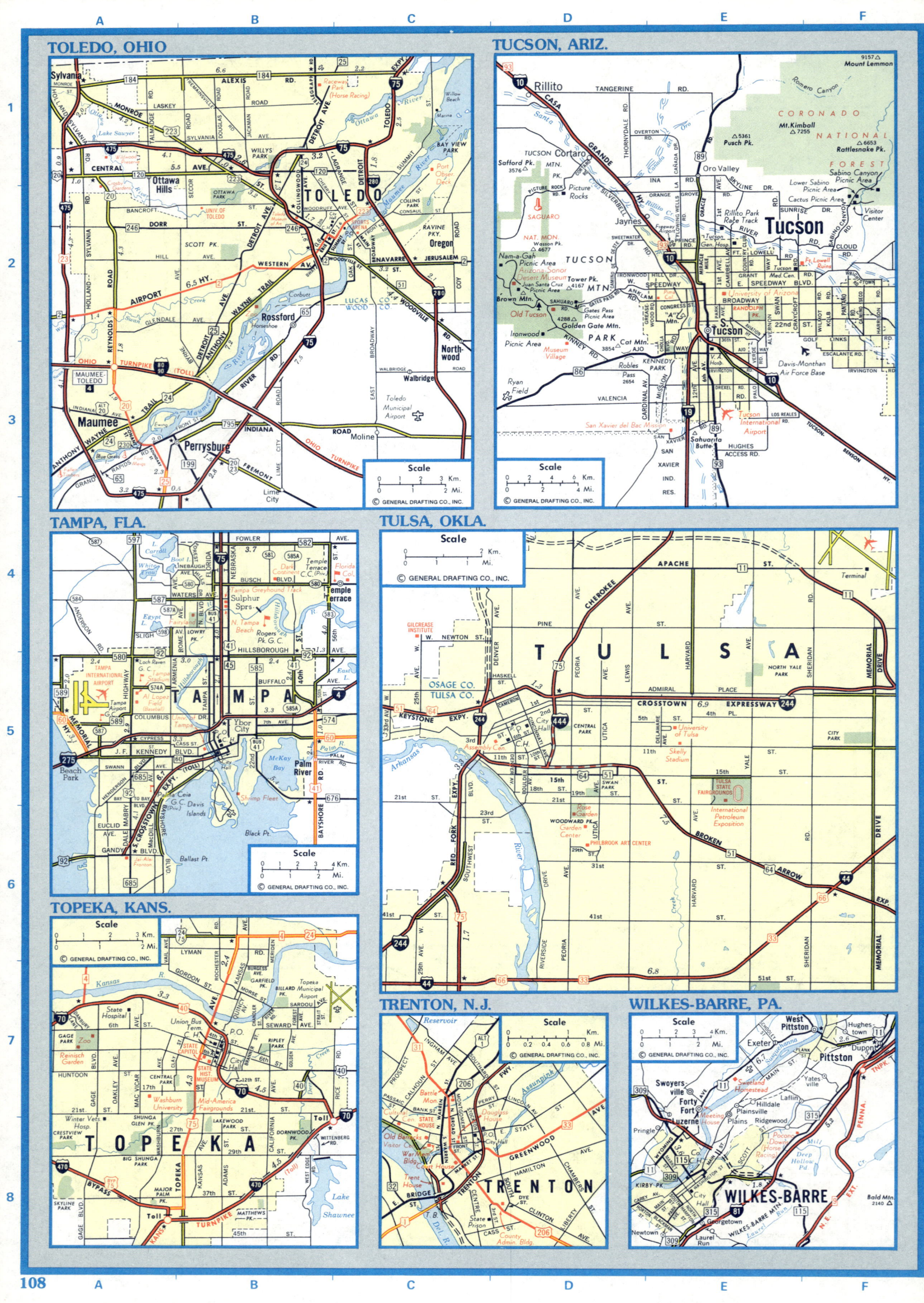

TOLEDO, OHIO

TUCSON, ARIZ.

TAMPA, FLA.

TULSA, OKLA.

TOPEKA, KANS.

TRENTON, N.J.

WILKES-BARRE, PA.

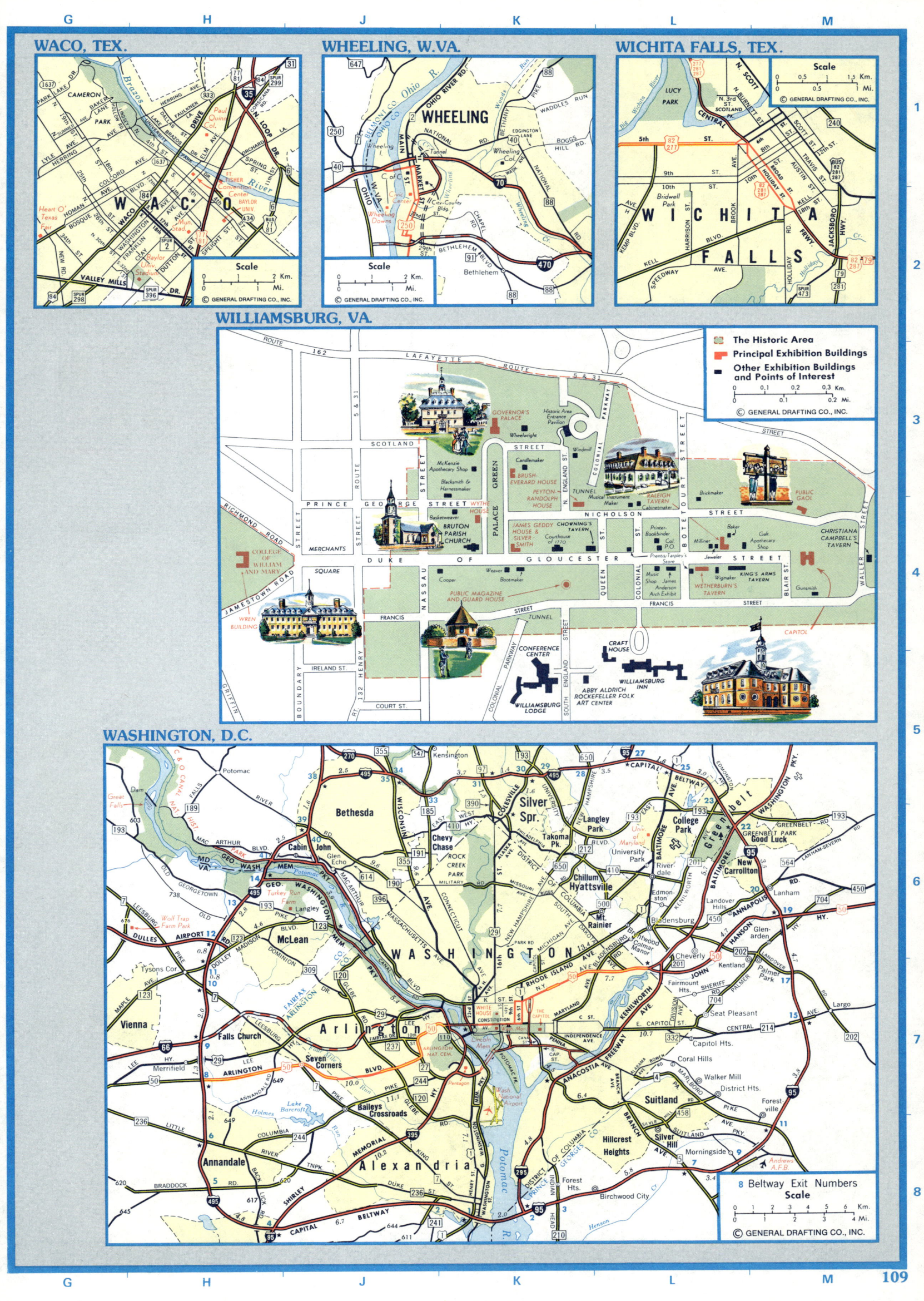

WACO, TEX.

WHEELING, W.VA.

WICHITA FALLS, TEX.

WILLIAMSBURG, VA.

The Historic Area
Principal Exhibition Buildings
Other Exhibition Buildings and Points of Interest

WASHINGTON, D.C.

8 Beltway Exit Numbers

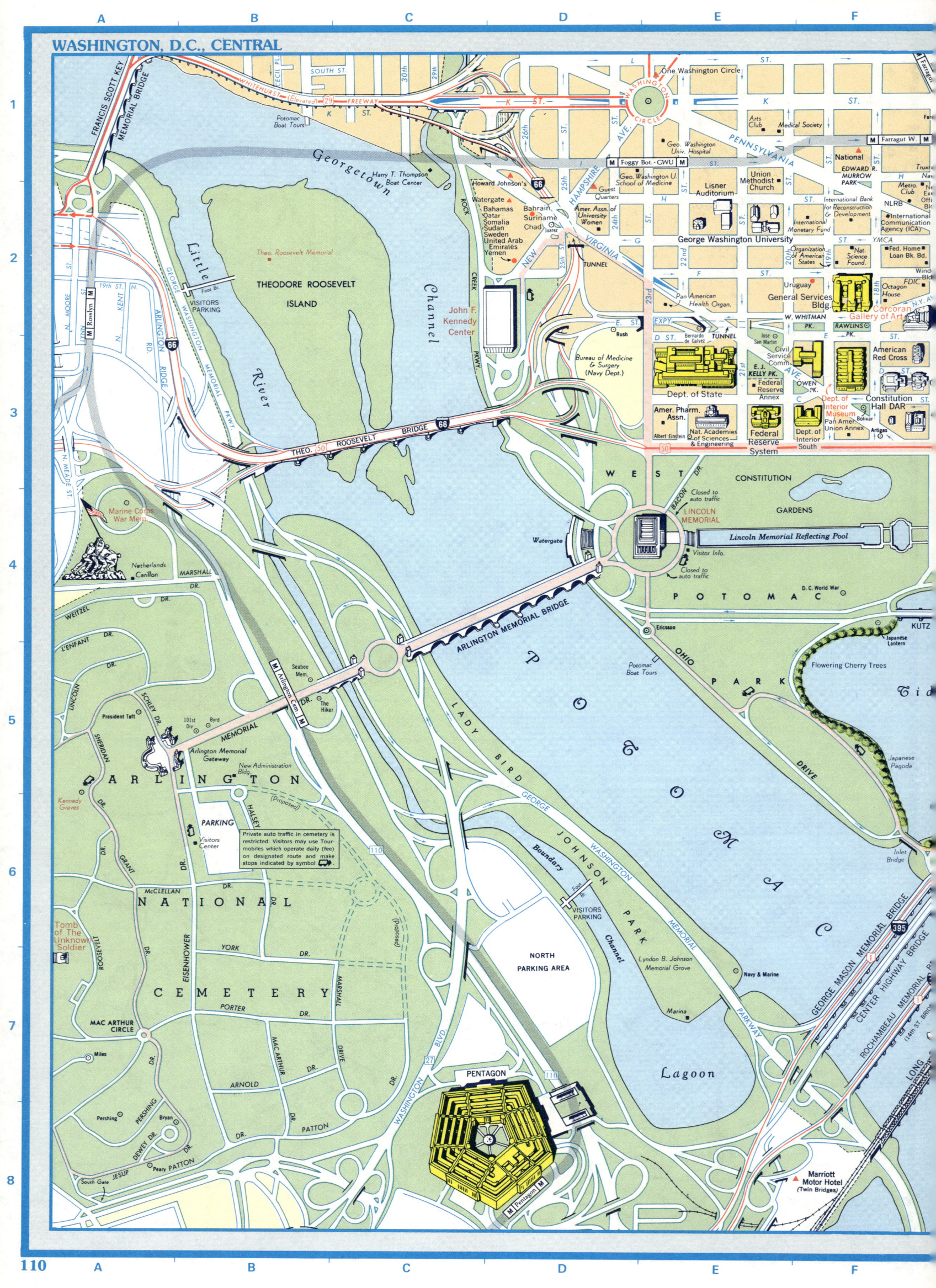

WASHINGTON, D.C., CENTRAL

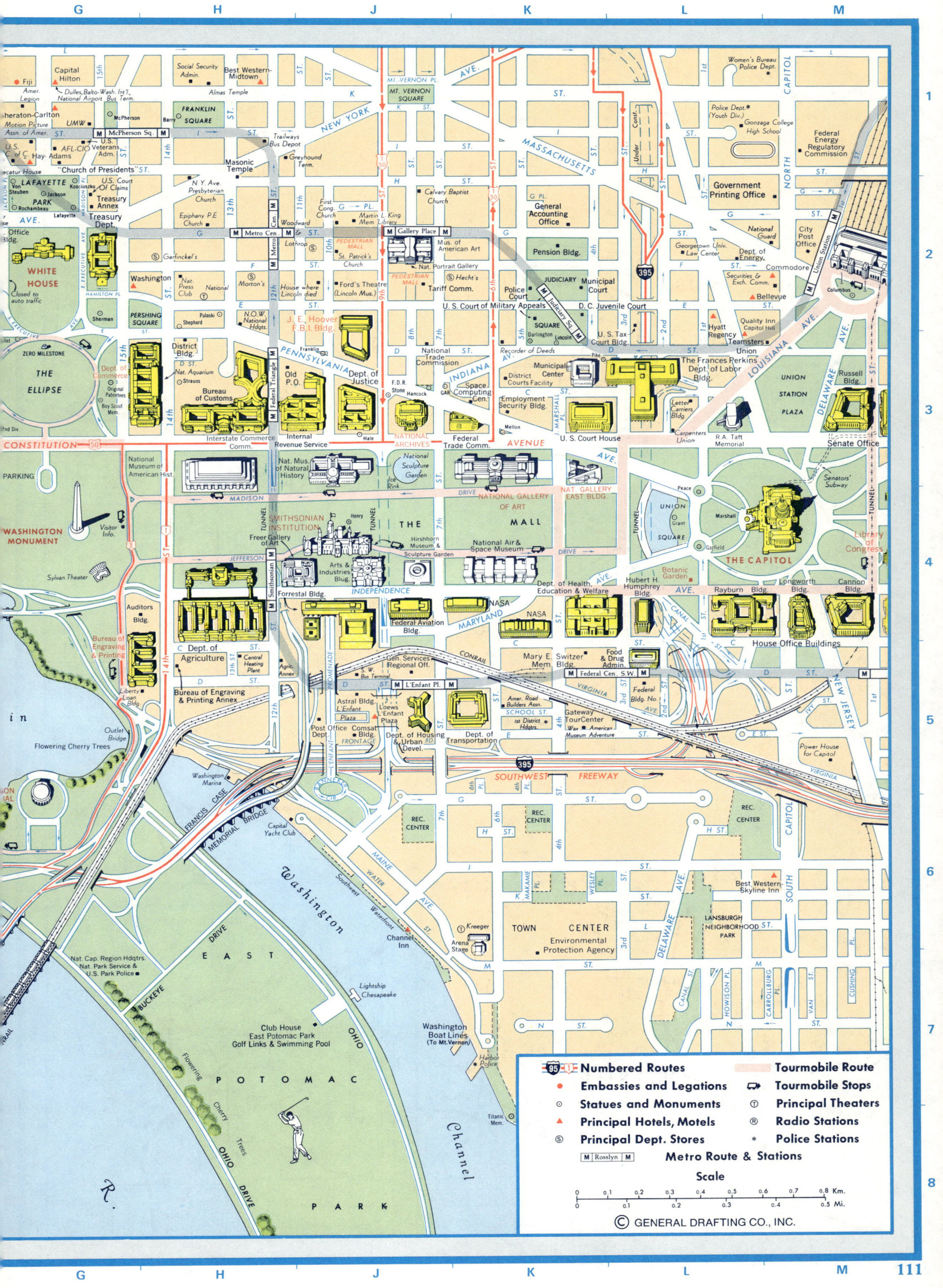

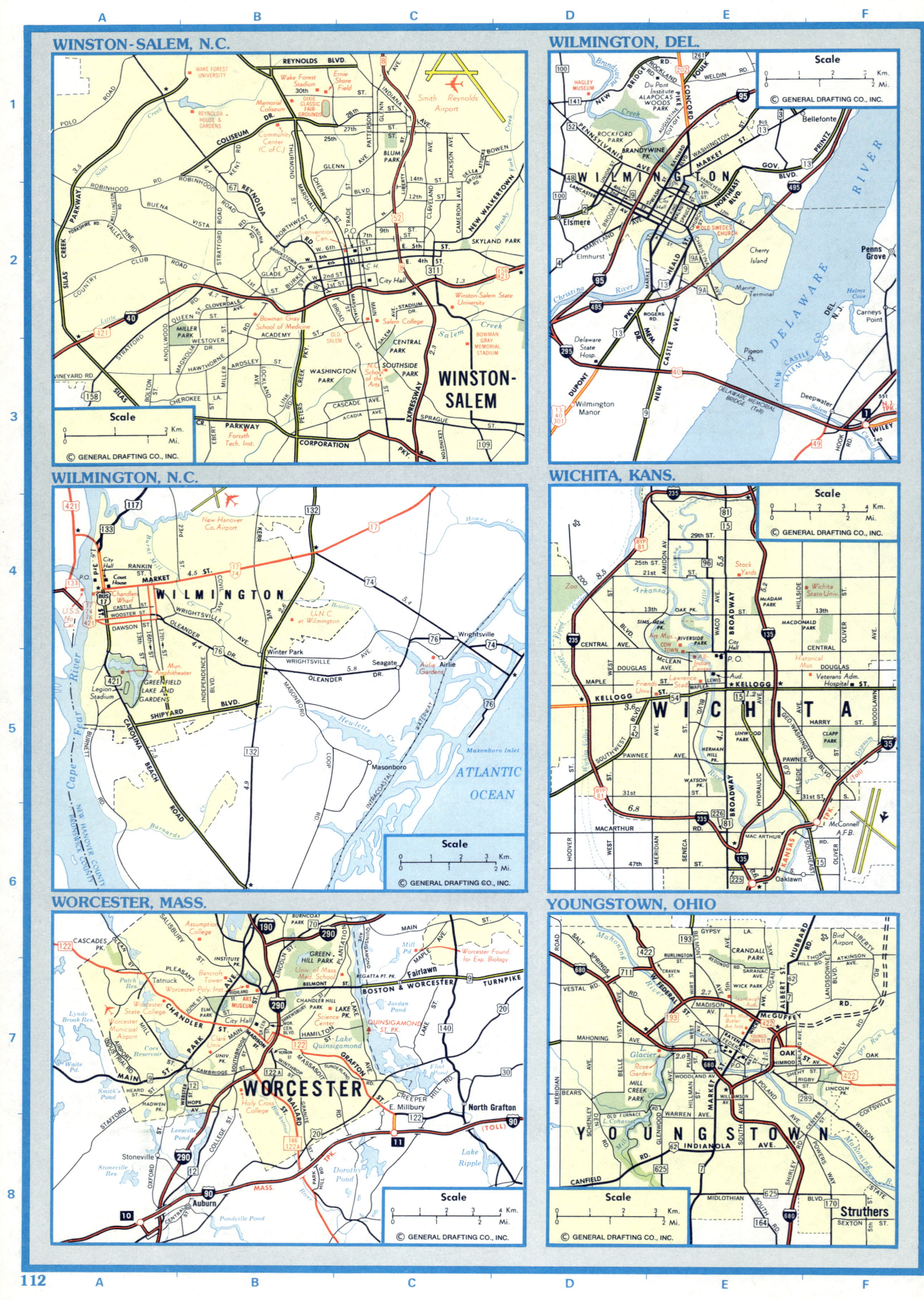

WINSTON-SALEM, N.C.

WILMINGTON, DEL.

WILMINGTON, N.C.

WICHITA, KANS.

WORCESTER, MASS.

YOUNGSTOWN, OHIO

CALGARY, ALTA.

EDMONTON, ALTA.

MONTRÉAL, QUÉ.

NIAGARA FALLS, ONT.

OTTAWA, ONT.

QUEBEC CENTRAL, QUÉ.

TORONTO (METROPOLITAN), ONT.

VANCOUVER, B.C.

MUNICIPALITY OF WEST VANCOUVER

MUNICIPALITY OF NORTH VANCOUVER

MT. SEYMOUR PROV. PK.

Horseshoe Bay
Eagle Harbour
Sherman
Wadsley
Dundarave
Hollyburn
LIGHTHOUSE PARK
FERRIES

North Vancouver

Deep Cove

MUNICIPALITY OF COQUITLAM

Capilano Suspension Bridge
Sky Ride
Chair Lift

Burrard Inlet

English Bay
Planetarium and Mus.
STANLEY PK.
Zoo & Aquarium

Burrard Inlet

SECOND NARROWS

Indian Arm

Buntzen L.

Sasamat L.

Ioco

MUN. OF BURNABY

Port Moody

Port Coquitlam

MUN. OF PITT MEADOWS

HASTINGS
EXHIBITION PK.
Empire Stadium
Broadway

Simon Fraser Univ.

Deer L.

N.W. MARINE DR.
BIANCA ST.
UNIV. OF BRITISH COLUMBIA
MARINE DR.
10th Ave.
16th
33rd
49th

Vancouver

Heritage Village

New Westminster

Pitt Meadows

Port Hammond

Strait of Georgia

VANCOUVER INT'L AIRPORT
Sea Island

LULU ISLAND

RICHMOND

Whalley

Port Kells

Newton

MUNICIPALITY OF SURREY

Langley

Steveston

MUN. OF DELTA

GEORGE MASSEY TUNNEL

Ladner

Cloverdale

Westham Island

Roberts Bank

VANCOUVER - BLAINE FREEWAY

Mud Bay

Crescent Beach

Ocean Park

White Rock

Boundary Bay

Semiahmoo Bay

Peace Arch Park

Tsawwassen
Boundary Bay

BRITISH COLUMBIA / WASHINGTON

CANADA / UNITED STATES

BRITISH COLUMBIA / WASHINGTON

Kilometers are shown in red along red roads.
Scale
© GENERAL DRAFTING CO., INC.

REGINA, SASK.

Scale
© GENERAL DRAFTING CO., INC.

R E G I N A

Pasqua Hospital
Saskatchewan House
TAYLOR FIELD (Sports)
Union Station P.O.
City Hall
Exhibition Grounds
Chamber of Commerce
Telecommunications Bldg. "Telorama" (College)
Mackenzie Art Gallery
Provincial Museum of Nat. Hist.
College of Educ.
Balfour Tech. School
General Hospital
Information Centre

KIWANIS PARK
Terminal Building
REGINA AIRPORT
KINSMEN PARK

Legislative Bldgs.
Trafalgar Fountain
LEGISLATIVE CENTRE GROUNDS
Provincial Admin. Bldg.
Provincial Health Bldg.
Diefenbaker Homestead

WASCANA LAKE
Centre of the Arts
WATER FOWL PARK
DOUGLAS PARK (Sports)

UNIVERSITY OF REGINA

TRANS CANADA HWY.
Plains Health Centre

WINNIPEG, MAN.

Scale
© GENERAL DRAFTING CO., INC.

KILDONAN PARK
Rainbow Stage
Lord Selkirk
West Kildonan
East Kildonan

WINNIPEG INTERNATIONAL AIRPORT
Terminal Building
Midland
City Hall
Centennial Centre
Upper Fort Garry Gate

Assiniboia (Race Track)
Stadium
Convention Cen.
MARION

PORTAGE
Assiniboine Park
ACADEMY
CORYDON
GRANT

Assiniboine Park

W I N N I P E G

Saint Boniface

Fort Garry

University of Manitoba

Saint Vital

Grande Pointe

PERIMETER

PORTAGE

115

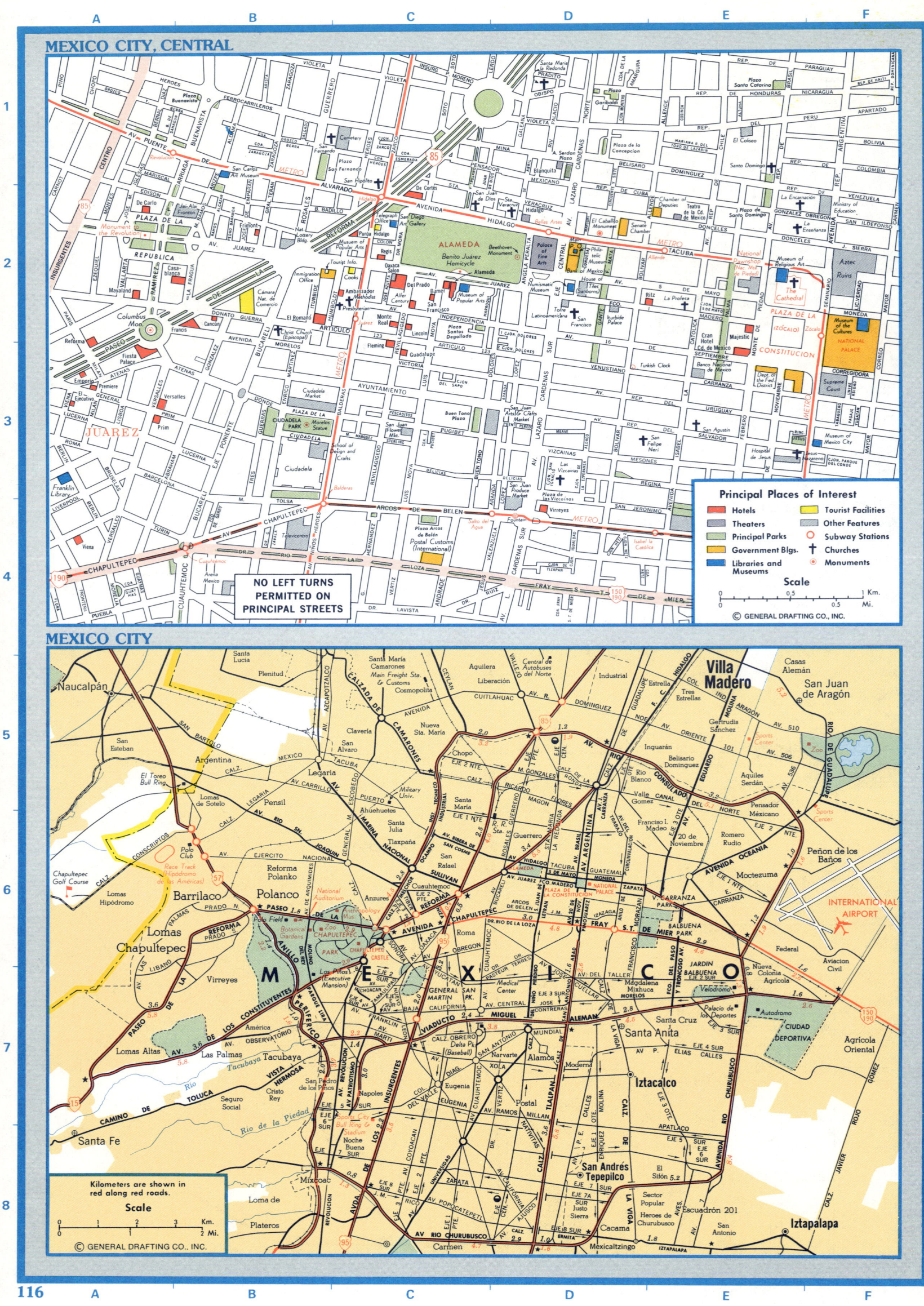

MEXICO CITY, CENTRAL

Principal Places of Interest

- Hotels
- Theaters
- Principal Parks
- Government Blgs.
- Libraries and Museums
- Tourist Facilities
- Other Features
- Subway Stations
- Churches
- Monuments

Scale

NO LEFT TURNS PERMITTED ON PRINCIPAL STREETS

© GENERAL DRAFTING CO., INC.

MEXICO CITY

Kilometers are shown in red along red roads.

Scale

© GENERAL DRAFTING CO., INC.

This country's richest stores of natural splendors and undisturbed wildlife are preserved in its national parks, located throughout the conterminous 48 states as well as Alaska, Hawaii and the U.S. Virgin Islands. These natural parks were set aside in order to preserve the many unusual features of the land. They have a unique character and afford visitors unmatched recreational opportunities in extraordinary settings. Sooner or later your vacation itinerary should include one or more of these parks.

The National Park Service was established in 1916 as a part of the Department of the Interior. The Service administers the National Park System—begun in 1872 when Congress set aside the Yellowstone wilderness for public recreation and natural preservation. The Antiquities Act has permitted presidents since then to place federal lands under Park Service protection for cultural, recreational, historical and scientific purposes. Since this legislation, the National Park System has grown to include more than 300 areas. The variety of natural and historic areas preserved by the National Park Service is sure to offer something of interest to any vacationer.

The Park System is comprised of approximately two dozen different types of areas. The most familiar are the national parks proper, followed by the recreation areas and monuments. In addition, the Park Service administers a variety of types of historic areas such as national cemeteries, historic sites, military parks and battlefields.

An entrance fee is charged at most National Park Service areas, although fees are waived if one possesses a Golden Eagle, Golden Age or Golden Access passport. Golden Eagle Passports cost $10 per calendar year and may be purchased by mail from National Park Service headquarters in Washington, D.C., or by mail or in person from regional offices. Golden Age

Passports are lifetime passes issued free to anyone 62 or older upon presentation of proof of age. Golden Access Passports are lifetime passes issued to the handicapped upon presentation of proof of disability. Both Golden Age and Golden Access passports may only be obtained in person at Park Service offices or at any federal fee area. All passports will also admit those accompanying the pass holder in a noncommercial vehicle. Campsite users, however, are charged a recreation use fee where applicable.

Sixteen of the most popular National Park Service areas are described on the following pages, accompanied by maps of each area. Among them are 13 national parks, a military park, a monument and a historical park.

AREA	PAGE
Acadia National Park	118
Crater Lake National Park	124
Craters of the Moon National Monument	122
Gettysburg National Military Park	120
Glacier National Park	123
Grand Canyon National Park	123
Grand Teton National Park	122
Great Smoky Mountains National Park	119
Hot Springs National Park	119
Independence National Historical Park	118
Mount Rainier National Park	124
Olympic National Park	124
Rocky Mountain National Park	120
Shenandoah National Park	120
Yellowstone National Park	121
Yosemite National Park	122

The map below locates and identifies the 37 national parks in the conterminous 48 states. For a more complete listing of National Park Service areas and facilities, refer to the charts on pages 125 and 126.

For further information, or to obtain a Passport, contact the National Park Service at their headquarters in Washington or at the regional office addresses given below. When seeking specific information on an area, write to the regional office serving the state in which that area is located as indicated by the listing in parenthesis.

National Park Service Headquarters, U.S. Department of the Interior, Washington, D.C. 20240

Alaska Regional Office: 540 W. Fifth Ave., Room 202, Anchorage, Alaska 99501 (Alaska)

Mid-Atlantic Regional Office: 143 S. Third St., Philadelphia, Pennsylvania 19106 (PA., MD., W. VA., DEL., VA.)

Midwest Regional Office: 1709 Jackson St., Omaha, Nebraska 68102 (OHIO, IND., MICH., WIS., ILL., MINN., IOWA, MO., NEBR., KANS.)

National Capital Regional Office: 1100 Ohio Drive, S.W., Washington, D.C. 20242 (D.C., MD., VA., W. VA.)

North Atlantic Regional Office: 15 State St., Boston, Massachusetts 02109 (ME., N.H., VT., MASS., R.I., CONN., N.Y., N.J.)

Pacific Northwest Regional Office: 601 Fourth & Pike Bldg., Seattle, Washington 98101 (IDAHO, OREG., WASH.)

Rocky Mountain Regional Office: Box 25287, 655 Parfait St., Denver, Colorado 80225 (MONT., N. DAK., S. DAK., WYO., UTAH, COLO.)

Southeast Regional Office: 75 Spring St., S.W., Atlanta, Georgia 30303 (KY., TENN., N.C., S.C., MISS., ALA., GA., FLA., Puerto Rico, Virgin Islands)

Southwest Regional Office: Box 728, Santa Fe, New Mexico 87501 (ARIZ., LA., TEXAS, OKLA., N. MEX.)

Western Regional Office: Box 36063, 450 Golden Gate Ave., San Francisco, California 94102 (CALIF., NEV., ARIZ., Hawaii, Guam)

NATIONAL PARKS

▲ NATIONAL PARKS

For locations of National Parks in Alaska and Hawaii, see page 52

INDEPENDENCE NAT. HISTORICAL PK., PA.

Legend:
- ■ Independence National Historical Park Points of Interest
- ■ Other Historical Points of Interest
- ■ Miscellaneous Features

Scale
0   0.1   0.2 Km.
0   0.1   Mi.

© GENERAL DRAFTING CO., INC.

INDEPENDENCE NATIONAL HISTORICAL PARK

One of the most historic sites in America is Independence National Historical Park in downtown Philadelphia. Within the revered shrines on these hallowed grounds was born American democracy. The properties preserved here are those most closely connected with the founding of the nation. Virtually the entire park lies within just a few blocks in the heart of the city. Surrounding the federal properties are numerous other structures of historical significance, such as the Tomb of the Unknown Soldier of the American Revolution, the Friends Meeting House and various colonial homes and churches.

Within the park are eight principal places of interest. Most famous of all is **Independence Hall,** where the Declaration of Independence was adopted and the Constitution written eleven years later. The Hall was formerly the location of the **Liberty Bell,** but now the symbol of our nation's freedom is in its own pavilion a short walk across the green. Next to Independence Hall sat the justices of the Supreme Court in the **Old City Hall.** To the west is **Congress Hall,** where the first U.S. Congress met prior to their move to Washington in 1800. Two blocks away the First Continental Congress convened in **Carpenters' Hall,** the home of the country's oldest builders' organization. **Franklin Court** contains five reconstructed dwellings, three of which belonged to Benjamin Franklin and have exhibits. While a delegate to the Continental Congress, Thomas Jefferson resided in the **Graff House,** where he drafted the Declaration of Independence. The **Second Bank of the United States** is a fine example of the Greek Revival style.

ACADIA NAT. PK., ME.

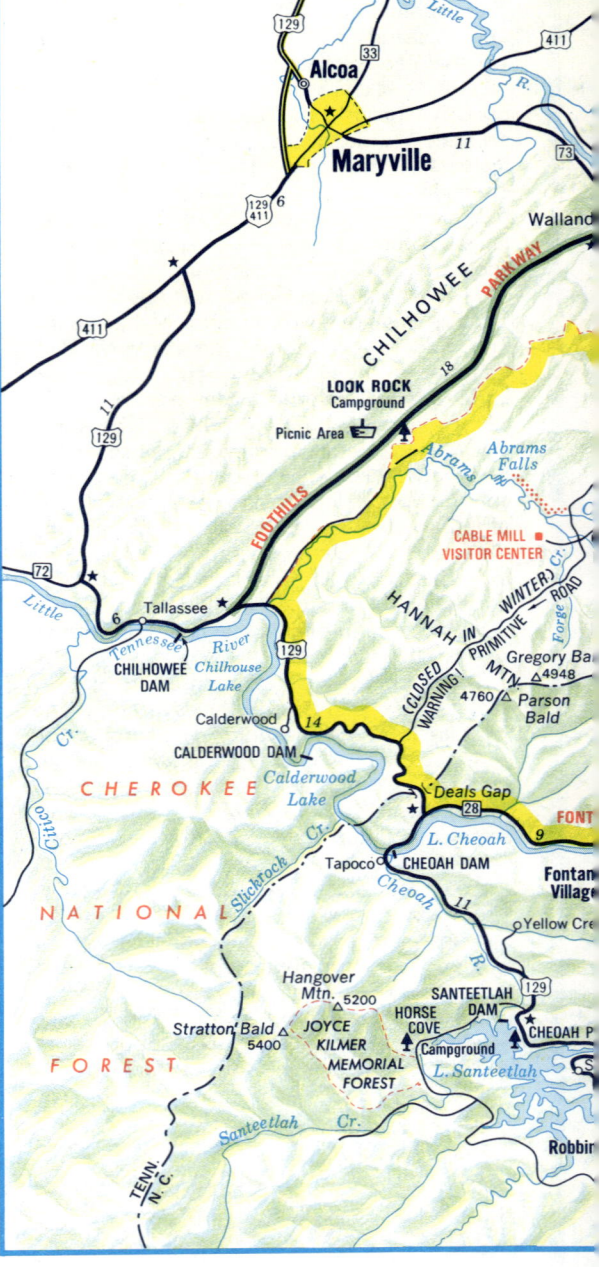

Scale
0 1 2 3 4 5 Km.
0   1   2   3 Mi.

© GENERAL DRAFTING CO., INC.

SEE SCALE AT LEFT

ACADIA NATIONAL PARK

This coastal preserve became the first national park east of the Mississippi and remains the Northeast's only National Park. It is special among all parks because of the beauty and character of its seascape, as well as its insularity and feeling of remoteness.

Originally scarcely more than 250 square feet and set aside by private individuals for public use, the small lot grew to thousands of acres and was given to the federal government. In time it changed from Sieur de Monts National Monument to Lafayette National Park and finally the present Acadia.

The shore was variously shaped into fiords, cliffs and caves by a pounding surf. Mt. Desert Island's Cadillac Mountain is the highest point on the Atlantic Coast.

Pine forests and wildflowers add color to the stony terrain where tidal pools capture tiny marine life for closer inspection by visitors. In addition to deer that inhabit the fields and woods, porpoises, seals, sea ducks, ospreys and egrets can be observed on tours led by park naturalists. The real wilderness is on Isle au Haut, which can be reached only by boat. Winter finds the park snowbound and its roads usually closed.

GREAT SMOKY MOUNTAINS NATIONAL PARK

A perennial winner as the most-visited of the national parks, the high peaks and misty, deep green valleys of the Great Smokies are preserved on 800 square miles that straddle the North Carolina-Tennessee border. Here is the largest remaining virgin forest of the wilderness that once was eastern America. Visitors exploring the park are surrounded by no less than 1,400 species of flowering plants.

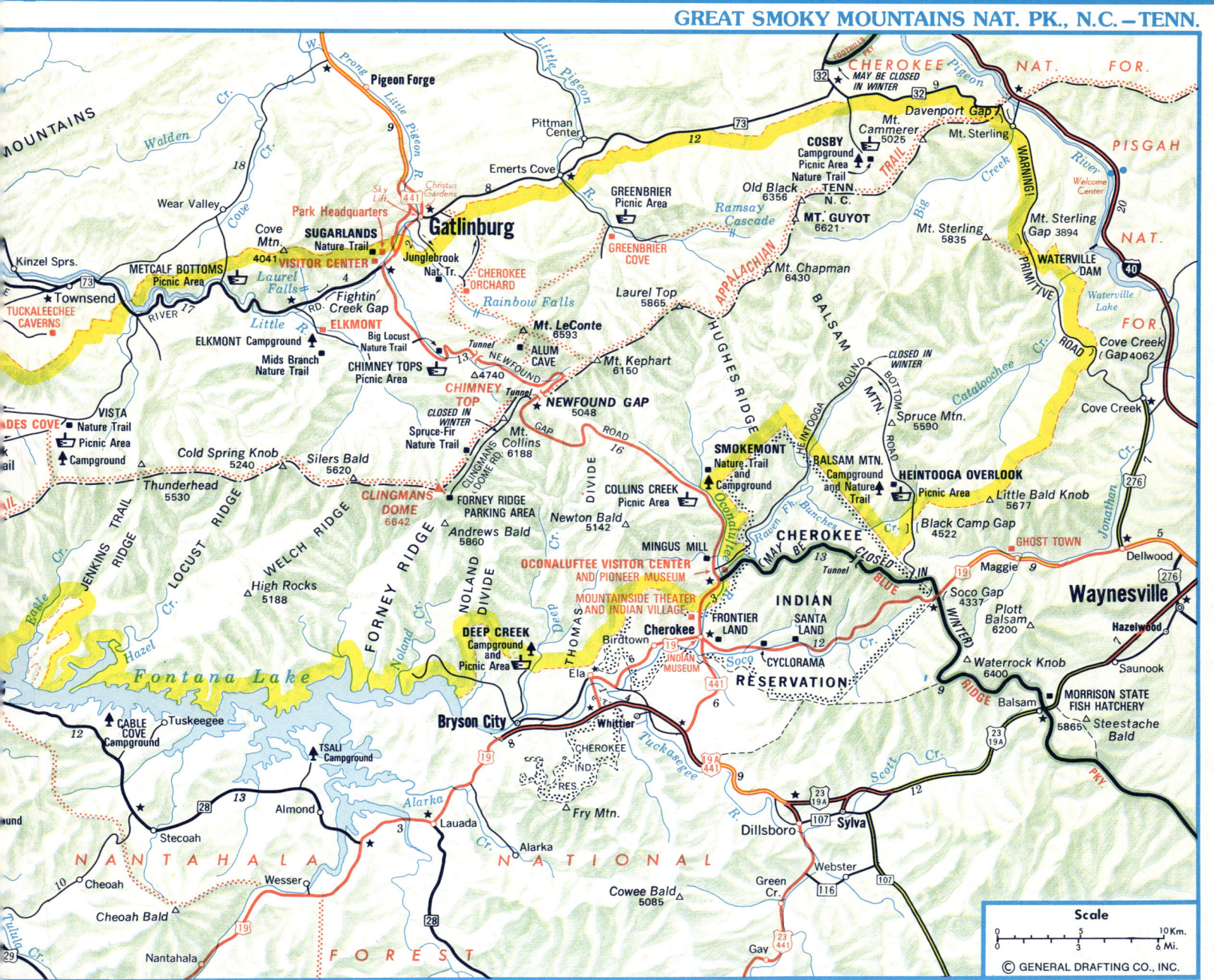

There are roughly 600 miles of tumbling streams for fishermen, and more than 800 miles of trails for horses and hikers, including the Appalachian Trail, which penetrates the park for 71 miles. Park wildlife includes almost 2,000 species of birds and dozens of kinds of mammals, ranging from the black bear to the tiny shrew.

The park not only preserves its natural assets, but also keeps alive the history of man's exploration and settlement of the wilderness. Cherokee Indians were first to inhabit the Smokies and their culture is preserved on the adjacent Qualla Boundary Reservation. Descendents of the first English and Scots-Irish frontiersmen and settlers retain their early crafts skills, music styles and distinctive patterns of speech.

National Park staff members are on hand to describe the park's features and offer an introduction to the area's history. There are visitor centers on the North Carolina and Tennessee sides of the park. The center at Oconaluftee, N.C., is a reconstructed pioneer homestead with authentic buildings and tools. On the Tennessee side is Cades Cove, with the farmhouses, fields and churches where the pioneers lived, worked and worshiped. At Cable Mill, corn is still stone-ground and sorghum is produced exactly as it was 150 years ago.

HOT SPRINGS NATIONAL PARK

Appreciation of the therapeutic worth of Arkansas' famed thermal springs dates to their discovery and use by early Indian tribes. Like the Indians, the U.S. government quickly recognized Hot Springs as a very special place. In 1832 it was set aside as a Federal Reservation and it became a national park in 1921. The park was expanded at various times to include thousands of acres of woodland and hills in addition to its peaceful town and Bathhouse Row.

Even today the hot mineral waters are sometimes of value for treatment of ailments and are used for bathing and drinking. Though all are admitted, a physician's recommendation is advisable. The 47 springs were sealed for indoor use and have never been affected by seasonal changes in temperature.

Winters are relatively mild and summer heat is moderated by the altitude. The foliage is quite colorful in both summer and autumn. Hiking, camping and boating on the nearby Ouachita River are popular activities.

HOT SPRINGS NAT. PK., ARK.

GETTYSBURG NAT. MILITARY PK., PA.

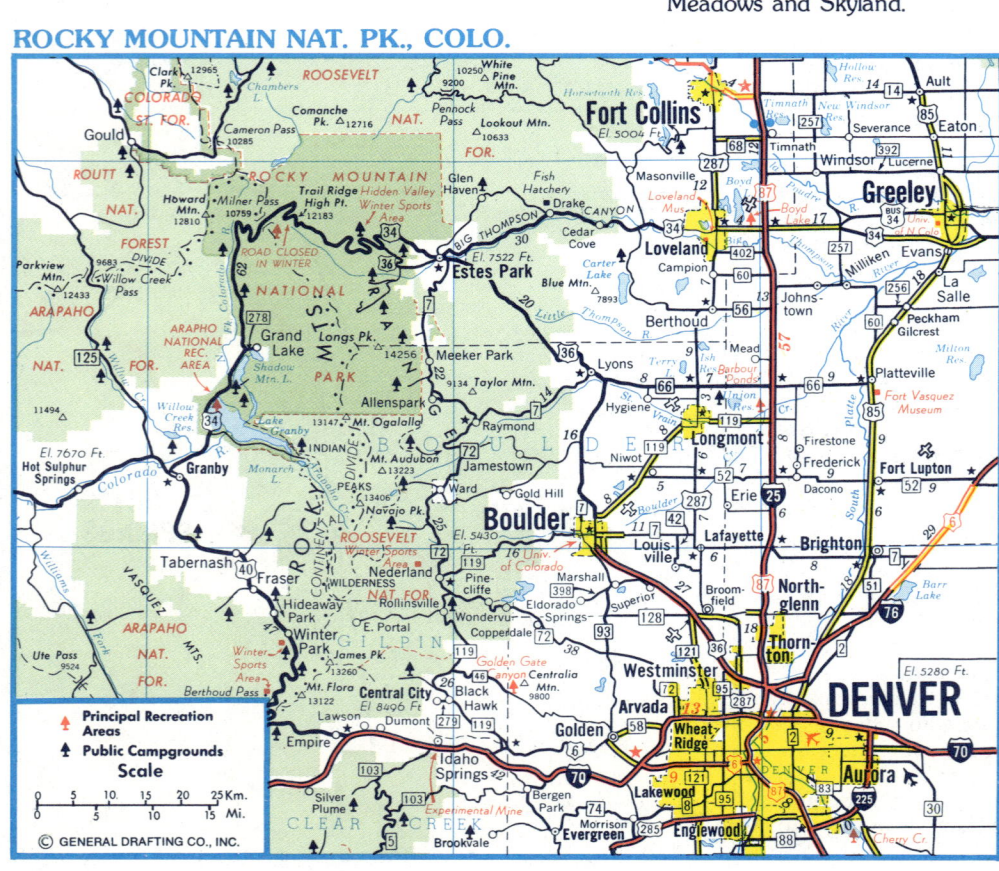

GETTYSBURG NAT. MILITARY PK.

The fields west of Gettysburg are the site of the greatest battle of the Civil War. The site is now a national shrine, as is the adjoining national cemetery, dedicated by President Lincoln in his "Gettysburg Address" some months after the battle.

In early July, 1863, the Confederacy's second and last offensive was turned back at the cost of tens of thousands of lives. When General Lee withdrew on July 4, the War Between the States had reached a turning point after which the fortunes of the South were never to recover.

More than 6,000 are buried in **Gettysburg National Cemetery.** Memorials and monuments dot the landscape. Roads connect such landmarks as Little Round Top and Cemetery Ridge. In addition to self-guiding trails, licensed park guides are available. The Cyclorama, exhibits and a museum aid visitors in understanding the course of events during the battle.

SHENANDOAH NATIONAL PARK

Stretching 80 miles along Virginia's Blue Ridge Mountains, this park was once home to descendents of America's first pioneers. In this century it has slowly returned to its original state of wilderness.

A segment of the Appalachian chain, these mountains are older than any in the west. Wildflowers bloom from mid-April to late summer, and dogwood, azalea and mountain laurel brighten the woods in May and June. In October the foliage changes to brilliant reds and gold.

The winding 105-mile Skyline Drive, traversing mountain crests, offers panoramic views of the magnificent scenery from 66 overlooks. It is the first section of the Blue Ridge Parkway, which follows the Blue Ridge Mountains into North Carolina and the Smoky Mountains.

The park derives its name from the river and valley that separate the Blue Ridge from the Alleghenies. The park is crossed by miles of hiking trails. Park ranger-naturalists conduct hikes daily during the summer. There are riding stables at Big Meadows and Skyland.

SHENANDOAH NAT. PK., VA.

ROCKY MOUNTAIN NAT. PK., COLO.

ROCKY MOUNTAIN NATIONAL PARK

Midway along the Front Range of the U.S. Rockies is a vast preserve of wildlife, meadows, forests and alpine tundra. Known for splendid scenery, Rocky Mountain National Park is a favored destination for novice and veteran mountain climbers, backpackers and tourists. Park elevations range from 7,640 feet above sea level to 14,256-foot Long's Peak.

Paved trails permit easier walks, and a popular way to tour the park is on horseback. First-time visitors can join hikes guided by rangers. The park offers year-round enjoyment, but like several other mountain parks closes its through roads in winter.

The range of elevations in the park provides three distinct climate zones. The pine-clad hills of the relatively warm and dry elevations are accented in autumn by the golden groves of aspens. Above this zone, stands of subalpine firs and spruce take over, broken by meadows of blue columbine. Above is the third zone, alpine tundra with strangely adapted dwarf grasses, shrubs and lichens surviving where trees cannot.

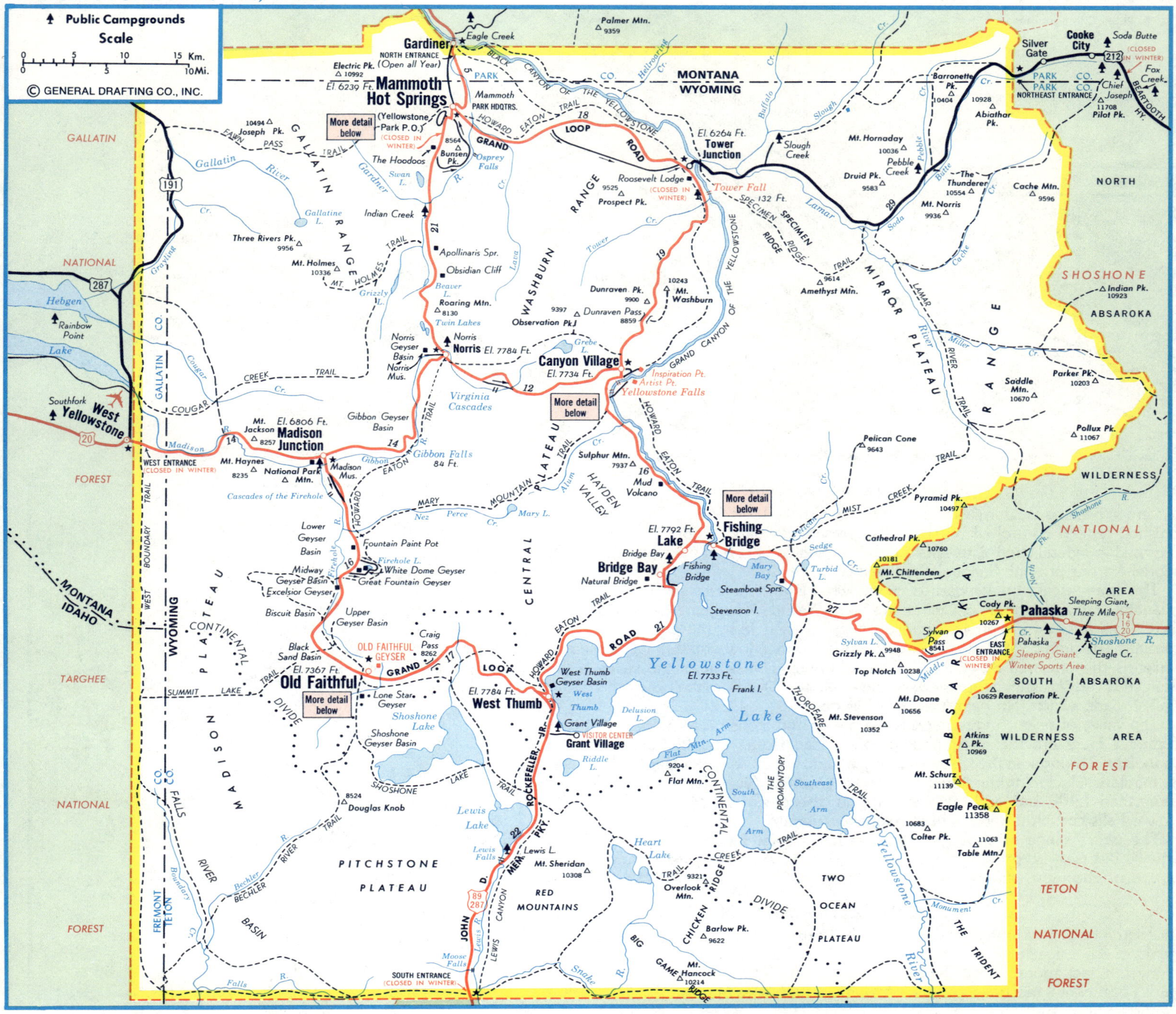

YELLOWSTONE NATIONAL PARK

America's first national park, Yellowstone is also one of the busiest, but its size permits uncrowded exploration. Established in 1872, the park comprises almost two and a quarter million acres in Wyoming, Montana and Idaho. It is best known for its principal geyser, Old Faithful, one of thousands of geysers and hot springs in the world's greatest geyser region.

While not the largest geyser, **Old Faithful** erupts with greatest regularity (at intervals of just over an hour), spraying more than 10,000 gallons of water 150 feet into the air. Such activity also takes place at **Mammoth Hot Springs,** where deposited minerals form ever-growing terraces.

The colorful **Grand Canyon of the Yellowstone** rises more than 1,000 feet above Yellowstone Lake. The beautiful **Falls** of the lower Yellowstone River are twice the height of Niagara.

As one of the largest wildlife sanctuaries in the U.S., the park is home to bear, bison, elk, moose, numerous birds and mountain sheep. Hundreds of species of wildflowers and other plants cover the Yellowstone plateau.

Trails vary from easy walks to rigorous climbs. Motorists can tour the park's 142-mile Grand Loop. The park can also be explored on horseback. In winter, the park is a snow-covered wilderness accessible only on cross-country skis or snowmobiles.

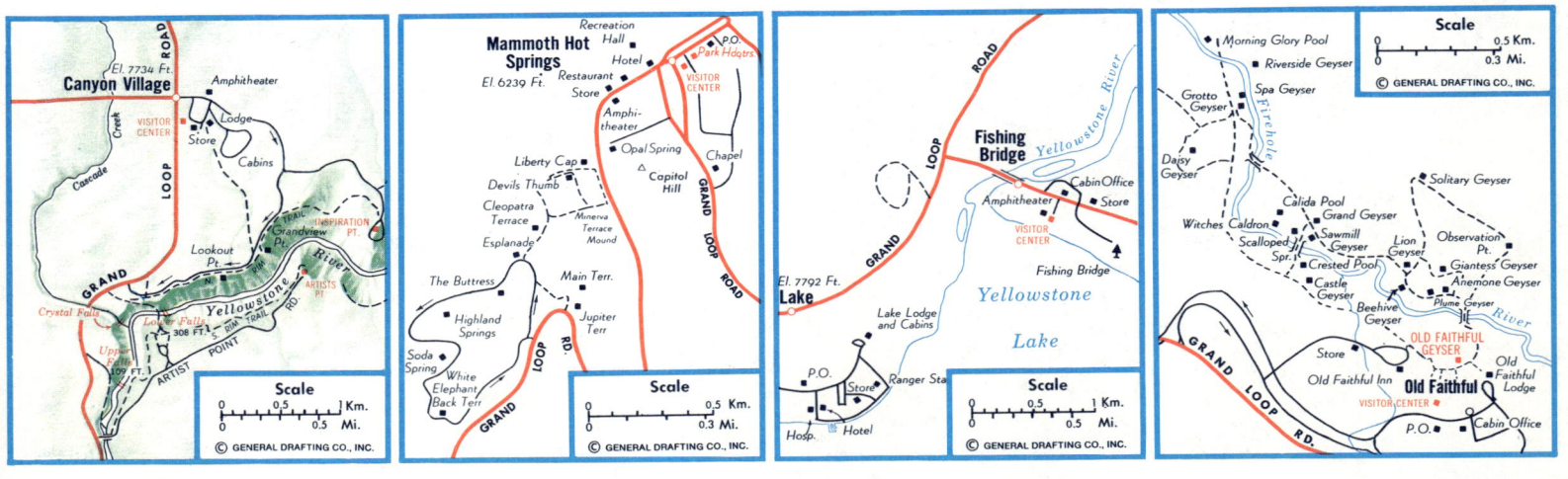

YOSEMITE NAT. PK. (YOSEMITE VALLEY), CALIF.

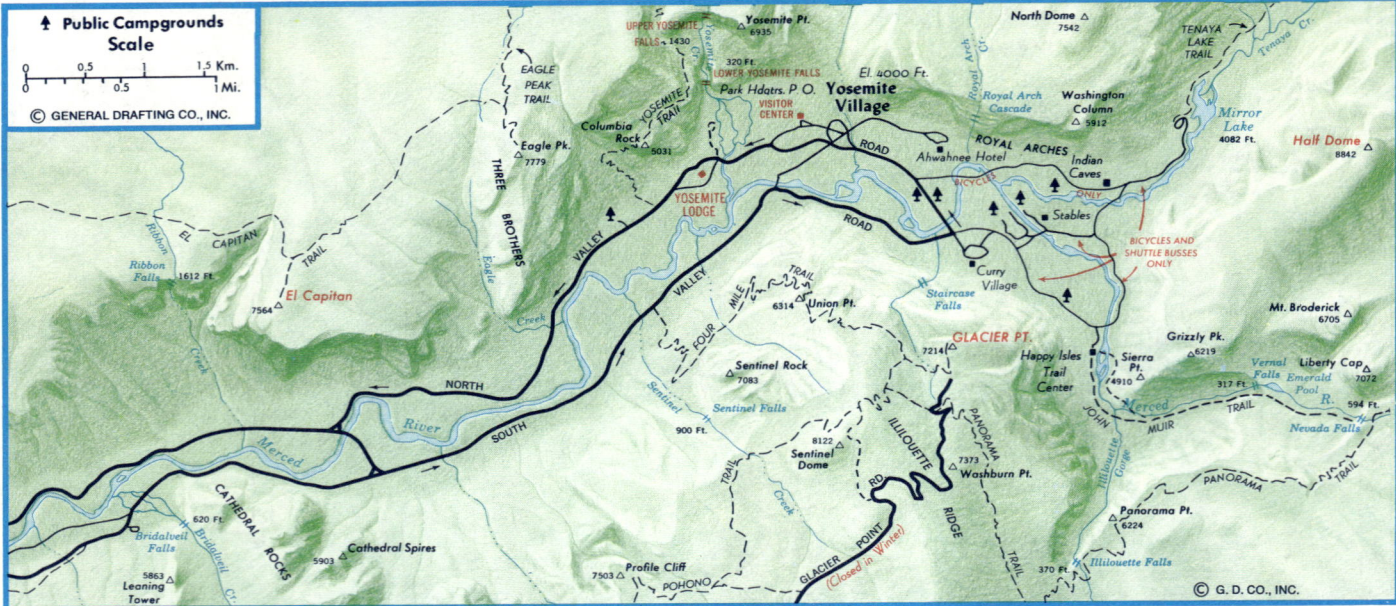

YOSEMITE NATIONAL PARK

Yosemite became California's first state park when, in 1864, President Lincoln relinquished federal control of much of the present parkland. Though subsequently returned to federal control, this first publicly protected tract became a precedent and model for our present National Park System.

This immensely popular park occupies what is undoubtedly some of the most beautiful and varied terrain of California's Sierra Nevada Mountains. California Route 120 slices the 3/4-million-acre park in two, entering from the east and passing through Tuolumne Meadows, the largest mountain meadow of the range. Yosemite Valley is the principal destination of most visitors. The heart of the valley is the Merced River, accessible from the west and easily surveyed from the North and South Valley roads, which join near Yosemite Village.

The valley of seven square miles is the site of numerous peaks, domes and bald summits. Here are the much-photographed Half Dome and El Capitan granite monoliths. Throughout the valley are the sparkling waterfalls of the Merced River and its tributaries, including the tiered falls of the Yosemite River. The Upper Falls drop more than 1,400 feet, and together with the cascades of the Middle Falls and the Lower Falls constitute the country's highest waterfall.

The park's range of elevations provides habitats for much wildlife. Plant life is equally diverse—from the ground cover so colorful in summer, to the ancient sequoias of enormous stature.

GRAND TETON NATIONAL PARK

Just south of Yellowstone National Park is the impressive Teton Range. The principal peaks of the range lie within Grand Teton National Park, and are among Wyoming's highest. In fact, more than 22 peaks rise above 10,000 feet, the tallest of which, the park's namesake, reaches more than two and a half miles into the sky. The park is almost entirely surrounded by National Forests, and is accessible from the north, south and east. To the north it is connected with its larger neighboring park by the scenic John D. Rockefeller Memorial Parkway.

Repeated glaciation has produced some of the most rugged scenery in the Rocky Mountain chain. It is also the origin of the park's beautiful lakes. The park is predominantly wooded with evergreens and there are stands of aspen and cottonwood whose yellow and gold of autumn make that season particularly attractive. The largest herd of North American elk winter in the park. .

CRATERS OF THE MOON NAT. MON., IDAHO

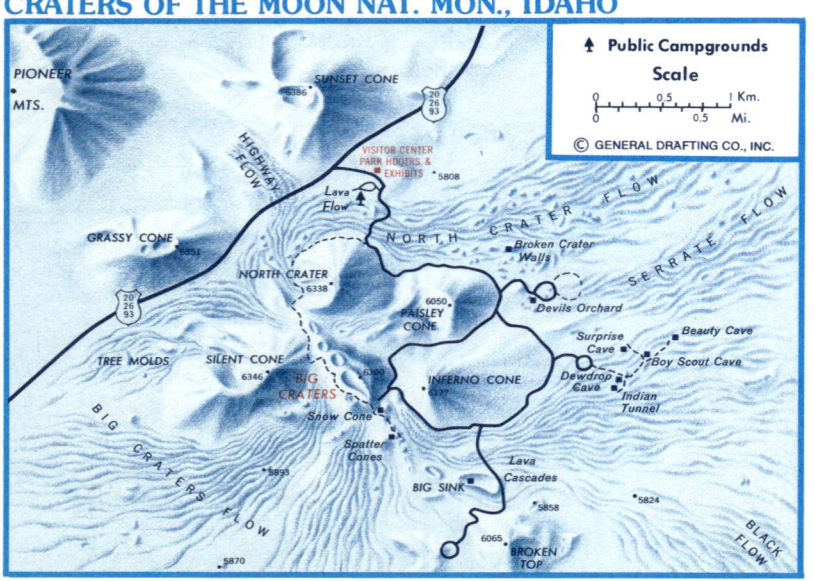

GRAND TETON NAT. PK., WYO.

CRATERS OF THE MOON NATIONAL MONUMENT

Numerous cinder cones with hollow or depressed centers are responsible for the name of this national monument that encompasses what is perhaps the eeriest landscape in North America. The rocky terrain of 83 square miles in south central Idaho is set aside to preserve a most peculiar natural formation. Visitors approaching through the surrounding wilderness observe a dark, seemingly dead and forbidding landscape. But the area is home to some wildlife and many plants.

The rocky terrain is a product of periodic volcanic activity, which last occurred about 2,000 years ago. The rock has taken many forms. Most notably there are ropy lengths that appear twisted or stranded, rivers of entire lava flows "frozen" in place. Other flows hardened above and emptied below to create tubes or caves. Other pieces of strangely shaped debris are known as "bombs."

Small mammals and deer wander across the area. In spring the flows come to life as wildflowers color the cinder gardens with silver and green, pink and yellow. Later, dwarf monkeyflowers cover the dark surface with deep red blooms.

A seven-mile loop provides a convenient tour for motorists, and in summer there are guided hikes and evening programs.

GLACIER NATIONAL PARK

A million acres of stunning scenery awaits visitors to our northernmost Rocky Mountain park, adjoined by Canada's Waterton Lakes National Park.

Unlike most parklands, Glacier's majesty is greater than the combined beauty of each of its features. First-time visitors, rather than exploring sections of the park, might do best to take in broad vistas. For this purpose a drive over Going-To-The-Sun Road is a must. It is a perfect interplay of mountains, glaciers, lakes, waterfalls, forests and skies that makes Glacier special.

The park claims no arches, geysers, caves or ancient trees: it is the main event, not a side show. There are vast opportunities for outdoorsmen, and such persons can enjoy a park relatively free of crowds owing to a lack of particular tourist attractions. This sky-high domain at the junction of three continental watersheds is perfect for boating, hiking, mountain climbing, observing nature or just dreaming.

The park's name is quite appropriate. Ancient frozen masses carved its lovely lakes and broad U-shaped valleys. Sheer rock faces of the mountains here are an imposing result of tremendous disturbances of the Earth during the area's billion-year history.

GRAND CANYON NATIONAL PARK

Though the average visit to this remarkable canyon may be only a matter of minutes, it is safe to say that no one leaves unimpressed. Nowhere else are the sudden and the gradual forces of nature more obvious.

Millions of years ago the earth's crust was thrust up a mile and a half, creating the mountains and plateaus of the Rockies. Ever since, the Colorado River has patiently undone this ancient work. For eons the unrelenting river has reduced elevations by almost two thirds. The canyon is a mile deep in places and four to 18 miles wide.

Incalculable erosion has made the canyon an open book of the Earth's history. Revealed are layers of landforms and lifeforms that appeared and disappeared through the ages.

The park stretches from Glen Canyon, near Utah, to Lake Mead Recreation Area, near Nevada. It may be seen from drives along the rim, upon sure-footed mules on trips into the canyon, on overnight hikes or even flights into and through the chasm.

The North and South rims have somewhat different climates owing to their respective altitudes. Both differ from the canyon floor. The North Rim is open between late spring and early fall. The South Rim is open year round.

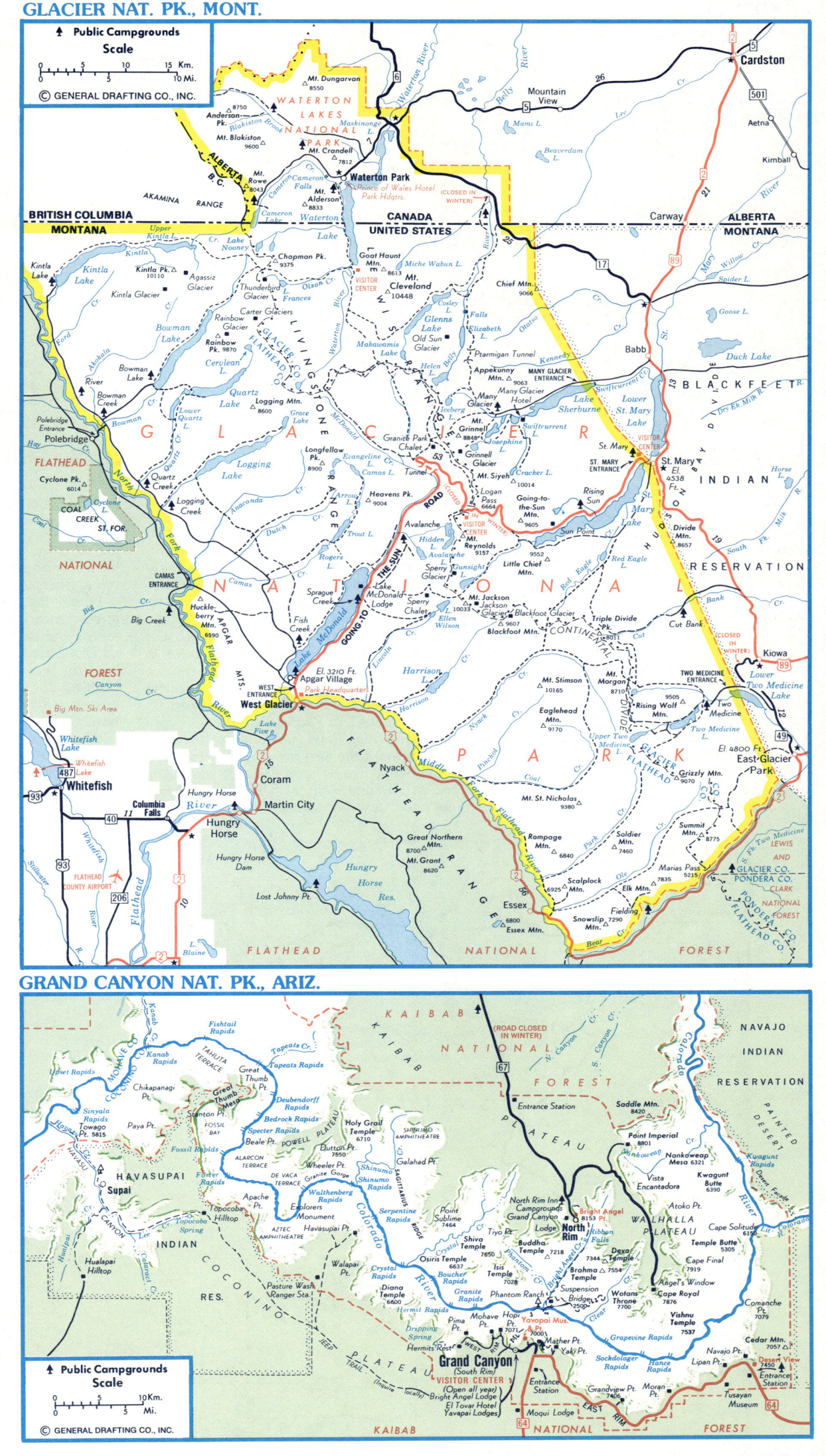

GLACIER NAT. PK., MONT.

GRAND CANYON NAT. PK., ARIZ.

CRATER LAKE NAT. PK., OREG.

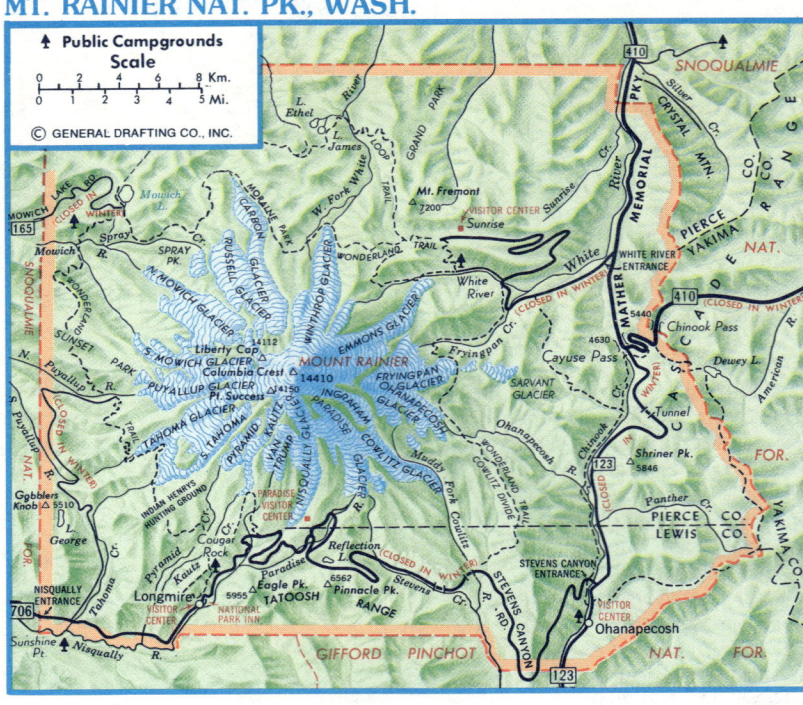

MT. RAINIER NAT. PK., WASH.

CRATER LAKE NATIONAL PARK

Talk of this park almost inevitably centers around the indescribable color and calm of the lake. Less obvious but equally amazing is the depth of the lake.

The centerpiece of the park is Mount Mazama, an ancient volcano believed to have risen to 12,000 feet. Centuries ago, cracks around the summit drained the molten core, collapsing the peak and creating what is now a water-filled crater. The wonderfully blue lake is almost 2,000 feet deep and lies below cliffs that rise just as high. In this hemisphere, only Canada's Great Slave Lake is deeper.

Wooded with fir trees and studded with wildflowers, the park is the protected abode of eagles, hawks, deer, bears and occasional bobcats, foxes, elk and coyote.

MOUNT RAINIER NATIONAL PARK

In one of the country's most beautiful mountain parks, Rainier's glacial system is the largest in the lower 48 states. Rainier rises above 14,000 feet, where winter is a permanent feature. The mountain can be seen from as far away as Oregon.

Its slopes are forested with evergreens, and the alpine meadows are a wash of color in spring. Mountain goats and deer roam the preserve, and the furry marmot's whistle might be heard, even if the creature isn't noticed.

The windward, or Pacific, slopes receive considerable rainfall (and tremendous snowfall) which supports the lush rain forests of towering fir and dense fern. In spite of heavy snows, the road to Paradise Park, at 5,400 feet, is often kept open to winter visitors.

OLYMPIC NATIONAL PARK

Probably no other U.S. National Park possesses the diversity of landscape and scenery contained within this park on Washington's enormous Olympic Peninsula.

Within the park's 1,400 square miles are forests, meadows, glaciers, river valleys, rainforests and more than 50 miles of coastline. In addition, most of the parkland is bounded by large tracts of Olympic National Forest. Opportunities for hikers and backpackers are numerous; they range from casual Pacific beachcombing to climbing alpine tundra.

A preserve of such variation in climate and terrain naturally supports a vast number of plant and animal species. Elk, deer, bear, otters, marmots, seals, shore birds and even migrating whales can all be seen.

OLYMPIC NAT. PK., WASH.

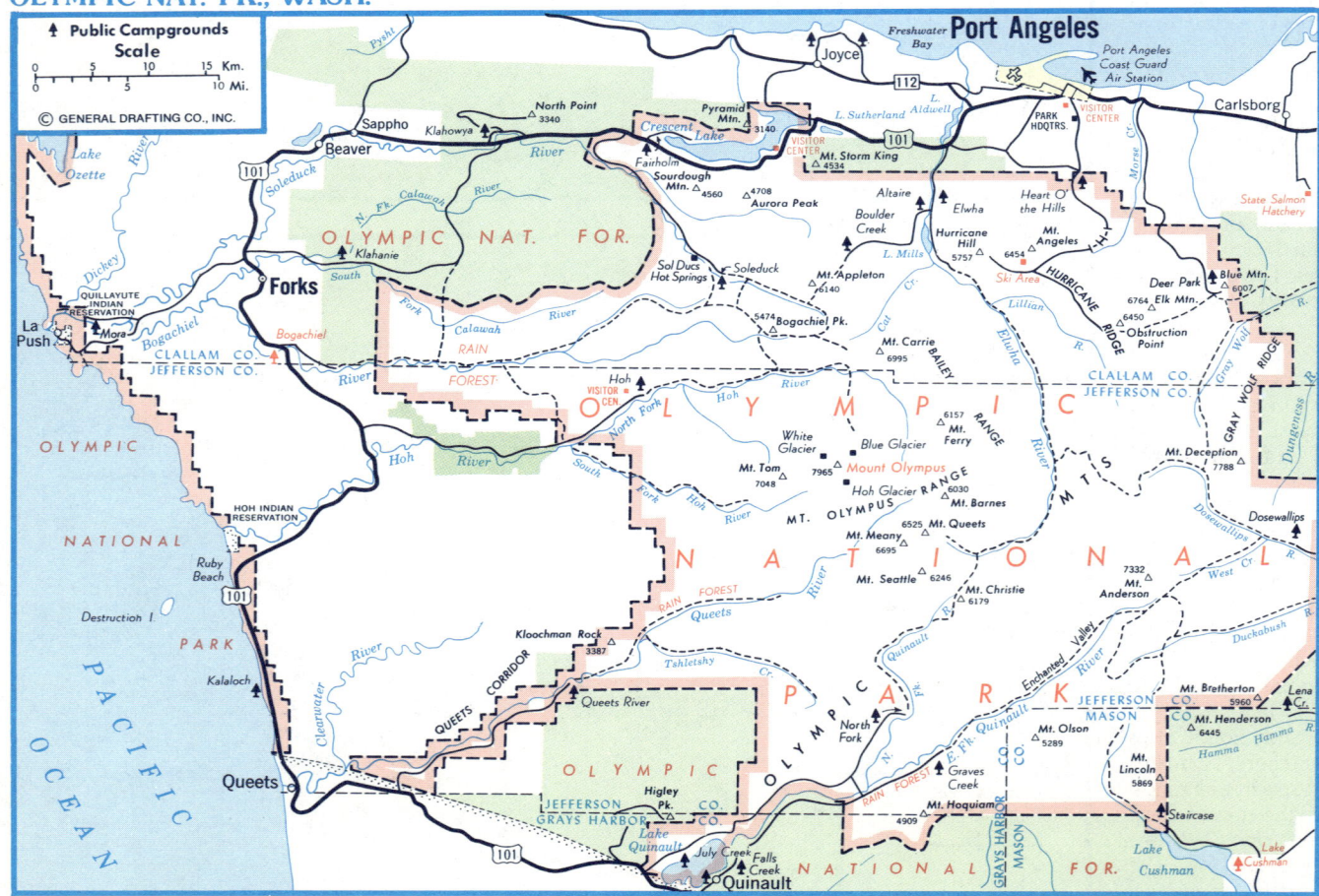

NATIONAL PARK AREAS

The following charts contain useful information about the facilities and recreational opportunities available within the U.S. National Park System. A green-shaded box on the chart indicates that the facility, service or recreational activity listed is available in that particular park.

Entrance Fees: Entrance fees ranging from $1-$3 per private passenger vehicle are required at many park areas. Free admission is given to the holders of Golden Eagle, Golden Age or Golden Access passports.

Exhibits: Many parks have some type of outdoor or indoor exhibit. An L printed in the green-shaded box indicates that the park presents a Living History program in which costumed park personnel re-enact certain events or periods in history.

Tours: An R printed in the green-shaded box indicates that a ranger-guided tour is available. The letter S indicates that self-guided tours are available, usually on clearly-marked trails.

Camping: Where camping is available, sites vary from primitive to fully equipped with hookups for recreational vehicles. Campsite users are charged recreation use fees ranging from $1-$4 at all campgrounds that have facilities and services. This charge is in addition to any park entrance fee. Most campsites are available on a first-come, first-served basis.

Food Service: Facilities are operated by concessionaires under contract with the Park Service. Services may range from snack bars to full-service restaurants.

Accommodations: From rustic cabins to comfortable hotels and motels, all park lodgings are concessionaire operated. Although the box under accommodations may indicate there are no lodgings on the park premises, there are often lodgings available just outside of the park proper, especially in urban areas.

Boating: Although boating may be permitted in a park, boats may not be available for rental on the premises. Where boats are available, a B is printed in the box. Where boat ramps are available, an R is indicated.

Swimming: Swimming is available at some parks.

Fishing: Contact park information offices to learn which varieties of fish are found in the park's lakes and streams.

Hiking: Where hiking is available, visitors may explore the park by following clearly marked trails.

Picnicking: Tables and fireplaces are available in designated areas.

Area Name and Mailing Address	Entrance Fee	Exhibit	Tours	Camping	Food Service	Accommodations	Boating	Swimming	Fishing	Hiking	Picnic Area
ALABAMA											
Horseshoe Bend Natl. Military Park (Daviston 36256)		L	R,S				R		●	●	●
Russell Cave Natl. Mon. (Bridgeport 35740)		L	S							●	
Tuskegee Institute Natl. Hist. Site (Tuskegee Institute 36088)			R								
ALASKA											
Aniakchak Natl. Mon. (Anchorage 99501)									●	●	
Cape Krusenstern Natl. Mon. (Anchorage 99501)										●	
Denali Natl. Park (Anchorage 99501)		●	R	●	●	●			●	●	
Gates of the Arctic Natl. Park (Anchorage 99501)						●			●	●	
Glacier Bay Natl. Park (Juneau 99802)		●	R,S	●	●	●	B		●	●	
Katmai Natl. Park (King Salmon 99613)		●	R,S	●	●	●	B		●	●	
Kenai Fjords Natl. Park (Anchorage 99501)									●	●	
Klondike Gold Rush Natl. Historical Park (Skagway 99840)		●	R,S	●	●	●			●	●	
Kobuk Valley Natl. Park (Anchorage 99501)									●	●	
Lake Clark Natl. Park (Anchorage 99501)						●	R			●	
Noatak Natl. Park (Anchorage 99501)							●		●	●	
Sitka Natl. Historical Park (Sitka 99835)		L	R,S		●	●			●		●
Wrangell-St. Elias Natl. Park (Anchorage 99501)							●		●	●	
ARIZONA											
Canyon de Chelly Natl. Mon. (Chinle 86503)		●	R,S	●	●	●				●	
Casa Grande Natl. Mon. (Coolidge 85228)	●	L	R,S								●
Chiricahua Natl. Mon. (Willcox 85643)	●	●	S	●						●	
Coronado Natl. Mem. (Hereford 85615)		L	S							●	
Fort Bowie Natl. Historic Site (Bowie 85605)		●	S								
Grand Canyon Natl. Park (Grand Canyon 86023)	●	●	R,S	●	●	●	●	●	●	●	●
Hubbell Trading Post Natl. Historic Site (Ganado 86505)		L	R,S								●
Montezuma Castle Natl. Mon. (Clarkdale 86324)	●	●	S								●
Navajo Natl. Mon. (Tonalea 86044)		●	R,S	●						●	●
Organ Pipe Cactus Natl. Mon. (Ajo 85321)		●	R,S	●						●	●
Petrified Forest Natl. Park (Petrified Forest Natl. Park 86028)	●	●	R,S		●					●	●
Pipe Spring Natl. Mon. (Moccasin 86022)	●	L	R,S		●						
Saguaro Natl. Mon. (Tuscon 85731)	●	●	R,S							●	●
Sunset Crater Natl. Mon. (Flagstaff 86001)		●	R,S	●							●
Tonto Natl. Mon. (Roosevelt 85545)	●	●	R,S								●
Tumacacori Natl. Mon. (Tumacacori 85640)	●	L	R,S								●
Tuzigoot Natl. Mon. (Clarkdale 86324)	●	●	S								
Walnut Canyon Natl. Mon. (Flagstaff 86001)	●	●	S							●	
Wupatki Natl. Mon. (Flagstaff 86001)		●	R,S							●	●
ARKANSAS											
Arkansas Post Natl. Mem. (Gillett 72055)		●	S						●	●	●
Buffalo Natl. River (Harrison 72601)		L	R,S	●	●	●	B,R	●	●	●	●
Fort Smith Natl. Historic Site (Fort Smith 72902)		●	R,S								
Hot Springs Natl. Park (Hot Springs 71901)		●	R,S	●						●	●
Pea Ridge Natl. Military Park (Pea Ridge 72751)		●	S							●	●
CALIFORNIA											
Cabrillo Natl. Mon. (San Diego 92106)		L	R,S		●				●		
Channel Island Natl. Park (Ventura 93003)		●	S	●		●	●	●	●	●	●
Death Valley Natl. Mon. (Death Valley 92328)		●	R,S	●	●			●		●	●
Devils Postpile Natl. Mon. (Three Rivers 93271)			R,S	●		●			●	●	●
Fort Point Natl. Historic Site (Presidio of San Francisco 94129)		●	R,S						●		
Golden Gate Natl. Recreation Area (San Francisco 94123)		L	R,S		●			●	●	●	●
John Muir Natl. Historic Site (Martinez 94553)	●	●	R,S								●
Joshua Tree Natl. Mon. (Twentynine Palms 92277)		●	R,S	●						●	●
Kings Canyon Natl. Park (Three Rivers 93271)	●	L	R,S	●	●	●			●		
Lassen Volcanic Natl. Park (Mineral 96063)	●	L	R,S	●	●	●	●	●	●	●	●
Lava Beds Natl. Mon. (Tulelake 96134)		L	R,S	●						●	●
Muir Woods Natl. Mon. (Mill Valley 94941)		●	S		●					●	
Pinnacles Natl. Mon. (Paicines 95043)	●	●	R,S	●						●	●
Point Reyes Natl. Seashore (Point Reyes 94956)		L	R,S	●	●	●		●	●	●	●
Redwood Natl. Park (Crescent City 95531)		L	R,S	●	●	●	R	●	●	●	●
Santa Monica Mountains Natl. Rec. Area (Woodland Hills 91364)										●	●
Sequoia Natl. Park (Three Rivers 93271)	●	L	R,S	●	●	●			●	●	●
Whiskeytown-Shasta-Trinity Nat. Rec. Area (Whiskeytown 96095)		●	S	●	●		B,R	●	●	●	●
Yosemite Natl. Park (Yosemite Natl. Park 95389)	●	L	R,S	●	●	●	●	●	●	●	●
COLORADO											
Bent's Old Fort Natl. Historic Site (La Junta 81050)		L	R,S								
Black Canyon of the Gunnison Natl. Mon. (Montrose 81401)	●	●	R,S	●	●				●	●	●
Colorado Natl. Mon. (Fruita 81521)	●	●	S	●						●	●
Curecanti Natl. Recreation Area (Gunnison 81230)		●	R	●	●		B,R	●	●	●	●
Dinosaur Natl. Mon. (Dinosaur 81610)		●	R,S	●			R		●	●	●
Florissant Fossil Beds Natl. Mon. (Florissant 80816)		●	R							●	●
Great Sand Dunes Natl. Mon. (Mosca 81146)	●	●	S	●					●	●	●
Hovenweep Natl. Mon. (Cortez 81321)			S	●	●					●	●
Mesa Verde Natl. Park (Mesa Verde Natl. Park 81330)	●	L	R,S	●	●	●					●
Rocky Mountain Natl. Park (Estes Park 80517)	●	●	R,S	●			●	●	●	●	●
DISTRICT OF COLUMBIA											
Ford's Theatre Natl. Historic Site (Washington, D.C. 20004)		L	R,S								
Frederick Douglass Home (Washington, D.C. 20020)		L	R,S								●
Kennedy Center for the Performing Arts (Washington, D.C. 20566)		L	R,S		●						
L. B. J. Mem. Grove on the Potomac (McLean, VA 22101)					●		●		●		●
Lincoln Memorial (Washington, D.C. 20242)			R,S								
National Mall (Washington, D.C. 20242)										●	●
National Visitor Center (Washington, D.C. 20002)		●	R		●						
Rock Creek Park (Washington, D.C. 20002)		L	R,S						●	●	●
Seawall-Belmont House Natl. Hist. Site (Washington, D.C. 20002)		●	R								
Theodore Roosevelt Island (McLean, VA 22101)			R,S						●	●	
Thomas Jefferson Memorial (Washington, D.C. 20242)			R,S								
Washington Monument (Washington, D.C. 20242)			R,S		●						
White House (Washington, D.C. 20242)		●	R,S								
FLORIDA											
Biscayne Natl. Park (Homestead 33030)		●	S	●			R	●	●		
Canaveral Natl. Seashore (Titusville 32780)		●	S		●		R	●	●	●	●
Castillo de San Marcos Natl. Mon. (St. Augustine 32084)	●	L	R								
DeSoto Natl. Mem. (Bradenton 33505)		L	S						●		
Everglades Natl. Park (Homestead 33030)	●	●	R,S	●	●	●	B,R		●	●	●
Fort Caroline Natl. Mem. (Jacksonville 32225)		●	S							●	●
Fort Matanzas Natl. Mon. (St. Augustine 32084)		●	R					●	●		
Gulf Islands Natl. Seashore (Gulf Breeze 32561)	●	L	R,S	●	●		●	●	●		●
GEORGIA											
Andersonville Natl. Historic Site (Andersonville 31711)		L									●
Chattahoochee River Natl. Recreation Area (Smyrna 30080)					●		●		●	●	●
Chickamauga & Chattanooga Nat. Mil. Pk. (Ft. Oglethorpe 30742)		L								●	
Cumberland Island Natl. Seashore (St. Marys 31558)		L	R	●				●	●	●	●
Fort Frederica Natl. Mon. (St. Simons Island 31522)		L	R								
Fort Pulaski Natl. Mon. (Tybee Island 31328)	●	L	R,S								●
Kennesaw Mountain Natl. Battlefield Park (Marietta 30061)		L	S							●	●
Ocmulgee Natl. Mon. (Macon 31201)		L	R,S						●		●
HAWAII											
Haleakala Natl. Park (Makawao 96768)		L	R,S					●	●	●	●
Hawaii Volcanoes Natl. Park (Hawaii Natl. Park 96718)	●	●	R,S					●	●	●	●
Pu'uhonua o Honaunau Natl. Hist Park (Honaunau 96726)	●	●	R,S	●	●					●	●
Puukohola Heiau Natl. Historic Site (Kawaihae 96743)		●	R,S						●	●	
IDAHO											
Craters of the Moon Natl. Mon. (Arco 83213)	●	●	R,S	●						●	●
Nez Perce Natl. Historical Park (Spalding 83551)		L	S								●
ILLINOIS											
Lincoln Home Natl. Historic Site (Springfield 62703)		●	R,S								
INDIANA											
George Rogers Clark Natl. Historical Park (Vincennes 47591)		L			●	●					
Indiana Dunes Natl. Lakeshore (Porter 46304)	●	●	R,S	●	●	●	●	●	●	●	●
Lincoln Boyhood Natl.Mem. (Lincoln City 47552)		L	R,S							●	
IOWA											
Effigy Mounds Natl. Mon. (McGregor 52157)		●	R,S							●	
Herbert Hoover Natl. Historic Site (West Branch 52358)		●	R,S								●
KANSAS											
Fort Larned Natl. Historic Site (Larned 67550)		L	R						●		
Fort Scott Natl. Historic Site (Fort Scott 66701)		●	R,S								
KENTUCKY											
Abraham Lincoln Birthplace Natl. Hist. Site (Hodgenville 42748)		●	R,S								●
Cumberland Gap Natl. Historical Park (Middlesboro 40965)	●	L	R,S	●	●	●			●	●	●
Mammoth Cave Natl. Park (Mammoth Cave 42259)			R,S	●	●	●	R	●	●	●	●
LOUISIANA											
Jean Lafitte Natl. Hist. Park and Preserve (New Orleans 70130)		L	R,S				●	●	●	●	
MAINE											
Acadia Natl. Park (Bar Harbor 04609)		●	R,S	●	●		R	●	●	●	●
MARYLAND											
Antietam Natl. Battlefield Site (Sharpsburg 21782)		L	R,S							●	●
Assateague Island Natl. Seashore (Berlin 21811)		●	R,S	●			●	●	●	●	●
Catoctin Mountain Park (Thurmont 21788)		L	R,S	●					●	●	●
Chesapeake and Ohio Canal Natl. Hist. Park (Sharpsburg 21782)		L	R, S	●	●		B, R		●	●	●
Clara Barton Natl. Historic Site (Glen Echo 20768)		●	R,S								
Fort McHenry Natl. Mon. and Historic Shrine (Baltimore 21230)		L	R,S								●
Fort Washington Park (Oxon Hill 20021)		L	R,S							●	●
Greenbelt Park (Greenbelt 20770)			R,S	●							●

Area Name and Mailing Address	Entrance Fee	Exhibit	Tours	Camping	Food Service	Accommodations	Boating	Swimming	Fishing	Hiking	Picnic Area
Hampton Natl. Historic Site (Towson 21204)			R,S								
Piscataway Park (Oxon Hill 20021)		L									
MASSACHUSETTS											
Adams Natl. Historic Site (Quincy 02269)			R								
Boston Natl. Historical Park (Boston 02109)		L	R,S								
Cape Cod Natl. Seashore (South Wellfleet 02663)			R,S								
John Fitzgerald Kennedy Natl. Historic Site (Brookline 02146)			S								
Longfellow Natl. Historic Site (Cambridge 02138)			R								
Lowell Natl. Historical Park (Lowell 01853)			R,S								
Minute Man Natl. Historical Park (Concord 01742)		L	R,S								
Salem Maritime Natl. Historic Site (Salem 01970)		L	R,S								
Saugus Iron Works Natl. Historic Site (Saugus 01906)		L	R,S								
Springfield Armory Natl. Historic Site (Springfield 01105)		L	R								
MICHIGAN											
Isle Royale Natl. Park (Houghton 49931)			R,S				B				
Pictured Rocks Natl. Lakeshore (Munising 49862)			R,S				R				
Sleeping Bear Dunes Natl. Lakeshore (Frankfort 49635)			R,S				B, R				
MINNESOTA											
Grand Portage Natl. Mon. (Grand Marais 55604)			R,S								
Pipestone Natl. Mon. (Pipestone 56164)		L	S								
Voyageurs Natl. Park (International Falls 56649)							B, R				
MISSISSIPPI											
Brices Cross Roads Natl. Battlefield Site (Tupelo 38801)											
Gulf Islands Natl. Seashore (Ocean Springs 39564)			R,S				R				
Natchez Trace Parkway (Tupelo 38801)		L	R,S				R				
Tupelo Natl. Battlefield (Tupelo 38801)											
Vicksburg Natl. Military Park (Vicksburg 39180)		L	S								
MISSOURI											
George Washington Carver Natl. Mon. (Diamond 64840)		L	R,S								
Jefferson Natl. Expansion Mem. Natl. Hist. Site (St. Louis 63102)		L	R								
Ozark Natl. Scenic Riverways (Van Buren 63965)		L	S				B,R				
Wilson's Creek Natl. Battlefield (Republic 65738)		L	S								
MONTANA											
Big Hole Natl. Battlefield (Wisdom 59761)			S								
Bighorn Canyon Natl. Recreation Area (Ft. Smith 59035)			R,S				B,R				
Custer Battlefield Natl. Mon. (Crow Agency 59022)			R,S								
Glacier Natl. Park (West Glacier 59936)		L	R,S				B,R				
Grant-Kohrs Ranch Natl. Historic Site (Deer Lodge 59722)		L	R,S								
NEBRASKA											
Agate Fossil Beds Natl. Mon. (Gering 69341)			S								
Homestead Natl. Mon. of America (Beatrice 68310)		L	R								
Scotts Bluff Natl. Mon. (Gering 69341)		L	R,S								
NEVADA											
Lake Mead Natl. Recreation Area (Boulder City 89005)			S				B,R				
Lehman Caves Natl. Mon. (Baker 89311)			R,S								
NEW HAMPSHIRE											
Saint-Gaudens Natl. Historic Site (Windsor, VT. 05089)		L	R,S								
NEW JERSEY											
Edison Natl. Historic Site (W. Orange 07052)		L	R								
Gateway Natl. Rec. Area-Sandy Hook Unit (Highlands 07732)		L	R,S								
Morristown Natl. Historical Park (Morristown 07960)		L	R,S								
NEW MEXICO											
Aztec Ruins Natl. Mon. (Aztec 87410)			S								
Bandelier Natl. Mon. (Los Alamos 87544)			R,S								
Capulin Mountain Natl. Mon. (Capulin 88414)			S								
Carlsbad Caverns Natl. Park (Carlsbad 88220)			R,S								
Chaco Culture Natl. Historical Park (Bloomfield 87413)			R,S								
El Morro Natl. Mon. (Ramah 87321)			R,S								
Fort Union Natl. Mon. (Watrous 87753)			S								
Gila Cliff Dwellings Natl. Mon. (Silver City 88061)			S								
Pecos Natl. Mon. (Pecos 87552)		L	R,S								
Salinas Natl. Mon. (Mountainair 87036)			R,S								
White Sands Natl. Mon. (Alamogordo 88310)			R,S								
NEW YORK											
Castle Clinton Natl. Mon. (N.Y.C. 10005)		L	R								
Eleanor Roosevelt Natl. Historic Site (Hyde Park 12538)		L	R								
Federal Hall Natl. Mem. (N.Y.C. 10005)		L	R,S								
Fire Island Natl. Seashore (Patchogue 11772)			R,S				B,R				
Fort Stanwix Natl. Mon. (Rome 13440)		L	R								
Gateway Natl. Recreation Area (Brooklyn 11234)		L	R,S				R				
General Grant Natl. Mem. (N.Y.C. 10005)			R								
Hamilton Grange Natl. Mem. (N.Y.C. 10031)			R								
Home of Franklin D. Roosevelt Natl. Hist. Site (Hyde Park 12538)			S								
Martin Van Buren Natl. Historic Site (Kinderhook 12106)			R								
Sagamore Hill Natl. Historic Site (Oyster Bay 11771)			R,S								
Saratoga Natl. Historical Park (Stillwater 12170)		L	S								
Statue of Liberty Natl. Mon. (N.Y.C. 10004)			R								
Theodore Roosevelt Birthplace Natl. Historic Site (N.Y.C. 10003)			R								
Theodore Roosevelt Inaugural Natl. Historic Site (Buffalo 14209)			R								
Upper Delaware Scenic & Rec. River (Narrowsburg 12764)			R				R				
Vanderbilt Mansion Natl. Historic Site (Hyde Park 12538)			R,S								
NORTH CAROLINA											
Blue Ridge Parkway (Asheville 28801)		L					B,R				
Cape Hatteras Natl. Seashore (Manteo 27954)		L	R,S				R				
Cape Lookout Natl. Seashore (Beaufort 28516)			R,S								
Carl Sandburg Home Natl. Historic Site (Flat Rock 28731)			R,S								
Fort Raleigh Natl. Historic Site (Manteo 27954)		L	S								
Guilford Courthouse Natl. Military Park (Greensboro 27408)		L	R,S								
Moores Creek Natl. Military Park (Currie 28435)		L	S								
Wright Brothers Natl. Mem. (Manteo 27954)		L									
NORTH DAKOTA											
Fort Union Trading Post Natl. Historic Site (Williston 58801)		L	R								
Knife River Indian Villages Natl. Historical Site (Stanton 58571)											
Theodore Roosevelt Natl. Park (Medora 58645)			R,S								
OHIO											
Cuyahoga Valley Natl. Recreation Area (Peninsula 44264)		L	R,S								
Mound City Group Natl. Mon. (Chillicothe 45601)			R,S								
Perry's Victory and International Peace Mem. (Put-In-Bay 43456)			R								
William Howard Taft Natl. Historic Site (Cincinnati 45219)			R								

Area Name and Mailing Address	Entrance Fee	Exhibit	Tours	Camping	Food Service	Accommodations	Boating	Swimming	Fishing	Hiking	Picnic Area
OKLAHOMA											
Chickasaw Natl. Recreation Area (Sulphur 73086)			R,S				R				
OREGON											
Crater Lake Natl. Park (Crater Lake 97604)			R,S								
Fort Clatsop Natl. Mem. (Astoria 97103)		L	S								
John Day Fossil Beds Natl. Mon. (John Day 97845)			R,S								
Oregon Caves Natl. Mon. (Cave Junction 97523)		L	S								
PENNSYLVANIA											
Allegheny Portage Railroad Natl. Historic Site (Cresson 16630)		L	R,S								
Delaware Water Gap Natl. Recreation Area (Bushkill 18324)		L	R,S				B,R				
Edgar Allen Poe Natl. Historic Site (Philadelphia 19106)			S								
Fort Necessity Natl. Battlefield (Farmington 15437)		L	R,S								
Gettysburg Natl. Military Park (Gettysburg 17325)		L	S								
Hopewell Village Natl. Historic Site (Elverson 19520)		L	R,S								
Independence Natl. Historic Park (Philadelphia 19106)		L	R,S								
Johnstown Flood Natl. Mem. (Cresson 16630)		L	R,S								
Thaddeus Kosciuszko Natl. Mem. (Philadelphia 19106)											
Valley Forge Natl. Historical Park (Valley Forge 19481)		L	S				R				
RHODE ISLAND											
Roger Williams Natl. Mem. (Providence 02901)											
SOUTH CAROLINA											
Cowpens Natl. Battlefield (Kings Mountain 28086)			R,S								
Fort Sumter Natl. Mon. (Sullivans Island 29482)		L	R,S								
Kings Mountain Natl. Military Park (Kings Mountain, NC 28086)		L	S								
Ninety Six Natl. Historic Site (Ninety Six 29666)			R,S								
SOUTH DAKOTA											
Badlands Natl. Park (Interior 57750)			R,S								
Jewel Cave Natl. Mon. (Custer 57730)			R								
Mount Rushmore Natl. Mem. (Keystone 57751)											
Wind Cave Natl. Park (Hot Springs 57747)		L	R,S								
TENNESSEE											
Andrew Johnson Natl. Historic Site (Greeneville 37743)			S								
Big South Fork Natl. River and Recreation Area (Oneida 37841)											
Fort Donelson Natl. Military Park (Dover 37058)		L	R,S								
Great Smoky Mountains Natl. Park (Gatlinburg 37738)		L	R,S								
Obed Wild and Scenic River (Oneida 37841)							R				
Shiloh Natl. Military Park (Shiloh 38376)		L	S								
Stones River Natl. Battlefield (Murfreesboro 37130)		L	R,S								
TEXAS											
Alibates Flint Quarries Natl. Mon. (Fritch 79036)			R								
Amistad Natl. Recreation Area (Del Rio 78840)			S				B,R				
Big Bend Natl. Park (Big Bend Natl. Park 79834)			R,S				R				
Chamizal Natl. Mem. (El Paso 79901)			R								
Fort Davis Natl. Historic Site (Fort Davis 79734)		L	S								
Guadalupe Mountains Natl. Park (Carlsbad, NM 88220)		L	R,S								
Lake Meredith Natl. Recreation Area (Fritch 79036)							R				
Lyndon B. Johnson Natl. Historic Site (Johnson City 78636)		L	R								
Padre Island Natl. Seashore (Corpus Christi 78418)			R,S								
San Antonio Missions Natl. Historical Park (San Antonio 78206)			S								
UTAH											
Arches Natl. Park (Moab 84532)		L	R,S								
Bryce Canyon Natl. Park (Bryce Canyon 84717)			R,S								
Canyonlands Natl. Park (Moab 84532)		L	R,S				B,R				
Capitol Reef Natl. Park (Torrey 84775)			R,S								
Cedar Breaks Natl. Mon. (Cedar City 84720)			S								
Glen Canyon Natl. Recreation Area (Page, AZ 86040)		L	R,S				B,R				
Golden Spike Natl. Historic Site (Brigham City 84302)		L	S								
Natural Bridges Natl. Mon. (Moab 84532)			R,S								
Rainbow Bridge Natl. Mon. (Page, AZ 86040)			R,S								
Timpanogos Cave Natl. Mon. (American Fork 84003)			R,S								
Zion Natl. Park (Springdale 84767)			R,S								
VIRGINIA											
Appomattox Court House Natl. Hist. Park (Appomattox 24522)		L	R,S								
Arlington House (McLean 22101)		L	R								
Booker T. Washington Natl. Mon. (Hardy 24101)		L	R,S								
Colonial Natl. Historical Park (Yorktown 23690)		L	R,S								
Fredericksburg Natl. Military Park (Fredericksburg 22401)		L	R,S								
G. Washington Birthplace Natl. Mon. (Wash. Birthplace 22575)		L	R,S								
G. Washington Mem. Parkway (McLean 22101)		L	R,S				R				
Manassas Natl. Battlefield Park (Manassas 22110)		L	R,S								
Petersburg Natl. Battlefield (Petersburg 23803)		L	R,S								
Prince William Forest Park (Triangle 22172)		L	R,S								
Richmond Natl. Battlefield Park (Richmond 23223)			S								
Shenandoah Natl. Park (Luray 22835)			R,S								
Wolf Trap Farm Park for the Performing Arts (Vienna 22180)			R,S								
WASHINGTON											
Coulee Dam Natl. Recreation Area (Coulee Dam 99116)		L	R,S				R				
Fort Vancouver Natl. Historic Site (Vancouver 98661)		L	R								
Klondike Gold Rush Natl. Historical Park (Seattle 98104)		L	S								
Lake Chelan Natl. Recreation Area (Sedro Woolley 98284)		L	R,S				B				
Mount Rainier Natl. Park (Ashford 98304)			R S								
North Cascades Natl. Park (Sedro Woolley 98284)											
Olympic Natl. Park (Port Angeles 98362)			S				B,R				
Ross Lake Natl. Recreation Area (Sedro Woolley 98284)			R,S				B,R				
San Juan Island Natl. Historical Park (Friday Harbor 98250)		L	R,S								
Whitman Mission Natl. Historic Site (Walla Walla 99362)		L	S								
WEST VIRGINIA											
Harpers Ferry Natl. Historical Park (Harpers Ferry 25425)		L	R,S								
WISCONSIN											
Apostle Island Natl. Lakeshore (Bayfield 54814)		L	R,S								
Lower St. Croix Natl. Scenic River (St. Croix Falls 54024)							B,R				
St. Croix Natl. Scenic River (St. Croix Falls 54024)			S				B,R				
WYOMING											
Devils Tower Natl. Mon. (Devils Tower 82714)			S								
Fort Laramie Natl. Historic Site (Fort Laramie 82212)		L	S								
Fossil Butte Natl. Mon. (Kemmerer 83101)			S								
Grand Teton Natl. Park (Moose 83012)		L	R,S				B,R				
John D. Rockefeller Jr. Mem. Parkway (Moose 83012)			R,S								
Yellowstone Natl. Park (Yellowstone Natl. Park 82190)		L	R,S				B,R				

HOTEL AND MOTEL CHAINS

How To Use The Accommodations Listings

Although there are over 60,000 motels, hotels, lodges and inns throughout the country, and new ones are opening each year, it is always a good idea to reserve a room prior to your arrival, particularly if you are traveling during peak season. Many of these accommodations are affiliated with major chains and generally offer consistency of facilities; you may obtain advance reservations by phone, writing or through toll-free reservation centers. To help simplify your travel plans, we have selected 27 major chains which provide these reservation services, listing their toll-free numbers where available. The following pages show the cities in which these hotels or motels are located. Although the listings are extensive, they are not to be considered a complete guide to all motels and hotels in the United States. There are many independent motels and hotels as well as other chains available to the traveler.

The following example illustrates how to use this Road Atlas chain listing information: **MARYLAND**

Aberdeen—BW, HI, QI
Annapolis—HH, HI, HJ
Annapolis Jct.—QI
Baltimore—HH, HI(4), RI, SH

By referring to the code letters shown after each city, and comparing these letters to the listings below, you can determine that in Baltimore, for instance, there is a Hilton Hotel, four Holiday Inns, a Ramada Inn and a Sheraton Hotel.

These listings were compiled from information obtained directly from the headquarters of each chain whose affiliates are listed on the next pages. Most of the chains offer a directory listing all of their locations, available by writing to the chain headquarters at the addresses given below or by stopping in at any of their properties.

HOTEL AND MOTEL CHAINS

AB ADMIRAL BENBOW INNS
c/o Morrison Incorporated
4721 Morrison Drive
Mobile, Alabama 36625
800-238-6877 in Cont'l U.S.A.
800-542-6844 in Tennessee
Children under 12 free*

BW BEST WESTERN INTERNATIONAL
Best Western Way
Box 10203
Phoenix, Arizona 85064
1-800-528-1234 in Cont'l U.S.A.
1-800-352-1222 in Arizona
　957-4200 in Phoenix
1-800-268-8993 in Canada
　485-2632 in Toronto

DI DAYS INN
2751 Buford Highway, N.E.
Atlanta, Georgia 30324
1-800-241-3400 in Ala., Fla., Ky.,
　Miss., N.C., S.C., Tenn.
1-800-241-7200 in Ark., Conn., Del.,
　Ill., Ind., Iowa, La., Md., Mass.,
　Mich., Mo., N.J., N.Y., Ohio, Okla.,
　Pa., R.I., Tex., Va., W.Va., Wis.
1-800-282-2424 in Georgia
　320-2000 in Atlanta
1-800-241-2345 in other Cont'l states
416-964-3434 in Canada call collect
　964-3434 in Toronto
Children under 18 $2.00 at
　most locations*

**DT DOWNTOWNER/ROWNTOWNER/
PASSPORT INNS**
5350 Poplar Ave.
Suite 518
Box 171807
Memphis, Tennessee 38117
800-238-6161 in Cont'l U.S.A.
800-582-6173 in Tennessee

**ET ECONO-TRAVEL MOTOR HOTELS/
ECONO LODGES**
Box 12188
Norfolk, Virginia 23502
1-800-446-6900 in Cont'l U.S.A.
1-800-582-5882 in Virginia
　461-6000 in Norfolk

FI FRIENDSHIP INNS
739 South 4th West Street
Salt Lake City, Utah 84101
801-532-1800 (toll call)
　in Cont'l U.S.A.
800-453-4511 in Cont'l U.S.A.
　8 a.m.-5 p.m. Mtn. Std. time.

HH HILTON HOTELS AND INNS
9880 Wilshire Blvd.
Beverly Hills, California 90210
Call local directory assistance or
　800-555-1212 for the number
　of the Hilton Reservation Service
　Office in your area.
Children regardless of age free*

HI HOLIDAY INNS
3796 Lamar Avenue
Memphis, Tennessee 38118
800-238-8000 in Cont'l U.S.A.
800-542-5270 in Tennessee
　363-3400 in Memphis
Children under 12 free*
Teens free at participating inns

HJ HOWARD JOHNSON'S
222 Forbes Road
Braintree, Massachusetts 02184
800-654-2000 in Cont'l U.S.A.
800-522-9041 in Oklahoma
1-800-268-4940 in Ont. and E. Can.
　363-7401 in Toronto
Children under 18 free*

HY HYATT HOTELS
9700 West Bryn Mawr Ave.
Rosemont, Illinois 60018
800-228-9000 in Cont'l U.S.A.
800-361-6172 in Canada

IM IMPERIAL '400' MOTELS
1830 N. Nash St.
Arlington, Virginia 22209
800-531-5300 in Cont'l U.S.A.
800-252-9649 in Texas
Family rates available*

LQ LA QUINTA MOTOR INNS
Marketing Division
The Centre
4100 McEwen, Suite 283
Dallas, Texas 75234
800-531-5900 in Cont'l U.S.A.
800-292-5200 in Texas
　349-4141 in San Antonio
Children under 12 free*

LO LOEWS HOTELS
666 Fifth Avenue
New York, New York 10019
For reservations call Loews
　Representation in your area.
Family rates available*

MM MARRIOTT HOTELS
Marriott Dr.
Washington, D.C. 20058
800-228-9290 in Cont'l U.S.A.
800-228-2180 in Alaska, Hawaii
402-571-5400 in Nebraska
800-268-8181 in Canada
　361-0408 in Toronto
Family rates available*

MH MASTER HOSTS INNS
see Red Carpet Inns

M6 MOTEL 6
51 Hitchcock Way
Santa Barbara, California 93105

QI QUALITY INNS
10750 Columbia Pike
Silver Spring, Maryland 20901
800-228-5151 in Cont'l U.S.A.
800-642-8700 in Nebraska
800-268-8990 in Canada
　485-2600 in Toronto
Children under 16 free*

RI RAMADA INNS
Box 590
Phoenix, Arizona 85001
800-228-2828 in Cont'l U.S.A.
800-642-9343 in Nebraska
1-800-268-8930 in Ontario or Quebec
1-800-268-8998 in other provinces
　485-2610 in Toronto
Family rates available*

**RD RED CARPET INNS/
MASTER HOSTS INNS**
2032 Hillview St.
Sarasota, Fla. 33579
800-327-9073 in Cont'l U.S.A.
800-432-9109 in Fla.
800-267-7187 in Canada
Children under 12 free*

RO RODEWAY INNS
2525 Stemmons Freeway
Dallas, Texas 75207
800-228-2000 in Cont'l U.S.A.
402-571-2000 in Alaska, Hawaii,
　Nebraska or Canada call collect
Children under 17 free*

SH SHERATON CORPORATION
Sixty State Street
Boston, Massachusetts 02109
800-325-3535 in Cont'l U.S.A.
1-800-392-3500 in Missouri
800-268-9393 in Eastern Canada
800-268-9330 in Western Canada
　and Newfoundland
　869-1414 in Toronto
Children under 18 free*

ST STOUFFER HOTELS
29800 Bainbridge Rd.
Solon, Ohio 44139
800-321-6888 in Cont'l U.S.A.
800-362-6100 in Ohio
216-248-4343 in Canada call collect
Children sharing Parents' room
and using existing beds are free

**SC SUSSE CHALET
MOTOR LODGES & INNS**
Two Progress Avenue
Nashua, New Hampshire 03060
1-800-258-1980 in Cont'l U.S.A.
1-800-572-1880 in New Hampshire

**TH THUNDERBIRD/RED LION
MOTOR INNS**
P.O. Box 1027
Vancouver, Washington 98666
800-547-8010 in Cont'l U.S.A.
800-452-0733 in Oregon
　221-1911 in Portland

TL TRAVELODGE INTERNATIONAL
250 TraveLodge Drive
El Cajon, California 92090
800-255-3050 in Cont'l U.S.A.
800-332-4350 in Kansas
800-255-6411 in Alaska, Hawaii
1-800-268-3330 in Canada
112-800-268-3330 in British Columbia
Children under 18 free*

TR TREADWAY INNS
140 Market Street
Paterson, New Jersey 07505
800-631-0182 in Cont'l U.S.A.
201-881-8483 in New Jersey call
　collect
Children 16 and under free*

WI WESTIN HOTELS
Marketing Division
The Westin Building
Seattle, Washington 98121
800-228-3000 in Cont'l U.S.A.
800-228-1212 in Alaska, Hawaii
800-268-8383 in Canada
　368-4684 in Toronto
Children 18 and under free*

*"Children free" plans require children to share same room with parents. These plans are available at participating properties.

127

CITIES AND TOWNS WITH CHAIN LODGINGS

ALABAMA

Anniston—DT
Attalla—HI
Auburn—BW
Bessemer—DI, HI, RI
Birmingham—BW(2), DT, ET(2), HI(4), HY, RD, RI(2), SH, TL(2)
Boaz—HI
Brewton—HI
Clanton—HI
Cullman—DI, HI
Daphne—HI
Decatur—HI, MH, RD, RI
Dothan—DI, HI, QI, RI, SH
Eufaula—HI
Evergreen—DI, HI
Florence—HI, HJ, TL
Fort Payne—BW, HI
Fultondale—RI
Gadsden—RI
Greenville—HI
Gulf Shores—DT, HI
Homewood—QI, RO
Hope Hull—DI
Huntsville—BW, HH, HI, HJ, RI, SH
Irondale—DI
Jasper—HI
Lanett—HI
Mobile—BW(2), DI, HH, HI(3), HJ, LQ, QI, RI(2) RO(2), SH, TL
Montgomery—BW(2), DI, DT, HI(3), HJ(2), M6, QI, RI(2), TL
Mountain Brook—SH
Northport—BW
Opelika—DI, HI, M6
Oxford—DI, HI, M6
Ozark—HI
Pell City—HI
Phenix City—ET, HI, RI
Prattville—HI, RI
Scottsboro—HI
Selma—HI
Troy—HI
Tuscaloosa—DI, HI(2), QI, RI, SH
Tuskegee—HI
Vestavia Hills—HI, HJ, RI, TL

ALASKA

Anchorage—BW, HH, HI, SH, TL
Homer—BW
Juneau—FI, HH

ARIZONA

Bullhead City—BW
Casa Grande—RI
Chambers—BW
Douglas—M6, TL
Eloy—BW
Flagstaff—BW(3), FL(4), HI, IM, LQ, M6, QI, RI(2), RO, TL(2)
Gila Bend—BW, FI, TL
Glendale—BW
Globe—FI
Goodyear—BW, RI
Grand Canyon—BW
Gray Mountain—BW
Holbrook—BW(2), FI(2), M6
Kayenta—HI
Kingman—BW(3), FI, HI, IM, LQ, M6, RI, RO, TL
Lake Havasu City—BW, HI, RO
Mesa—BW(3), FI, QI, RO, TL
Miami—BW
Nogales—BW(2), M6
Page—BW, HI

Patagonia—BW
Peoria—FI
Phoenix—BW(3), DI(2), DT, FI(3), HI(2), HY, LQ, M6(3), QI(3), RI(2), RO(3), SH(2), TL(3), WI
Prescott—BW, M6
Safford—BW(2), FI
Scottsdale—BW, HH, HI, MM, M6, QI, RI(2), RO, SH, TL
Sedona—BW(2), FI, QI
Show Low—BW(2)
Sierra Vista—M6
Tempe—FI, HI, HJ, TL
Tombstone—BW
Tucson—BW(6), FI, HH, HI(2), HJ, IM, LQ, MM, M6(2), QI(2), RI, RO(2), SH, TL(2)
Wickenburg—BW
Wilcox—BW(2), IM, M6, TL
Williams—BW, FI, RI, TL(2)
Winslow—BW, IM, M6, TL
Youngtown—M6
Yuma—BW(3), FI(2), HI, M6, QI, RI, RO, TL

ARKANSAS

Arkadelphia—BW, HI
Batesville—BW
Beebe—FI
Benton—HI
Bentonville—HI
Blytheville—BW, DI, FI, HI, RI
Brinkley—BW, FI
Carlisle—BW
Clarksville—FI
Conway—BW, FI, HI, M6, RI
Crossett—RI
Dardanelle—BW
El Dorado—BW, HI
Eureka Springs—BW(2)
Fayetteville—HI, RI
Forest City—BW, HI
Fort Smith—BW(2), FI, HI, M6, RI, SH
Hardy—BW
Harrison—BW, HI, RI
Helena—HI
Heth—BW
Hope—BW, FI, HI, QI
Hot Springs—BW(2), DT, HI(2), MH, RI
Jacksonville—HI, RI
Jonesboro—BW, M6, RI
Little Rock—BW(2), HH, HI(2), IM, LQ(2), MH, M6, QI, SH, TL
Magnolia—BW
Marion—BW
Marked Tree—BW
Monticello—BW
Morrilton—BW
Mountain Home—BW, HI, RI
Newport—BW, TL
N. Little Rock—BW, DI(2), DT, RI
Osceola—BW
Pine Bluff—AB, BW, HI
Pocahontas—BW
Rogers—BW
Russellville—BW, HI, M6, RI
Searcy—BW
Springdale—BW, HI
Texarkana—BW(2), HI, RI, RO
Walnut Ridge—BW
West Memphis—HI, RI

CALIFORNIA

Alameda—TL
Alturas—BW

Anaheim—BW(7), FI(8), HI, HJ, HY, M6, MM, QI, SH, TL(5)
Anderson—BW
Arcadia—BW, M6, RI
Arcata—M6, RI
Atascadero—M6
Auburn—BW, FI
Baker—FI
Bakersfield—BW(2), FI(3), HH, HI, IM, M6(2), RI, RO, TL
Baldwin Park—HJ
Barstow—BW, FI, HI, HJ, IM, M6, TL
Beaumont—BW
Belmont—HI
Berkeley—BW(2), FI, MM, TL
Beverly Hills—HH, HI, RI
Big Bear City—M6
Bishop—BW, FI, TL
Blythe—BW(2), FI(2), M6, RO, TL
Bodega Bay—BW
Brawley—TL
Bridgeport—BW, FI
Buellton—BW, M6
Buena Park—BW, FI(2), HI, QI, TL(2)
Burbank—FI, TL
Burlingame—HY, RI, SH
Buttonwillow—M6
Camarillo—M6
Cambria—BW
Canoga Park—BW
Carlsbad—BW, TL
Carmel—BW(3), FI, HI
Carpinteria—FI, M6
Carson—RI
Chico—HI, IM, M6, TL
Chino—BW
Chula Vista—BW, RI, TL
Claremont—HJ, RO
Coalinga—M6
Colton—BW, HJ, TL
Commerce—HY
Compton—BW
Concord—HI, SH
Corning—FI
Corona—BW, TL
Corte Madera—BW
Costa Mesa—BW(2), HI, LQ, RO, WI
Crescent City—BW, FI
Culver City—HJ, RI
Dana Point—BW
Davis—BW, M6
Delano—FI
Desert Hot Sprs.—BW
Dublin—HJ
Dunnigan—BW
Dunsmuir—BW, FI(2), TL
El Cajon—BW, TL(3)
El Centro—BW, HI, M6, TL(2)
El Monte—M6, TL
Emeryville—HI
Encino—TL
Escondido—BW, M6, TL(2)
Eureka—BW, FI(2), IM, TH, TL
Fairfield—FI, HI, M6
Fontana—M6, TL
Fort Bragg—BW, FI
Fremont—BW, TI
Fresno—BW(3), FI(2), HH, HI, IM, M6(2), QI, RI, TL(3)
Fullerton—HI, TL
Garden Grove—TL
Gilroy—M6
Glendale—BW, FI, HI, TL
Goleta—HI, M6
Half Moon Bay—BW
Harbor City—TL
Hayward—FI, TL
Healdsburg—FI
Hemet—BW
Hollister—BW

Hollywood—BW(2), FI(2), HI, HY, TL(3)
Imperial Beach—TL
Indio—BW, FI, M6(2), QI, TL
Inglewood—BW, FI(2)
Irvine—BW
Jackson—BW
King City—FI, M6
Kings Beach—FI
Laguna Hills—HI, TL
La Habra—BW
La Jolla—BW, TL(2)
Lancaster—BW
Lee Vining—BW
Livermore—HI
Livingston—FI
Lodi—BW
Lompoc—BW, M6
Lone Pine—BW
Long Beach—BW(3), FI, HH, HI, HY, IM, RI, TL(2)
Los Angeles—BW(5), FI(2), HH(2), HI(6), HY(3), IM, MM, QI, RI, SH(3), TL(4), WI(2)
Los Banos—TL
Lost Hills—M6
Madera—BW(2)
Mammoth Lakes—BW, FI, M6, TL
Manhattan Bch.—BW
Manteca—FI
Marina Del Rey—BW, MM
Martinez—FI
Marysville—FI, IM, TL
Merced—BW(2), FI, M6, TL
Millbrae—BW, TL
Mill Valley—HJ, TL
Milpitas—TL
Mission Hills—BW
Modesto—BW, FI, HI, M6, TL
Mojave—FI, IM, M6, TL
Monrovia—HJ
Montebello—HI
Monterey—BW(3), FI(2), HH, HI, HY, M6, RI, TL(3)
Monterey Park—BW
Monte Rio—FI
Morro Bay—BW(2), FI, M6, TL
Mountain View—BW(2)
Mt. Shasta—BW, FI(2)
Napa—HI, M6
Needles—BW, FI, IM, M6, TL
Newark—M6
Newport Beach—BW, FI, MM, SH, TL
North Highlands—RO
N. Hollywood—BW, HJ, SH
Oakland—BW(2), FI, HH, HI, HY, M6, TL
Oceanside—BW(2), FI, HH, HI, HY, M6, TL
Oceanside—BW(2), M6, TL
Ontario—HI, M6, TL
Orange—BW, HI, RO, TL
Oroville—M6
Oxnard—BW, FI, HH, QI
Pacific Grove—FI
Palmdale—HI, M6
Palm Springs—BW(4), HH, M6, RI, RO, SH(2), TL
Palo Alto—BW, FI, HI, HY(2), M6, TL
Pasadena—FI, HH, HI, IM, SH, TL
Paso Robles—BW, FI, TL
Perris—TL
Petaluma—BW, M6
Pico Rivera—IM, TL
Pismo Beach—BW, FI
Pittsburg—M6
Pleasanton—BW
Pomona—TL
Porterville—M6

Rancho Cordova—M6
Rancho Mirage—MM
Red Bluff—BW, FI, M6, TL
Redding—BW(3), FI, HI, IM, M6, QI, TH, TL
Redlands—BW, M6, TL
Redwood City—BW, HJ
Richmond—TL
Ridgecrest—M6
Riverside—BW, HI, HJ, M6(2), RI, TL
Rohnert Park—BW
Rosemead—M6
Roseville—BW
Sacramento—BW(3), FI(4), HI(3), HJ, IM, M6, QI(2), TH(2), TL(2)
Salinas—BW, FI(2), M6, QI, TL
San Bernardino—BW, HH, HI, IM, M6, TL(2)
San Bruno—FI
San Carlos—TL
San Clemente—BW
San Diego—BW(4), FI(3), HH, HI(3), HJ, HY, IM(2), M6, QI(3), RI, RO, SH(2), TL(18)
San Francisco—BW(5), DT(2), FI(6), HH, HI(5), HJ, HY(2), QI, RI, RO, SH(2), TL(9), WI(2)
S.F. Intl. Arpt.—HH
San Jose—BW(3), HI(2), HJ, HY, M6, QI, TL
San Juan Capistrano—BW
San Luis Obispo—BW(3), FI(2), HJ, M6, TL
San Pedro—BW, IM
San Rafael—FI, HI
San Simeon—BW(2), FI
Santa Ana—BW, RI, TL
Santa Barbara—BW(3), MM, M6(2), QI, SH, TL(2)
Santa Clara—BW, FI, HJ, MM, M6, RI
Santa Cruz—BW, FI, HI, TL(2)
Santa Fe Sprs.—BW
Santa Maria—BW, FI, HI, HJ, M6, TL
Santa Monica—HI, SH, TL
Santa Nella—BW, HI, M6
Santa Rosa—BW(2), FI(3), HI, M6, TL(2)
San Ysidro—BW, FI, M6, TL(2)
Saugus—BW
Seaside—BW, FI
Sepulveda—FI
Sherman Oaks—FI, HH
Shingle Springs—BW
Simi Valley—M6
Smith River—BW
Solvang—BW(2)
Sonora—M6
S. Lake Tahoe—BW(2), M6, TL(3)
S. San Francisco—BW, HI, IM, TL
Stanton—M6
Stockton—BW, FI, HH, HI, M6, TL
Sunnyvale—BW, HH, HI, M6, SH
Susanville—BW
Tahoe City—TL
Tehachapi—BW
Thousand Oaks—BW, HI, HJ, M6
Three Rivers—BW
Torrance—FI, HI, RI
Tracy—M6
Tujunga—TL
Tulare—M6, TL(2)
Turlock—M6
Tustin—FI
Ukiah—BW(2), M6, TL
Upland—BW
Vacaville—BW, M6
Vallejo—BW, M6, QI, TL(2)

Van Nuys—HI
Ventura—HI, M6, TL
Victorville—BW, HI, M6, TL
Visalia—BW, FI, HI, TL
Vista—BW
Walnut Creek—M6
Wasco—BW
West Covina—HI
Westminster—M6
West Sacramento—RO, TL
Whittier—BW, M6, TL
Willits—BW
Willows—BW
Woodland—M6
Woodland Hills—BW, HI
Yountville—BW
Yreka—BW, FI, M6, TL
Yuba City—BW, M6

COLORADO

Alamosa—BW
Aspen—HI
Aurora—FI, HI, QI
Basalt—BW
Boulder—BW(2), HH, HI, RO, TL
Breckenridge—QI
Brush—BW
Buena Vista—BW
Burlington—BW, FI, RI
Canon City—BW, RI
Colorado City—BW
Colorado Sprs.—BW(2), FI(3), HH, HI(2), HJ, IM(2), M6, RI(2), RO, SH, TL(2)
Commerce City—BW
Cortez—BW(2), FI(2), RI
Craig—BW
Delta—BW
Denver—BW(6), FI(3), HH(2), HI(6), HJ(3), LQ(3), MM(2), M6, QI, RI(4), RO, SH(2), ST, TL(3)
Dillon—BW
Durango—BW(4), FI, HI, QI, TL
Eads—BW
Eagle—BW
Estes Park—BW(2), FI(3), HI
Evans—BW, FI, M6, RI
Fort Collins—BW(3), FI, HI, M6, RI
Fort Morgan—BW, FI, RI
Fowler—FI
Frisco—BW, FI, HI
Georgetown—BW
Glenwood Sprs.—BW(2), FI(2), HI, RI
Golden—FI, HI
Granby—BW
Grand Junction—BW(3), HI, HJ, M6, RI, TL
Greeley—HI, TL
Greenwood Vill.—RO
Gunnison—BW, FI
Julesburg—BW
La Junta—BW
Lakewood—FI, M6, RI(2)
Lamar—BW, FI, TL
Las Animas—BW
Leadville—BW
Longmont—BW, FI
Loveland—BW(2), FI
Monte Vista—BW
Montrose—BW, FI, MH
Monument—BW
Northglenn—DI
Ouray—BW, FI(2)
Pagosa Springs—BW, FI
Pueblo—BW(2), FI(2), HI, RI ,TL
Rocky Ford—BW
Salida—BW, FI
Silverthorne—RI
Snowmass Vill.—BW
Steamboat Sprs.—BW(2), FI(2), HI, RI, SH
Sterling—BW(2), FI, HI, RI
Stratton—BW
Thornton—BW
Trinidad—BW, FI, RI

Vail—BW, HI
Walsenburg—BW, FI
Watkins—FI
Wheat Ridge—QI, RI
Winter Park—BW, FI

CONNECTICUT

Bethel—BW
Bridgeport—HI
Danbury—HH, HI, HJ, RI
Darien—HI, HJ
East Hartford—HI, HJ, IM, RI
East Lyme—HJ
East Windsor—RI
Greenwich—SH
Groton—HI
Hamden—HJ
Hartford—HH, HI, HJ, SC, SH
Meriden—HI
Milford—BW, HI, HJ(2)
Mystic—HJ, RI
New Britain—HI
New Haven—HI(2), HJ, SH
New London—HI
Niantic—SC, TL
North Haven—RI
Norwalk—HI
Norwich—SH
Old Saybrook—HJ
Plainville—HI
Rocky Hill—HJ
Southington—HJ, SC
Stamford—HJ, MM, RI
Stratford—HJ
Vernon—HJ, QI
Waterbury—HI, HJ
Wethersfield—RI
Windsor—SH
Windsor Locks—HJ, RI(2)

DELAWARE

Dover—BW, HI, QI, SH
Newark—HI, HJ, SH, TL
New Castle—HJ, QI(2), RI
Wilmington—HH, HI, SH

DISTRICT OF COLUMBIA

Washington—BW(5), HH(2), HI(7), HJ, HY, LO, MM, MH, QI(3), RI, SH(2), TL, WI

FLORIDA

Alachua—HI
Altamonte Sprs.—BW, DI(2), HI
Apollo Beach—HI
Arcadia—BW
Bal Harbour—BW, SH
Belle Glade—HI
Boca Raton—BW, HI(2), HJ
Bonifay—BW
Bonita Springs—BW
Boynton Beach—BW
Bradenton—DI
Brooksville—DI, HI
Bushnell—BW
Callahan—QI
Carol City—HI
Clearwater—BW, DI, FI, HI, HJ, QI, RI(2), TL
Clearwater Bch.—BW, FI, HH, HI, SH
Clermont—DI, HI, HJ, RI, SH
Cocoa—BW, DI, HI, HJ, QI
Cocoa Beach—BW, ET, HI, HJ, M6, QI
Coral Gables—HI(2), HJ, QI
Crestview—HI
Crystal River—HI
Cutler Ridge—HI
Dania—HJ, TL

Davenport—BW, HJ, RI, SC
Daytona Beach—BW(2), DI, ET(2), FI, HI(4), HJ(3), QI, RI(2), RO, TL
Daytona Bch. Shores—BW, DI(2), HH, QI, SH, TL
Deerfield Beach—DI, FI, HI, HJ, RI
De Funiak Sprs.—RI
Deltona—BW
Dundee—HI
Englewood—DI
Ft. Lauderdale—BW(2), DI, HH, HI(4), HJ(3), MM, M6, QI(2), RI(2), SH(2), ST(2), TL
Ft. Myers—DI, HI, RI, RO, SH, TL
Ft. Myers Beach—BW, HI
Ft. Pierce—HI, HJ, QI, RI
Fort Walton Bch.—DI, ET, HI, HJ, RI, SH, TL
Gainesville—BW, DI, ET(2), HH, HI(2), HJ(2), QI, RI, TL
Gulf Breeze—HI
Haines City—BW, HI
Hallandale—RI
Hialeah—HI
Highland Beach—HI
Hobe Sound—DI
Holly Hill—TL
Hollywood—HI(3), HJ(2)
Homestead—HI, HJ, RI
Homosassa Sprs.—SH
Indialantic—HI
Inverrary—HH
Islamorada—FI, HI, HJ
Jacksonville—AB, DI(3), ET(2), HH, HI(6), HJ(2), LQ, QI(2), RD, RI(4), RO, SH, TL
Jacksonville Bch.—FI, HI, HJ, RI, SH
Jasper—BW
Jennings—HI
Jensen Beach—HI
Juno Beach—HJ
Jupiter—HH
Kendall—HJ, RI
Key Biscayne—HJ, SH
Key Largo—FI, HI, HJ
Key West—BW, DI, HI, HJ, MM, QI, RI, TL
Kissimmee—BW(2), DI(4), HH, HI(3), HJ(2), HY, QI, RD(2), RI, RO, SH
L. Buena Vista—BW, HJ, TL
Lake City—BW, DI, HI, HJ, QI, RO
Lakeland—BW, DI, HH, HI(3), HJ, RD, RI
Lake Park—TL
Lake Placid—HI
Lake Wales—ET, HJ
Lake Worth—FI
Lauderdale/By-The-Sea—HI, HJ, TL
Leesburg—HI
Lido Beach—BW, HI, RO, SH
Longboat Key—HH, HI
Longwood—QI
Madeira Beach—HI
Marathon—HI, HJ
Marco Island—MM
Marianna—HI, QI
Marineland—QI
Melbourne—BW, DI, ET, HI(2), HJ, RI
Merritt Island—HI
Miami—BW(2), FI, HI(4), HJ(4), MH, MM, QI(2), RI(2), SH
Miami Beach—BW, HH, HI(3), HJ, SH
Miami Springs—BW, HI, FI
Micanopy—DI
Naples—BW, DI, HI, HJ, RI(2), SH
Navarre—HI
Neptune Beach—BW
New Port Richey—HI

New Smyrna Beach—BW
North Bay Village—BW
N. Ft. Myers—BW, DI, ET
North Miami—HI(2), HJ(2), RI
N. Palm Beach—ET
Ocala—BW, DI, ET, HI, HJ, QI(2), RD, RI, SH, TL
Ocoee—HI, RI
Orange Park—BW, DI, HI
Orlando—BW(3), DI(7), ET(2), FI, HH(2), HI(5), HJ(7), LQ(2), MM, QI(3), RD, RI(2), SH(5), TL(2)
Ormond Beach—ET, HI, HJ, QI(2)
Osprey—HI
Palatka—HI
Palm Beach—HH, HJ
Palm Beach Gardens—HI
Palm Beach Shores—BW
Palm Coast—SH
Panama City—DI, HI, HJ, RI, TL
Panama City Bch.—DT, HI, HJ, SH
Pensacola—BW, DI, HI, HJ, M6, RI, RO(2), SH, TL
Pensacola Beach—HI, HJ
Perry—HI, HJ
Pinellas Park—DI
Plantation—HI
Plant City—DI, HI
Pompano Beach—BW, DI, HI, HJ
Port Charlotte—RI
Port Richey—DI
Punta Gorda—HI, HJ
Ruskin—QI
St. Augustine—BW(2), DI(2), ET(2), HI(2), HJ(2), QI, RD, RI, TL
St. Augustine Bch.—FI, HI, RI, SH
St. Petersburg—BW, FI(2), HI(2), HJ(2), SH
St. Petersburg Bch.—HH, HI, HJ
Sanford—DI, HI(2)
Sanibel Island—BW, RI
Sarasota—BW(3), DI, FI, HI, HJ(2), HY, QI, RI, TL
Seffner—DI
Silver Springs—BW, HI, HJ
Singer Island—HH, HI, SH
South Bay—DI
Starke—BW, HI
Stuart—HI, HJ, QI
Tallahassee—BW, DI(2), ET, HH, HI(2), HJ(2), LQ, MH, QI, RI(2), TL
Tamarac—HI
Tampa—AB, BW(2), DI(2), ET, HH, HI(6), HJ(2), LQ, MH, QI(2), RI(3), RO, SH, TL(2)
Tarpon Springs—BW, DI, HJ
Tavares—BW
Titusville—DI, ET, HI, HJ, QI, RI
Treasure Island—HJ, QI, RI, TL
Venice—BW, M6
Vero Beach—DI, HI(2), HJ, M6, SH
Weeki Wachee—HI
West Palm Beach—DI, HI(2), HJ, RI, SH
Wildwood—DI(2), HI
Winter Garden—AB
Winter Haven—BW(2), HI, HJ, QI(2), RI
Winter Park—BW, QI
Yeehaw Jct.—HI
Yulee—BW
Zephyrhills—BW, DI

GEORGIA

Adel—BW, QI
Albany—DT, HI, M6, QI(2), RI, SH

Americus—BW
Arabi—BW
Ashburn—QI
Athens—BW, DI, DT, HI, HJ, RI, SH
Atlanta—AB, BW(2), DI(5), DT, HH, HI(4), HJ(5), HY(2), MM(3), MH, M6, QI(2), RI(3), RO(2), SH(2), TL(3), WI
Atlanta Intl. Arpt.—QI
Augusta—DI, ET, HH, HI, HJ, RI(2)
Austell—RD
Bainbridge—HI
Bremen—HI
Brunswick—BW(2), DI, HI(2), HJ, QI(2), RI(2)
Buford—ST
Byron—DI
Calhoun—BW, DI, HI, RI
Cartersville—BW, DI, HI, HJ, QI, RI
Cecil—QI
Chamblee—DI(2)
Chula—DI
College Park—AB, BW, LQ, MH, MM
Columbus—HI(2), LQ, M6, QI, RI
Commerce—HI, QI
Conyers—HI
Cordele—DI, HI, RI(2)
Dalton—BW, DT, HI, HJ, QI, RO
Decatur—BW, DI, HI, MH, RI, SH
Dillard—BW
Doraville—BW, HI
Douglas—HI
Dublin—HI
East Point—HI, HJ, M6, SH
Folkston—QI
Forest Park—DI(2), RD
Forsyth—BW, DI, HI, HJ. QI, RD, RI
Gainesville—DI, HI
Griffin—DI, HI
Hahira—DI
Hapeville—HH, HI, RI
Hazlehurst—FI
Jekyll Island—HH, HI, RI, SH
Jesup—HI
Jonesboro—HI, TL
La Grange—DI
Lake Park—BW, DI
Lavonia—DI
Locust Grove—FI, HI
Mableton—BW
Macon—BW, DI, FI, HH, HI, HJ(2), M6, QI(2), RI(2), SH
Madison—HI, QI
Marietta—BW, HI, MH, RI
Martinez—HI
McDonough—DI, HI, QI, RD
Milledgeville—DI, FI, HI
Newnan—HI
Perry—HI, HJ, QI, RD
Pinehurst—DI
Richmond Hill—DI, ET, HI, RI
Ringgold—DI
Rome—HI, SH
Savannah—BW(3), DI, DT, HH, HI(2), HJ(2), HY, MH, QI(3), RI, SH, TL
Smyrna—SH
Statesboro--HI, MH, QI
Stockbridge—BW
Suwanee—DI, HI
Sylvania—DI
Thomasville—DI, HI
Tifton—DI, HI, HJ, QI, RI, TL
Townsend—BW, DI
Tucker—HH, SH
Tybee Island—BW, DI
Unadilla—DI
Valdosta—BW, ET, HI, HJ, QI(2), RI, SH
Warner Robins—HI, RI
Waycross—DI, HI, QI

HAWAII

Hawaii Volcanoes Nat. Pk.—SH
Hilo—SH, TL
Honolulu—BW(2), HH(2), HI(3), HY, RI, SH(5), TL, WI
Kaanapali (Lahaina P.O.)—SH
Kahuku—HY
Kailua-Kona—HH
Kamuela (Waimea)—WI
Kepuhi (Maunaloa P.O.)—SH
Kihei—BW
Lahaina—TL
Poipu (Koloa P.O.)—SH
Wailea (Kihei P.O.)—WI
Waipouli (Kapaa P.O.)—HI

IDAHO

Blackfoot—BW
Boise—BW(2), HI, M6, QI, RI, RO, TH(2), TL(2)
Burley—BW, FI
Chubbuck—M6, RO
Coeur d'Alene—BW, FI, HI, M6
Driggs—BW
Idaho Falls—BW(2), FI, M6
Ketchum—BW(2)
Lewiston—BW, M6
Montpelier—BW, FI
Moscow—BW, M6, TL
Nampa—BW
Pocatello—BW(2), FI, HH, HI, IM, TL
Rexburg—BW(2)
St. Anthony—BW
Twin Falls—BW(3), HI, IM, M6, TL

ILLINOIS

Alsip—HI
Altamont—BW
Alton—RI, TL
Arcola—BW
Arlington Hts.—BW, HH
Aurora—BW
Belleville—IM
Benton—HI
Bloomington—FI, HI, HJ, RI
Bradley—HI
Broadview—BW
Burbank—BW
Burr Ridge—QI
Carbondale—HI, RI
Champaign—BW, HI, HJ, RI
Charleston—HI
Chicago—BW, HH(3), HI(5), HY(2), MM(2), WI, RI, RO, SH TL(3), WI
Chicago Hts.—HI
Chicago O'Hare Intl.—HH
Collinsville—BW(2), HI, HJ
Danville—BW, HI, RI, SH
Decatur—HI, SH
De Kalb—HI, M6
Des Plaines—BW, HI, RI, TL
Dolton—FI
East Peoria—HI, M6
Edwardsville—HI
Effingham—BW, DI, HI, RI, TL
Elgin—FI, HI, HJ, RI
Elk Grove—BW, HI, HY
Elmhurst—HI
Evanston—HI
Fairview Hts.—RI
Freeport—HI
Galesburg—HI, HJ, SH, TL
Glen Ellyn—BW, HI
Granite City—TL
Gurnee—HI
Harrisburg—HI
Harvey—HI
Havana—BW
Highland Park—HI

Highwood—RI
Hillside—HI
Homewood—SH
Itasca—HI
Jacksonville—HI, M6
Joliet—HI, HJ, QI
Kankakee—HJ, IM, RI
La Grange—BW(2)
Lansing—HI
La Salle—HI, HJ
Libertyville—BW
Lincoln—HI
Lincolnshire—MM
Lincolnwood—HY
Litchfield—BW
Lombard—FI
Lyons—BW
Macomb—HI, TL
Marion—HI
Matteson—FI
Mattoon—HI
Melrose Park—HI
Moline—HI, HJ, LQ
Morris—HI
Mt. Prospect—HI
Mt. Vernon—BW, HI, TL
Mundelein—HI
Nauvoo—BW
Niles—BW, FI, M6, TL
Normal—M6
North Aurora—HH, HI
Northbrook—HI, SH
North Chicago—HI
Oak Brook—HY, SH, ST
Oakbrook Ter.—HI
Oak Lawn—BW, HI
Olney—HI
Ottawa—RI
Palatine—BW, HJ
Pekin—HI
Peoria—DI, HH, HI, HJ, IM, RI
Peru—M6
Pontiac—BW
Princeton—HI
Quincy—HI, RI, TL
Rantoul—HI(2)
Rock Falls—RI
Rockford—BW, HI, HJ, IM, M6, RI
Rock Island—SH
Rolling Meadows—HI
Rosemont—BW, HI, HY, SH
St. Charles—BW
Salem—HI
Schaumburg—SH
Schiller Park—HI, HJ, QI
Skokie—HH, HI, HJ
South Beloit—BW
Springfield—BW(3), DI, HH, HI(2), HJ(2), M6, RI, SH, TL
Urbana—BW, HI, HJ, M6, TL
Vandalia—BW, HI, TL
Watseka—FI
Waukegan—BW, SH, TL
Willowbrook—HI
Zion—HI

INDIANA

Anderson—HI, M6, SH
Angola—BW, HI
Bloomington—BW, HI, HJ, M6, RI, TL
Clarksville—BW, HI, MM
Cloverdale—HI
Columbus—HI, IM, MH
Connersville—HI
Corydon—BW
Crawfordsville—HI
East Gary—RI
Elkhart—BW(2), DI, HI, RI
Evansville—HI, RI, SH, TL
Fort Wayne—BW, DI, ET, HH, HI(2), HJ, MH, MM, M6(2), RI, SC, TL
French Lick—SH
Gary—SC, SH, TL
Goshen—BW, HI
Greenwood—RI
Hammond—HI(2), HJ(2)
Huntingburg—BW

Indianapolis—BW(4), DI(2), FI, HH(2), HI(7), HJ(4), HY, LQ, MM, M6, QI(2), RI(4), RO(2), SC, SH(2), ST, TL(2)
Jasper—HI
Jeffersonville—DI, HH
Kokomo—BW, DI, HJ, RI, SC
Lafayette—HI(2), HJ(2), RI
La Porte—HI
Lebanon—HI
Logansport—HI
Marion—HI, SH
Merrillville—HI, LQ
Michigan City—HI, HJ, TL
Mishawaka—HI
Muncie—HI, QI, TL
Nashville—RI
New Albany—HI
New Castle—FI, HI
Plymouth—HI, M6
Portage—HI, HJ, M6
Princeton—HI
Remington—BW, DI
Richmond—BW, HI, HJ
Scottsburg—RI
Sellersburg—DI
Seymour—BW, DI, HI
Shelbyville—HI
South Bend—DI, HI, HJ, M6, QI, RI, TL
Tell City—BW
Terre Haute—BW(2), FI, HI, HJ, MH, QI, RI, SH, TL
Valparaiso—HI
Vincennes—HI, TL
Wabash—HI
Warsaw—HI
West Lafayette—TL

IOWA

Adair—BW
Albia—FI
Ames—BW(2), HI, RI, TL
Ankeny—BW
Atlantic—BW(2)
Avoca—FI
Bettendorf—FI, HI
Bloomfield—BW
Boone—BW
Burlington—BW, HI
Carroll—BW
Cedar Falls—BW, HI, HJ, M6
Cedar Rapids—BW(3), HI, HJ, RI, SH, ST
Cherokee—BW
Clear Lake—BW, DI
Clinton—BW, IM, TL
Clive—BW, SH
Coralville—BW(2), FI, HI, M6
Council Bluffs—BW(2), HI, HJ, M6
Davenport—BW, FI, HI, M6, QI, RI
Denison—BW
Des Moines—BW(3), FI, HH, HI(3), HJ(2), HY, MM, M6, RI(2), RO, TL
Dubuque—BW(3), HI
Fort Dodge—BW, HI, TL
Fort Madison—HI
Glenwood—BW
Grinnell—BW
Iowa City—FI, HJ
Keokuk—HI
Marshalltown—BW(2), TL
Mason City—HI, SH, TL
Muscatine—BW, HI
Newton—BW, HI
Okoboji—BW
Osceola—BW
Oskaloosa—BW(2), FI
Ottumwa—HI
Red Oak—HI
Sergeant Bluff—M6
Sioux City—BW, FI, HH, HI, HJ, RO
Spencer—BW
Spirit Lake—FI
Storm Lake—BW

KANSAS

Abilene—BW(2)
Arkansas City—BW
Atchison—BW
Belleville—BW, FI
Beloit—BW
Chanute—BW, HI
Coffeyville—BW
Colby—BW(2), RI
Concordia—BW
Dodge City—BW, TL
El Dorado—BW(2), FI
Emporia—BW, HI, RI, TL
Eureka—BW
Fort Scott—BW
Garden City—BW(3), HH
Goodland—BW, HI, M6
Great Bend—BW, FI, HI
Greensburg—BW, FI
Hays—BW, FI(2), HI, M6
Hiawatha—BW
Holton—BW
Hutchinson—BW, HI, IM, RI
Independence—BW
Iola—BW
Junction City—BW(2), HH, RI
Kansas City—BW(3), HI, TL
Kingman—BW
Lawrence—BW(2), HI, RI, TL
Leavenworth—BW, RI
Lenexa—DI, HI, HJ, LQ
Liberal—BW(2), FI(2), HI, TL
Manhattan—BW, HI, M6, RI
Marysville—BW, FI
McPherson—BW
Medicine Lodge—BW
Merriam—BW
Newton—BW, FI
Norton—BW
Oakley—BW, TL
Olathe—BW
Ottawa—BW, FI
Overland Park—BW, HI, RI, RO
Parsons—BW(2)
Phillipsburg—FI
Pittsburg—BW, HI
Pratt—BW, FI, TL
Russell—RI
Salina—BW(2), FI, HH, HI, QI, RI, TL
Scott City—BW, FI
Seneca—FI
Sharon Springs—FI
Smith Center—BW, FI
Topeka—BW(2), HI(3), HJ, M6(2), RI(2)
WaKeeney—BW, FI, TL
Wellington—HI
Wichita—BW(3), HH, HI(3), HJ, LQ, M6, RI(2), SH, TL

KENTUCKY

Bardstown—BW, HI
Benton—FI
Berea—ET
Bowling Green—BW(2), DI, ET, FI, HI(2), M6, QI, RD, RI, TL
Carrollton—HI
Cave City—BW(2), HI, QI, RI
Corbin—DI, HI, HJ, QI
Covington—HI, QI
Danville—HI

Elizabethtown—BW, DI, FI, HI(2), HJ, M6, RI
Erlanger—BW, HJ
Florence—HI, RI
Ft. Mitchell—DI, HI, RI
Frankfort—DI, HI, QI, TL
Fulton—BW
Georgetown—BW, DI
Gilbertsville—HI
Glasgow—HI
Henderson—HI, RI
Hopkinsville—HI
Horse Cave—RO
Jeffersontown—HI
La Grange—DI
Lexington—BW(2), DI, HH, HI(2), HJ(2), HY, MH, QI, RI(2), SH
London—RI
Lousia—BW
Louisville—AB, HI(7), HJ, HY, LQ, M6, RI, RO, SH, ST, TL
Madisonville—BW, ET, RI
Mayfield—HI
Middlesboro—BW
Middletown—BW
Morehead—HI
Mt. Sterling—DI
Murray—HI
Newport—TL
Owensboro—DI, HI, M6
Paducah—BW, DI, HI, TL
Pikeville—BW, HI
Richmond—BW, DI, HI
Richwood—BW, DI
St. Matthews—HJ
Shepherdsville—BW, DI, QI
Shively—HI
Williamsburg—HI
Winchester—BW, HI, RI

LOUISIANA

Alexandria—BW, FI, HI, HJ, IM, RI, RO, SH, TL
Baton Rouge—AB, BW(3), DI, FI, HH, HI(2), HJ, RI, RO, SH, TL
Bossier City—BW, DI, HH, HI, SH
Covington—BW
Franklin—BW
Gretna—HI, RO, SH
Hammond—HI, RI
Harvey—TL
Houma—BW, HI, RI
Jennings—HI
Kenner—BW, DI, HH, HI, RI, RO, SH, TL
Lafayette—BW, FI, HI(2), HJ, IM, RI, RO, SH, TL(2)
Lake Charles—BW, DI, DT, FI, HI, HJ, IM, SH, TL
Leesville—HI, MH
Metairie—BW, HI, HJ, LQ, RI
Minden—RI
Monroe—HI, HJ, RI, TL
Morgan City—HI
Natchitoches—HI
New Iberia—HI, MH, RI
New Orleans—BW(3), DT, FI, HH, HI(4), HJ(2), HY, LQ, MH(2), MM, QI(3), RI(2), RD, RO, TL(2)
Port Allen—DI, HI, HJ, RI
Ruston—BW, HI
St. Francisville—HI
Shreveport—BW, DI, HI(2), HJ, M6, RD, RI, SH
Slidell—BW, DI, ET, HI, RI
Sulphur—HI
Thibodaux—HI, SH
Vinton—BW
Winnfield—BW
West Monroe—RO

MAINE

Auburn—HI
Augusta—BW, HI, HJ, SC
Bangor—BW, HH, HI(2), HJ, QI, RI, SC

Bath—SH
Brunswick—HI
Ellsworth—HI
Lewiston—RI
Portland—HI(2), HJ, RI, SC
Rumford—FI
S. Portland—BW, QI, SH
Trenton—BW
Waterville—HI, HJ

MARYLAND

Aberdeen—BW, HI, QI
Annapolis—HH, HI, HJ
Annapolis Jct.—QI
Baltimore—HH, HI(4), RI, SH
Beltsville—RI
Bethesda—HI, MM, RI
Bowie—HI
Cambridge—QI
Camp Springs—HI
Catonsville—BW, HI, HJ, QI
Cheverly—HJ
Chevy Chase—HI
Cockeysville—BW
College Park—BW, HI(2), QI
Cumberland—HI
Easton—ET, HI
Edgewood—BW
Elkridge—ET
Frederick—BW, HI, SH
Gaithersburg—HI, MM, QI
Glen Burnie—HI(2)
Hagerstown—BW, FI, HI, QI, RI, SH
Hanover—HJ
Hunt Valley—MM
Jessup—ET, HI
Joppa—FI
Lanham—RI
La Plata—HJ
Laurel—BW, HI, HJ
La Vale—BW
Linthicum—HI
New Carrollton—SH
Ocean City—BW, QI(3), SH
Odenton—RD
Pikesville—HH, HI, QI
Pocomoke City—HI, QI
Rockville—RI, SH
Salisbury—BW, DI, FI, HI, HJ, SH
Silver Hill—HJ
Silver Spring—HI, QI, SH
Towson—HI(2), QI
Waldorf—BW
Wheaton—HJ
Williamsport—DI

MASSACHUSETTS

Amesbury—SC
Andover—SH
Auburn—SH
Bedford—TL
Boston—HH, HI, HJ(3), QI, RI(2), SC, SH
Boxborough—SH
Braintree—HJ, SC, SH
Brockton—HI
Brookline—TL
Burlington—HI, HJ
Cambridge—HI, HJ, HY, SC, SH
Chelmsford—HJ
Chicopee—BW, SC
Concord—HJ
Danvers—HJ, QI
Dedham—HI
Eastham—SH
Falmouth—BW, HI, SH
Framingham—HI, HJ, SH
Gloucester—HH
Great Barrington—FI
Greenfield—HJ, MH
Hadley—HJ
Haverhill—HJ
Holyoke—HI, SC
Hyannis—BW, HI, HJ, QI, SH
Kingston—HJ

Lawrence—HI
Lenox—HI, SC
Leominster—HI, SC
Lexington—SH
Mansfield—SH
Marlborough—HI
Methuen—HJ
Natick—HH, TL
New Ashford—QI
New Bedford—HI
Newton—HI, HJ, MM, SC
Northampton—HH
Norwood—HJ
Peabody—HI
Pittsfield—HH, TL
Provincetown—BW, HI, QI
Randolph—HI
Revere—HJ
Seekonk—HJ, QI, RI, SC
Somerset—HJ
Somerville—HI
South Attleboro—HI
South Deerfield—M6
Springfield—HI, MM, RI, RO, TL
Sturbridge—BW, QI, SH, TR
Teaticket—BW
Tewksbury—HI
Wakefield—BW, HH
Waltham—BW, HI
Wellesley—TR
West Boylston—HJ
Westfield—HJ
Westminster—BW
West Springfield—BW, HJ, SH
Woburn—HI, HJ, RI
Worcester—DI, HI, HJ, SH

MICHIGAN

Adrian—HI
Albion—BW, HI
Allen Park—RI
Alma—TL
Alpena—BW, HI
Ann Arbor—BW, HH, HI(2), HJ, MM, SH
Battle Creek—HI, HJ, RO, TL
Bay City—BW, FI(2), HI, IM
Bellaire—HH
Belleville—HJ
Benton Harbor—HI, HJ, RI
Big Rapids—BW
Bloomfield Hills—HI
Clare—BW
Coldwater—BW
Dearborn—HI, HY, TL
Detroit—HJ, QI, TL, WI
East Lansing—BW
Escanaba—BW
Farmington Hills—HI
Flint—BW(2), HI, HJ, IM, SH
Gaylord—BW, HI
Grand Haven—BW
Grand Rapids—HI, HJ, TL
Grayling—HI
Hancock—BW
Hazel Park—HI
Holland—DI, HI
Houghton—BW
Howell—HI
Inkster—FI
Iron Mountain—BW
Ironwood—BW
Jackson—HI, QI, SH
Kalamazoo—HH, HI(2), HJ, SH, TL
Kentwood—BW(2), HH, HI, HJ, MM, M6
Lansing—BW, HH, HI(2), HJ, M6, RI
Livonia—HI, QI
Ludington—HI
Mackinaw City—BW, FI, RI, TL
Manistee—BW
Manistique—BW, FI, RI
Marquette—BW(2), FI, HI, RI
Marysville—BW
Midland—HI, RI(2)
Monroe—HI, HJ

Mount Clemens—HI
Mount Pleasant—HI
Munising—BW
Muskegon Hts.—DI, HI
Niles—HI
Petoskey—BW, FI, HI
Plymouth—HH
Pontiac—SH
Portage—BW, RI
Port Huron—HI, HJ
Romulus—HH, HI, RI, SH
Roosevelt Park—RI
Saginaw—HI(2), RI
St. Ignace—BW, QI
St. Joseph—BW, HI
Sault Ste. Marie—BW, RI
Southfield—HI, HJ, RI, SH, ST, WI
Spring Lake—HI
Sturgis—FI, HI
Taylor—HI
Three Rivers—FI
Traverse City—BW(2), DI, FI, HH, HI, TL
Trout Lake—BW
Troy—HH(2), HI
Walker—M6
Warren—BW, HI
Woodhaven—SH
Wyoming—HI

MINNESOTA

Albert Lea—BW(2), FI, HI
Alexandria—HI
Anoka—HI
Austin—BW
Bemidji—HI
Bloomington—BW, FI, HI (3), HJ(2), MM, RI, RO, SH
Brainerd—HI
Brooklyn Center—HI
Brooklyn Park—SH
Burnsville—HJ
Cambridge—FI
Cannon Falls—BW
Crookston—BW
Detroit Lakes—BW, HI
Duluth—BW(4), HI
Elk River—BW
Eveleth—HI
Fairmont—HI
Faribault—BW
Fergus Falls—HI
Golden Valley—BW
Grand Marais—BW, FI
Grand Rapids—HI
Hastings—BW
Hibbing—BW
Hinckley—BW
Hopkins—BW
Internatl. Falls—BW, HI
Jackson—BW
Lutsen—BW
Luverne—BW
Mankato—HI(2)
Maplewood—HI
Marshall—BW
Minneapolis—BW(3), HH, HI, IM, SH
Monticello—BW
Moorhead—HI, RI
Morris—BW
New Ulm—BW, HI
Olivia—BW
Owatonna—BW, HI
Plymouth—RA
Redwood Falls—BW
Rochester—BW(4), FI, HI(2), HJ, M6, RI
Roseville—HI
St. Cloud—BW(3), HI
St. Louis Park—BW(2)
St. Paul—BW, HI, RI, TL
Thief R. Falls—BW
Virginia—BW
Wadena—BW
Wilmar—BW, HI
Winona—BW, HI
Woodbury—HJ
Worthington—BW, HI

MISSISSIPPI

Bilox—AB, BW(3), HH, HI, HJ, QI, RI, RO

Brookhaven—BW, HI
Clarksdale—HI
Cleveland—HI
Columbus—DT, FI, HH, HI, RI
Corinth—HI, RI
Forest—RI
Greenville—BW, HI, RI
Greenwood—HI, RI
Grenada—BW, DI, HI
Gulfport—BW, HI, SH
Hattiesburg—BW, DI, HI, HJ, M6, RI
Jackson—AB, BW, DI(2), DT(2), HH, HI(4), HJ, LQ, M6, RI(2), RO, SH, TL(2)
Laurel—HI, MH
Long Beach—RI
McComb—HI
Meridian—DI, HI(2), HJ, RI
Natchez—DI, HI, RI
Oxford—HI, RI
Pearl—DI
Sardis—DI, HI
Senatobia—RI
Starkville—BW, HI, RI
Tupelo—HI, RI, TL
Vicksburg—BW, DT, HI, RI

MISSOURI

Belton—RI
Berkeley—RI
Bethany—BW, FI
Blue Springs—BW
Branson—BW(4), DT, FI(2), HI, QI, RO
Bridgeton—BW(2), HI(2), SH
Cameron—BW
Cape Girardeau—BW, FI, HI, HJ, RI
Chillicothe—BW
Clarksville—BW
Clayton—HI, RI
Clinton—BW
Columbia—BW(2), HH, HI(2), HJ, MH, M6, RI
Concordia—BW
Creve Coeur—HI, RI, SH(2)
Edmundson—MM
Eureka—HI
Flat River—BW
Grandview—BW
Hannibal—BW(2), FI, HI
Harrisonville—BW, FI
Hazelwood—HI, HJ, LQ
Higginsville—BW
Independence—HJ, RI
Jefferson City—BW, HI, RI, RO
Joplin—BW, HI, HJ, QI, RI, SH
Kansas City—BW(4), HH(2), HI(7), HJ, HY, MM, QI, RI(3), SH(2), TL(2)
Kimberling City—BW, DT
Kirksville—BW
Kirkwood—HJ
Lake Ozark—HI, HJ
Lamar—BW
Lebanon—BW(2), FI
Lee's Summit—BW
Lemay—HJ, RI
Lexington—BW
Liberty—BW
Macon—BW
Marston—BW
Mehlville—HI
Mexico—BW
Moberly—RI
Mt. Vernon—BW
Neosho—RI
Nevada—BW
N. Kansas City—RO
Osage Beach—BW(2), MM
Platte City—BW
Poplar Bluff—HI
Rock Port—BW
Rolla—BW, HI, HJ
St. Charles—BW, HJ
St. Joseph—BW(2), HI, RI
St. Louis—BW, FI, HH, HI(2), HJ, MM, QI, RI, RO, SH, ST, TL

Sedalia—BW, RI
Sikeston—BW, ET, HI, RI, TL
Spanish Lake—BW
Springfield—BW(4), FI, HH, HI, HJ, B6, RI, SH, TL
Sullivan—BW, HI
Sunset Hills—BW, HI
Villa Ridge—BW
Warrensburg—BW
Waynesville—BW, FI, RI
Wentzville—BW
West Plains—HI, RI
Woodson Terrace—DI, HH, M6, RO

MONTANA

Bigfork—BW
Billings—BW(2), FI(2), HI(2), IM, M6, RI, SH, TL
Bozeman—BW, FI, HI, IM, RI, TL
Butte—BW, M6, RI, TL
Cooke City—FI
Deer Lodge—BW
Dillon—BW
Forsyth—BW
Glendive—BW, FI
Great Falls—BW(2), HI, IM, SH
Hamilton—BW
Hardin—FI
Havre—FI
Helena—BW(2), IM, M6, TL
Kalispell—BW, M6, TH
Laurel—BW
Lewistown—BW
Libby—BW
Livingston—BW, FI
Miles City—BW, M6
Missoula—BW(3), HI, TH(2), TL
Polson—BW
Red Lodge—BW, FI
W. Yellowstone—BW(3), FI(2)

NEBRASKA

Alliance—BW
Auburn—FI
Aurora—BW
Beatrice—BW, FI
Bellevue—BW
Big Springs—BW
Chadron—BW, FI
Columbus—FI, HI
Cozad—BW
Fremont—BW(2)
Geneva—FI
Grand Island—BW, FI, HI(2), RI, TL
Hastings—BW, FI, HI
Henderson—BW
Holdrege—FI
Kearney—BW(2), FI(2), HI, RI
Kimball—BW, FI
Lexington—BW(2), FI
Lincoln—BW(3), DI, FI(3), HH, HI(2), M6, RI
McCook—BW
Norfolk—BW, FI
North Platte—BW(2), FI(2), HI, HJ, M6, RI, TL
Ogallala—FI, HI ,RI
Omaha—BW(7), FI, HI(2), HJ, IM, M6, RI(2), RO, SH, TH, TL
O'Neill—BW
Scottsbluff—BW
Sidney—FI
South Sioux City—BW, FI, TL
Valentine—BW, FI
York—BW(2)

NEVADA

Battle Mountain—BW
Boulder City—FI

131

Carson City—BW(3), M6, TL(2)
Elko—BW(3), FI, HI, TH
Ely—BW(2), M6
Fallon—BW, FI, TL
Hawthorne—BW, FI
Henderson—BW
Incline Village—HY
Las Vegas—BW(8), DT, FI(3), HH(2), HI(3), IM, LQ, M6, RO(2), TL(6)
Laughlin—BW
Reno—BW(5), FI(3), HI(2), M6(3), RI, TL
Sparks—FI, QI
Tonopah—BW(2), FI
Wells—BW, FI, M6
Winnemucca—BW(2), M6, TH

NEW HAMPSHIRE

Campton—HI
Concord—HJ, RI
Dover—RI
Exeter—BW
Franconia—BW
Hampton—SH
Keene—BW, RI
Littleton—MH
Manchester—HI, HJ, SH
Merrimack—HH
Nashua—HI, HJ, SC, SH
North Conway—BW
Portsmouth—HI, HJ
Seabrook—BW
West Lebanon—SH

NEW JERSEY

Atlantic City—BW
Bellmawr—BW
Bordentown—HI, HJ, QI, SH
Bridgewater—HI
Burlington—BW
Cape May—BW
Carteret—HI
Cherry Hill—BW, HI, HY, SH
Clark—HJ, RI
Clifton—HJ, RI
Cranford—BW
Deepwater—HI
East Brunswick—RI, SH
East Hanover—RI, RO
East Orange—HI
East Windsor—HH
Edison—HI, RI
Elizabeth—HI, SH
Fort Lee—HI
Freehold—SH
Gloucester City—HI
Hasbrouck Hts.—QI, SH
Hazlet—SH
Jersey City—HI
Kenilworth—HI
Lakewood—BW, HI
Ledgewood—HI
Linden—FI
Livingston—HI
Lyndhurst—HI
Middletown—HJ
Montvale—RI
Moorestown—QI
Morristown—BW
Mount Holly—FI, HJ
Mt. Laurel—HH, TL
Neptune—HJ
Newark—HH, HI, HJ
New Brunswick—HJ
North Bergen—HI
North Brunswick—HI
North Plainfield—HJ
Paramus—HI(2), HJ, RD
Parsippany—HI, HJ
Penns Grove—HJ
Phillipsburg—HI, HJ
Piscataway—SH
Princeton—HI, TR
Ramsey—HI, HJ
Ridgefield Park—HJ
Rochelle Park—RI
Rockaway—HJ
Runnemede—RI

Saddle Brook—HI, HJ, MM
Secaucus—HH, HJ
Somerset—MM, TL
South Plainfield—HI, HJ
Springfield—HJ
Tinton Falls—HH
Toms River—HI, HJ, TL
Trenton—HJ
Vineland—HI, IM
Wayne—HI, HJ, RI
Westfield—BW
W. Long Beach—HI
Whippany—HJ
Wildwood Crest—HI
Woodcliff—HH

NEW MEXICO

Alamogordo—BW, FI, HI, TL
Albuquerque—BW(6), DI, FI(4), HH, HI, HJ(2), IM, LQ, M6(2), RI(2), RO, SH, TL(2)
Carlsbad—BW, HI, M6, RO, TL
Chama—BW
Cimarron—BW
Clovis—HI, M6, RO
Deming—BW, HI, M6, RI
Española—FI
Farmington—BW, HI, M6, TL
Gallup—BW(2), RI(2), HI, M6, RI, TL(2)
Grants—BW, FI, HI, M6, TL
Hobbs—BW, FI, HI, M6, RI
Las Cruces—BW(2), FI, HI, HJ, M6, QI, RO, TL
Las Vegas—BW, FI
Lordsburg—BW, FI
Raton—BW, FI(2), HI, MH, M6
Rociada—BW
Roswell—BW(3), RI, TL(2)
Ruidoso—BW, HI
Santa Fe—BW(2), DT, FI, HH, HI, M6, RI, RO, SH, TL
Santa Rosa—BW, FI(2), HI, M6, TL
Socorro—BW
Taos—BW, HI
Truth or Consequences FI
Tucumcari—BW(2), FI, HI, RI, TL
Vaughn—BW
Whites City—BW

NEW YORK

Albany—BW(2), HJ, QI, RI, SH, TL
Amherst—HI
Amsterdam—HI
Armonk—RI
Auburn—HI, TL
Batavia—HI, TR
Bath—RI
Binghamton—BW, HI(2), HJ, QI, RI
Boonville—BW
Buffalo—BW(2), HH, HI, TL
Canandaigua—SH
Centereach—HI
Cheektowaga—HI(2), HJ(2), SH
Colonie—BW, HI, MH, SC, SH
Commack—HJ
Corning—HH
Cortland—HI, IM
Dunkirk—QI
East Elmhurst—HI, SH
East Syracuse—HI, HJ, MM, RO
Elmira—HI
Elmsford—HI, HJ, RI
Endicott—BW
Fishkill—HI
Flushing—BW, FI
Fredonia—HI
Geneseo—HI

Glens Falls—FI(2), HJ, SC, SH
Grand Island—HI
Hamburg—HI
Harrison—ST
Happauge—HI, RI
Hawthorne—QI
Hempstead—HI
Henrietta—BW, MM, QI
Horseheads—HI, HJ, QI
Huntington Sta.—HJ
Ithaca—HI, HJ, RI, SH
Jamaica—HH, HI, HJ, TL
Jamestown—HI
Jericho—HJ
Johnson City—BW
Johnstown—HI
Kingston—HI, HJ, RI
Lake George—BW, HI, RI
Lake Placid—BW, FI, HH, HI, HJ, RI, SH
Latham—HI, HJ
Liberty—HI, HJ
Little Falls—BW
Liverpool—SH
Lockport—SH
Massapequa Pk.—BW
Menands—HI
Middletown—HI, HJ
Mount Kisco—HI
Nanuet—SH
Newark—DT
Newburgh—HI, HJ, IM, RI
New Hartford—RI, TR
New Rochelle—SH
New York—BW(2), HH(3), HI, HJ(2), HY(2), LO(3), MM, QI, RI, SH(5), WI
Niagara Falls—BW, HI, HI, HJ(2), MH, RI, TL
North Syracuse—BW, HI, HJ
Norwich—HJ
Ogdensburg—QI
Olean—BW, HI
Oneonta—HI
Orangeburg—HI
Oriskany—RI
Owego—TR
Painted Post—BW, HI
Plainview—HI, HJ
Plattsburgh—ET, HI, HJ
Port Jervis—HI
Poughkeepsie—BW, HI
Riverhead—BW, HI
Rochester—DT, HH, HI(2), HJ, MM, RI, SH, TL(3)
Rockville Cen.—HI
Rome—BW, HI
Rye—HH
Saratoga Sprs.—BW, HI
Saugerties—HJ
Schenectady—HI, IM, RI, TL
Spring Valley—HI, HJ
Staten Island—HI
Suffern—HI
Syracuse—BW, HH, HI(2), RI, TL(2), TR
Tarrytown—HH, MM
Troy—HI
Uniondale—SH
Utica—HJ, QI, SH, TL
Vestal—HJ
Waterloo—HI
Watertown—HI, HJ, RI
Weedsport—BW
Westbury—HI
Williamsville—HJ
Woodbury—QI, RI
Yonkers—HI

NORTH CAROLINA

Apex—RI
Asheboro—ET, HI
Asheville—BW, DI, DT, ET(2), HH, HI(3), HJ, QI, RI, RO, SH
Atlantic Beach—RI
Banner Elk—HI
Battleboro—BW, DI, ET, HI, IM, QI, RI
Benson—BW, DI
Boone—HI, RI
Burlington—BW, HI, RI

Candler—DI
Chapel Hill—HI
Charlotte—BW(3), DI(3), DT, ET(2), HI(5), HJ(3), M6(2), QI(2), RI(3). RO(2), SH
Cherokee—HI
Clemmons—RI
Concord—DI, HI, QI
Dunn—HI, HJ, RI
Durham—BW, DI, DT, ET, HH, HI, HJ, M6, RI(2)
Elizabeth City—HI
Fayetteville—BW, DT, HI, M6, QI(2), RI, SH
Fletcher—DI
Franklin—QI
Gastonia—DI, HI, HJ, RI
Goldsboro—DI, ET, HI, M6, QI
Greensboro—AB, BW, DI, DT, ET, HH(2), HI(2), HJ(3), M6, QI(2), RI, TL
Greenville—ET, HI, RI
Henderson—DI, HI, HJ
Hendersonville—HI, RI
Hickory—HI, HJ, RI, SH
High Point—HI, HJ, M6, TL
Jacksonville—HI, MH, RD
Kernersville—QI
Kill Devil Hills—HI, QI
Kinston—ET, HI, HJ
Laurinburg—HI
Lenoir—HI
Lexington—HI
Lumberton—BW, DI, HI, HJ, M6, RI, SC
Monroe—FI, HI
Morehead City—HI
Morganton—DI, HI
Nags Head—QI
New Bern—FI, HI, QI, RI
Oxford—HI
Pilot Mountain—HI
Raleigh—BW, DI, DT, ET, HH, HI(2), HJ(2), QI, RI, SH(2)
Reidsville—HI
Roanoke Rapids—HI, HJ, QI
Rocky Mount—HI, HJ, QI(2)
Rowland—DI
Salisbury—DI, ET, HI, HJ
Sanford—ET,HI
Selma—DI, ET, HI
Shelby—HI
Smithfield—HJ
Southern Pines—HI, HJ, QI, SH
Statesville—DI, HI, MH, RI
Wade—DI
Washington—ET, HI
Waynesville—HI
Weldon—ET
Wilkesboro—HI
Williamston—HI
Wilmington—BW, DI, ET, HH, HI, M6, QI, RI, RO
Wilson—HI, QI
Winston-Salem—HH, HI(3), HJ, HY, M6, RI(2), SH
Wrightsville Bch.—HI

NORTH DAKOTA

Belfield—BW
Beulah—BW
Bismarck—BW(3), FI, HI, M6, RI
Bowman—BW
Devils Lake—BW, M6
Dickinson—BW(2), FI(2), HI
Fargo—BW(2), HI, M6, QI
Grand Forks—BW(2), HI, RI, TL
Jamestown—BW, FI, HI, M6, RI
Lakota—BW
Mandan—BW
Minot—BW, HI, RI
Williston—BW(2), M6, RI

OHIO

Akron—HH, HI(3), IM, RI
Aurora—SH, TR
Austinburg—HI, TL
Austintown—DI, HI
Beachwood—HI, MM, SH
Bedford Heights—RI
Bellefontaine—HI
Bellevue—BW
Boston Heights—HI
Botkins—BW
Bowling Green—BW, HI, HJ
Brook Park—HH, HI, HJ, RI, SH
Bucyrus—HI
Cambridge—BW, HI, TL
Canton—HI, SH, TL
Chillicothe—HI
Cincinnati—BW, HH(2), HI, MH, ST, TL, TR, WI
Cleveland—HI(2), HJ, MM, SH, ST, TL(3)
Columbus—BW, HH(2), HI(5), HJ(3), HY, IM, MH(2), MM(2), QI, RI, RO, SH(3), ST, TL
Conneaut—RI
Cuyahoga Falls—TL
Dayton—DI, HI(2), HJ, LQ(2), MH, RI(2), SH, ST
Defiance—HI
Delaware—HI
Elyria—BW, HI, TL
Englewood—HI
Euclid—HJ, SH
Fairborn—HI
Fairfield—HI
Fairlawn—HH, HI, RI
Findlay—ET, HI, MH
Fremont—HI, RI
Gallipolis—BW, ET, HI
Geneva—HJ
Girard—DI, ET, HI, M6
Grove City—DI, ET, HI, HJ, RI
Independence—HI
Jeffersonville—DI
Kent—FI, HI
Lancaster—HI
Lebanon—BW
Lemoyne—HI
Lima—BW, DI, HJ, RI, SC, TL
Mansfield—BW, FI, TL, QI
Marietta—BW, HI, RI
Marion—IM, SC
Mason—BW(2), DI, HI
Massillon—HI, TL
Maumee—HI, HJ
Mayfield—HI
Medina—DI, HI
Miamisburg—DI, MH
Middleburg Hts.—HI, RI
Middletown—HI, HJ
Monroe—DI
Montgomery—RI
Montpelier—FI, HI
Moraine—DI, HI, Q
Napoleon—HI
Newark—HI, HJ, SH
New Philadelphia—HI, M6
Niles—BW
North Canton—HI, MH
North Jackson—HI
North Randall—HI, HJ
Northwood—QI
Norwich—FI
Norwood—QI
Perrysburg—DI, HI(2), RI
Portsmouth—HI, RI
Reynoldsburg—HI, LQ, RI
Richfield—HI
Rocky River—SH
St. Clairsville—BW, FI, HI, SH
Sandusky—BW(2), DI, HI, RI
Shaker Heights—ST
Sharonville—BW, HH, HI, HJ, LQ, MM, RI
Sidney—DI, HI, MH
South Point—BW, HI, RI
Springdale—BW, RO
Springfield—BW, HI(2), RI, TL

Steubenville—HI
Strongsville—FI, HI, HJ, QI
Toledo—HI(2), RI, SH
Troy—HI, MH
Van Wert—HI
Wapakoneta—HI
Warren—ST, TL
Wauseon—BW
West Chester—BW
Wheelerburg—DI
Whitehall—HI
Wickliffe—HI, QI, RI
Willoughby—SH
Wooster—RI
Worthington—HH
Youngstown—HJ, QI(2), RI
Zanesville—BW, HI, HJ, QI, TL

OKLAHOMA

Ada—HI
Alva—FI
Anadarko—BW
Ardmore—BW, HI, RI
Bartlesville—BW, HI, RI
Blackwell—FI
Calumet—BW
Chickasha—BW, FI
Claremore—BW
Clinton—BW, FI, HI, RI, TL
Duncan—HI
Durant—HI
Elk City—BW(2), FI, QI, RI
El Reno—BW, FI
Enid—HI, RI
Erick—BW
Guymon—BW(2), FI
Henryetta—HI
Idabel—HI
Lawton—BW, FI, HI, RI
McAlester—BW, FI, HI
Miami—BW, FI
Midwest City—HI
Moore—BW
Muskogee—BW, FI, HI
Norman—HI, HJ, RI
Oklahoma City—BW(4), DI(2), DT, FI, HH(2), HI(4), HJ(2), LQ, M6, QI(2), RD, RI(4), SH(2), TL
Okmulgee—FI
Pauls Valley—BW
Perry—BW
Ponca City—BW(2), HI
Seminole—RI
Shawnee—BW(2), HI
Stillwater—HI
Tulsa—BW(3), DI, DT(2), HH, HI(3), HJ, LQ, QI(2), RD, RI, SH(2), TL, WI
Weatherford—BW
Woodward—BW

OREGON

Albany—BW, FI
Ashland—BW
Astoria—TH
Baker—BW
Beaverton—BW, QI
Bend—BW, FI, TH(2)
Biggs—BW, FI
Boardman—BW, FI
Brookings—BW, FI
Burns—BW
Cannon Beach—BW
Coos Bay—BW, TH
Corvallis—BW
Eugene—BW(2), HI, M6, QI, TH, TL
Gladstone—FI
Gold Beach—BW, FI
Grants Pass—BW, FI, M6, TL
Halsey—BW
Klamath Falls—FI, M6, TH, TL
La Grande—BW, FI
Lake Oswego—BW, M6
Lakeview—BW, FI
Lincoln City—FI
Madras—BW, FI

McNary—BW
Medford—BW(2), FI, HI, M6, RO, TH
Newport—BW, HH
Ontario—BW(2), FI, M6
Pendleton—BW, IM, M6
Portland—BW(4), FI(3), HH, HI(2), IM, MM, M6, RI, RO, SH(2), TH(4), TL(2), WI
Prineville—BW
Roseburg—BW(4), FI, TL
Salem—BW(2), FI, HI, M6, RO, TL
Seaside—BW(2)
Springfield—RI, RO
Sutherlin—FI
The Dalles—BW
Tigard—BW, RO
Tillamook—BW
Tualatin—RI, TL
Welches—TH
Wilsonville—HI

PENNSYLVANIA

Allentown—DI, HI, QI, SH
Allison Park—RI
Altoona—BW, HI, SH
Bartonsville—HI
Beaver Falls—BW, HI
Bedford—HI, QI
Belle Vernon—HI
Berkeley Hills—SH
Bethlehem—HI, HJ
Bird-in-Hand—FI
Bloomsburg—QI
Bradford—HI
Breezewood—BW, HI, HJ, QI, RI
Bristol—IM
Brookville—HI, HJ
Burnham—HI
Butler—BW, HI
Carlisle—BW, HJ, QI
Chambersburg—BW, HI, HJ, TL
Chester—HJ
Clarion—HI, SH
Clarks Summit—RI
Clearfield—HI
Concordville—RI
Connellsville—FI
Coraopolis—HH, HI, HJ, RI, SH
Cornwells Hts.—HI
Danville—HI, HJ, SH
Del. Water Gap—HJ
Denver—HI, HJ
Dickson City—TR
Downingtown—RI
Du Bois—BW, HI, SH
Dunmore—HI
Easton—SH
E. Stroudsburg—HI
Ebensburg—FI
Edinboro—HI
Ephrata—BW
Erie—HH, HI(2), HJ, QI(2), RI, TL
Essington—HI, RI
Fairview—BW
Ft. Washington—HI, RI
Gettysburg—BW(2), FI, HI, HJ, QI(2), SH, TL
Gibsonia—BW, HJ
Grantville—TR
Greensburg—FI, SH
Green Tree—BW, HI, MM
Harmarville—HI
Harrisburg—BW, DI, ET, FI, HJ, MM, QI, SH(2), TL
Hazleton—BW, HI
Hershey—BW
Huntingdon—BW, HI
Indiana—HI, SH
Intercourse—BW
Irwin—BW, HI
Johnstown—HI, SH
King of Prussia—HH, HI, HJ, SH, ST
Kittanning—FI, QI
Kulpsville—HI
Lamar—HI
Lancaster—HI(2), HJ, QI, SH, TL, TR

Lebanon—TR
Levittown—HI
Ligonier—HI
Linglestown—BW
Lionville—HI
Mars—BW, M6
McKeesport—SH
Meadville—BW, HI
Mechanicsburg—BW
Media—HI
Mercer—HJ
Milesburg—SH
Monroeville—BW, HI, HJ, MM
Morgantown—ET
New Cumberland—QI
New Hope—HI
New Kensington—HI
New Stanton—BW, HI, HJ, QI
North East—BW
North Versailles—QI
Oakdale—HI
Oil City—HI
Paradise—BW
Philadelphia—BW, HH(2), HI(5), HJ, MM, SH(3), TR)2)
Pittsburgh—BW, HH, HI, HJ, HY, QI
Pittston—HJ
Pleasant Hills—HJ
Pottstown—HI
Quakertown—BW
Reading—ET, HI
Sayre—BW
Scranton—HI, SH
Sewickley—HI
Shippensburg—BW
Somerset—HI, RI
State College—BW, HI, IM, SH
Stroudsburg—HH, SH
Summerdale—QI
Sunbury—HI
Tannersville—BW
Trevose—HH
Uniontown—HI
Upper St. Clair—SH
Warren—HI
Warrendale—HI
Washington—HH, HI, HJ, QI, RI
Waynesboro—BW
West Chester—TR
West Middlesex—HI, SH
West Mifflin—HI
West Reading—FI
White Haven—HI, HJ
Wikes-Barre—BW, HI, MH, TR
Wilkinsburg—HI
Williamsport—HI, RI
Wyomissing—SH
York—HI(2), HJ, QI, RI, TL

RHODE ISLAND

Charlestown—FI
Cranston—HH·
Middletown—HJ
Newport—SH, TR
North Kingstown—BW
Pawtucket—HJ
Portsmouth—RI
Providence—HI, MM
S. Kingstown—HI
Warwick—HJ, SH

SOUTH CAROLINA

Aiken—HI, RI
Allendale—QI
Anderson—BW, DI, HI, HJ
Beaufort—RI
Cayce—HJ
Charleston—BW(2), ET, HI(3), M6, SH
Clemson—HI
Clinton—HI
Columbia—BW, DT, ET, HI, HJ, LQ, MH, QI
Dentsville—DI
Dillon—DI, HI, HJ

Florence—DI, DT, HI(2), HJ, QI, RD, RI, SH
Fort Mill—HI
Gaffney—HI
Georgetown—BW, HI
Greenville—BW, DI, ET, HI(2), HJ(2), QI(2), RI, RO, SH(2)
Greenwood—HI
Hardeeville—DI, ET, HI, HJ, RI
Hilton Head I.—HI, HY, MM
Lugoff—HI
Manning—DI, HJ, QI
Myrtle Beach—BW(3), DT, HH, HI, HJ, QI(2), SH
Newberry—BW
N. Charleston—BW(2), DI, ET(2), HI, HJ, MH, RI, SH
N. Myrtle Beach—ET, HI
Orangeburg—BW, QI, RD
Pawleys Island—QI
Ridgeland—BW
Rock Hill—HI, HJ, RI
St. George—HI
Santee—BW, DI, HI, QI, RI, SH
Spartanburg—BW, HI, HJ, RI, SH, TL
Summerton—BW, HI
Sumter—HI
Surfside Beach—HI
Turbeville—HI
Walterboro—BW, ET, HI, HJ, RI
West Columbia—BW, DI, ET, HI, RI
Yemassee—HI

SOUTH DAKOTA

Aberdeen—HI, SH
Belle Fourche—BW
Brookings—BW, FI, HI
Canistota—BW
Chamberlain—BW, FI
Custer—BW, FI(2)
Deadwood—BW, FI
Hill City—BW
Hot Springs—BW, FI
Kadoka—BW, FI
Keystone—BW, FI
Mitchell—BW, FI, HI, M6
Murdo—BW, FI
Pierre—BW, HI, M6
Rapid City—BW(2), FI(2), HI, HJ, M6, RI, TL
Rockerville—FI
Sioux Falls—BW(3), FI(2), HI(2), HJ, M6, RI, TL
Spearfish—BW(2), FI, HI
Sturgis—BW
Vermillion—BW
Wall—BW, FI, MH
Watertown—BW, HI

TENNESSEE

Alcoa—HI, QI
Athens—SH
Bristol—BW, HI(2)
Camden—DI
Caryville—HI, QI
Chattanooga—AB, BW(2), DT, ET, HH, HI(2), HJ, QI, RI, SH
Clarksville—DT, HI, M6
Cleveland—FI, HI(2), QI, RI, TL
Columbia—HI
Concord—DI
Cookeville—BW, DI, HI, HJ, RD
Crossville—HI
Cumberland Gap—HI
Dickson—HI
Dyersburg—BW, HI
East Ridge—DI, HI, HJ, QI(2), RI, RO, SH
Franklin—BW(2), HI
Gatlinburg—BW(4), DT, ET, HI, HJ, MH, QI, RI, RO, SH, TL
Goodlettsville—DT
Greeneville—ET, HI
Harriman—HI

Hermitage—DI
Hurricane Mills—BW, QI
Jackson—DI, FI, HI, QI, RI
Jellico—BW, DI, QI
Johnson City—HI
Kingsport—BW, ET, HI, RI
Kingston—BW
Knoxville—BW(2), ET, HI(4), HJ(3), HY, RI, RO, SH(2)
Lebanon—BW, DI, HI
Lenoir City—RD
Manchester—BW, DI, HI, QI
Maryville—BW
Memphis—AB(3), BW(4), DI(3), DT, HH, HI, HJ, HY, LQ, MH, QI(3), RI(3), RO, SH(3), TL(4)
Milan—HI
Monteagle—HI
Morristown—HI
Murfreesboro—BW, DI, HI, HJ, B6, QI, RI
Nashville—BW(6), DI(3), DT, HH, HI(5), HJ(2), HY, LQ, MH(2), M6, QI(2), RI(4), RO, SH(2), TL
Newport—BW, HI
Oak Ridge—BW, HI
Pigeon Forge—BW(2), RD, RI
Pulaski—BW
Sevierville—BW, TL
Sweetwater—QI, RI
Tiftonia—HI
Wildersville—BW

TEXAS

Abilene—BW(2), FI, HI, LQ, M6, QI, RI, SH
Alice—HI, RO
Alpine—RI
Alvin—HI
Amarillo—BW(3), FI(2), HH, HI, HJ(2), M6, QI, RI, RO, TL(3)
Arlington—BW(2), DI, FI, HI, M6, QI, RI, RO
Atlanta—BW
Austin—BW(4), FI, HH, HI(2), HJ, IM, LQ(3), MM, M6(2), QI, RI(4), RO(2), SH, TL(2)
Bay City—BW
Baytown—HI, RI
Beaumont—BW, DI, HI, HJ, LQ, M6, RD, RI, RO, SH
Bedford—BW
Belton—BW
Big Spring—BW, HI, RI
Brookshire—DI
Brownsville—BW, DT, HI, LQ, M6, QI, RI, RO
Brownwood—HI
Bryan—HI, RO
Cameron—FI
Carrollton—HI
Carthage—BW
Center—BW
Childress—BW
Cleburne—FI
Clute—LQ, M6
College Station—HI, LQ, M6, RI
Colorado City—BW
Columbus—FI, HI
Gonroe—HI
Conway—FI
Corpus Christi—BW, HH, HI, LQ(2), M6, QI(2), RI, SH, TL
Corsicana—HI
Dalhart—BW(2), FI
Dallas—BW(6), DI(3), HH(3), HI(5), HJ(3), HY, LO, LQ(5), MM(2), M6(2), QI, RD, RI(2), RO(3), SH(4), TL(3)
Del Rio—BW(2), FI, HI, M6, RI, RO
Denison—FI, HI
Denton—HI, LQ, M6, RI
Dumas—BW

Duncanville—HI
Eastland—RI
El Paso—BW(3), FI(2), HH, HI(3), HJ, IM, LQ, M6(2), QI, RI, RO, SH, TL(3)
Ennis—BW
Euless—RO
Farmers Branch—DI, LQ, RI
Fort Stockton—BW, FI, HI, M6, RI
Fort Worth—BW(5), DI, FI, HH, HI, HJ, HY, M6(3), RI(2), RO, TL
Fredericksburg—BW
Galveston—HI, LQ, MM, M6
Garland—DI, FI, HI, LQ
Gatesville—BW
Grand Prairie—FI, LQ, SH
Greenville—HI, RI
Groves—HI
Haltom City—M6
Harlingen—HI, M6, RI, RO, SH
Hearne—BW
Henderson—FI, HI
Hereford—BW
Hillsboro—BW
Houston—BW(3), DI(2), HH(6), HI(12), HJ(4), HY, LQ(7), MM(4), M6, QI(2), RD(6), RI(6), RO(3), SH(4), ST, TL(2), WI
Humble—RO
Huntsville—BW, HI, M6
Hurst—HI
Hutchins—BW
Irving—BW, DI(2), HI(3), LQ, QI, RI
Jacksonville—HI
Kerrville—BW, FI, HI
Killeen—BW, HI, LQ
Kingsville—BW, HI, RO
Lake Jackson—HH, HI, HI
Lamesa—FI
Lampasas—FI
Laredo—BW(2), HH, HI, LQ, M6, RI, RO, SH, TL
Lewisville—HI, RI
Livingston—HI
Longview—BW, HI, IM, M6, RI, TL
Lubbock—BW(4), FI(2), HH, HI, HJ, LQ, M6, QI, RI, RO, TL
Lufkin—HI, RI
Marshall—HI, RI
McAllen—HH, HI(2), LQ, M6, RO, SH
McKinney—FI
Memphis—BW
Mercedes—RO
Mesquite—BW, DI, HI, M6
Midland—BW, HH, HI(2), M6, RO, SH
Mineral Wells—HI
Monahans—BW
Mt. Pleasant—BW, FI, HI
Nacogdoches—HI, QI
Nederland—BW
New Braunfels—BW, HI
N. Richland Hills—LQ
Odessa—FI(2), HI, M6, RI, RO, TL
Orange—BW, HI, RI
Ozona—BW
Palestine—HI
Pampa—BW
Paris—BW, QI, RI
Pasadena—RI, RO
Pecos—BW, FI, HI, M6
Pharr—QI
Plainview—HI, TL
Plano—LQ
Port Arthur—IM, MH
Port Lavaca—BW
Rio Grande City—BW
Rosenberg—HI
Round Rock—BW
San Angelo—BW(2), HI, LQ, M6, RO
San Antonio—BW(6), DI(2), FI, HH, HI(3), HJ, LQ(7), MH, MM, M6(2), RI(2), RO(4), SH, TL(4)
San Augustine—BW

San Marcos—HI, M6, RI
Seguin—BW, HI
Shamrock—BW(2), FI, TL
Sherman—HI, RI, RO
Snyder—FI, TL
Sonora—FI
South Padre I.—HH, HI
Spearman—BW
Stephenville—BW
Sulphur Springs—RI
Sweetwater—HI
Temple—BW(2), HI, M6, RI
Terrell—BW
Texarkana—FI, HJ, M6, SH
Texas City—HI, LQ
Tyler—BW(2), HI, M6, RI, RO, SH
Uvalde—BW, RI
Van Horn—BW(2), HI, RI, RO
Vega—BW
Vernon—BW(2), RI
Victoria—BW, HI, M6, RI, RO
Waco—BW(2), HI, LQ, QI, RI, SH
Webster—DI
Weslaco—FI
Westfield—HI
White Settlement—DI
Wichita Falls—BW(2), HI(2), LQ, M6

UTAH

Beaver—BW, TL
Blanding—BW
Brigham City—BW
Bryce Canyon—BW, FI
Cedar City—BW(2), FI, IM, RO, TL(2)
Fillmore—BW, FI(2)
Green River—BW, FI, M6
Hatch—BW, FI
Heber City—BW, FI(2)
Hurricane—BW
Kanab—BW, FI
Logan—BW(2)
Midvale—BW, LQ, RO
Moab—BW, FI(2), RI, TL
Monticello—BW, FI
Mt. Carmel Jct.—BW, FI
Murray—QI
Nephi—BW, FI
Ogden—BW(2), FI, HI, M6, RI, TL
Orem—TL
Panguitch—BW, FI(2)
Park City—HI
Parowan—BW
Price—BW, FI
Provo—BW(2), FI, HI, IM, M6, QI, RO, TL
Richfield—BW(2), FI, RO
Roosevelt—FI
St. George—BW(3), FI(2), HH, M6, RO, TL(2)
Salina—BW, FI, QI
Salt Lake City—BW(2), FI(2), HH(3), HI(2), HJ, IM, MM, M6(2), RI, RO, TL(4)
Springdale—BW
Vernal—BW(2), FI
Wendover—BW, M6

VERMONT

Bennington—BW, RI
Brattleboro—HI, SC
Jeffersonville—BW
Killington—FI
Newport—BW
Rutland—HI, HJ, TL
S. Burlington—BW, ET, HI, HJ, RI, SH
Springfield—HJ
Waterbury—HI
White R. Jct.—HI, HJ, SC

VIRGINIA

Abingdon—MH
Alexandria—BW, DI, ET, HI(3), HJ, IM, QI, RI(2)

Arlington—BW(2), ET, HI(3), HJ, HY, IM, MM(3), QI(3), SH, ST
Ashland—BW, DI, ET, HI
Blacksburg—ET, HI, MM, SH
Bristol—ET, HI, HJ
Cape Charles—HI
Carmel Church—DI, RI
Charlottesville—BW(2), ET(2), HI(2), HJ, RI, SH
Chesapeake—ET(3)
Chester—DI, HI, HJ
Chilhowie—BW
Christiansburg—DI, ET
Cloverdale—FI
Collinsville—ET, HI
Colonial Hts.—DI
Covington—HI
Culpeper—ET, HI
Danville—BW, DT, ET, HI
Doswell—BW
Dulles Intl. Airport—MM
Dumfries—ET, HI
Emporia—DI, HI, QI(2)
Fairfax—HI, QI
Falls Church—BW(2), MM, QI, RI
Franklin—BW
Fredericksburg—BW(2), DI, ET(2), HI(2), HJ, RI, SH
Front Royal—QI
Hampton—DI, ET(2), HI, QI, SH
Harrisonburg—ET, HI, HJ, RD, SH
Hollins—HJ
Leesburg—QI
Lexington—BW, DI, ET, FI, HI, HJ
Luray—HI
Lynchburg—ET, HH, HI, RI, SH
Madison Hts.—HI, HJ
Manassas—HI, RI
Marion—HI, QI
McKenney—FI
McLean—HI
Mt. Jackson—BW
New Market—QI
Newport News—BW, ET(2), HI, RI
Norfolk—BW, ET(5), HI(3), QI(2), RI(2), SH
Petersburg—BW, HI(2), HJ, QI, RI
Portsmouth—HI, IM, QI
Pulaski—RD
Reston—SH
Richmond—BW, DT, ET(2), HI(8), HJ(2), HY, QI(3), RI
Ridgeway—TL
Roanoke—BW, DI, ET(3), HI(4), RI, TL
Salem—ET, HI, SH
Sandston—DI, ET, HI, RI
Skippers—BW
South Hill—ET, HI
Springfield—BW, HH, HI, HJ
Staunton—DI, ET, HI(2), MH, RD
Sterling—HI
Suffolk—ET, HI
Triangle—QI, RI
Troutville—HJ
Verona—ET
Virginia Beach—BW(3), ET(5), FI, HH, HI(4), HJ, QI, RI(2), SH
Warrenton—HJ
Waynesboro—BW, HI, RD
Williamsburg—BW(4), DI, ET(5), FI, HH, HI(3), HJ, M6, QI(5), RI(2), RO SH(2)
Winchester—BW, ET(2), HI(2), HJ, QI(2)
Woodbridge—ET(2)
Wytheville—ET, HI, HJ

WASHINGTON

Aberdeen—TH
Auburn—BW, TL

Bellevue—BW, HH, HI, TH, TL
Bellingham—BW, HI, M6, TL
Blaine—IM
Bremerton—BW, HI
Burlington—FI
Centralia—BW, FI
Chehalis—BW
Clarkston—BW
Coulee Dam—FI
E. Wenatchee—BW
Ellensburg—HI, M6
Ephrata—TL
Everett—HI, IM, M6, RO, TL
Fife—HI, M6
Issaquah—HI, M6
Kelso—BW, M6, TH
Kennewick—QI, TL
Kent—BW
Kirkland—RI
Lakewood Cen.—BW
Lynnwood—BW
Mercer Island—TL
Moses Lake—BW, IM, M6, TL
Ocean Shores—RO
Olympia—BW
Omak—TL
Othello—BW
Pasco—BW, M6, TH, TL
Port Angeles—FI, TH
Pullman—TL
Quincy—FI
Rendton—SH
Richland—FI, HI, IM, TH
Seattle—BW(3), FI, HH(3), HI(2), HY, IM(2), MM, M6, QI, RI, RO, TH, TL(4), WI
Spokane—BW(3), FI(2), HI(2), M6, RI, SH, TH, TL
Sunnyside—BW, TL
Tacoma—FI, RO, RI(2)
Tumwater—B6
Vancouver—BW, TH, TL
Walla Walla—IM, TL
Wenatchee—FI, IM, TH, TL
Yakima—BW(2), HI, IM, M6, TH(2), TL

WEST VIRGINIA

Barboursville—HI
Beckley—BW, DI, HI, RI
Bluefield—ET, HI, RD, SH
Bridgeport—HI
Burnsville—FI
Charleston—HI(4), RI
Clarksburg—SH
Davis—BW
Fairmont—FI, HI
Grafton—BW
Harpers Ferry—BW
Huntington—BW, FI, HI(2), RI
Lewisburg—FI
Martinsburg—HI
Morgantown—HI, RI
Nitro—BW
Parkersburg—BW, HI, SH
Princeton—HI
Ripley—BW
St. Marys—BW
Triadelphia—HI, RI
Weirton—HI
Wheeling—HJ, QI
White Sulphur Sprs.—BW
Williamstown—DI

WISCONSIN

Appleton—BW
Ashland—BW, FI
Baraboo—FI
Beaver Dam—BW
Black River Falls—BW
Brookfield—MM
Camp Douglas—FI
De Pere—HJ
Dodgeville—BW
Eau Claire—BW(2), HI, HJ, RI

Fond Du Lac—HI, HJ, IM, M6
Fort Atkinson—BW
Glendale—HH, HI
Green Bay—BW(2), HI(2), IM, M6, RI(2)
Hurley—HI
Janesville—BW, HI, HJ
Kenosha—BW, HI, HJ
La Crosse—BW(2), FI, HI, RI
Ladysmith—BW
Lake Delton—BW
Lake Geneva—HH
Lancaster—BW
Madison—BW(4), FI, HI(2), HJ(2), M6, QI(2), RI, SH, TL
Marinette—HI
Marshfield—BW, FI, HI
Mauston—FI
Menomonee Falls—HI
Menomonie—BW
Merrill—BW
Milwaukee—BW(3), HI(2), HJ(2), HY, M6, RI(2)
Mishicot—BW
New Lisbon—BW
Oshkosh—HI, HJ, M6
Osseo—BW
Platteville—BW
Plover—BW
Portage—BW
Port Washington—BW
Prairie Du Chien—FI
Racine—HI
Reedsburg—BW
Rhinelander—BW, HI
Rice Lake—BW
Ripon—BW
St. Croix Falls—BW
Sheboygan—BW
Stevens Point—BW, HI
Superior—BW(2), HI
Tomah—BW, HI
Watertown—BW
Waukesha—BW, HI
Wausau—BW, HI, HJ
Wauwatosa—BW, HI, HJ, RI, SH
West Bend—HI
Wisconsin Rapids—BW

WYOMING

Afton—BW, FI
Alpine—BW
Buffalo—BW, FI(2)
Casper—BW(2), FI, HH, HI, IM, LQ, RI
Cheyenne—BW(2), FI(2), M6, RI, TL
Cody—BW(3), FI(3), HI, MH
Douglas—BW, FI(2)
Dubois—FI
Evanston—BW, FI(2), RI
Gillette—BW, FI, HI, RI
Greybull—BW
Jackson—BW(2), FI(4), MH, M6, RI
Kemmerer—BW, FI
Lander—BW
Laramie—BW(3), FI(2), HI, M6, RI, TL
Little America—BW
Lovell—FI
Lusk—BW, FI
Newcastle—BW, FI
Pine Bluffs—FI
Powell—BW
Ranchester—BW
Rawlins—BW, FI, HI, QI, RI, TL
Riverton—BW(2), FI
Rock Springs—BW(2), FI, HI
Shell—BW
Sheridan—BW, MH
Sundance—BW, FI
Teton Village—BW
Thermopolis—BW, FI, HI
Torrington—BW
Upton—FI
Wapiti—FI
Wheatland—BW, FI(2)
Worland—BW

A Handy Source For
Planning Your Next Camping Trip

Every year millions of Americans get a taste of the Great Outdoors by enjoying a camping vacation. Camping offers a reasonably priced way to see the country and get closer to nature. If you are among those planning a camping vacation, you may find it a good idea to reserve your campsite in advance, especially during the busy seasons. To assist you in making your plans, we have compiled a list of selected privately owned and operated campgrounds throughout the United States, using information provided to us by Woodall Publishing Company. (Should you desire total listings, obtain Woodall's Campground Directory, available at most bookstores or RV dealers.)

The illustration below details how to interpret the information from the charts on the following pages. Under each state name is an alphabetical listing of towns. Following the town name is the name, location and telephone number of a campground that is in or near that town. The total number of available sites in that campground is also listed.

Almost all campgrounds differ in terms of facilities, ranging from natural settings where you can tent by a stream and cook over an open fire, to sites that provide you with nearly all the comforts of home. Most of the campgrounds listed here have facilities which include reasonable access and site size, level sites, lighted areas, well-maintained hookups, hot showers and flush toilets, laundry, public phone, a small store and ice. Some campgrounds also have recreation halls, boat docks, nature and hiking trails, swimming pools, bike trails, horses for rent and more.

If you have never considered a camping vacation and would like to try something different this year, you can rent tents, fully equipped pop-up campers, or trailers at many locations. Renting allows you to determine whether camping is the right vacation for you before investing in equipment.

There are many fine government-managed public campgrounds located in state and national forests and parks in addition to the privately owned and operated facilities listed here. For information on public campgrounds, write the state offices of tourism at the addresses listed on page 7. For national park camping information, see the charts on pages 125-126 or contact the National Park Service at the addresses given on page 117.

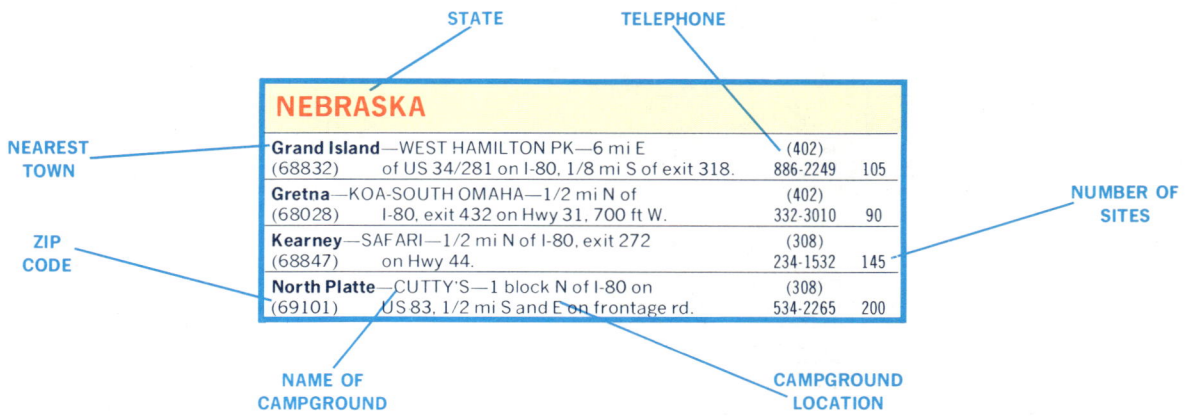

CITY—CAMPGROUND—LOCATION	PHONE	SITES
ALABAMA		
Decatur—POINT MALLARD (CITY PARK)—2 mi S of Alt	(205)	225
(35601) US 72 on US 31, 3 mi E on 8th Ave, S.E.	350-3003	
Eufaula—LAKEPOINT RESORT STATE PARK—7 mi N of	(205)	190
(36027) US 82 on US 431.	687-6676	
Gulf Shores—GULF STATE PARK—3 mi E of Hwy 59	(205)	468
(36542) on Hwy 182.	968-7544	
Guntersville—RIVERVIEW—3 mi W of Hwy 431 on Hwy	(205)	175
(35976) 69, 1 mi N on Cha-la-kee Rd.	582-3014	
Jemison—PEACH QUEEN—1/4 mi E of jct of I-65	(205)	106
(35094) and CR 42 on Mineral Springs Rd.	688-2573	
Leeds—HOLIDAY TRAVEL PARK—At I-20 and US 411.	(205)	137
(35094)	640-5300	
Lillian—KOA-PERDIDO BAY—1½ mi S of US	(205)	85
(36549) 98 on CR 99.	962-2727	
Lincoln—TALLADEGA TRAVEL TRAILER—1 mi W of I-20,	(205)	260
(35096) exit 168 on McCaig St.	763-2599	
Mobile—KOA/STYX RIVER—9 mi E of Hwy 59 on I-10,	(205)	180
(36567) to Wilcox Rd. exit, 1 1/4 mi E on service rd.	964-5998	
Montgomery—KOA—7 1/2 mi S of I-85 on I-65, 1/4 mi S	(205)	150
(36043) on US 31.	288-0728	
Rogersville—JOE WHEELER STATE PARK—1 mi W of	(205)	116
(35672) town on US 72, 3 mi S on Park Rd.	247-1184	
Valley Head—KOA-SEQUOYAH CAVERNS—1 mi E of I-59	(205)	85
(35989) on Hwy 40, 6 mi N on US 11, 1 mi NW on CR	635-2311	
ALASKA		
Glenn Hwy—KROA—At Milepost 153 on	(907)	200
(99588) Glenn Hwy.	822-3346	
McKinley Park—TEKLANIKA RIVER (Denali Nat. Pk.)—	(907)	110
(99755) 28 mi W of entrance station.	683-2294	
ARIZONA		
Apache Junction—LOST DUTCHMAN—1/4 mi N of jct on	(602)	726
(85220) US 60/89 Hwy 88, 1/4 mi N on N Plaza Dr.	982-4173	
Apache Junction—ROCK SHADOWS—1/4 mi SE of Hwy	(602)	680
(85220) 88 on US 60/80/89, 1/4 mi S on Idaho Rd.	982-0450	
Flagstaff—BLACK BART'S RV LODGE—3 mi E of I-17 on	(602)	175
(86001) I-40, 1/4 mi SE on Butler Ave.	774-1912	
Florence—CALIENTE CASA DE SOL—2 1/2 mi E of Hwy	(602)	250
(85232) 287 on US 89.	868-5520	
Grand Canyon—TRAILER VILLAGE—3/4 mi	(602)	80
(86023) E of Visitor Center, S rim.	638-2401	
Mesa—FIESTA—5 mi E of Hwy 87/93 on US 60/80/89,	(602)	343
(85205) 1/2 mi N on Val Vista Dr, 1/4 mi E.	832-6490	

CITY—CAMPGROUND—LOCATION	PHONE	SITES
Mesa—TRAILER VILLAGE—4 1/4 mi E of Hwy 87 on US	(602)	1679
(85203) 60/80/89.	832-1770	
Mesa—VAL VISTA VILLAGE—5 mi E of Hwy 87 on US	(602)	788
(85203) 60/80/89, 1/4 mi N on Val Vista Dr.	832-2547	
Mesa—VENTURE OUT AT MESA—6 3/4 mi E of Hwy	(602)	1289
(85206) 87/93 on US 60/80/89.	832-0200	
Phoenix—ROYAL PALM—6 1/4 mi N of US 60/89 Hwy	(602)	124
(85021) 93 on I-17, 3/4 mi E on Dunlap Ave.	943-5833	
Scottsdale—ROADRUNNER—3/4 mi S of US 60/89/Hwy	(602)	627
(85256) 93 on I-17, 13 mi E on McDowell, 1/2 mi S.	945-0787	
Tempe—CASA FIESTA—3/4 mi E of Hwy 360 on I-10	(602)	555
(85283) mi E on Baseline Rd.	839-1052	
Tempe—TEMPE VILLA—3 1/2 mi E of I-10 on Hwy 360,	(602)	160
(85281) 1/2 mi N on McClintock, E on US 60/80/89.	968-1411	
Topock—5 MILE LANDING (HAVASU WILDLIFE REFUGE)—	(602)	75
(86436) 7 mi N of I-40, exit 1 on Hwy 95.	768-2350	
Tucson—FAR HORIZONS—2 3/4 mi N of I-19 on I-10, 9 1/4	(602)	516
(85710) mi E on Speedway, 1/2 mi S on Pantano.	296-1234	
Tucson—RINCON COUNTRY—15 3/4 mi SE of I-19 on	(602)	450
(85730) I-10, 9 3/4 mi N on Houghton Rd., 2 1/2 mi W.	886-8431	
Two Guns—KOA-TWO GUNS—1/4 mi S of I-40, exit 230.	(602)	81
(86047)	289-3939	
Wickenburg—WESTPARK—3 1/4 mi W of US 89,	(602)	60
(85358) Hwy 93 on US 60.	684-2210	
Yuma—SUNI-SANDS—1 3/4 mi SE of US 95 on I-8,	(602)	312
(85364) 1/2 mi N on Ave 3E, 1 mi W.	726-5941	
ARKANSAS		
Alma—KOA-FORT SMITH—2 1/2 mi N of I-40 on US 71.	(501)	60
(72921)	632-2704	
Arkadelphia—KOA-LAKE DEGRAY—900 feet N of I-30	(501)	70
(71923) on Hwy 7, 1 mi NE on service rd.	246-4922	
Benton—KOA—1 3/4 mi E of I-30 on service rd.	(501)	87
(72015)	778-1244	
Benton—YOGI BEAR'S JELLYSTONE PK—1/4 mi N of I-30	(501)	113
(72015) exit 121 on Alcoa Rd.	794-2950	
Cotter—KOA-WHITE RIVER—100 yds W of city limits	(501)	110
(72626) on US 62, 1/8 mi N on entry rd.	453-2299	
Eureka Springs—KOA—4 1/4 mi NW of Hwy 23 on	(501)	100
(72632) US 62, 3/4 mi SW on Hwy 187.	253-8036	
Harrison—OZARK SAFARI—7 mi S of US 62/65	(501)	75
(72601) on Hwy 7, 1 mi E on Hwy 206.	743-2343	
Hot Springs—KOA—1/4 mi E of east city limits on	(501)	115
(71901) US 70.	624-5912	
Lake Village—LAKE CHICOT STATE PK—8 mi N of town	(501)	127
(71653) on Hwy 144.	265-5480	

CITY—CAMPGROUND—LOCATION	PHONE	SITES
Morrilton—KOA/CONWAY—100 feet N of I-40, exit 107	(501)	68
(72110) on Hwy 95.	354-8262	
Mountain Home—WILDERNESS POINT—10 1/2 mi E	(501)	230
(72544) of Hwy 5 on US 62, 1 mi S on Wilderness Pt.	488-5340	
North Little Rock—KOA—1 mi SW of I-40, exit 148 on	(501)	100
(72115) Crystal Hill Rd.	758-4598	
Rogers—BEAVER LAKE CAMPGROUND—5 1/2	(501)	66
(72756) mi E of US 62/71 on Hwy 94.	925-1265	
Texarkana—BEST WESTERN-MASTER HOST—2 mi S	(501)	32
(75501) of I-30 on Hwy 245, 2 mi SW on US 67.	773-3161	
CALIFORNIA		
Anaheim—KOA—3 1/4 mi NW of Hwy 22 on	(714)	221
(92803) I-5, 1 block N on Harbor, 1/2 mi W on Ball.	533-7720	
Anaheim—VACATION LAND—3 1/4 mi NW of Hwy 22	(714)	406
(92802) on I-5, 1 block N on Harbor, 1/2 mi W on Ball.	533-7270	
Arroyo Grande—LOPEZ RECREATION AREA—2 mi NE of	(805)	221
(93420) town on Huasna Rd, 4 mi E on Lopez Dr.	489-2095	
Bakersfield—BUENA VISTA AQUATIC REC AREA—2 mi W	(805)	112
(93276) of I-5 on Hwy 119, S on Hwy 43 to gate.	763-1526	
Bradley—SAN ANTONIO LAKE—12 mi W of town	(805)	600
(93426) on Nacimiento, 9 mi W on CR G-14, 4 mi NE.	472-2311	
Buellton—FLYING FLAGS—1/4 mi W of US 101 on Hwy	(805)	200
(93427) 246, 1 block S on Avenue of the Flags.	688-3716	
Carpinteria—CARPINTERIA STATE BEACH—In town on	(805)	160
(93013) Palm Ave.	684-2811	
Cathedral City—DE ANZA PALM SPRINGS OASIS—9 1/4	(714)	140
(92234) mi E of Hwy 62 on I-10, 4 mi N on Date Palm.	328-4813	
Chula Vista—KOA-SAN DIEGO METROPOLITAN—1 1/4	(714)	183
(92010) mi E of I-5 on "E" St, 1 mi N on 2nd Ave.	427-3601	
Coarsegold—KOA—8 mi S of Hwy 49 on Hwy 41.	(209)	215
(93614)	683-7855	
Coloma—CAMP COLOMA—1/4 mi E of Hwy 49 on Camp	(916)	125
(95613) Coloma Rd.	622-6700	
Crescent City—KOA-CAMP LINCOLN—1 mi N of US 199	(707)	120
(95531) on US 101.	464-5744	
Del Mar—SURF & TURF—1 block W of I-5 on Via De La	(714)	234
(92014) Valle Rd, 1 block S on Turf Rd.	755-5400	
Desert Hot Springs—SKY VALLEY—6 1/4 mi E of Hwy 62	(714)	
(92240) on I-10, 3 mi N on Palm, 8 mi E.	329-7415	
El Cajon—CIRCLE RV RANCH—1 block N of I-8 on	(714)	180
(92021) Greenfield Dr., 1/4 mi E on Main St.	440-0040	
El Cajon—KOA—1/4 mi N of I-8 on Hwy 111, W on	(714)	165
(92243) Ross Rd.	353-1051	
Eureka—KOA—4 mi N of town on US 101.	(707)	166
(95501)	822-4243	

U.S. CAMPGROUNDS

CITY—CAMPGROUND—LOCATION	PHONE	SITES
Garberville—BENBOW VALLEY RESORT—2 mi S of (95440) town center on US 101.	(707) 923-2777	102
Hemet—GOLDEN VILLAGE—2 1/4 mi W of CR R 3 on (92343) Hwy 74/79.	(714) 925-2518	1041
Hemet—ROADRUNNER—1 1/2 mi W of CR R 3 on (92343) Hwy 74/79, 1/4 mi S on Kirby, 2 blocks W.	(714) 925-2515	358
Idyllwild—WILDERNESS PINES—8 mi N of Hwy 74 on (92349) Hwy 243, 1/2 mi E on Marion Ridge Dr.	(714) 659-2123	283
Indio—HAPPY WANDERER RV PK—1 block N of I-10, (92201) Hwy 111 exit on Van Buren, 1/4 mi E.	(714) 347-7749	459
Isleton—THE LIGHTHOUSE—9 mi W of I-5 on Hwy 12, (95641) 1/4 mi NE, 2 1/4 mi S on Brannon Is. Rd.	(916) 777-6681	270
Jamul—PIO PICO PK—2 mi W of Hwy 94 on Otay Lakes (92035) Rd.	(714) 421-2424	234
Julian—BUTTERFIELD RANCH—12 mi E of Hwy 79 on (92036) Hwy 78, 12 mi S on CR S2.	(714) 765-1463	288
June Lake—PINE CLIFF RESORT—1 mi SW of US 395 on (93529) Hwy 158, 1 mi W on Ohio Ridge Rd.	(714) 648-7558	170
La Grange—FLEMING MEADOWS—7 1/2 mi E of CR J-59 (95329) on Hwy 132, 3 mi N on Bond's Flat Rd.	(209) 852-2396	236
Lakehead—ANTLER'S TRAILER PK—1 1/4 mi S of I-5 on (96051) Antler's Rd.	(916) 238-2322	118
Lebec—KOA-FORT TEJON/LEBEC—9 mi N of Hwy 138 (93243) on I-5, 2 1/2 mi N on Lebec Rd.	(805) 248-6145	85
Lewiston—PINE COVE TRAILER PK—6 mi N of Rush (96052) Creek Rd. on Trinity Dam Blvd.	(916) 778-3838	72
Lodi—TOWER PK—5 mi W of I-5 on Hwy 12. (95240)	(209) 369-1041	252
Los Banos—KOA-LAKE SAN LUIS—2 mi W of I-5 on Hwy (93635) 152.	(209) 826-6298	108
Manteca—OAKWOOD LAKE—5 mi W of Hwy 99 on Hwy (95336) 120, 1 mi S on McKinley, 3/4 mi W.	(209) 239-9566	320
Mt. Shasta—KOA—3 blocks E of I-5 on Lake St, 1/2 mi N (96067) on North Mt. Shasta Blvd.	(916) 926-4029	110
Olema—OLEMA RANCH—1/4 mi S of Sir Francis Drake (94950) Blvd on Hwy 1.	(415) 663-1363	150
Orange—ORANGELAND—1 3/4 mi N of I-5 on Hwy 57, (92667) 1/2 mi E on Katella, 100 feet on Struck.	(714) 633-0414	212
Palm Springs—HAPPY TRAVELER—5 mi S of I-10 on (92262) Indian, 1 block W on Valmonte, 1/2 mi S.	(714) 325-8518	140
Pauma Valley—RANCHO CORRIDO—5 mi N of CR S6 on (92061) Hwy 76.	(714) 742-3755	149
Petaluma—KOA—1/4 mi W of US 101 on Petaluma Blvd, (94952) 1/4 mi N on Stony Point Rd, 1/4 mi W.	(707) 763-1492	247
Plymouth—FAR HORIZONS 49ER TRAILER VILLAGE—1 (95669) mi N of Hwy 16 on Hwy 49.	(209) 245-6981	300
Porterville—KOA-YOKUT—5 mi E of Hwy 65 on Hwy 190. (93257)	(209) 784-2123	120
Red Bluff—RIO VISTA—1 mi E of I-5 on Hwy 36, 3/4 mi N (96080) on Chestnut, 1/4 mi W on Paynes Cr.	(916) 527-2793	30
Redding—MARINA RV PK—1 1/2 mi W of I-5 on Hwy (96001) 299, 1 mi S on Park Marina Drive.	(916) 241-4396	85
Salton City—SALTON CITY SPA & RV PK—3 1/2 mi E of (92274) city on S Marina Dr, 3/4 mi S on Sea View Dr.	(714) 394-4133	189
San Diego—BORDER GATE RV PK—1 block E of I-5 on (92173) Dairymart Rd, 1 block N on San Ysidro Blvd.	(714) 428-4411	179
San Francisco—SAN FRANCISCO RV PK—1 block NE (94105) of I-280, 4th St. exit on Berry St.	(415) 986-8730	200
San Juan Bautista—MISSION FARM—3 mi. E of US 101 on (95045) Hwy 156, 1 block S on Alameda, 1/4 mi E.	(408) 623-4456	140
San Rafael—MARIN RV PK—1/4 mi E of US 101 on Lucky (94904) Dr, 1/2 mi N on Redwood Hwy.	(415) 461-5199	171
Scotts Valley—HOLIDAY HOST—5 1/4 mi N of Hwy 1 on (95066) Hwy 17, 1/4 mi N on Santa's Village Rd.	(408) 438-1600	152
Smith River—SALMON HARBOR RESORT—2 3/4 mi N (95567) of town on US 101.	(707) 487-3341	84
South Lake Tahoe—TAHOE VALLEY—1/4 mi SW of Hwy (95731) 89 on US 50.	(916) 541-2222	287
Temecula—BUTTERFIELD COUNTRY & VAIL LAKE RV (92390) RESORT—9 mi E of I-15 on Hwy 79.	(714) 676-5694	467
Trinity Center—WYNTOON PK—3/4 mi S of north city (96091) limits on Hwy 3.	(916) 266-3337	217
Victorville—KOA—1 mi N of Hwy 18 on I-15, (92392) 1/4 mi S on Stoddard Wells Rd.	(714) 245-6867	131
Wallace—SOUTH CAMANCHE SHORE—6 mi NE of Hwy 12 (95254) on S Camanche Pkwy.	(209) 763-5178	728
Watsonville—KOA-SANTA CRUZ—5 mi N of Hwy 152 on (95076) Hwy 1, 3 1/2 mi SW on San Andreas, 500 ft E.	(408) 722-0551	253
West Sacramento—KOA—1/2 mi W of I-880 on (95691) I-80, 1 block S on Enterprise, 1/4 mi E.	(916) 371-6771	176
Willits—KOA-KAMP SUNDOWN—1 1/2 mi W of US 101 on (95490) Hwy 20.	(707) 459-6179	65

COLORADO

CITY—CAMPGROUND—LOCATION	PHONE	SITES
Alamosa—KOA-SIERRA VISTA—3 1/4 mi E of Hwy 17 on (81101) US 160.	(303) 589-9757	87
Buena Vista—KOA—1 1/4 mi E of US 24 on US 24/285, (81211) 1/4 mi N on CR 303.	(303) 395-8318	100
Canon City—RV STATION—2 mi E of Hwy 115 on US 50. (81212)	(303) 275-4576	66
Coaldale—CUTTY'S OF HAYDEN CREEK—3 1/2 mi S of US (81222) 50 on Hayden Creek Rd.	(303) 942-3455	233
Colorado Springs—PEAK VIEW—1/2 mi E of I-25 on (80907) Garden of the Gods Rd, 1/4 mi N.	(303) 598-1434	135
Cortez—KOA-CORTEZ/MESA VERDE—1/2 mi E of Hwy 145 (81321) on US 160.	(303) 565-9301	79
Cotopaxi—KOA-COTOPAXI—1 1/2 mi E of town center on (81223) US 50.	(303) 275-9972	83
Denver—KOA—15 mi N of I-70 on I-25, 100 yards E of exit (80020) 229 on Hwy 7.	(303) 452-4120	141
Durango—UNITED CAMPGROUNDS—1 1/2 mi N of city (81301) limits on US 550, 1/8 mi S on frontage rd.	(303) 247-3853	196
Estes Park—SPRUCE LAKE—1/4 mi W of US 34/36 on Bus (80517) US 34, 1 1/2 mi SW on US 36, 1 block S.	(303) 586-2889	110
Fountain—BONANZA COACH STOP—100 ft E of (80817) I-25 on Hwy 16, 1 mi S on Bonanza Tr.	(303) 382-7575	301
Grand Junction—KOA—1/2 mi SE of I-70, exit 37 (81520) on Business Loop I-70.	(303) 434-6644	134
Greeley—GATEWAY—1/2 mi E of US 85 on Byp (80631) US 34, 900 feet W on frontage rd.	(303) 353-6476	98
Gunnison—PARADISE PARK—2 1/4 mi N of US 50 on Hwy (81230) 135, 600 feet E.	(303) 641-2927	109
Leadville—SUGAR LOAFIN'—2 mi SE of Hwy 91 on (80461) US 24, 3 1/2 mi NW on CR 4.	(303) 486-1031	95
Loveland—JOHNSON CORNER—300 feet E of I-25, exit 254 (80537) on Campion, 1/2 mi S.	(303) 669-8400	266
Montrose—KOA-WESTERN HIGHLANDER—3/4 mi E of US (81401) 550 on US 50, 1/2 block N on Cedar Ave.	(303) 249-9177	80
Monument—YOGI BEAR'S JELLYSTONE—I-25, exit (80132) 163 County Line Rd, S on Monument Hill Rd.	(303) 481-2336	75
Ouray—4 J + 1 + 1—3 blocks W of US 550 on 7th Ave. (81427)	(303) 325-4418	105
Pueblo—KOA-WEST PUEBLO—16 1/2 mi W of I-25 (81007) on US 50.	(303) 547-2101	103
South Fork—KOA-AQUA RAMON—4 mi E of Hwy 149 on (81132) US 160.	(303) 873-5500	104
Steamboat Springs—SKI TOWN CAMPGROUND—2 mi W (80477) of W city limits on US 40.	(303) 879-0273	100
Strasburg—KOA-DENVER EAST—200 ft N of I-70 on (80136) Strasburg Rd, 1/4 mi E on frontage rd.	(303) 622-9274	68
Woodland Park—SOLAR RV PK—1/2 mi N of US 24 on (80863) Hwy 67.	(303) 687-9518	53

CONNECTICUT

CITY—CAMPGROUND—LOCATION	PHONE	SITES
Baltic—SALT ROCK FAMILY—2 mi N of Hwy 207 on Hwy (06330) 97.	(203 822-8728	130
Clinton—RIVERDALE FARM—2 mi W of (06417) Hwy 81 on I-95, 2 mi N on River Rd.	(203) 669-5388	250
East Haddam—WOLFS DEN FAMILY CAMPSITES—3 mi E (06423) of Hwy 149 on Hwy 82.	(203) 873-9681	220
Jewett City—ROSS HILL PARK—1/2 mi N of Hwy 52 on (06351) Hwy 12, 1/2 mi W on Hwy 138, 1 1/2 mi N.	(203) 376-9606	250
Kent—TREETOPS—5 1/2 mi E of US 7 on Hwy (06757) 341, 1/2 mi N on Kenico Rd.	(203) 927-3555	262
North Stonington—HIGHLAND ORCHARDS—1/4 (06359) mi N of I-95 on Hwy 49.	(203) 599-5101	200
Norwich—ACORN ACRES—6 1/2 mi W of Hwy 52 on Hwy (06415) 2, 2 mi S of Hwy 2, exit 22 on Scott Hill Rd.	(203) 859-1020	185
Old Mystic—SEAPORT—1 1/4 mi N of I-95, exit (06372) 90 on Hwy 27, 1/2 mi E on Hwy 184.	(203) 536-4044	130
Salem—WITCH MEADOW LAKE—1/8 mi E of Hwy 11, exit (06415) 5 on Witch Meadow Rd.	(203) 859-1542	225
Warrenville—BRIALEE—1 mi N of US 44 on Hwy 89, 1/2 (06278) mi W on Perry Hill, 3/4 mi N on Laurel Lane.	(203) 429-8359	175
West Willington—KOA-GOOSEMEADOW—4 1/2 mi E of (06279) I-86, exit 100 on US 44, 1 mi N.	(203) 429-7451	160
Winsted—WHITE PINES—1 3/4 N of US 44 on Hwy 8, 1/4 (06098) mi E on Hwy 20, 3/4 mi S on Old North.	(203) 379-0124	185

DELAWARE

CITY—CAMPGROUND—LOCATION	PHONE	SITES
Dagsboro—TUCKAHOE ACRES—6 1/4 mi E of US 113 on (19939) Hwy 26, 2 1/2 mi N on CR 346.	(302) 539-9841	540
Fenwick Island—TREASURE BEACH—1 1/2 mi W of Hwy 1 (19975) on Hwy 54.	(302) 436-8001	725
Georgetown—HOMESTEAD—6 1/2 mi E of US (19947) 113 on US 9 and Hwy 18, 3/4 mi N on CR 254.	(302) 684-4278	136
Rehoboth Beach—3 SEASONS—1 1/2 mi W (19971) of Hwy 1 on CR 273.	(302) 227-2564	225

FLORIDA

CITY—CAMPGROUND—LOCATION	PHONE	SITES
Apopka—YOGI BEAR'S JELLYSTONE PARK—2 1/2 mi S (32703) of US 441 on Hwy 435.	(305) 889-3048	300
Arcadia—KOA-PEACE RIVER—2 mi W of US 17 on Hwy 70. (33821)	(813) 494-0214	150
Bonita Springs—IMPERIAL BONITA ESTATES—1 mi E of (33923) Old US 41 on Dean St.	(813) 992-0511	300
Bradenton—SUGAR CREEK—1 mi N of Hwy 684 on US (33508) 41, 2 mi E on 26th Ave.	(813) 747-6331	244
Brooksville—CLOVER LEAF FOREST—1 mi N of (33512) US 98 on US 41.	(904) 796-8016	300
Chokoloskee—BAY VIEW TRAVELERS PK—8 mi S of US (33925) 41 on Hwy 29 (on Chokoloskee Island).	(813) 695-2881	250
Clearwater—TRAVEL TOWNE TRAVEL TRAILER RESORT— (33515) 5 mi N of Hwy 60 on US 19.	(813) 784-2500	360
Clermont—SUNSHINE HOLIDAY CAMPER RESORT—14 (32711) mi S of Hwy 50 on US 27.	(305) 394-4041	393
Clewiston—KOA—2 1/2 mi W of west city limits on US (33440) 27, 100 feet N on Hwy 720.	(813) 983-7078	120
Crystal River—CRYSTAL RIVER VACATION CLUB—4 mi (32629) W of US 19 on Hwy 44.	(904) 795-3411	300
Dade City—BLUE JAY MOBILE HOME PK—2 mi S of city (33525) limits on US 301 & 98, 1/2 mi SE on US 98.	(904) 567-7117	190
Davenport—THREE WORLDS CAMP — 3 mi N of Hwy 547 (33837) on US 17-92.	(813) 424-1286	185
Daytona Beach—HOLIDAY INN TRAVEL PARK—5 mi (32010) N of US 1 on I-95.	(904) 672-8122	205
Destin—HOLIDAY TRAVEL PARK—10 mi E (32541) of Destin Br on US 98.	(904) 837-6334	253
Dunedin—DUNEDIN BEACH—3/4 mi N of Hwy 586 on Alt (33528) US, 19	(813) 784-3719	195
Englewood—HOLIDAY TRAVEL PARK—1 mi E of Hwy 775 (33533) On Hwy 776 and Alt 45.	(813) 474-5078	450
Estero—SHADY ACRES TRAILER PK—5 3/4 mi S of Hwy (33928) 865 on US 41.	(813) 481-3361	198
Fort Lauderdale—TWIN LAKES TRAVEL PK—1500 feet E (33314) of Fla Tpk, exit 12 on Hwy 84.	(305) 587-0101	371
Fort Myers—FORT MYERS CAMPGROUND—2 mi S of (33908) Hwy 865 on US 41.	(813) 481-1007	357
Fort Myers—WOODSMOKE—5 1/2 mi S of Hwy 865 on US (33908) 41.	(813) 992-1772	300
Fort Myers Beach—INDIAN CREEK PK—2 mi S of (33931) Hwy 867 on Hwy 865.	(813) 481-4546	845
Fort Pierce—BRYN MAWR OCEAN RESORT—2 mi N of (33450) Hwy 70 on US 1, 6 mi NE on Hwy A1A.	(305) 465-1003	319
Haines City—CENTRAL PK—1/2 mi N of US 92 (33844) on US 27.	(813) 422-5322	352
Holiday—HOLIDAY TRAVEL PK—At US 19 and Alt (33590) US 19.	(813) 934-6782	525
Hollywood—HOLIDAY TOWERS—1/4 mi W of (33021) I-95 on Hwy 824.	(305) 962-7400	176
Homestead—ROYAL COLONIAL— 2 mi N of city on US 1, (33032) 2 blocks NE on Waldin Dr.	(305) 247-3387	104
Homosassa Springs—NATURE'S CAMPGROUND— 1 1/4 (32647) mi W of US 19 on Hwy 490A.	(904) 628-2892	287
Jacksonville—KOA-ST. AUGUSTINE—20 mi S of I-10 on (32224) I-95, 1 block E on Hwy 210.	(904) 824-8309	120
Jennings—OUTDOOR RESORTS OF AMERICA—500 feet (32053) W of I-75 on Hwy 143.	(904) 938-3321	140
Jensen Beach—OUTDOOR RESORTS AT NETTLES (33457) ISLAND—2 mi N of Hwy 707A on A1A.	(305) 229-1300	1576
Key Largo—BYRN MAWR—5 mi S of city limits on US 1 (33037) (Milepost 95).	(305) 852-3011	232
Key West—VENTURE OUT AT CUDJOE CAY—23 mi NE of (33042) town on US 1, 1/4 mi E on Spanish Main Dr.	(305) 745-3233	659
Kissimmee—CAMPING WORLD—4 mi E (32741) of I-4 on US 192.	(305) 846-3424	509
Kissimmee—SHERWOOD FOREST—2 mi E of I-4 (32741) on US 192.	(305) 846-7431	376
Lady Lake—BLUE PARROT—1 1/2 mi N of US 27 on Alt (32659) US 27/441.	(904) 753-2026	327
Lake Buena Vista—FORT WILDERNESS—1 mi W of I-4 on (32830) US 192, N to the park admission gate.	(305) 824-8000	825
Lake Buena Vista—YOGI BEAR'S JELLYSTONE PK—4 mi (32741) W of I-4 on US 192.	(305) 423-4751	703
Lake Wales—SADDLEBAG LAKE—8 mi E of US 27 on Hwy (33853) 60.	(813) 696-1115	790
Lakeland—SANLAN RANCH—9 mi S of I-4 on US 98. (33801)	(813) 665-1726	320
Largo—RAINBOW VILLAGE—1/3 mi W of US 19 (33541) on Hwy 688, 3/4 mi S on 66th St. N.	(813) 536-3545	190
Largo—YANKEE TRAVELER—2 mi W of US 19 on Hwy (33541) 688.	(813) 531-7998	200
Leesburg—HOLIDAY TRAVEL PK—1/4 mi W of US 27 on (32748) Hwy 33.	(904) 787-5151	371
Long Key—OUTDOOR RESORTS OF AMERICA—On US 1 (33001) at Milepost 66.	(305) 664-4415	408
Marathon—SUNSHINE KEY TRAV-L-PARK—9 mi (33043) S of town on US 1.	(305) 872-2217	400
Marianna—ARROWHEAD—2 mi N of I-10 on Hwy 71, 1/4 (32446) mi W on US 90.	(904) 482-5583	280
Melbourne Beach—OCEAN HOLIDAY—4 mi S of US 192 (32951) on A1A.	(305) 724-2600	576
Milton—EAST PENSACOLA SAFARI—200 feet N of I-10 (32570) on Hwy 87.	(904) 623-3936	129
Naples—KOA—7 1/2 mi SE of Hwy 84 on US 41, 3/4 mi S (33942) on Hwy 951, 1/2 mi W on TV Tower Rd.	(813) 774-5455	174
Naples—MARCO NAPLES HITCHING POST—5 miles SE (33942) of Hwy 84 on US 41.	(813) 774-1259	310
Navarre Beach—KOA-ISLAND VIEW—1/4 mi E of Hwy 87 (32561) on US 98, 4 1/2 mi SW on Hwy 399.	(904) 932-2530	250
Ocala—HOLIDAY TRAVEL PARK—1000 feet W of (32670) I-75 on Hwy 40.	(904) 622-5330	156
Okeechobee—CRYSTAL LAKES RV RESORTS—3 mi S (33472) of Hwy 70 on US 98/441.	(813) 763-0231	810
Orlando—OUTDOOR RESORTS OF AMERICA—6 1/2 mi W (32711) of I-4 on US 192.	(305) 422-7461	980
Orlando—YOGI BEAR'S—3 mi W of Fla Tpk on I-4, 50 ft W (32809) on Hwy 528, 1 mi S on access rd.	(305) 351-4394	455
Ormond Beach—SUNSHINE HOLIDAY—1/2 mi NW of (32074) I-95 on US 1.	(904) 672-3045	210
Palm Harbor—CYPRESS POINT ON LAKE TARPON— (33563) 2 1/2 mi N of Hwy 584 on US 19.	(813) 938-1966	400
Palmetto—WINTERSET TRAVEL TRAILER PK—2 1/2 mi N (33561) of US 19 on US 41.	(813) 722-4884	217
Panama City Beach—KOA—1/4 mi W of Hathaway Br on (32407) US 98, 1 mi S on Hwy 757.	(904) 234-5032	179
Panama City Beach—VENTURE OUT—1/2 mi W of (32407) Hathaway Br on US 98, 3 1/2 mi S.	(904) 234-2247	735
Pompano Beach—HIGHLAND PINES—1 mi E of I-95 (33064) on Hwy 834, 1 mi N on Hwy 811, 1 block W.	(305) 421-5372	351
Port Richey—SON-MAR TRAVEL PK—10 mi N of town on (33568) US 19.	(813) 868-2285	315
Ruskin—HAWAIIAN ISLES—2 1/2 mi S of Hwy 674 on (33570) US 41, 1 mi W on Bay Rd.	(813) 645-1098	625
St. Augustine—COOKSEY'S—2 1/2 mi E of US 1 on Hwy (32084) 312, 1 mi S on Hwy 3.	(904) 829-3297	250
St. James City—LAKE WOOD TRAVEL RESORT—6 mi S of (33956) Hwy 78 on Hwy 767.	(813) 283-2415	360
St. Petersburg—KOA—2 1/2 mi S of Hwy 694 on Alt US (33708) 19, 1/2 mi E on 95th St N.	(813) 392-2233	438
Sanford—12 OAKS—2 mi W of I-4 on Hwy 46. (32771)	(305) 323-0880	250
Sarasota—KOA—1 mi E of Us 41 on Hwy 778. (33580)	(813) 355-8585	153
Sarasota—SUN-N-FUN—6 mi E of US 301 on Hwy 780. (33582)	(813) 371-2505	1061
Seminole—HOLIDAY—1 mi E of Alt US 19 on Hwy 694. (33543)	(813) 391-4960	515
Silver Springs—SILVER SPRINGS CAMPERS GARDEN— (32688) 1 block E of Hwy 35 on Hwy 40.	(904) 236-3700	139
Tampa—SPANISH MAIN—6 mi N of I-4 on US 301. (33617)	(813) 986-2415	330
Titusville—SPACE CENTER—3 mi S of Hwy 50 on US 1. (32780)	(305) 269-0947	248
Venice—RAMBLERS REST—6 mi E of Byp US 41 on (33595) Center Rd, 1 mi S on River Rd.	(813) 485-4004	465
Weeki Wachee—CAMP-A-WYLE—3 mi N of Hwy 50 on US (33512) 19.	(904) 596-2139	200
Weeki Wachee—HOLIDAY INN TRAV-L-PARK— 7 mi (33512) S of Hwy 50 on US 19.	(904) 683-0034	248

CITY—CAMPGROUND—LOCATION	PHONE	SITES
West Palm Beach—VACATION INN—2 mi E of Fla Tpk on (33407) Hwy 704, 4 1/2 mi N on Hwy 809.	(305) 848-6166	300
Winter Garden—KOA-ORLANDO—2 mi W of Fla Tpk, exit (32787) 80 on Hwy 50.	(305) 656-1415	350
Winter Garden—STAGE STOP SAFARI—2 mi W of Fla Tpk, (32787) exit 80 on Hwy 50.	(305) 656-8000	248
Winter Haven—HOLIDAY INN TRAV-L-PARK—1 mi W (33880) of US 27 on Hwy 540.	(813) 324-7400	200

GEORGIA

CITY—CAMPGROUND—LOCATION	PHONE	SITES
Adairsville—SAFARI CAMP—1/4 mi W of I-75 on Hwy 140, (30103) 1/10 mi S on county rd.	(404) 773-7320	89
Atlanta—STONE MOUNTAIN PARK (STATE PARK)—5 1/2 (30086) mi S of I-85 on I-285, 7 1/2 mi E on US 78.	(404) 469-9831	461
Cartersville—ALLATOONA CAMPGROUND—1 1/2 mi E (30120) of I-75, exit 122 on Emerson/Allatoona Rd.	(404) 974-3182	140
Cartersville—YOGI BEAR'S JELLYSTONE PK—1/4 mi W (30120) of I-75, exit 126 on Hwy 411.	(404) 382-6216	112
Cleveland—MOUNTAIN SHADOWS JELLYSTONE—2 1/2 (30528) mi N of US 129 on Hwy 356.	(404) 865-4742	226
Douglas—HOLIDAY BEACH—5 mi W of US 441/221 on (31533) Hwy 158.	(912) 384-6099	300
Flippen—KOA-ATLANTA SOUTH—1/10 mi W of I-75, (30253) and Hwy 351 (exit 72).	(404) 957-2610	265
Forsyth—YOGI BEAR'S ATLANTA SOUTH—11 (30204) mi N of Hwy 83 on I-75, 1/10 mi W.	(404) 358-2205	120
Lake Park—KOA—1/10 mi E of I-75 and (31636) Hwy 376 on Twin Lakes Rd, 1/2 mi S.	(912) 559-5192	350
Marietta—KOA-ATLANTA NORTH—1 mi W of I-75, exit (30144) 116 on Roberts Rd, 1/2 mi N on US 41.	(404) 427-2406	225
Perry—PERRY OVERNIGHT PARK—1/4 mi E of I-75 (31069) on US 341, 1/10 mi N on Perimeter Rd.	(912) 987-3371	66
Ringgold—KOA-CHATTANOOGA SOUTH—1/10 mi W of (30736) I-75, exit 141 on Hwy 2.	(404) 937-4166	157
Rossville—HOLIDAY TRAV-L-PARK—1/4 mi W of I-75 (37412) on US 41, 1/2 mi S on Mack Smith Rd.	(404) 891-9766	171
Savannah—SAFARI BELLAIRE WOODS—2 3/4 mi W of (31405) I-95, exit 16 on Hwy 204.	(912) 748-9560	122
Trenton—KOA-MOUNTAIN SHADOWS—2 1/2 mi W of (30752) I-59, exit 3 on Slygo Rd.	(404) 657-6815	120
Unadilla—KOA—1/10 mi N of I-75 on US 41, 1/10 mi E (31091) on Speeg Rd, 1/2 mi S on E Railroad.	(912) 627-3255	120

IDAHO

CITY—CAMPGROUND—LOCATION	PHONE	SITES
Boise—KOA—1/2 mi E of I-84, exit 57 on Federal Way. (83706)	(208) 342-9714	162
Coeur d'Alene—KOA—1/4 mi S of I-90, exit (83814) 22 on Hwy 97, 1/4 mi E on Wolf Lodge Bay.	(208) 664-9307	120
Declo—KOA-SNAKE RIVER—1/4 mi N of I-84, (83323) exit 216 on Hwy 25.	(208) 654-6691	156
Eden—ANDERSON'S CAMP—100 yds N of I-84 (83325) exit 182 on Hwy 150, 1/2 mi E on Tipperary.	(208) 733-6756	157
Idaho Falls—KOA—1/4 mi SE of I-15, exit 119 on US 20, (83401) 1/2 mi N on Lindsay Blvd.	(208) 523-3362	175
Lava Hot Springs—KOA—1/4 mi E of town on US 30. (83246)	(208) 776-5295	100
Sandpoint—KOA — SANDPOINT—6 mi S of US 2 on (83864) US 95, 2 blocks E on Garfield Rd.	(208) 263-4824	100

ILLINOIS

CITY—CAMPGROUND—LOCATION	PHONE	SITES
Alpha—SHADY LAKES—1/2 mi S of Hwy (61465) 17 on Hwy 150, 4 mi W.	(309) 667-2709	290
Benton—CAMPGROUND BENTON—1/4 mi E of I-57, (62897) exit 77 on Hwy 183, 1/2 mi S on Hwy 37.	(618) 435-3401	120
Casey—KOA—1/4 mi N of I-70 on Hwy (62420) 49, 1/4 mi W on county rd.	(217) 932-5319	85
Galena—PALACE—1 mi W of west edge of town on US 20. (61036)	(815) 777-2466	200
Jacksonville—CRAZY HORSE—8 mi N of N Main St on (62612) Hwy 78, 2 1/2 mi E of Literberry, 1 1/2 mi S.	(217) 886-2523	220
Litchfield—RAINMAKER—5 1/2 mi E of I-55 on Hwy 16, 4 (62056) mi NW on county rd.	(217) 532-6370	230
Marengo—KOA-CHICAGO NORTHWEST—5 mi E of Hwy (60180) 23 on US 20, 1/4 mi N on Union Rd.	(815) 923-4206	120
Peru—KOA—7 mi E of US 51 on I-80, 1 1/2 mi N on Hwy (61373) 178, 1/2 mi W on a county rd.	(815) 667-4988	135
Petersburg—YOGI BEAR'S NEW SALEM—1 3/4 mi E of (62675) Hwy 97 on Hwy 123, 1 1/2 mi N and E.	(217) 632-7517	359
Pittsfield—PINE LAKES RESORT—3 blocks W of Hwy 107 (62363) on US 36, 1 mi N on Memorial St.	(217) 285-6719	181
Round Lake—FISH LAKE BEACH—5 mi W of Hwy 134 on (60073) Hwy 120, 1 block N on Hwy 12.	(312) 546-2228	500
Springfield—SAFARI—2 mi S of I-55, exit 88. (62629)	(217) 483-9998	147

INDIANA

CITY—CAMPGROUND—LOCATION	PHONE	SITES
Cicero—HIDDEN BAY—1/2 mi W of city limits on 236th (46034) St.	(317) 984-5329	196
Columbus—KOA—4 1/2 mi S of Hwy 46 on I-65, 1/2 mi W (47201) on Hwy 58, 1 mi S on CR 300W.	(812) 342-6229	101
Elkhart—KOA—1/4 mi N of I-80/90, exit 9, 1/4 mi (46514) E on CR 4.	(219) 264-2914	365
Greenfield—ACRES OF FUN—1/2 mi N of I-70 on Hwy 9, (46140) 3 mi W on CR 300 N.	(317) 326-3171	206
Howe—TWIN MILLS RESORT—2 1/2 mi S of I-80/90, exit (46746) 11 on Hwy 9, 1 3/4 mi W on Hwy 120.	(219) 562-3212	335
Indianapolis—KAMPER KORNER—1 1/2 mi (46217) S of I-465 on Hwy 37.	(317) 788-1488	160
La Porte—CUTTY'S OF NORTHERN INDIANA—1/4 mi S (46350) of I-80/90, exit 7 on Hwy 39.	(219) 362-5111	400
Michigan City—KOA—1 1/2 mi S of I-94, (46360) exit 34A on US 421.	(219) 872-7600	205
Monticello—INDIANA BEACH RESORT—3 1/2 mi N of US (47960) 24 on Shafer Rd.	(219) 583-8306	1156
Nashville—KOA-BROWN COUNTY—2 1/2 mi E of town (47448) center on Hwy 46/135.	(812) 988-4475	138
Pierceton—YOGI BEAR'S JELLYSTONE PK—4 mi N of US (46562) 30 on Hwy 13, 1 mi E on CR 200 N.	(219) 594-2124	1200

CITY—CAMPGROUND—LOCATION	PHONE	SITES
Portage—YOGI BEAR'S JELLYSTONE PK—1 block S of (46368) I-94 on Hwy 249, 1 mi W on US 20.	(219) 762-7758	930
Terre Haute—KOA—1 block S of I-70 on Hwy 46, 1/2 mi (47802) E on county rd.	(812) 232-2457	76
Thorntown—THE OLD MILL RUN PK—3/4 mi W of Hwy 75 (46071) on Hwy 47, 1 mi N on CR 825W.	(317) 436-7190	500

IOWA

CITY—CAMPGROUND—LOCATION	PHONE	SITES
Council Bluffs—KOA-OMAHA—1/4 mi W of I-29, exit 66. (51526)	(712) 545-3202	110
Des Moines—ADVENTURELAND—5 mi E of US 69 on I-80, (50316) 1/8 mi S on US 65.	(515) 265-7384	283
Des Moines—CUTTY'S—1/2 mi N of I-80, exit 127 (50111) on Hwy 141, 3/4 mi E on NW 54th Ave.	(515) 986-3929	479
Onawa—KOA—1/2 mi W of I-29, exit 112 on Hwy 175, (51040) 1 1/2 mi N on county rd.	(712) 423-1633	109
Stockton—KOA-DAVENPORT—1/4 mi S of (52769) I-80, exit 280 on CR Y-30.	(319) 284-6881	64
Story City—KOA-WHISPERING OAKS—1/4 mi W of I-35 on (50248) Story City Rd, 1/2 mi N on Timberland Dr.	(515) 733-2521	85

KANSAS

CITY—CAMPGROUND—LOCATION	PHONE	SITES
Abilene—SAFARI—6 mi E of Hwy 15 on (67410) I-70, 1 block N of exit 281 on Hwy 43.	(913) 598-2215	100
Goodland—MID/AMERICA CAMP INN—1/4 mi S (66735) of I-70 on Hwy 27.	(913) 899-5431	129
Lawrence—KOA—1/2 mi N of I-70 on US 59/40, 1/4 mi E (66044) on US 40.	(913) 842-3877	115
Melvern—MELVERN LANE—3 1/2 mi W of town on Hwy (66510) 174.	(913) 528-4900	310
Salina—KOA—1 3/4 mi E of I-70 on I-135, 1/2 (67401) block N on Bus US 81, 1/2 mi W on Diamond.	(913) 827-3182	145
Wakeeney—KOA—1/2 block S of I-70, exit 127 on (67672) US 283, 1/4 mi E on service rd.	(913) 743-5612	80

KENTUCKY

CITY—CAMPGROUND—LOCATION	PHONE	SITES
Aurora—SPORTSMAN'S—3 mi NW of Hwy 80 on US 68. (42025)	(502) 354-8493	65
Campbellsville—KOA-GREEN RIVER LAKE—9 mi S of US (42718) 68 on Hwy 55.	(502) 465-3916	100
Cave City—YOGI BEAR'S JELLYSTONE PK—3/4 mi W of (42127) I-65, exit 53 on Hwy 70.	(502) 773-3840	218
Corbin—KOA-LAUREL RIVER—1/4 mi W of I-75 and US (40701) 25 E on Hwy 770, follow signs 1/4 mi S.	(606) 528-1534	85
Crittenden—KOA-CINCINNATI SOUTH—1/4 mi E of I-75, (41030) exit 166 on Hwy 491, 2 mi S on US 25.	(606) 428-2000	100
Eddyville—DAYTONA SHORES—3 1/2 mi S of US 62 on (42038) Hwy 93, 100 yds W on Hwy 293, 1 mi S.	(502) 388-7709	255
Elizabethtown—GLENDALE—5 1/2 mi S of US 31 (42701) on I-65, 1/4 mi E on Hwy 222, 3/4 mi N.	(502) 369-7755	108
Georgetown—KOA-BLUEGRASS—1 mi NE of I-75, exit 126 (40324) on US 62.	(502) 863-1205	100
Horse Cave—KOA—100 yards W of I-65, exit 58 on (42749) Hwy 218.	(502) 786-2819	135
Owensboro—DIAMOND LAKE—11 1/2 mi W (42377) of Byp US 60 on Hwy 56, 1000 ft SW.	(502) 229-4146	450
Park City—KOA-MAMMOTH CAVE—1 1/4 mi NW of I-65, (42160) exit 48 on Hwy 255.	(502) 749-4400	204
Russell Springs—KOA—2 1/2 mi SE of Hwy 80 on Hwy (42642) 910, 4 1/2 mi S on Hwy 76, 1 1/2 mi SW.	(502) 866-5616	179
Shepherdsville—KOA-LOUISVILLE SOUTH—1 1/2 mi E of (40165) I-65 on Hwy 44.	(502) 543-2041	220
Williamsburg—WILLIAMSBURG TRAVEL TRAILER PK—50 (40769) yards W of I-75, exit 11 on Hwy 92.	(606) 549-2300	87

LOUISIANA

CITY—CAMPGROUND—LOCATION	PHONE	SITES
Abita Springs—MONEY HILL PLANTATION—7 mi NE of (70431) Hwy 59 on Hwy 435, 2 mi N on access rd.	(504) 892-0388	800
Baton Rouge—KOA—1/2 mi S of I-12, exit 10 on Hwy (70726) 3002, 1/2 mi W on Hwy 1034.	(504) 664-7281	125
Hammond—YOGI BEAR'S JELLYSTONE PK—10 mi S of (70455) I-55 on I-12, 3 mi N on Hwy 445.	(504) 542-1507	217
Henderson—FRENCHMAN'S WILDERNESS—6 3/4 mi E of (70517) Hwy 347 on I-10, 3/4 mi S on Hwy 3177.	(318) 228-2616	140
Lafayette—KOA—5 1/4 mi W of US 167 on I-10, (70501) 1/2 mi S of exit 97 on Hwy 93.	(318) 235-2739	150
Laplace—LAPLACE TRAILER PK—4 1/4 mi S of I-10 on US (70068) 51, 1 1/2 mi E on US 61.	(504) 652-9086	100
New Orleans—KOA—2 1/2 mi S of I-10 on Hwy 49, 3/4 (70123) mi E on US 48.	(504) 721-0246	133
New Orleans—NEW ORLEANS TRAVEL PK—7 blocks E of (70126) I-10, exit 241 on US 90.	(504) 242-7795	155
New Orleans—PARC D'ORLEANS I—8 blocks E of I-10 on (70126) US 90 (7676 Chef Menteur Hwy.)	(504) 241-3167	74
St. Joseph—KOA-LAKE BRUIN—1/2 mi SE of US 65 on (71366) Hwy 605.	(318) 766-3334	87
Shreveport—KOA-BOSSIER—1/2 mi S of I-20, exit 10 (71129) on Pines Rd, 1 mi W on Hwy 511.	(318) 687-4567	95
Slidell—KOA-NEW ORLEANS EAST—3 mi SW of I-12 on (70458) I-10, 1/2 mi E on Hwy 433.	(504) 643-3850	132
Vinton—KOA-LAKE CHARLES—1/10 mi N of I-10, (70668) exit 8 on Hwy 108, 1/10 mi W on Goodwin St.	(318) 589-2300	100

MAINE

CITY—CAMPGROUND—LOCATION	PHONE	SITES
Boothbay—SHORE HILLS—2 mi N of town on Hwy 27. (04537)	(207) 633-4782	100
Damariscotta—LAKE PEMAQUID—1 mi E of Hwy 130 on (04543) Bus US 1, 2 mi S on Biscay, 1/4 mi E.	(207) 563-5202	200
Kennebunk—YANKEELAND—2 mi NW of US 1 on Hwy 35, (04043) bear left at fork, 2 1/2 mi.	(207) 985-7576	130
Medway—KOA-KATAHDIN SHADOWS—1 1/2 mi W of I-95 (04430) on Hwy 157.	(207) 746-9933	108
Naples—FOUR SEASONS REC AREA—2 1/2 mi NW on Hwy (04055) 11/114 on US 302.	(207) 693-6797	125
North Monmouth—BEAVER BROOK—2 1/2 mi E of Hwy (04265) 106 on US 202, 3 mi NW on Wilson Pond Rd.	(207) 933-2108	200
North Waterford—PAPOOSE POND—3 1/4 mi E of Hwy 35 (04267) on Hwy 118.	(207) 583-4470	170

CITY—CAMPGROUND—LOCATION	PHONE	SITES
Old Orchard Beach—POWDER HORN CAMPING—1 mi NW (04064) of Hwy 5 on Hwy 98.	(207) 934-4733	364
Portland—WASSAMKI SPRINGS—2 1/4 mi W of I-95, (04092) exit 8 on Hwy 25, 3 mi S on Saco St.	(207) 839-4276	150
Scarborough—BAYLEY'S—3 mi S of Hwy 207 on (04074) US 1/Hwy 9, 2 1/2 mi SE on Hwy 9.	(207) 883-6043	299
South Casco—POINT SEBAGO CAMP—2 mi NW of city on (04015) US 302, 1 1/2 mi SW on Point Sebago Rd.	(207) 655-3821	470
South Lebanon—THE KING'S AND QUEEN'S COURT—2 (04027) mi N of US 202 on River Rd.	(207) 339-9580	450
Wells—BEACH ACRES—1 1/2 mi E of I-95 on (04090) Hwy 109, 2 1/2 mi S on US 1, 1/4 mi E.	(207) 646-5612	230

MARYLAND

CITY—CAMPGROUND—LOCATION	PHONE	SITES
Annapolis—KOA-CAPITOL—1/2 mi N of Hwy (21108) 32 on Hwy 3, 1/2 mi W on Hog Farm Rd.	(301) 923-2771	119
Freeland—MORRIS MEADOWS REC FARM—4 mi W of (21053) I-83, exit 36 on Freeland Rd.	(301) 357-8689	250
Greensboro—HOLIDAY PARK—1 mi NE of Hwy 313 & 314 (21639) on Boyce Mill Rd, 2 mi N on Drapers Mill Rd.	(301) 482-6797	140
Hagerstown—SAFARI—1 1/2 mi S of I-70 on I-81, 1 mi S (21795) on US 11, 2 1/2 mi E on Hwy 68.	(301) 223-7117	100
Leonardtown—LA GRANDE ESTATE—4 mi S of Hwy 245 (20650) on Hw 5.	(301) 475-8550	79
Lothian—KOA—1/2 mi E of US 301 on Hwy 4, (20820) 1/2 mi E on Hwy 408.	(301) 627-3909	300
Ocean Pines—WHITE HORSE—1 mi N of Hwy 90 (21811) on Hwy 589, 2 mi E on Beauchamp Rd.	(301) 641-1102	175
Perryville—RIVERSIDE—3 1/4 mi E of US 222 on US (21903) 40, 1 mi E on Hwy 7, 3 1/2 mi S on CR 267.	(301) 642-3431	300
Princess Anne—PRINCESS ANNE CG—1/2 mi N of (21853) Hwy 362 on US 13.	(301) 651-1520	70
Smithsburg—RAVEN ROCK—4 mi NE of Hwy 64 on Hwy (21783) 491.	(301) 824-7101	102
Woodbine—RAMBLIN' PINES—2 1/2 mi N (21797) of I-70 on Hwy 97, 1/2 mi NW on Hoods Mill.	(301) 795-5161	97

MASSACHUSETTS

CITY—CAMPGROUND—LOCATION	PHONE	SITES
Bellingham—CIRCLE CG FARM—1 mi S of I-495 on (02019) Hwy 126, 1/2 mi SE on N. Main St.	(617) 966-1136	126
Bourne—BAY VIEW—1 mi S of US 6 on Hwy 28. (02532)	(617) 759-7610	286
Brewster—SPRAWLING HILLS PK—3 1/4 mi N of US 6 on (02631) Hwy 124, 1/2 mi W on Tubman Rd.	(617) 896-3939	330
Cedarville—INDIANHEAD RESORT—2 mi N of Hwy 3, (02360) exit 2 on Hwy 3A.	(617) 888-3688	180
Foxboro—NORMANDY FARM FAMILY—1 mi N of I-495 on (02035) US 1, 1 1/2 mi E on Thurston-West Sts.	(617) 543-7600	200
Granville—SUNLEDGE—4 mi W of Hwy 189 (01034) on Hwy 57.	(413) 357-6494	150
Middleboro—KOA-PLYMOUTH ROCK—2 3/4 mi E of Hwy (02346) 25 on US 44.	(617) 947-6435	267
Monson—PARTRIDGE HOLLOW—1/4 mi E of Hwy 32 on (01057) Wales Rd, 2 1/2 mi NE on Munn Rd.	(413) 267-5122	125
Orange—WAGON WHEEL—2 mi N of Hwy 2 on Hwy (01364) 122, 3 mi W on Hwy 2A, 2 mi N.	(617) 544-3425	108
Rochester—KOA-CAPE COD—100 feet N of I-195 on (02770) Hwy 105, 4 mi NE on CR, 1 1/2 mi W on High.	(617) 763-5911	200
Sandwich—PETERS POND PK—2 mi S of US 6, exit 2 on (02563) Hwy 130, 2 mi SE on Cotuit Rd.	(617) 477-1775	417
Savoy—SHADY PINES—3 mi E of Hwy 8A on Hwy 116. (01256)	(413) 743-2694	125
Shelburne Falls—SPRINGBROOK—5 1/2 mi W of I-91 on (01370) Hwy 2, 1 1/2 mi N on Little Mohawk, 1 mi NW.	(413) 625-6618	100
Sturbridge—YOGI BEAR'S JELLYSTONE PK—1 mi W of (01566) US 20 on I-86, 3/4 mi E of exit 3 on River Rd.	(617) 347-9570	200
Vineyard Haven—MARTHA'S VINEYARD FAMILY—1/4 (02568) mi S of ferry on Water St.	(617) 693-3772	180
Washington—SKYVIEW CAMPSITES—10 mi N of US 20 (01235) on Hwy 8, 1 3/4 mi E on Summit Hill Rd.	(413) 623-8821	104
Wrentham—KOA-BOSTON HUB—1/4 mi S of I-495 on (02093) Hwy 1A.	(617) 384-8930	110

MICHIGAN

CITY—CAMPGROUND—LOCATION	PHONE	SITES
Brighton—LAKE CHEMUNG—3 mi NW of I-196, exit 145 (48843) on Grand River Rd, 2 mi N on Hughes Rd.	(517) 546-6361	340
Cambridge Junction—KOA-IRISH HILLS—5 mi W of Hwy (49233) 50 on US 12.	(517) 592-6751	130
Cedar—LEELANAU PINES—1/2 mi E and N of CR 651 on (49621) CR 645, 3 mi E and N on CR 643.	(616) 228-5742	150
Champion—MICHIGAMME SHORES—2 1/2 mi W of (49814) west town limits on US 41/Hwy 48.	(906) 339-2116	76
Coldwater—WAFFLE FARM—2 3/4 mi W of I-69, exit 16 (49036) on Jonesville Rd, 3/4 mi N on Union City Rd.	(517) 278-4315	290
Coloma—KOA-BENTON HARBOR/ST JOSEPH—4 mi N of (49084) I-94 on I-196, 1/4 mi E on Coloma-Riverside.	(616) 849-3333	123
Decatur—TIMBER TRAILS—7 mi S of I-94 on Hwy 51, (49045) 2 mi W on Hwy 51, 3/4 mi N on 47 1/2 St.	(616) 423-7311	107
Dundee—KOA-MONROE COUNTY—9 mi S of Hwy 50 on (49270) US 23, 200 yds SE on Summerfield Rd.	(313) 856-4972	199
Fitchburg—WILDERNESS—1/4 mi W of Hwy 106 on Hwy (49285) 52, 6 mi W on Fitchburg.	(517) 565-3200	200
Gaylord—KOA—1/2 mi W of I-75 on Hwy 32. (49735)	(517) 732-4126	136
Grayling—YOGI BEAR'S JELLYSTONE CAMP RESORT—4 (49738) 1/2 mi E of I-75, exit 251 on Four Mile Rd.	(517) 348-6431	268
Holly—HOLLY HILLS—100 yds E of I-75 on Fenton- (48442) Grange Hall Road.	(313) 634-8621	147
Indian River—YOGI BEAR'S JELLYSTONE PK—4 1/2 mi (49749) E of I-75, Indian River exit on Hwy 68.	(616) 238-8259	173
Iron Mountain—RIVERS BEND—1/2 mi W of Hwy 95 on (49801) US 2, 1/2 mi SW on Pine Mtn, 1/4 mi NW.	(906) 774-9817	130
Jackson—KOA—5 1/4 mi E of US 127 on I-94, 1/4 mi S on (49240) Race, 1/4 mi E on Ann Arbor, 3/4 mi E.	(517) 522-8459	145
Mackinaw City—MACKINAW—3 mi SE of I-75 on US 23. (49701)	(616) 436-5584	600
Newaygo—WOODS & WATER—4 1/2 mi N of Hwy 82 on (49349) Hwy 37, 1 1/2 mi E on 40th, 1/2 mi S.	(616) 689-6701	280

CITY—CAMPGROUND—LOCATION	PHONE	SITES
Newberry—KOA—1/4 mi E of Hwy 123 on Hwy 28. (49868)	(906) 293-3452	148
North Branch—SUTTER'S—1 3/4 mi E of Hwy 24 on Hwy (48461) 90, 1/2 mi S on McKibben, 1/2 mi W on Tozer.	(313) 688-3761	140
Northport—TIMBER SHORES RESORT—2 mi SE of Hwy (49670) 201 on Hwy 22.	(616) 386-5191	561
Oscoda—KOA—1 1/2 mi S of city limits on US 23, 3/4 mi (48750) W on Johnson Rd, 3/4 mi S on Forest Ave.	(517) 739-5115	100
Port Huron—CRAZY HORSE—3 mi W of I-94 on Hwy 21, (48060) 1/2 mi N on Wadhams, 1/4 mi E on Lapeer.	(313) 987-4070	160
Powers—TAMU SAFARI CAMP—1/2 mi S of US 2 on US (49874) 41, 1 mi E on paved rd.	(906) 497-5457	100
Roscommon—ROSCOMMON-HIGGINS LAKE SAFARI— (48653) 1 1/4 mi W of I-75, exit 244 on Old Hwy 76.	(517) 275-8151	100
Sand Lake—PARADISE COVE—1 mi E of US 131 on Sand (49343) Lake Rd.	(616) 866-1415	135
Saugatuck—KOA-WEST WIND RESORT—1/2 mi W of (49453) I-196, exit 41 on CR A-2	(616) 857-2528	150
Six Lakes—PLEASURE POINT—1 mi E of Hwy 46/66 on (48886) Bridge St-Fleck St, 1/2 mi N on Musson Rd.	(517) 365-3133	155
St. Ignace—KOA—3 mi W of I-75 on US 2. (49781)	(906) 643-9305	200
Topinabee—KOA-INDIAN RIVER—1 1/2 mi NE of I-75 (49791) exit 313 on Hwy 27.	(616) 238-7733	150
Traverse City—YOGI BEAR'S—E of Hwy 37 on US (49684) 31, 2 mi S on 4 Mile Rd, 2 mi E on Hammond.	(616) 947-2770	221
Vicksburg—OAK SHORES RESORT—2 mi E of east city (49097) limits on W Ave, 1/2 mi N on 28th St.	(616) 649-1310	117
West Branch—KOA—3 mi S of I-75, exit 215 on (48661) Cook, 1 mi W on Channel, 1/4 mi W.	(517) 345-1203	100

MINNESOTA

CITY—CAMPGROUND—LOCATION	PHONE	SITES
Bemidji—GULL LAKE—11 mi N of US 2 (56683) on US 71, 5 mi N on CR 23 and 29.	(218) 586-2842	129
Cannon Falls—KOA—2 1/2 mi E of US 52 on Hwy 19, (55009) 2/10 mi S on Oak Lane.	(507) 263-3145	150
Detroit Lakes—RIDGEWOOD—2 1/4 mi S of US 10 on US (56501) 59 & Hwy 34, 1 1/4 mi W on CR 6.	(218) 847-8436	180
Garrison—KOA-MILLE LACS—1/2 mi N of Hwy 18 on US (56450) 169 and Hwy 18.	(612) 692-4587	100
Hinckley—ST CROIX HAVEN—23 1/2 mi E of I-35 on Hwy (55037) 48, 1 mi N on CR 173.	(612) 655-7989	106
Lincoln—KOA-FISHTRAP LAKE—1 mi S of town limits on (56443) US 10, 1 1/4 mi E on Fishtrap Lake Dr.	(218) 575-2485	150
Minneapolis—KOA—3 mi N of I-494 on I-94, 2 mi W on CR (55374) 30, 1 mi N on Hwy 101.	(612) 420-2255	160
Onamia—RUM RIVER VILLAGE—9 mi S (56401) of Hwy 27 on US 169.	(612) 532-3166	220
Owatonna—KOA—1 1/2 mi E of I-35, exit 40 on US 14/218, (55060) 1/2 mi S on CR 45, 1 1/2 mi W on CR 18.	(507) 451-8050	185
Park Rapids—VAGABOND VILLAGE—7 1/2 mi N of Hwy 34 (56470) on US 71, 6 mi E on CR 40.	(218) 732-5234	86
Rochester—SILVER LAKE—2 3/10 mi N of US 14 (55901) on US 63.	(507) 289-6412	85
Sturgeon Lake—TIMBERLINE—2 1/4 mi W of I-35 (55783) on CR 46, 3/4 mi N on access road.	(218) 372-3272	110
Theilman—WHIPPOORWILL—2 1/2 mi S of village (55978) limits on CR 4.	(507) 534-3569	209
Waseca—KIESLER'S CLEAR LAKE—1 1/2 mi E of Hwy 13 (56093) on US 14.	(507) 835-3179	200
Waterville—KAMP DELS—1 mi N of Hwy 60 on Hwy 13, (56096) 1/2 mi E on CR 131.	(507) 362-8616	200

MISSISSIPPI

CITY—CAMPGROUND—LOCATION	PHONE	SITES
Bay St. Louis—BUCCANEER STATE PK—10 mi W of town (39756) on US 90.	(601) 467-3822	104
Gautier—KAMPERS KOVE—6/10 mi E of Van Cleave Rd (39553) on US 90, 2/10 mi S on campground rd.	(601) 497-4186	70
Gulfport—GAYWOOD—4 mi E of US 49 on (39501) US 90, 1 1/2 mi N on Cowan Rd.	(601) 896-9831	65
Lucedale—KOA-MEADOWLAWN—1 mi S of Hwy 98 on (39452) Hwy 613, 1/2 mi S on Mill St. ext.	(601) 947-8193	56
Meridian—NANABE CREEK—5 mi E of town (39301) on I-59/20, 1 mi N of Russell exit.	(601) 485-4711	66
Ocean Springs—KOA—1/2 mile N of I-10 on Hwy 57. (39564)	(601) 875-2100	183
Picayune—YOGI BEAR'S JELLYSTONE—1 mi E of (39466) I-59, exit 2 on Hwy 43, 1/2 mi S.	(601) 798-2239	193

MISSOURI

CITY—CAMPGROUND—LOCATION	PHONE	SITES
Allenton—SAFARI ST. LOUIS SOUTHWEST—1/2 mi W of (63069) I-44, Pacific-Allenton exit on N Fox Creek.	(314) 587-3300	125
Branson—YOGI BEAR'S JELLYSTONE PK—7 mi S of Hwy (65616) 76 on US 65, 1 mi W on Hwy P.	(417) 334-4131	328
Columbia—LEWIS AND CLARK—100 yds S of (65279) I-44, Richeport exit on Hwy BB.	(314) 364-5490	266
Cuba—YOGI BEAR'S JELLYSTONE PK—3 mi E of Hwy 19 (65453) on I-44, 1 1/4 mi N on Hwy UU.	(314) 885-2541	380
Eureka—TOP NOTCH CAMP—1 1/4 mi W of (63025) I-44 on N Outer Rd, 1/4 mi N.	(314) 587-3100	300
Joplin—KOA—1/8 mi S of I-44, exit 4 on Hwy 43. (64801)	(417) 623-2246	245
Kimberling City—SAFARI OF TABLE ROCK—1 1/2 mi W of (65686) of Hwy 13 on Hwy 00.	(417) 739-4958	100
Lake Ozark—LAZY "K"—1 1/2 mi N of (65049) Bus US 54 on US 54, 1 1/2 mi E on Hwy V.	(314) 365-2374	100
Lebanon—BENNETT SPRING STATE PARK—12 mi W of (65536) town on Hwy 64.	(417) 532-4338	235
Lesterville—TWIN RIVERS LANDING—1/4 mi E of (63654) east city limits on Hwy 49/72, 1/2 mi SW.	(314) 637-2274	108
Pevely—SAFARI ST. LOUIS SOUTH—50 yards W of I-55 (63070) on Hwy Z, 3/4 mi S on Weier Rd.	(314) 479-4433	90
Theodosia—FT. COOK RV PK—1/2 mi E of (65761) east city limits on US 16D.	(417) 273-4444	98
Villa Ridge—DEANS RESORT—100 yards W of I-44 on (63089) US 50, 1 mi NE on CR "AT."	(314) 742-2633	250
Wentzville—SAFARI CAMP ST. LOUIS WEST—7 mi E of (63366) I-70 on US 40/61.	(314) 441-0123	135

MONTANA

CITY—CAMPGROUND—LOCATION	PHONE	SITES
Alder—KOA—1/4 mi E of village center on Hwy 287. (59710)	(406) 842-5677	89
Big Sky—CAMPER VILLAGE—4 mi W of town center on (59716) Spur 191.	(406) 995-4407	172
Billings—KOA-METRO LINDE'S LANDING—1 block (59101) S of I-90 on Hwy 3, 1/2 mi SE on Garden.	(406) 252-3104	171
Bozeman—KOA—1 mi S of Hwy 84 on US 191. (59715)	(406) 587-3030	150
Butte—KOA—1 block N of I-90 and I-15 (59701) on Montana St, 1/4 mi E on George St.	(406) 792-0663	100
Clinton—THE ELKHORN GUEST RANCH—4 mi S of (59825) I-90, exit 126 on Rock Creek Rd.	(406) 825-3220	150
Hungry Horse—FLATHEAD RIVER RANCH—3/4 (59919) mi W of town on US 2, 1/2 mi S.	(406) 387-5482	84
Kalispell—GLACIER PINES—3 1/2 mi E of (59901) US 93 on US 2.	(406) 257-6760	149
Missoula—KOA-EL MAR—1 1/2 mi S of I-90, exit 101 on (59801) Reserve St, 1/4 mi NW on Tina St.	(406) 549-0881	208
Polson—KOA-POLSON—4 1/2 mi E of US 93 on Hwy 35. (59860)	(406) 883-2151	130
Three Forks—KOA—1 mi S of I-90, exit 274S on Hwy 287. (59752)	(406) 285-3611	90
West Yellowstone—KOA—6 mi W of US 287 on US (59758) 191/20.	(406) 646-7607	233

NEBRASKA

CITY—CAMPGROUND—LOCATION	PHONE	SITES
Grand Island—WEST HAMILTON PK—6 mi E (68832) of US 34/281 on I-80, 1/8 mi S of exit 318.	(402) 886-2249	105
Gretna—KOA-SOUTH OMAHA—1/2 mi N of (68028) I-80, exit 432 on Hwy 31, 700 ft W.	(402) 332-3010	90
Kearney—SAFARI—1/2 mi N of I-80, exit 272 (68847) on Hwy 44.	(308) 234-1532	145
North Platte—CUTTY'S—1 block N of I-80 on (69101) US 83, 1/2 mi S and E on frontage rd.	(308) 534-2265	200

NEVADA

CITY—CAMPGROUND—LOCATION	PHONE	SITES
Boulder City—LAKESHORE—2 mi NE of US 93 on Hwy 41. (89005)	(702) 293-2540	75
Ely—KOA—3 mi SE of town center on US 6/50/93. (89305)	(702) 289-3413	100
Las Vegas—HACIENDA CAMPERLAND—1/2 mi E of I-15 on (89119) Tropicana Ave, 3/4 mi S on Las Vegas Blvd.	(702) 739-8214	451
Las Vegas—KOA—4 mi E of I-15 on Sahara (89121) Ave, 1 1/2 mi S on Boulder Hwy.	(702) 451-5527	320
Las Vegas—STARDUST CAMPERLAND—3/4 mi E of I-15 (89109) on Sahara Ave, 2/3 mi S on Las Vegas Blvd.	(702) 732-6564	400
Reno—MGM RENO—1/2 mi S of I-80 (89502) on US 395, 1 block E on Glendale Rd.	(702) 789-2147	462
Wells—KOA-CRESTED ACRES—5 mi W of US 93 on I-80. (89835)	(702) 752-3557	250

NEW HAMPSHIRE

CITY—CAMPGROUND—LOCATION	PHONE	SITES
Ashland—YOGI BEAR'S—3/4 mi SE of I-93 (03217) on US 3, 2 3/4 mi S on Hwy 3-B.	(603) 968-3654	220
Barrington—LEN-KAY—1 1/2 mi S of Hwy 9 on Hwy 125, (03825) 1 mi W on Beauty Hill Rd, 1 mi S on Hall Rd.	(603) 664-9333	120
Center Ossipee—TERRACE PINES—1/8 mi S of Hwy 25 (03814) on Hwy 16, 4 1/2 mi SW on Don Hole Pond.	(603) 539-6210	130
Conway—PINE-KNOLL CAMPING—5 mi S of US 302 on (03818) Hwy 16.	(603) 447-8982	131
Gorham—TIMBERLAND—5 mi E of wy 16 on US 2. (03581)	(603) 466-3872	120
Goshen—RAND POND—4 mi S of Hwy 11 on Hwy 10, 2 (03752) 1/2 mi E on Rand Pond Rd.	(603) 863-3350	115
Hampton Falls—WAKEDA—1/4 mi E of I-95 on Hwy 107, (03844) 3 mi N on US 1, 3 1/2 mi N on Hwy 88.	(603) 772-5274	320
Hudson—TUCKAWAY—1 1/2 mi N of Hwy 3A on Hwy 102, (03051) 1 mi E on Old Derry Rd.	(603) 882-2056	120
Laconia—WEIRS BEACH—6 mi N of Hwy 11 on US 3. (03246)	(603) 366-4747	148
Lancaster—ROGER'S—1 1/2 mi E of US 3 on US 2. (03584)	(603) 788-4885	350
Lee—SIESTA SHORES—4 1/2 mi W of Hwy 108 on Hwy (03857) 152.	(603) 659-3852	88
Milton—MI-TE-JO—8 mi N of Spaulding Tpk on Hwy 16, 1 (03851) mi E of Townhouse Rd, 1/2 mi N.	(603) 652-4297	210
New Hampton—CLEARWATER—3 mi E of I-93, exit 23 (03824) on Hwy 104.	(603) 279-7761	140
North Conway—SACO RIVER—1/2 mi N (03860) of US 302 on Hwy 16.	(603) 356-3360	148
North Woodstock—LOST RIVER VALLEY—3 1/2 mi W of (03262) I-93 on Hwy 112.	(603) 745-8321	109
Ossipee—BEAVER HOLLOW—1 mi S of Hwy 28 on Hwy 16. (03864)	(603) 539-4800	120
Raymond—PINES ACRES FAMILY—3/4 mi S of Hwy 107 (03077) on Hwy 101, 1/2 mi SW on Prescott Rd.	(603) 895-2519	350
Richmond—SHIR-ROY—1 mi S of Hwy 119 on Hwy 32. (03470)	(603) 239-4768	118
Sanbornville—LAKE FOREST RESORT—3 mi E of Hwy 16 (03872) on Hwy 153, 1 1/2 mi E on Acton Ridge.	(603) 522-3306	130
West Ossipee—TOTEM POLE PARK—1/2 mi NE of Hwy 16 (03818) on Hwy 41, 4 mi SE on Ossipee L. Rd.	(603) 539-4420	400

NEW JERSEY

CITY—CAMPGROUND—LOCATION	PHONE	SITES
Branchville—KYMER'S—4 8/10 mi N of US 206 (07826) on CR 519, 1 mi W on Kymer Rd.	(201) 875-3167	230
Buena—BUENA VISTA CAMPING WORLD—1/10 (08310) mi E of Hwy 54 on Hwy 40.	(609) 697-2004	500
Cape May—LAKE LAURIE—2 1/2 mi S of Hwy 47 on US 9. (08204)	(609) 884-3567	750
Cape May Court House—BIG TIMBER LAKE—3 mi N of (08210) CR 585/657 on US 9, 1 mi W.	(609) 465-4456	540
Cape May Court House—DRIFTWOOD—4 1/2 (08210) mi N of CR 585/657 on US 9.	(609) 263-2677	500
Egg Harbor City—KOA-ATLANTIC CITY WEST—1 mi W of (08215) Hwy 50 on US 30, 1 mi N on Heidleburg.	(609) 965-1944	465
Elmer—TALL PINES—6 2/10 mi S of US 40 on Daretown (08318) Rd, 3/10 mi E on Beal Rd.	(609) 451-7479	225
Glassboro—LAKE KANDLE—2 2/10 mi N of US 322 on (08080) Hwy 47, 1 1/2 mi E on Chapel Heights Rd.	(609) 589-2158	160
Hope—TRIPLEBROOK—3 3/10 mi SW of CR 521 on High (07825) St, 1 1/10 mi NW, 4/10 mi NE on Honeyrun.	(201) 459-4079	187
Jackson—BUTTERFLY—3 mi E of CR 57 on CR 528, (08527) 2 1/4 mi E on CR 527/528, 1/2 mi N.	(201) 928-2107	135
Mays Landing—WINDING RIVER—4 mi W of Hwy 50 on (08330) CR 559.	(609) 625-3191	133
Ocean View—OCEAN VIEW—2 1/4 mi S of Hwy 50 on US (08230) 9.	(609) 263-8382	1502
Pleasantville—PLEASANTVILLE CAMPGROUND NO. 2— (08232) 6/10 mi N of Atlantic City Expwy on US 9.	(609) 641-3176	300
Port Republic—CHESTNUT LAKE—1/2 mi W of US 9 on (08241) CR 575.	(609) 652-7251	200
Sussex—PLEASANT ACRES FARM—4 3/4 mi N of (07461) Hwy 284 on Hwy 23, 1 mi E on Dewitt Rd.	(201) 875-4166	300
Toms River—KOA-SURF & STREAM—1 1/2 mi W of US 9 (08753) on CR 571.	(201) 349-8919	250
Tuckerton—KOA-ATLANTIC CITY NORTH—1 mi S of (08087) CR 539 on US 9, 2 1/2 mi W.	(609) 296-9163	150
West Milford—KOA—4 1/4 mi E of Hwy 23 on Echo Lake, (07480) 2 mi N on Macopin, 1 1/4 mi E on Westbrook.	(201) 697-8400	150
Wildwood—HOLLY SHORES—1 1/4 mi S of Hwy 47 (08204) on US 9.	(609) 886-1234	300
Williamstown—HOSPITALITY CREEK—5 mi E of Hwy 42 (08094) on US 322, 1/4 mi S on CR 538.	(609) 629-5140	215

NEW MEXICO

CITY—CAMPGROUND—LOCATION	PHONE	SITES
Alamogordo—KOA—3 3/4 mi N of US 70/82 on US 54, 1 (88310) 1/2 blocks E on 24th St.	(505) 437-3003	82
Albuquerque—KOA—4 1/2 mi W of I-25 on I-40, 1 block N (87120) on Coors Rd, 1 block W on Ouray Rd.	(505) 831-1911	91
Albuquerque—KOA—8 3/4 mi E of I-25 on I-40, 1/2 mi SW (87123) on Central Ave, 1 block N on Figueroa St.	(505) 296-2729	200
Bernalillo—KOA-ALBUQUERQUE-NORTH—1/2 block W of (87004) I-25, exit 240, 1 mi N on frontage rd.	(505) 867-5227	82
Clayton—KOA-MEADOWLARK—3/4 mi S of US 64 on US (88415) 87, 4 blocks E on Spruce St.	(505) 374-9508	91
Deming—KOA-ROADRUNNER—1 block S of I-10, exit 82A (88030) on Hwy 11, 2 mi E on Pine St.	(505) 546-9035	74
Edgewood—RED ARROW—1 block S of I-40 and Hwy 344, (87015) 1/2 mi E on frontage rd.	(505) 281-3721	102
Gallup—KOA—3 1/4 mi W of US 666/Hwy 32 on US 66. (87301)	(505) 863-5021	182
Las Cruces—KOA—1 3/4 mi E of (88001) exit 135 on US 70.	(505) 526-9030	102
Las Vegas—KOA-RANCHERO—1 block SE of I-25, exit 339 (87701) on US 84, 1/2 mi SW on frontage rd.	(505) 454-0180	58
Prewitt—KOA-GRANTS WEST/PREWITT—1/4 mi N of I-40, (87045) exit 63 on Hwy 412, 1/4 mi E on frontage rd.	(505) 876-2662	70
Ruidoso Downs—CIRCLE B—1 1/2 mi E of city limits on (88346) US 70.	(505) 378-4990	211
Santa Fe—KOA—3 mi E of US 85 on US 84-285, 9 mi N (87501) on I-25, 1 mi N and E of exit 290.	(505) 983-3482	108
Tucumcari—KOA—3 1/2 mi E of Hwy 18 on US 66. (88401)	(505) 461-1841	153

NEW YORK

CITY—CAMPGROUND—LOCATION	PHONE	SITES
Ausable Chasm—KOA-AUSABLE CHASM—1/4 mi E of US (12911) 9 on Hwy 373.	(518) 834-9990	120
Austerlitz—WOODLAND HILLS—3 mi N of Hwy 203 on (12017) Hwy 22, 6/10 mi W on Middle, N. on Fog Hill.	(518) 392-3557	128
Bath—HICKORY HILL—1 mi N of Hwy 17, exit 38 on (14810) Hwy 54, 2 mi N on Haverling.	(607) 776-4345	125
Bergen—KOA-SOUTHWEST ROCHESTER—2 mi N of I-90, (14416) exit 47 on Hwy 19.	(716) 494-1550	149
Canandaigua—KOA—4 mi S of I-90, exit 44 on Hwy 332, (14424) follow signs.	(716) 398-3582	120
Chautauqua—CAMP CHAUTAUQUA—2 3/4 mi E of east (14785) town limits on Hwy 394.	(716) 789-3435	450
Copake—OLEANA FAMILY CAMPGROUND—1 1/2 mi W (12593) of town limits on CR 7A.	(518) 329-2811	320
Corinth—ALPINE LAKE—1 1/2 mi S of town limits (12822) on Hwy 9N.	(518) 654-6260	290
Corning—FERENBAUGH—6 mi N of Hwy 17 on Hwy 414. (14830)	(607) 962-6193	173
Dansville—SKY BROOK—1 mi W of Hwy 36 on Hwy 414. (14437) 1 mi SW on Ossian, 2 3/4 mi S on McCurdy.	(716) 335-3904	225
Darien Center—SKYLINE RESORT—4 mi E of Hwy (14040) 77 on Hwy 20, 1 mi S on Townline Rd.	(716) 591-2021	225
Ellenville—BIRCHWOOD ACRES—8 mi W of US 209 on (12789) Hwy 52, 1/2 mi S on Martinfeld.	(914) 434-4743	200
Ellicottville—YOGI BEAR'S JELLYSTONE PK—6 mi N of (14729) Hwy 242 on US 219, 3 1/2 mi W.	(716) 699-2351	659
Forestport—FAIRWOOD EVERGREEN—2 mi N of Hwy 12 (13338) on Hwy 28, follow signs 3 mi S, 1/2 mi E.	(315) 831-5077	124
Gainesville—WOODSTREAM—1 mi S of Hwy 78 on (14066) Hwy 19, 3/4 mi E on Lamont.	(716) 493-5643	200
Gilboa—NICKERSON PK—5 mi N of Hwy 23 on Hwy 30, (12076) follow signs.	(607) 588-7327	300
Herkimer—HERKIMER DIAMOND—7 mi N of Hwy 5 (13350) on Hwy 28.	(315) 891-7355	89
Lake George—LAKE GEORGE RV PK—3/4 mi N of I-87, (12845) exit 20 on Hwy 9, 1/4 mi E on Hwy 149.	(518) 792-3775	350
Lake Placid—KOA-L. PLACID—10 mi E of (12946) town limits on Hwy 86.	(518) 946-7878	162
Lakeville—CONESUS LAKE—5 3/4 mi S of US 20A (14435) on E Lake Rd.	(716) 346-5472	125
Massena—KOA-MASSENA—3 3/4 mi E of Hwy 420 on (13662) Hwy 37, 1/2 mi NE.	(315) 769-9483	115
Mayville—SAFARI CAMP-CHAUTAUQUA LAKE—6 mi S of (14728) Hwy 394 on Hwy 17, follow signs.	(716) 386-3804	100
Middleburg—TWIN OAKS—2 3/4 mi N of Hwy 30 on (12122) Hwy 145, 1 mi W on CR 41.	(518) 827-5641	125
Montezuma—HEJAMADA—3 1/2 mi N of Hwy 5/20 on (13117) Hwy 90, 1/4 mi SE on Fuller, E on McDonald.	(315) 776-5887	155

CITY—CAMPGROUND—LOCATION	PHONE	SITES
Natural Bridge—KOA—1 mi E of town limits on Hwy 3. (13665)	(315) 644-4880	117
Old Forge—KOA—1 mi N of town center on Hwy 28. (13420)	(315) 369-6011	160
Richburg—SUNNY HILL—3 mi N of Hwy 417 on Hwy 275, 1 mi N on CR 40, 1 mi N on CR 5B. (14774)	(716) 928-1053	180
Roscoe—RUSSELL BROOK—3/4 mi N of Hwy 17, exit 93 on Russell Brook Rd. (12776)	(607) 498-5416	160
Saratoga Springs—INTERLAKEN PARK—1 1/4 mi E of I-87 on Hwy 9P. (12866)	(518) 587-1444	240
South Glens Falls—JELLYSTONE PK—1/4 mi N of I-87, exit 17N on US 9, follow signs. (12831)	(518) 792-0485	225
Summitville—WANDERWOOD—1 mi W of Hwy 17 on old Hwy 17, 4 1/2 mi N on New Rd. (12781)	(914) 888-2161	200
Swan Lake—SWAN LAKE—1/2 mi E of Hwy 17, exit 102 on Old Hwy 17, 2 mi N on CR 75, 3/4 mi NW. (12734)	(914) 292-4781	135
Wading River—WILDWOOD STATE PK—3 mi E of town on Hwy 25A. (11792)	(516) 929-4314	322
Watkins Glen—KOA—4 1/2 mi S of Hwy 14 on Hwy 414. (14891)	(607) 535-7404	106
Westfield—KOA-LAKE ERIE—1/4 mi N of I-90, exit 60 on Hwy 17, 1 mi E on Hwy 5. (14787)	(716) 326-3573	116

NORTH CAROLINA

CITY—CAMPGROUND—LOCATION	PHONE	SITES
Belhaven—RIVERSIDE—10 mi E of Bus US 264, on US 264, 2 mi NW on CR 1300. (27810)	(919) 943-2849	100
Boone—KOA—3 mi N of US 221/421 on Hwy 194, 1 mi NW on Ray Brown Rd. (28607)	(704) 264-7250	125
Charlotte—FROG CREEK—13 mi S of I-85 on I-77, 1 mi W on Carowinds Blvd. (28224)	(704) 588-3363	207
Cherokee—FLAMING ARROW—2 mi S of US 19 on US 441. (28719)	(704) 497-5451	115
Cherokee—HAPPY HOLIDAY FAMILY—4 1/2 mi E of US 441 on US 19. (28719)	(704) 497-7250	326
Denver—CROSS COUNTRY—1 mi E of Hwy 16 on Hwy 150. (28037)	(704) 483-5897	400
Emerald Isle—HOLIDAY TRAV-L PARK—1 1/2 mi W of town center on US 58. (28557)	(919) 354-2250	300
Enfield—KOA—3/4 mi W of I-95, exit 154 on Hwy 481. (27823)	(919) 445-5925	120
Greensboro—KOA—1/4 mi S of I-85 and Hwy 6 on East Lee St, 1 mi W on Sharpe Rd. (27406)	(919) 274-4143	96
Hatteras—SURF AND SOUND—3 1/2 mi N of town center on Hwy 12. (27943)	(919) 986-2505	237
Hendersonville—JAYMAR TRAVEL PARK—3 mi E of I-26 on US 64. (28739)	(704) 685-3771	117
Holden Beach—SAND-N-SEA—5 mi W of Hwy 130 on Ocean Blvd. (28462)	(919) 842-6306	167
Kitty Hawk—OCEAN BEACH—5 mi N of US 158 and Bus US 158 on CR 1200. (27949)	(919) 261-2200	200
Laurinburg—SANDHILL—14 mi N of US 401 on US 501, 2 mi E of CR 1400. (28396)	(919) 281-4444	89
Linville Falls—BEAR DEN—7 mi S of US 221 on Blue Ridge Pky, 1/2 mi E on Bear Den Mountain Rd. (28777)	(704) 765-2888	138
Littleton—AMERICAN HERITAGE—1 1/2 mi N of town center on CR 1352, 2 mi NW on CR 1365. (27850)	(919) 586-4121	300
Lumberton—SAFARI—At US 74 and I-95, exit 14 on service rd. (28358)	(919) 739-4372	80
Manteo—SANDPIPER'S TRACE—3 1/2 mi W of town center on US 64/264. (27954)	(919) 473-3471	500
Morehead City—PENDER PARK CAMPSITES—8 mi W of US 70 on Hwy 24. (28570)	(919) 726-4902	150
Morganton—KOA—3 mi W of US 64 on I-40, 100 feet E on access rd. (28655)	(704) 584-1733	106
Rodanthe—KOA-HOLIDAY—1/2 mi S of town center on Hwy 12. (27968)	(919) 987-2307	255
Salter Path—ARROWHEAD—1/2 mi W of town center on Hwy 58. (28557)	(919) 726-7974	158
Salter Path—SALTER PATH—3/4 mi W of town center on Hwy 58. (28557)	(919) 726-2710	200
Selma—KOA-LAKEVIEW—1 mi NE of US 70A on I-95, 1/4 mi NE on frontage rd. (27576)	(919) 965-5923	93
Statesville—KOA—6 mi S of I-40 on I-77, 1/2 mi N on E frontage rd. (28677)	(704) 873-5560	88
Williamston—GREEN ACRES—5 mi S of US 64 on US 17, 1 mi W on Green Acres Rd. (27892)	(919) 792-3939	175
Wilmington—WILMINGTON SAFARI—4 mi NE of US 74 on US 17. (28405)	(919) 686-7705	100

NORTH DAKOTA

CITY—CAMPGROUND—LOCATION	PHONE	SITES
Bismarck—DAKOTA HILLS—1 mi N of I-94, exit 37 on county rd. (58501)	(701) 255-0873	107
Grafton—SHADY OAK—1/2 mi N of Hwy 17 on US 81, 1 mi E on 5th St, 1/4 mi N. (58237)	(701) 352-1005	48
Grand Forks—KOA-GRAND FORKS—3/4 mi S of I-29, exit 138 on W frontage rd. (58201)	(701) 775-2814	101
Williston—KOA-BUFFALO TRAILS—2 1/2 mi N of US 2 and 85 on US 2/85, 1/2 mi N. (58801)	(701) 572-3206	120

OHIO

CITY—CAMPGROUND—LOCATION	PHONE	SITES
Aurora—YOGI BEAR'S JELLYSTONE PK—4 mi E of Hwy 306 on Hwy 82. (44255)	(216) 562-6500	500
Berlin Heights—PIN OAK LAKE PK—3 1/2 mi E of Hwy 61 on Hwy 113. (44814)	(419) 588-2052	225
Bluffton—TWIN LAKES—1/10 mi S of I-75 on Hwy 235, 1/2 mi E on Twp Rd 234. (45817)	(419) 477-9109	84
Brookville—KOA/TALL TIMBERS—1/2 mi W of I-70 on Brookville-Salem Rd, 1/4 mi N on Wellbaum. (45309)	(513) 833-3888	140
Canal Fulton—CLAY'S PK—2 mi S of Hwy 21 on Hwy 93, 1/2 mi E on Patterson Rd. (44614)	(216) 854-3961	600
Circleville—AIRY ACRES—2 mi E of Hwy 56 on US 22, 1 1/2 mi N on Bolender Pontious, 1/2 mi E. (43113)	(614) 477-1909	250
Geneva—KOA—3 mi S of US 20 on Hwy 534, 1 1/2 mi E on Hwy 307. (44041)	(216) 466-4909	99
Kings Mills—KINGS ISLAND—1/2 mi E of I-71 on Kings Mills Rd, 1/2 mi S. (45034)	(513) 398-2901	350
Marlboro—SUNSET TRAILER PK—1/4 mi E of Hwy 44 on Hwy 619. (44641)	(216) 935-2733	250
Milan—TRAV-L-Park—250 yards N of Ohio Tpk, exit 7 on US 250. (44846)	(419) 499-4627	130
Mount Gilead—KOA-COLUMBUS NORTH—7 mi E of US 42, Hwy 61 on Hwy 95. (43338)	(419) 768-3428	147
North Baltimore—KOA—1/5 mi S of I-75, exit 168 on Grant, 3/4 mi E on Eagleville, 1 1/2 mi N. (45872)	(419) 257-2667	169
Salem—TIMASHAMIE TRAVEL TRAILER PK—5 3/4 mi SW of Byp Hwy 45 on CR 400. (44460)	(216) 525-7054	175
Sandusky—CEDAR POINT CAMPER VILLAGE—1 mi E of US 250 on US 6, 4 mi N on Causeway Dr. (44870)	(419) 626-0830	700
Shreve—WHISPERING HILLS—2 mi S of Hwy 226 on Hwy 514. (44676)	(216) 567-2137	340
Van Buren—PLEASANT VIEW—1 mi E of I-75 on Hwy 613, 1/2 mi SE on Twp Rd 218. (45889)	(419) 299-3897	255
Vanlue—KOA—1/2 mi N of Hwy 15 on Hwy 330, 1 block SE, 3/4 mi E on Twp Rd 171. (45890)	(419) 387-7738	190
Wilmington—HILLSIDE HAVEN—1/4 mi E of I-71 on Hwy 73, 2 1/2 mi S on Hwy 380, 1/4 mi E. (45177)	(513) 382-8591	327
Zanesville—KOA-ZANE GREY—3/4 mi S of I-70 and Hwy 93 on S Pleasant Grove Rd. (43701)	(614) 452-5025	125

OKLAHOMA

CITY—CAMPGROUND—LOCATION	PHONE	SITES
Checotah—KOA-EUFAULA LAKE—10 mi W of US 69 on I-40, 1/2 mi W of Pierce exit. (74426)	(918) 473-6511	469
Elk City—KOA-CLINTON LAKE—11 1/2 mi E of Hwy 6 on I-40 to Clinton Lake Road exit. (73626)	(405) 592-3256	129
Oklahoma City—RTA CAMPGROUNDS—1 mi S of I-44 on I-35/US 66. (73131)	(405) 478-0278	209
Salisaw—KOA—1/2 mi S of I-40 on US 59. (74955)	(918) 775-2792	78
Tulsa—KOA—1/4 mi N of I-44, exit 193 on East Ave, 1/4 mi E. (74015)	(918) 266-4227	200
West Siloam Springs—DRIPPING SPRINGS SAFARI—5 mi W of US 59 on Hwy 33. (73761)	(918) 422-5372	111

OREGON

CITY—CAMPGROUND—LOCATION	PHONE	SITES
Ashland—KOA-"GLENYAN"—3 1/4 mi SE of I-5 on Hwy 66. (97520)	(503) 482-4138	84
Cascade Locks—KOA-BRIDGE OF THE GODS—2 1/2 mi E of town center on Forest Lane. (97014)	(503) 374-8668	90
Creswell—KOA-SHERWOOD FOREST—1/2 block W of I-5, Creswell exit on Oregon Av. (97426)	(503) 895-4110	121
Gold Beach—INDIAN CREEK—1 mi N of town center on US 101, 1/2 mi E on Jerry's Flat Rd. (97444)	(503) 247-7704	126
Gold Hill—KOA-GOLD'N ROGUE—1/4 mi N of I-5 on Hwy 234, 1/4 mi E on Blackwell Rd. (97502)	(503) 855-7710	96
Joseph—WALLOWA LAKE STATE PK—6 mi S of town on Hwy 82. (97846)	(503) 432-5181	210
La Pine—LA PINE STATE REC AREA—7 mi N of town on US 97, follow signs. (97739)	(503) 536-2428	145
Madras—COVE PALISADES STATE PK—15 mi SW of town on US 97. (97741)	(503) 546-3591	186
Nehalem—NEHALEM BAY TRAILER PK—1 mi NW of town limits on US 101. (97131)	(503) 368-5180	100
Newport—BEVERLY BEACH STATE PK—7 mi N of town on US 101. (97365)	(503) 265-7655	278
Rogue River—VALLEY OF THE ROGUE STATE PK—3 mi E of town on I-5. (97537)	(503) 582-1312	174
Salem—KOA—1/4 mi E of I-5, exit 253 on Hwy 22. (97301)	(503) 581-6736	203
Tualatin—TRAILER PARK OF PORTLAND—3/4 mi N of I-205 on I-5, 1/4 mi E of exit 289 on Hwy 212. (97062)	(503) 638-7304	120

PENNSYLVANIA

CITY—CAMPGROUND—LOCATION	PHONE	SITES
Beaver Springs—GRAY SQUIRREL—4 1/2 mi N of US 522 on Hwy 235, 2 mi E on LR 54029, 1/2 mi N. (17813)	(717) 837-0333	230
Bellefonte—KOA—2 mi N of I-80, exit 24 on Hwy 26. (16823)	(814) 355-7912	142
Bowmansville—LAKE-IN-WOOD—3/4 mi E of Hwy 625 on Maple Grove, 1/2 mi S on Oaklyn, 1 1/2 mi E. (17555)	(215) 445-5525	200
Bushkill—KEN'S WOODS—1 1/2 mi W of US 209 on Bushkill Falls Rd, 4 mi N, 2 1/2 mi W. (18324)	(717) 588-6381	100
Carlisle—SAFARI-CARLISLE—1 mi S of Pa Tpk, exit 16 on US 11. (17013)	(717) 249-4563	115
Coatesville—BEECHWOOD—2 mi E of Hwy 82 on US 30, 1 1/2 mi N of Vet Hosp exit on Reeseville Rd. (19320)	(215) 384-1457	400
Denver—KOA-LANCASTER/READING—1/8 mi S of Pa Tpk, exit 21 on Hwy 272. (17517)	(215) 267-2112	120
Dillsburg—KOA-HARRISBURG SOUTH—3 1/2 mi S of Hwy 74 on US 15, 1/2 mi SE. (17019)	(717) 432-4523	114
Donegal—MAPLE GROVE—2 mi E of Pa Tpk, exit 9 on Hwy 31, 2 mi S. on Hwy 711. (15622)	(412) 455-3300	350
Ephrata—STARLITE—5 mi W of Hwy 272 on US 322, 1 mi N on Clay Rd, 1 1/4 mi E on Hopeland Rd. (17578)	(717) 733-9655	200
Gettysburg—BATTLEFIELD CAMP RESORT—4 mi S of US 30 on Bus US 15. (17325)	(717) 334-1577	250
Gettysburg—DRUMMER BOY—100 yards E of Byp US 15 on Hwy 116, 1/8 mi N on Rocky Grove Rd. (17325)	(717) 334-3277	300
Gettysburg—GETTYSBURG—3 mi W of US 30 on Hwy 116. (17325)	(717) 334-3304	250
Hawley—SAFARI-LAKE WALLENPAUPACK AREA—1 1/2 mi W of US 6 on Hwy 590, 1 mi N. (18428)	(717) 226-3317	146
Hershey—DECK ACRES—6 1/4 mi E of Hwy 743 on US 322, 3/4 mi S on Mt Pleasant Rd. (17042)	(717) 272-4677	110
Honey Brook—HONEY BROOK—3 1/2 mi S of US 322 on Hwy 10, 1/2 mi E on Hill Rd. (19344)	(215) 273-3152	180
Jim Thorpe—KOA—1/4 mi N of Hwy 903 on US 209, 2 mi W on Broadway. (18229)	(717) 325-2644	135
Knox—KOA-WOLF'S—1/8 mi N of I-80, exit 7 on Hwy 338. (16232)	(814) 797-1103	478
Lancaster—CIRCLE M—2 1/4 mi W of Hwy 283 on US 30, 5 1/2 mi S on Millersville-Rohrerstown Rd. (17603)	(717) 872-4651	217
Lenhartsville—ROBIN HILL—1/2 mi S of I-78, US 22 on Hwy 143, 1 1/2 mi E, follow signs. (19534)	(215) 756-6117	238
Manheim—DISTELFINK—1/4 mi N of Pa Tpk, exit 20 on Hwy 72. (17545)	(717) 665-9561	350
Meadville—PLAYLAND—4 mi E of Hwy 77 on Hwy 27, 3 1/2 mi SE on Wayland Rd. (16335)	(814) 425-7314	150
Mercer—SKY VIEW—3 mi N of US 19 on Hwy 258, 1/2 mi W on Skyline Rd. (16137)	(412) 662-5317	375
Mercersburg—SAUNDEROSA PK—4 1/2 mi W of Hwy 75 on Hwy 16, 2 1/2 mi S on Hwy 456. (17236)	(717) 328-2216	200
Morgantown—WARWICK WOODS—1/8 mi S of Pa Tpk on Hwy 10, 7 mi E on Hwy 23, 1/2 mi N. (19470)	(215) 286-9655	200
New Columbia—NITTANY MOUNTAIN—5 mi W of I-80, exit 30. (17856)	(717) 568-5541	350
New Holland—SPRING GULCH—4 1/2 mi S of Hwy 23 on Hwy 897. (17557)	(717) 354-3100	200
New Stanton—FOX DEN ACRES—1/2 mi N of Pa Tpk, exit 8 and I-70 on US 119, follow signs. (15672)	(412) 925-7054	300
Portland—DRIFTSTONE ON THE DELAWARE—4 mi S of Hwy 611 on River Rd. (18343)	(717) 897-6859	190
Sandy Lake—KOA-GODDARD—4 1/2 mi N of US 62 on Hwy 173, 4 mi W on Goddard St. Pk. Rd. (16145)	(412) 253-4645	400
Shartlesville—MOUNTAIN SPRINGS—1/2 mi N of I-78, US 22, exit 23 on Shartlesville Rd, 1/2 mi E. (19554)	(215) 488-6859	300
Stroudsburg—OTTER LAKE—4 mi N of I-80, exit 52 on US 209, 300 feet N on Hwy 402, 7 mi W. (18335)	(717) 223-0123	250
Stroudsburg—TIMOTHY LAKE—1/2 mi N of Hwy 402 on US 209, 2 mi W on Magic Valley Rd, 2 mi S. (18301)	(717) 588-6631	245
Tamaqua—ROSEMOUNT—4 mi SW of Hwy 309 on US 209, 2 1/2 mi S on New Ringgold Rd. (18252)	(717) 668-2580	150
Tunkhannock—KOA-SHADOW BROOK—2 mi E of Hwy 309 on US 6, 1/4 mi S. (18657)	(717) 836-4122	121
Upper Black Eddy—COLONIAL WOODS—1/2 mi N of town on Hwy 32, 4 mi W on Bridgeton Hill. (18972)	(215) 847-5808	185
Waymart—KEEN LAKE—1 1/2 mi E of Hwy 296 on US 6. (18472)	(717) 488-6161	240
Yocumtown—PARK AWAY PARK—At I-83, exit 14 and Hwy 392. (17319)	(717) 938-1686	110

RHODE ISLAND

CITY—CAMPGROUND—LOCATION	PHONE	SITES
Chepachet—BOWDISH LAKE—5 1/2 mi W of Hwy 102 on US 44. (02814)	(401) 568-8890	327
Foster—GINNY-B—3 1/4 mi W of Hwy 94 on US 6, 3 1/2 mi S on Cucumber Hill, 1/2 mi E. (02825)	(401) 397-9477	275
Hope Valley—WHISPERING PINES—2 1/2 mi W of Hwy 3 on Hwy 138, 1/2 mi N on Saw Mill Rd. (02832)	(401) 539-7011	166

SOUTH CAROLINA

CITY—CAMPGROUND—LOCATION	PHONE	SITES
Anderson—KOA-LAKE HARTWELL—1 mi SE of I-85 on Hwy 187. (29621)	(803) 287-3161	140
Clover—LITTLE MOUNTAIN—6 mi E of US 321 on Hwy 55, 1/4 mi N on Hwy 46-152. (29710)	(803) 631-5612	142
Eutawville—ROCKS POND—5 mi E of Hwy 45 on Hwy 6, follow signs 2 mi N. (29048)	(803) 492-7711	569
Florence—KOA—1/10 mi S of I-95, exit 169 on TV Rd, 1 mi N on Frontage Rd. (29501)	(803) 665-7007	134
Hilton Head Island—OUTDOOR RESORTS—11 mi SE of bridge on US 278, 1/4 mi N. (29928)	(803) 785-7699	401
Myrtle Beach—KOA-NORTH SHERWOOD FOREST—11 1/2 mi N of US 501 on US 17. (29582)	(803) 272-6420	525
Myrtle Beach—LAKE ARROWHEAD—8 mi N of US 501 on US 17, follow signs 3/4 mi E. (29577)	(803) 449-5626	1300
Myrtle Beach—LAKEWOOD—5 3/4 mi S of US 501 on US 17. (29577)	(803) 238-5161	1900
Myrtle Beach—MYRTLE BEACH TRAVEL PARK—9 3/4 mi N of US 501 on US 17, follow signs 3/4 mi E. (29577)	(803) 449-3714	1000
Myrtle Beach—PIRATELAND—5 mi S of US 501 on US 17. (29577)	(803) 238-5155	1100
Point South—KOA—300 feet N of I-95, exit 33 on US 17, 1/10 mi S. (29945)	(803) 726-5733	78
Rock Hill—SAFARI PINE TREE—1/2 mi W of I-77, exit 72 on Gold Hill Rd. (29715)	(803) 548-0216	209
Santee—SHAWNEE CAMPGROUND—3 mi E of I-95, exit 98 on Hwy 6, 3/4 mi N. (29148)	(803) 854-2136	250
Spartanburg—KOA—2 mi N of I-85 exit 69 on Rd 41, 1/4 mi E on Rd 42/219. (29303)	(803) 576-0970	107

SOUTH DAKOTA

CITY—CAMPGROUND—LOCATION	PHONE	SITES
Custer—KOA-BLACK HILLS-MT RUSHMORE—3 1/4 mi W of US 16A on US 16. (57730)	(605) 673-4304	164
Hill City—RAFTER J-BAR RANCH—At US 16 and Hwy 87. (57745)	(605) 574-2527	200
Interior—KOA-BADLANDS—4 mi SE of Hwy 377 on Hwy 44. (57750)	(605) 433-5337	125
Mitchell—GOLDIE'S CAMPING—3/4 mi N of I-90 on Bus I-90, 1/2 mi W on S. Havens St. (57301)	(605) 996-1181	125
Murdo—KOA-I-90/MIDWAY—1/2 mi N of I-90 on Hwy 63. (57552)	(605) 344-2247	165
Rapid City—HAPPY HOLIDAY—8 1/2 mi S of I-90 on US 16. (57701)	(605) 342-7365	208
Rapid City—HOLIDAY TRAVEL PARK—1/4 mi S of I-90 on LaCrosse St. (57709)	(605) 342-3611	204
Rapid City—THE BERRY PATCH—1/2 mi S of I-90 on Bus US 16. (57701)	(605) 341-5588	130
Salem—KAMP AMERICA—1 1/2 mi N of I-90 on US 81. (57058)	(605) 425-9085	66
Sioux Falls—KOA—1/4 mi N of I-90 on US 77, 1/4 mi E on local rd. (57101)	(605) 332-9987	110
Spearfish—KOA—100 yds S of I-90 and US 85N on Bus Loop 90, 1/4 mi W. (57783)	(605) 642-4633	131

TENNESSEE

CITY—CAMPGROUND—LOCATION	PHONE	SITES
Athens—KOA-ATHENS—2 blocks E of I-75, exit 49 on Hwy 30. (37303)	(615) 745-9199	74
Columbia—COLUMBIA SAFARI—1/2 mi E of I-65 on Hwy 99. (38401)	(615) 381-9545	100
Dickson—KOA—1/4 mi N of I-40 on Hwy 46. (37055)	(615) 446-9925	97
Gatlinburg—OUTDOOR RESORT—10 mi E of US 441 on Hwy 73. (37738)	(615) 436-5861	376
Gatlinburg—SMOKY MOUNTAIN CRAZY HORSE—12 1/2 mi E of US 441 on Hwy 73. (37738)	(615) 436-4434	225
Hurricane Mills—LORETTA LYNN'S DUDE RANCH—8 mi N of I-40 on Hwy 13. (37078)	(615) 296-7700	191

CITY—CAMPGROUND—LOCATION	PHONE	SITES
Lake City—LINDSAY MILL RESORT—1 1/4 mi SE of (37769) I-40 on US 441, 2 1/2 mi E on Oak Grove Rd.	(615) 426-2032	168
Lebanon—KOA-NASHVILLE-LEBANON—2 mi S (37087) of I-40 on US 231.	(615) 449-0000	147
Lexington—KOA—1 1/4 mi N of I-40 on Hwy 22. (38351)	(901) 968-9551	122
Nashville—FIDDLERS INN—5 1/2 mi N of I-40 on Briley (37214) Pkwy, 1/4 mi W on McGavock Pike.	(615) 885-1440	154
Nashville—HOLIDAY NASHVILLE—5 1/2 mi N of I-40 on (37214) Briley Pkwy, 1/4 mi W on McGavock Pike.	(615) 889-4225	338
Nashville—KOA-OPRYLAND—5 1/2 mi N of I-40 on Briley (37204) Pkwy, 1/4 mi W on McGavock, 2 mi N.	(615) 889-0282	416
Nashville—YOGI BEAR'S JELLYSTONE PK—10 mi N of (37072) I-24 on I-65, 2 mi N on US 31W.	(615) 859-0348	100
Pigeon Forge—KOA-GATLINBURG—1/4 mi E of US 441 (37863) on Silver Dollar City Rd, 1000 feet N.	(615) 453-7903	159
Pigeon Forge—SAFARI CAMPING RESORT—1/2 mi W of (37862) US 441 on Wears Valley Rd.	(615) 453-2607	170
Rockwood—CAMP'N AIRE PINEY CREEK—800 feet W of (37854) I-40 on Westel Rd.	(615) 354-3280	169
Smyrna—KOA-NASHVILLE I-24—1 1/2 mi E of I-24, exit (37167) 66B on Sam Ridley Pkwy, 1 1/2 mi S.	(615) 459-5818	140
Townsend—SAFARI SUNDOWN RESORT—2 mi E of town (37882) on Hwy 73.	(615) 448-6936	100

TEXAS

CITY—CAMPGROUND—LOCATION	PHONE	SITES
Alamo—MORNINGSIDE TRAVEL PARK—1 1/2 mi W of FM (78516) 907 on US 83, 1 block S on Morningside Rd.	(512) 787-5784	460
Amarillo—A & A LONGHORN TRAILER INN—10 mi W of (79106) I-27 on I-40, 1 1/2 mi W on S service rd.	(806) 359-6302	100
Arlington—KOA—1 3/4 mi N of I-20, exit (76015) 449 on FM 157.	(817) 461-0601	190
Brownsville—AUTUMN ACRES—2 mi S of FM 802 on US (78521) 77/83, 3 1/2 mi E on Hwy 4.	(512) 546-4979	140
Brownsville—CROOKED TREE CAMPLAND—1/4 mi W of (78520) US 77/83 on FM 802.	(512) 546-9617	200
Brownsville—SAFARI—6 mi E of US 77/83 on FM 802, 1/2 (78521) mi S on Vermillion.	(512) 831-4427	123
Denton—KOA-DALLAS—9 mi S of I-35W on I-35E, 1 mi S (76201) of Corinth exit on service road.	(817) 497-3353	120
Donna—BIT-O-HEAVEN—1 mi S of FM 1423 on US 83, 1/2 (78537) mi W on Bus US 83, 1/4 mi S on Whalen Rd.	(512) 464-5191	128
Edinburg—ORANGE GROVE RV PK—3 1/2 mi E of US 281 (78539) on Hwy 107.	(512) 383-7931	480
El Paso—EL PASO ROADRUNNER RV PK—1/4 mi E (79907) of I-10, exit 28B on Gateway East.	(915) 598-4469	130
El Paso—KOA—23 mi W then N of US 54 on I-10, 50 feet (79998) W on FM 1905, 1/2 mi S on service rd.	(915) 886-2397	102
Falfurrias—COUNTRYSIDE RV PK—1 mi N of US 285 (78355) on US 281, 1 block E of Magnolia Hts.	(512) 325-5353	200
Grand Prairie—TRADERS VILLAGE TRAILER PK—6 1/4 (75051) mi S of I-30 on Hwy 360, 1/2 mi E.	(214) 647-2331	255
Harlingen—EMERALD GROVE TRAVEL TRAILER PK—4 mi (78550) W of US 77/83 on Bus US 83.	(512) 425-2111	170
Harlingen—PARADISE PARK-HARLINGEN—2 mi N of US (78550) 83 on US 77, 1 mi S on service rd.	(512) 425-6881	294
Harlingen—SUNSHINE—2 1/2 mi N of US 83 on US 77, 1 (78550) block E on Primera, 1/4 mi N.	(512) 428-4137	694
Houston—KOA-HOUSTON NORTH—3 1/2 mi E of I-45 on (77039) N. Belt Dr, 1 mi S on Aldine-Westfield.	(713) 442-3700	131
Houston—KOA-HOUSTON WEST—1 mi S of I-10 (77079) on Hwy 6.	(713) 493-2391	125
Kemp—EVENING SHADOWS—14 3/4 mi SE of US 175 on (75143) Hwy 274, 1/2 mi E on Mason Lane.	(214) 432-3455	120
Kerrville—AMERICAMP LEISURE RESORT—500 ft N of (78028) I-10 on Hwy 16, 1/2 mi W on Benson Dr.	(512) 896-6052	150
Livingston—KOA—15 mi W of US 59 (77360) on US 190.	(713) 646-3824	137
Lone Oak—PAWNEE'S WIND POINT PK—1/2 mi S of US (75453) 69 on FM 513, 4 1/2 mi W on FM 1571.	(214) 662-5331	212
Mc Allen—PARADISE PARK—1 1/2 mi N of US 83 on Hwy (78501) 336, 3/4 mi E on Hackberry St.	(512) 682-8081	140
Mc Allen—VALLEY GRANDE TRAILER VILLAGE—3 1/4 mi (78501) W of Hwy 336 on US 83, 1 mi E.	(512) 686-1305	120
Mercedes—LLANO GRANDE LAKE PK—1 mi W of FM 491 (78570) on US 83, 2 mi S on Mile 2 W Road.	(512) 565-2638	276
Mission—CIRCLE T PK—1 mi N of US 83 on FM 1016, 1 (78572) 1/2 mi W on Bus US 83.	(512) 585-5381	250
Mission—EL VALLE DEL SOL—2 1/2 mi E of FM 1016 on (78572) US 83, 1 mi N on FM 494, 1/4 mi E.	(512) 585-5704	450
Mission—OLEANDER ACRES—2 1/2 mi E of FM (78572) 1016 on US 83, 1 mi N on FM 494.	(512) 585-4587	234
Orange—KOA-OAK LEAF—1/5 mi N of I-10, exit (77630) 873 on local road.	(713) 886-4082	120
Pharr—PARADISE PARK—1 1/4 mi S of US 83 on US 281, (78577) 1 1/2 mi W on Kelly Ave.	(512) 787-1521	240
Port Isabel—OUTDOOR RESORTS—1 mi S of Hwy (78597) 100 at Garcia St.	(512) 943-6449	455
Rockport—LAKEWOOD SAFARI CAMP—6 3/4 mi N of FM (78382) 881 on Hwy 35.	(512) 729-7317	118
Rockport—WOODY ACRES—4 mi N of FM 881 on Hwy 35, (78358) 1/2 mi W on Mesquite Rd.	(512) 729-5636	136
San Antonio—GREENTREE VILLAGE NORTH—2 mi N of (78233) Loop 410 & BR 81 on I-35, 1/2 mi W.	(512) 655-3331	250
San Antonio—KOA-ALAMO—1/2 mi S of I-35 on Coliseum (78219) Rd, 1 mi E on Gembler Rd.	(512) 224-9296	400
San Benito—FUN N SUN TOO—1/2 mi SW of US 77/83 (78586) on Helen Moore Rd.	(512) 399-5125	1341
South Padre Island—CAMERON COUNTY'S SEA RANCH— (78578) 1 mi S of Hwy 100 on Park Rd 100.	(512) 943-2585	120
South Padre Island—ISLA BLANCA—3/4 mi S (78578) of Hwy 100 on Park Rd 100.	(512) 943-2585	400
Weslaco—PINE TO PALMS MOBILE HOME—1 mi S of US (78596) 83 on FM 88, 3 mi E on Bus US 83.	(512) 968-5760	236
Weslaco—RIO VALLEY ESTATES—1 mi W of FM 88 on US (78596) 83, 500 feet S on Mi 6 W Rd.	(512) 968-2708	200
Zapata—4 SEASONS TRAILER COURT—2 1/2 mi SW of (78076) US 83 on FM 496, 1/2 mi S on FM 3074.	(512) 765-4241	200

UTAH

CITY—CAMPGROUND—LOCATION	PHONE	SITES
Kanab—CRAZY HORSE—1/2 mi E of Alt US 89 on US 89. (84741)	(801) 644-2782	70
Kaysville—CHERRY HILL—1/4 mi E of I-15, Kaysville exit (84037) on 2nd St. N, 2 mi S on Main.	(801) 451-5379	250
Moab—SLICKROCK COUNTRY—3/4 mi S of Hwy 128 on (84532) US 163.	(801) 259-7660	108
Provo—SILVER FOX—1/2 block N of I-15 on University (84601) Ave.	(801) 377-0033	118
St. George—REDLANDS—1 block E of I-15, Washington (84780) exit, 1 block N on frontage rd.	(801) 673-9700	204
Salt Lake City—KOA—1/4 mi N of I-80 (84116) on Hwy 68, 1/4 mi E on N. Temple.	(801) 355-1192	300
Vernal—CAMPGROUND DINA—1 mi N of US 40 on Hwy (84078) 44.	(801) 789-2148	100

VERMONT

CITY—CAMPGROUND—LOCATION	PHONE	SITES
Alburg—ALBURG TRAVEL TRAILER PK—2 mi E of US 2 (05440) on Hwy 78, 1/2 mile S on Blue Rock Rd.	(802) 796-3733	152
Andover—HORSESHOE ACRES—1 mi W of town limits on (05143) Andover-Weston Rd.	(802) 875-2960	100
Arlington—CAMPING ON THE BATTENKILL—3/4 mi N of (05250) Hwy 313 on US 7.	(802) 275-6663	95
Bomoseen—LAKE BOMOSEEN—5 mi N of US 4 on (05732) Hwy 30.	(802) 273-2061	85
Colchester—LONE PINE CAMPSITE—1 mi W of (05446) US 2/7, on Hwy 127.	(802) 878-5447	255
Georgia Center—HOMESTEAD CAMPGROUNDS—1/4 mi S (05468) of I-89 on US 7.	(802) 524-2356	144
Orleans—WILL-O-WOOD—6 1/4 mi E of I-91 on Hwy 58, 1/2 (05860) mi S on Hwy 5A.	(802) 525-3575	75
Randolph—MOBILE ACRES—2 mi N of Hwy 12 on Hwy 12A (05060)	(802) 728-5548	93
Salisbury—LAKE DUNMORE KAMPERSVILLE—1 1/2 mi E (05753) of US 7 on Hwy 53.	(802) 352-4501	160
South Hero—KOA-CEDAR RIDGE—1 mi E of town on US (05486) 2, 1/2 mi N on Kibbe Point Rd.	(802) 372-5070	75
Stowe—GOLD BROOK CAMPGROUND—2 mi S of (05672) Hwy 108 on Hwy 100.	(802) 253-8906	82
Townshend—BALD MOUNTAIN—1 1/4 mi S of Hwy 35 on (05353) Hwy 30, 3/4 mi W on CR, 1 1/4 mi NW.	(802) 365-7510	170
Townshend—CAMPERAMA VACATION—3/4 mi S of Hwy (05353) 35 on Hwy 30, 1/4 mi W on Depot St.	(802) 365-4451	225

VIRGINIA

CITY—CAMPGROUND—LOCATION	PHONE	SITES
Bowling Green—KOA-BOWLING GREEN—3 mi S of Hwy 2 (22427) on US 301.	(804) 633-7592	100
Bracey—AMERICAMPS LAKE GASTON—5 mi E of I-85 (23919) on Hwy 637.	(804) 636-2668	407
Burgess—GLEBE POINT—2 mi S of US 360 (22432) on Hwy 200.	(804) 453-3440	95
Charlottesville—SAFARI MONTICELLO-SKYLINE—1 mi W (22901) of I-64 on US 250.	(703) 456-6409	90
Cheriton—KOA-CHERRYSTONE—1 1/2 mi W of Byp (23316) US 13 and CR 680 on county rds.	(804) 331-3063	575
Chincoteague—CAMPER'S RANCH—11 mi E of US 13 on (23336) Hwy 175, 1 mi S on Main, E on Bunting.	(804) 336-6371	500
Chincoteague—TOM'S COVE—11 mi E of US 13 on Hwy (23336) 175, 1 1/4 mi S on Main, 1/2 mi E on Beebe.	(804) 336-6498	830
Front Royal—NORTH FORK CAMPSITES—2 mi W of US (22630) 340 on Hwy 55.	(703) 636-9949	275
Gasburg—CAMPTOWN—2 1/2 mi W of Hwy 46 (23857) on CR 626.	(804) 577-2423	320
Gloucester—GLOUCESTER POINT—6 mi S of Byp (23072) US 17 on US 17, 3 mi E on Hwy 636/656.	(804) 642-4316	230
Gwynn Island—CAMPERS HAVEN—5 mi E of Hwy 198 on (23066) Hwy 223, 1/2 mi S on gravel road.	(804) 725-5700	120
Hartfield—BUSH PARK—2 mi E of Hwy 3 on Hwy 33, (23071) 1 1/2 mi N on CR 628, 1 mi E on CR 7G2.	(804) 776-6750	400
Haymarket—GREENVILLE FARM—4 mi N of Hwy 55 on US (22069) 15, E on Hwy 234, 1 mi E on Hwy 601.	(703) 754-7944	175
Lynchburg—KOA-JAMES RIVER—4 mi E of US 501 on US (24505) 460.	(804) 845-6127	100
Petersburg—HOLIDAY INN TRAV-L-PARK—4 mi S of I-85 (23803) on US 1.	(804) 861-2616	150
Saluda—KOA-CYPRESS SHORES—5 mi E of US 17 on Hwy (23061) 198.	(804) 693-3792	450
Stafford—SAFARI WASHINGTON-FREDERICKSBURG— (22554) 200 yards E of I-95 on Hwy 610, 1/4 mi N.	(703) 659-3447	120
Talleysville—KOA-WOODBOURNE—3 mi N of I-64 on Hwy (23124) 609, 3 mi NW on Hwy 606, 1/2 mi NW.	(804) 932-4776	135
Union Hall—PELICAN POINT—1 1/2 mi N of Hwy 40 on (24176) Hwy 945, 3 1/4 mi NE on Hwy 663.	(703) 576-2019	120
Verona—KOA-SHENANDOAH—1 mi W of I-81 on Hwy 612, (24482) 1/2 mi N on US 11, 1 mi W on Hwy 781.	(703) 248-2746	130
Virginia Beach—HOLIDAY TRAV-L-PARK—1 mi S of US 58 (23451) on Pacific Ave, 2 mi S on Gen Booth Blvd.	(804) 425-0249	1019
Virginia Beach—KOA—1 mi S of US 58 on Pacific Ave, 2 (23451) 1/2 mi S on Gen Booth Blvd.	(804) 428-1444	760
Waverly—LOG AND LANTERN—2 1/2 mi SE of (23890) Hwy 40 on US 460, 1/4 mi N on Hwy 614.	(804) 834-8454	135
Williamsburg—JAMESTOWN BEACH CAMPSITES—4 mi W (23185) of I-64 on Hwy 199, 4 mi S on Hwy 31.	(804) 229-8366	600
Williamsburg—KOA-HOLIDAY TRAV-L-PARK—4 1/2 mi (23185) W of I-64 on Hwy 199, 8 mi W on Hwy 5.	(804) 229-1453	600
Williamsburg—SAFARI COLONIAL—1 mi NE of I-64 on (23185) Hwy 646.	(804) 229-2734	200
Williamsburg—YOGI BEAR'S JELLYSTONE PK—1/10 mi (23185) W of I-64 on Hwy 607, 1/4 mi N on Hwy 168.	(804) 564-9808	233
Wytheville—KOA—1/2 mi S of I-81 on Hwy 758. (24382)	(703) 228-2601	103

WASHINGTON

CITY—CAMPGROUND—LOCATION	PHONE	SITES
Bothell—KOA-SEATTLE NORTH—1/4 mi S of I-405 on (98011) Hwy 527, 1/4 mi E on 228th St, 100 yards S.	(206) 486-1972	154
Chelan—LAKE CHELAN TR PK & MARINA (CHELAN (98816) PUBLIC PK)—1 mi W of US 97 on Hwy 150.	(509) 682-5031	160
Coulee City—SUN LAKES PARK RESORT, INC—4 mi S of (99115) US 2 on Hwy 17, 1 1/4 mi E on St Pk Rd.	(509) 632-5583	110
Ellensburg—KOA—1/2 block S of I-90, exit 106 (98926) on Thorp Rd.	(509) 925-9319	140
Gig Harbor—KOA-TACOMA GIG HBR.—6 mi N of bridge on (98335) Hwy 16, 1 1/4 mi SE on Burnham.	(206) 858-8138	140
Kent—KOA-SEATTLE SOUTH—2 1/2 mi SE (98031) of I-5, exit 152 on Orillia Rd.	(206) 852-8652	140
Lynden—KOA—3 mi E of Hwy 539 on Hwy 546, 1/2 mi S (98264) on Line Rd.	(206) 354-4772	175
Monroe—THUNDERBIRD PARK—1 1/4 mi S of US 2 on (98272) Hwy 203, 5 mi E on Ben Howard Rd.	(206) 794-8987	120
Mossyrock—KOA-PLANT'S PARADISE—14 1/4 mi E of I-5 (98585) on US 12, 2 1/2 mi N.	(206) 985-2121	160
Ocean Park—OCEAN PARK TOURIST CAMP—2 blocks (98640) E of Hwy 103 on 259th Place.	(206) 665-5485	160
Olympia—KOA-AMERICAN HERITAGE—1/4 mi E of (98502) I-5, exit 99 on Lathrop Rd, 1/4 mi S.	(206) 943-8778	109
Snoqualmie—SAFARI SEATTLE—1/4 mi N of I-90 on Echo (98065) Glen Rd, 1/8 mi E on 99th St.	(206) 888-1324	118
Spokane—TRAILER INNS—3 mi E of US 2/95 on I-90, (99206) 1 block S on Havana, 1/4 mi E on 4th Ave.	(509) 535-1811	158
Spokane—UNITED CAMPGROUND—1/4mi N of I-90. (99016) exit 293 on Baker Rd.	(509) 928-3300	143
Vantage—KOA—1/2 mi N of I-90, exit 136 on Old Vantage (98950) Hwy.	(509) 856-2230	160
Yakima—TRAILER INNS—3 mi N of Hwy 24 on I-82, 1/8 (98901) mi S on N First St.	(509) 452-9561	101

WEST VIRGINIA

CITY—CAMPGROUND—LOCATION	PHONE	SITES
Ashton—ASHTON RIVERVIEW—1 mi S of town on US 2. (25503)	(304) 576-9074	97
Berkeley Springs—SAFARI TRI-LAKE PK—12 mi S of Hwy (25411) 9 on US 522, 2 1/2 mi E on Fish Hatchery Rd.	(304) 258-1331	225
Elkins—REVELLE'S MOUNTAINAIRE—8 1/2 mi E of US (26241) 219 on US 33, 1/4 mi S on Bowden Rd.	(304) 636-0023	112
Harpers Ferry—KOA—1 mi W of Park Entrance on US (25425) 340, 1/4 mi S on local road.	(304) 535-6895	300
Milton—FOX FIRE—3 mi W of I-64, exit 28 on US 60. (25541)	(304) 743-5622	203
Summersville—MOUNTAIN MANOR—1 mi S of town on (26651) US 19, 2 mi W on Airport Rd.	(304) 872-4220	200

WISCONSIN

CITY—CAMPGROUND—LOCATION	PHONE	SITES
Bagley—YOGI BEAR'S JELLYSTONE PK—1 mi N of town (53801) center on CR X.	(608) 966-2201	206
Bancroft—VISTA ROYALLE—3/4 mi E OF US 51 on CR W, 1 (54921) mi N on Radcliffe Ave.	(715) 366-6860	144
Big Flats—PINELAND CAMPING PK—7 mi N of Hwy 21 on (54613) Hwy 13.	(608) 564-7818	150
Caledonia—YOGI BEAR'S JELLYSTONE PK—2 mi E of I-94 (53108) on Seven Mile Rd, 1/3 mi N on Hwy 38.	(414) 835-2565	232
Denmark—DEVILS RIVER CAMPERS PK—5 1/4 mi S of (54227) Hwy 96 on US 141.	(414) 863-2812	142
Eagle River—PINE AIRE—3 mi N of town center on US 45, (54521) 1/8 mi E on Chain-O-Lakes Rd.	(715) 479-9208	132
Edgerton—HICKORY HILLS—1 mi N of I-90 on Hwy 73, (53534) 3/4 mi E on Hwy 106, 3/4 mi N on Hillside Rd.	(608) 884-6327	265
Egg Harbor—KOA-DOOR COUNTY—3 mi S of city limits (54209) on Hwy 42, 1/4 mi E on Sunny Point Rd.	(414) 868-3151	168
Elkhart Lake—PLYMOUTH ROCK LTD—3 mi S of CR A (53020) on Hwy 67.	(414) 892-4252	440
Fort Atkinson—YOGI BEAR'S—5 mi S of US 12 on (53538) Hwy 26, 3/4 mi W on Koshkonong L. Rd.	(414) 563-5714	275
Fremont—YOGI BEAR'S JELLYSTONE PK—1 1/2 mi W of (54940) city limits on US 10.	(414) 446-3420	124
Greenbush—WESTWARD HO CAMP RESORTS—5 mi SW (53023) of Hwy 20 on CR T.	(414) 526-3407	250
Green Lake—GREEN LAKE—3 1/2 mi W of Hwy 49 on Hwy (54941) 23.	(414) 294-3543	240
Knowlton—CAMP DUBAY SHORES—2 1/2 mi S of US 51 (54455) on Hwy 34, 600 feet E on Dubay Dr.	(715) 457-2484	191
Milton—LAKELAND—2 1/4 mi E of I-90, exit 163 on Hwy (53563) 59.	(608) 868-4255	450
Montello—WILDERNESS—8 mi S of Hwy 23 on Hwy 22. (53949)	(414) 297-2002	350
Pardeeville—INDIAN TRAILS—1 1/4 mi N of Hwy 44 (53954) on Hwy 22, 1 mi W of Haynes Rd.	(608) 429-3244	300
Rio—WILLOW MILL CAMPSITES—1 mi E of town center (53960) on Rio St, 3 mi N on CR SS.	(414) 992-5355	149
Sturgeon Bay—YOGI BEAR'S—6 1/2 mi W on CR C, (54235) 1 1/4 mi N on May Rd, 3/4 mi E.	(414) 743-9001	290
Tomahawk—THE OUTPOST—1/2 mi N of US 8 on US (54487) 51, 1 mi E on CR N.	(715) 453-3468	263
Warrens—YOGI BEAR'S JELLYSTONE PK—1/2 mi E of (54666) I-94, exit 135 on CR E.	(608) 378-4303	220
Wisconsin Dells—ARROWHEAD—1 mi NW of I-90/94 on (53965) US 12/16, 1 1/2 mi W on Arrowhead Rd.	(608) 254-7344	228
Wisconsin Dells—DELL BOO—1 1/4 mi E of I-90/94, exit 92 (53965) on US 12, 3/4 mi W on Shady Lane.	(608) 356-5898	138
Wisconsin Dells—YOGI BEAR'S—1/2 mi W of I-90/94, exit (53965) 92 on US 12, 1 mi NW on Gasser Rd.	(608) 254-2568	240
Woodruff—INDIAN SHORES—4 3/4 mi SE of US 51 on (54568) Hwy 47.	(715) 356-5552	210

WYOMING

CITY—CAMPGROUND—LOCATION	PHONE	SITES
Buffalo—INDIAN—1 1/2 mi S of I-90 on I-25, 100 feet W (82834) on US 16.	(307) 684-9601	130
Cody—KOA—3 mi E of Hwy 120 on US 16/14/20. (82414)	(307) 587-2369	200
Dubois—CIRCLE UP CAMPER COURT—1 block S (82513) of town center on Riverton St.	(307) 455-2238	100
Fishing Bridge—FISHING BRIDGE RV PARK—1 (82190) mi E of US 89 on US 14/16.	(307) 242-7231	358
Moran Junction—COLTER BAY TRAILER VILLAGE—10 (83001) 1/2 mi N of US 26/89/287 on US 89/287.	(307) 543-2811	112
Sundance—KOA-SUNDANCE—1/2 mi W of I-90 on Bus (82729) I-90, 1/4 mi N on county rd.	(307) 283-1557	114

On the following pages you will find discount coupons for over one hundred outstanding attractions in every region of the United States. You can reduce your family's travel expenses by using them. The management of each of the places of interest included on these pages would like to help you save money. Just clip out the coupon and present it at the ticket window; you will receive the reduced rate of admission indicated on the coupon.

These coupons are valuable because of the money they save you plus the attractions you visit will provide you with many fun-filled hours. You and your family can explore natural wonders, relive American history by visiting historic sites, or take in a museum exhibit. You can wander through a haunted ghost town or, if it's more excitement you want, have fun at an amusement park.

Before presenting the coupon to a participating facility, be certain to read and understand all the statements on the back of the coupon that relate to the discount being offered, as well as any restrictions and limitations to the coupon use. **Responsibility:** We assume no responsibility for changes in terms or discounts established by participating facilities, nor for failure of any facility to honor a coupon.

20% off	$1.00 off	15% off
Alabama Space & Rocket Center Huntsville, Alabama	**Marine World/Africa U.S.A.** (on San Francisco Bay) Marine World Pky. Redwood City, California	**Sea World** Mission Bay San Diego California
AMERICAN EXPRESS TRAVEL & LEISURE ROAD ATLAS	AMERICAN EXPRESS TRAVEL & LEISURE ROAD ATLAS	AMERICAN EXPRESS TRAVEL & LEISURE ROAD ATLAS
10% off	$2.00 off	10% off
U.S.S. Alabama Battleship U.S. 90-98 East Mobile, Alabama	**Marriott's Great America** Great America Parkway Santa Clara, California	**Winchester Mystery House** 525 S. Winchester Blvd. San Jose, California
AMERICAN EXPRESS TRAVEL & LEISURE ROAD ATLAS	AMERICAN EXPRESS TRAVEL & LEISURE ROAD ATLAS	AMERICAN EXPRESS TRAVEL & LEISURE ROAD ATLAS
50¢ off	10% off	50¢ off
Meteor Crater I-40 40 miles east of Flagstaff, Arizona	**Palm Springs Aerial Tramway** Palm Springs, California	**Argo Gold Mill** 2350 Riverside Dr. Idaho Springs, Colorado
AMERICAN EXPRESS TRAVEL & LEISURE ROAD ATLAS	AMERICAN EXPRESS TRAVEL & LEISURE ROAD ATLAS	AMERICAN EXPRESS TRAVEL & LEISURE ROAD ATLAS

Clip along dashed lines

$1.00 off	$1.25 off	$AVING$
Dogpatch USA Dogpatch, Arkansas	**R.M.S. Queen Mary-Tour** Long Beach, California	**Pikes Peak Ghost Town** U.S. 24, 21st St. Exit Colorado Springs, Colorado
AMERICAN EXPRESS TRAVEL & LEISURE ROAD ATLAS	AMERICAN EXPRESS TRAVEL & LEISURE ROAD ATLAS	AMERICAN EXPRESS TRAVEL & LEISURE ROAD ATLAS
50¢ off	10% off	$AVING$
Miles Musical Museum U.S. 62 West in Eureka Springs, Arkansas	**San Diego Harbor Excursion** Foot of Broadway at Harbor Dr. San Diego, California	**Mystic Seaport** Mystic, Connecticut
AMERICAN EXPRESS TRAVEL & LEISURE ROAD ATLAS	AMERICAN EXPRESS TRAVEL & LEISURE ROAD ATLAS	AMERICAN EXPRESS TRAVEL & LEISURE ROAD ATLAS

Clip along dashed lines

50¢ off	10% off	10% off
Ozark Folk Center Mountain View, Arkansas	**San Diego Wild Animal Park** San Diego, California Code: PI-442	**Steam Train and Riverboat** Railroad Ave. Essex, Connecticut
AMERICAN EXPRESS TRAVEL & LEISURE ROAD ATLAS	AMERICAN EXPRESS TRAVEL & LEISURE ROAD ATLAS	AMERICAN EXPRESS TRAVEL & LEISURE ROAD ATLAS
10% off	10% off	50¢ off
Lion Country Safari 8800 Irvine Center Drive Laguna Hills, California	**San Diego Zoo** Balboa Park San Diego, California Code: AI-442	**National Historical Wax Museum** 333 E. St., S.W. Washington, D.C. 20024
AMERICAN EXPRESS TRAVEL & LEISURE ROAD ATLAS	AMERICAN EXPRESS TRAVEL & LEISURE ROAD ATLAS	AMERICAN EXPRESS TRAVEL & LEISURE ROAD ATLAS

Responsibility: We assume no responsibility for changes in terms or discounts established by participating facilities, nor for failure of any facility to honor a coupon.

For your convenience, there is an indication on the front of the coupons of the percentage or cash amount off the normally applicable rate. The special rates and conditions of coupon validity vary with each participant, and the coupon may not be honored on certain "special event" days. Be sure to read and understand all statements on the back of the coupon before presenting it.

Sea World
Features Shamu, the 4-ton killer whale; five major shows; over 25 exhibits; Cap'n Kids' World and the new live shark display.
With this voucher: 15% discount. Voucher may not be used in combination with any other discount program. Prices subject to change.
Voucher limit: 6 family members. Expires June 30, 1983.   A:164   C:165

Marine World/Africa U.S.A.
Performing sea mammals and African land animals in 6 different shows. West Coast's only water ski and boat show.
With this voucher: $1.00 discount. Open daily Memorial Day-Labor Day; Wed.-Sun., Apr.-May, Sept.-Oct.; weekends and holidays Dec.-Mar.
Voucher limit: 8 family members. Expires Dec. 31, 1982.

Alabama Space & Rocket Center.
Explore earth's largest space museum. See and touch the whole story of man's adventure into space. Price includes tour of museum and NASA's Marshall Space Flight Center.
With this voucher: 20% off.
Voucher may not be used in combination with any other discount program. Closed Christmas Day only.
Voucher limit: family members.

— — — Clip along dashed lines — — — —

Winchester Mystery House
Discover the world's largest oddest dwelling - a 160-room bizarre Victorian mansion built by Winchester Rifle heiress, Sarah Winchester to appease evil spirits!
With this voucher:
Deluxe tour: 10% discount. (Offers reverts to 50¢ off house tour rate when weather closes grounds tour portion.) Closed Christmas. Not valid Sundays July 1-Labor Day.
Voucher limit: 6 family members. Expires Dec. 31, 1983.

Marriott's Great America
Northern California's largest family attraction; 45 miles south of San Francisco on U.S. 101.
With this voucher:
$2.00 off current rate. Voucher may not be used in combination with any other discount program. Open daily in summer, weekends spring and fall.
Voucher limit: 6 family members.

U.S.S. Alabama Battleship
W.W. II battleship and submarine; military displays representing every service branch; rose garden, picnic area.
With this voucher:
10% discount.
Closed Christmas.
Voucher limit: family members.

Argo Gold Mill
Includes Clear Creek Historic Mining and Milling Museum.
With this voucher:
50¢ discount.
Open April -Oct. (Closed rest of year.)
Voucher limit: 6 family members.

Palm Springs Aerial Tramway
Longest single-span lift carries visitors from 2,643-ft. Valley Station to 8,516-ft. Mountain Station and threshold of Mt. San Jacinto Wilderness and State Pk.
With this voucher:
10% discount. No credit cards on discounts. Closed Sept.
Voucher limit: 6 family members.

Meteor Crater
Training site for Apollo astronauts; Astronaut Hall of Fame; Museum of Astrogeology; film and lecture presentation; gift and rock shop; snack bar.
With this voucher:
50¢ discount.
Voucher limit: 5 family members.

Pikes Peak Ghost Town
Authentic Old West town.
With this voucher:
One adult admission free with one or more paid adult admissions.
Open daily May 1-Oct. 15.
Voucher limit: 1 person

R.M.S. Queen Mary-Tour
World-famous 1930s transatlantic ocean liner complete with exhibits, unique entertainment, food, hotel and more.
With this voucher:
$1.25 discount. Voucher may not be used in combination with any other discount program. Open daily, 10 a.m.-5 p.m.
Voucher limit: entire party.

Dogpatch USA
Theme park featuring rides, music shows, Li'l Abner characters, trout fishing.
With this voucher:
$1.00 discount.
Daily Memorial Day-late Aug. Weekends only May, Sept. Coupon not good with any other special offer.

— — — Clip along dashed lines — — — —

Mystic Seaport
Nation's largest maritime museum dedicated to preserving the ships, artifacts and craftsmanship of 19th-century America.
With this voucher:
1 to 4 persons at group rates. Call (203) 536-2631 for further information.
Closed Christmas.

San Diego Harbor Excursion
With this voucher:
10% discount.
Open all year.
Voucher: Limit: 6 family members.

Miles Musical Museum
Miniature mechanical circus; chapel; Gay 90's musical instruments, clocks, paintings, carvings; conducted tours.
With this voucher: 50¢ discount on adults admission only. Voucher may not be used in combination with any other discount program. Daily 9 a.m.-5p.m. May 1-Oct. 31. Phone (501) 253-8961.
Voucher limit: 5 family members.

Steam Train and Riverboat
Turn-of-the-century adventure: steam train excursion, closed and open cars, first-class parlor car (additional fare); connecting riverboat cruises.
With this voucher:
10% discount on all trips. Daily mid-June to 1st week of Sept. (Weekends only May-June, Sept.-Oct. & Dec.)
Voucher limit: 4 family members.

San Diego Wild Animal Park
1,800-acre wildlife preserve and authentic African-style village.
With this coupon:
10% discount on ticket package: admission to Nairobi Village, Wgasa Bush Line, Kilimanjaro Hiking Trail. Open daily 9 a.m.-dusk; mid-June thru Labor Day, 9 a.m.-9 p.m.
Coupon limit: entire party.

Ozark Folk Center
State park featuring live demonstrations of crafts; music and lore of the Ozarks.
With this voucher:
50¢ discount.
May 1-Nov. 1. Closed Mon. & Tues. in May and Sept. Closed Mondays in Oct.
Voucher limit: 1 person.

National Historical Wax Museum
Scenes from American history; stirring animated Hall of Presidents finale.
With this voucher:
50¢ discount. Closed New Year's, Thanksgiving and Christmas. Voucher not to be used in conjunction with any other discount program.
Voucher limit: 4 family members.

San Diego Zoo
3,400 animals representing 800 species of wildlife.
With this coupon:
10% discount on ticket package: admission to Zoo, guided bus tour, admission to Children's Zoo, Skyfari aerial tram.
Open daily 9 a.m. -dusk.
Coupon limit: entire party.

Lion Country Safari
African wildlife preserve; 20 mi. S. of Disneyland.
With this voucher:
10% discount for tour of preserve, shows, rides, parking. Voucher may not be used in combination with any other discount program. Prices subject to change without notice.
Voucher limit: 6 family members.

20% off	**$AVING$**	**$1.00 off**
Florida Cypress Gardens Fla. 540 (30 minutes west of Disney World) Cypress Gardens, Florida	**Crystal Lake Cave** U.S. 52 Dubuque, Iowa	**Edaville Railroad** Mass. 58 South Carver, Massachusetts
AMERICAN EXPRESS TRAVEL & LEISURE ROAD ATLAS	AMERICAN EXPRESS TRAVEL & LEISURE ROAD ATLAS	AMERICAN EXPRESS TRAVEL & LEISURE ROAD ATLAS
10% off	**25% off**	**25¢ off**
Florida's Silver Springs Fla. 40,1 mile east of Ocala Silver Springs, Florida	**Agricultural Hall of Fame** 630 N. 126 St. Bonner Springs, Kansas	**National Basketball Hall of Fame** 460 Alden St. Springfield, Massachusetts
AMERICAN EXPRESS TRAVEL & LEISURE ROAD ATLAS	AMERICAN EXPRESS TRAVEL & LEISURE ROAD ATLAS	AMERICAN EXPRESS TRAVEL & LEISURE ROAD ATLAS

- - - Clip along dashed lines - - - -

10% off	**$AVING$**	**$AVING$**
Florida's Weeki Wachee U.S. 19 at Fla. 50 Weeki Wachee, Florida	**Historic Wondering Woods** Ky. 70 Wondering Woods, Kentucky	**Arcadian Copper Mine Tours** Mich. 26, Ripley Hancock, Michigan
AMERICAN EXPRESS TRAVEL & LEISURE ROAD ATLAS	AMERICAN EXPRESS TRAVEL & LEISURE ROAD ATLAS	AMERICAN EXPRESS TRAVEL & LEISURE ROAD ATLAS
20% off	**$1.00 off**	**$AVING$**
Marineland of Florida Fla. A1A St. Augustine, Florida	**Horse Cave Theatre** 129 1/2 Main Street Horse Cave, Kentucky	**Quincy Mine Hoist** U.S. 41, Quincy Hill Hancock, Michigan
AMERICAN EXPRESS TRAVEL & LEISURE ROAD ATLAS	AMERICAN EXPRESS TRAVEL & LEISURE ROAD ATLAS	AMERICAN EXPRESS TRAVEL & LEISURE ROAD ATLAS
50¢ off	**25¢ off**	**$AVING$**
MGM's Bounty Exhibit 345 2nd Ave., N.E. St. Petersburg, Florida	**Pontalba Historical Puppetorium** 514 St. Peter St. New Orleans, Louisiana	**Lumbertown, U.S.A.** Brainerd, Minnesota
AMERICAN EXPRESS TRAVEL & LEISURE ROAD ATLAS	AMERICAN EXPRESS TRAVEL & LEISURE ROAD ATLAS	AMERICAN EXPRESS TRAVEL & LEISURE ROAD ATLAS

- - - Clip along dashed lines - - - -

10% off	**50¢ off**	**$AVING$**
Silver Springs Wild Waters Fla. 40,1 mile east of Ocala Silver Springs, Florida	**ARGO Cruises** Pier 6, Fisherman's Wharf Boothbay Harbor, Maine	**Beauvoir** U.S. 90 Biloxi, Mississippi
AMERICAN EXPRESS TRAVEL & LEISURE ROAD ATLAS	AMERICAN EXPRESS TRAVEL & LEISURE ROAD ATLAS	AMERICAN EXPRESS TRAVEL & LEISURE ROAD ATLAS
15% off	**25% off**	**$AVING$**
White Water Raft Trip Salmon River Salmon, Idaho	**Casco Bay Lines** Custom House Whart Portland, Maine	**Missouri Botanical Garden** 2101 Tower Grove Ave. St. Louis, Missouri
AMERICAN EXPRESS TRAVEL & LEISURE ROAD ATLAS	AMERICAN EXPRESS TRAVEL & LEISURE ROAD ATLAS	AMERICAN EXPRESS TRAVEL & LEISURE ROAD ATLAS

- - - Clip along dashed lines - - - -

$1.00 off	**50¢ off**	**10% off**
Amish Acres 1600 W. Market St. Nappanee, Indiana	**Chesapeake Bay Maritime Museum** Navy Point St. Michaels, Maryland	**Six Flags-St. Louis** 1-44 and Allenton Exit Eureka, Missouri
AMERICAN EXPRESS TRAVEL & LEISURE ROAD ATLAS	AMERICAN EXPRESS TRAVEL & LEISURE ROAD ATLAS	AMERICAN EXPRESS TRAVEL & LEISURE ROAD ATLAS
$AVING$	**$2.00 off**	**$AVING$**
Conner Prairie Pioneer Settlement 13400 Allisonville Road Noblesville, Indiana	**Bay State-Spray & Provincetown Cruises** 20 Long Wharf Boston, Massachusetts 02110	**Frontier Town** U.S. 12 West of Helena, Montana
AMERICAN EXPRESS TRAVEL & LEISURE ROAD ATLAS	AMERICAN EXPRESS TRAVEL & LEISURE ROAD ATLAS	AMERICAN EXPRESS TRAVEL & LEISURE ROAD ATLAS

Edaville Railroad
A 5½-mile steam train ride through an 1,800-acre cranberry plantation.
With this voucher:
$1.00 savings on each combination (train & museum) ticket.
Not valid Sundays in Dec.
Voucher limit: 6 family members.

Crystal Lake Cave
With this voucher:
One child under 12 admitted free with each 2 adult admissions paid.
Open May 1-Oct. 31. Weekends only May, Sept.-Oct. (Closed rest of year.) Prices subject to change.
Voucher limit: 6 family members.

Florida Cypress Gardens
The world's most beautiful gardens and finest Water Ski Revue; many specialty shops and an Exotic Bird Show. Now 14 new acres of family attractions in two themed areas - Southern Crossroads and the The Living Forest.
With this voucher:
20% discount.

National Basketball Hall of Fame
Honors Court; original Gym replica; movies; game displays.
With this voucher:
25¢ discount.
Closed New Year's, Thanksgiving, Christmas.
Voucher limit: 6 family members.

Agricultural Hall of Fame
Exhibits trace the development of farm equipment. Seasonal demonstrations.
With this voucher:
25% discount.
Closed Thanksgiving and Christmas.
Voucher limit: family members.

Florida's Silver Springs
Underwater panorama viewed through glass-bottom boats. Price includes Glass-bottom Boat Ride, Jungle Cruise Ride, Deer Park and Reptile Institute, Cypress Point and Antique Car Collection.
With this voucher: 10% discount. Prices subject to change without notice. Voucher not to be used in conjuntion with any other offer.
Voucher limit: 4 family members.

Arcadian Copper Mine Tours
Underground tour of a real copper mine.
With this voucher:
One free admission for one paid admission of the same type.
Daily June 1-Oct. 15. (Closed rest of year.)
Voucher limit: 6 persons.

Historic Wondering Woods
With this voucher:
A $avings discount.
Open daily Memorial Day-Labor Day.
Weekends only May-Sept.

Florida's Weeki Wachee
World's only underwater magic show; wilderness river cruise; birds of prey show; exotic bird show; tropical rain forest and more.
With this voucher:
10% discount. Voucher may not be used in combination with any other discount program.
Voucher limit: 4 family members.

Quincy Mine Hoist
World's largest steam hoist. One of the wonders of the old mining world.
With this voucher:
One free admission for one paid admission of the same type.
Daily June 15-Sept. 1. (Closed rest of year.)
Voucher limit: 6 persons.

Horse Cave Theatre
Professional repertory theatre.
With this voucher:
$1.00 discount.
Five plays alternating during season; closed on Mondays. Closed Oct.-May.
Voucher limit: entire party.

Marineland of Florida
"The world's original oceanarium." Florida's only oceanfront resort/attraction. Marine life shows plus complete resort facilities.
With this voucher:
20% discount.
Voucher limit: family members.

— Clip along dashed lines —

Lumbertown, U.S.A.
With this voucher:
One free admission for one paid admission of the same type.
Open daily late May-Labor Day.
Voucher limit: 6 family members.

Pontalba Historical Puppetorium
Puppets portray the legend of Jean Lafitte and the Battle of New Orleans.
With this voucher:
25¢ discount.

MGM's Bounty Exhibit
With this voucher:
50¢ discount.
Open daily 9 a.m.-10 p.m.
Voucher limit: 6 persons.

Beauvoir
Historic last home of Jefferson Davis. Home, pavilions, two museums and 74 acres of gardens and trails; Confederate Veterans Cemetery and Tomb of the Unknown Soldier of the C.S.A.
With this voucher:
$1.00 off adult admission, children 8-17, 25¢ off.
Closed Christmas.
Voucher limit: 5 family members.

ARGO Cruises
Boat cruises along Maine Coast.
With this voucher:
50¢ discount. Not valid for supper cruises. Daily from May 25-Oct. 20. (Closed rest of year.)
Voucher limit: 6 family members.

Silver Springs Wild Waters
Six exciting water flume rides, wave pool, water fort, miniature golf.
With this voucher:
10% discount.
Daily Apr.-1st week of Sept.
Voucher limit: 4 family members.

— Clip along dashed lines —

Missouri Botanical Garden
National Historic Landmark; contains indoor and outdoor horticultural displays and the largest Japanese garden in North America.
With this voucher:
One free admission with one paid adult admission.
Closed Christmas.

Casco Bay Lines
Cruises along picturesque waterway with 365 islands.
With this voucher:
25% off on all cruises, except city tour No. 7.
Voucher limit: 6 family members.

White Water Raft Trip
Exciting two-day white water raft trip on the Salmon River and one night camping along its banks.
With this voucher:
15% off total price. A minimum of 3 people is required before discount applies. For reservations write Idaho Adventures, Box 834-EXA, Salmon, Idaho 83467 or call (208) 756-2986.
Voucher limit: 5 family members.

Six Flags-St. Louis
SAVE 10% ON A ONE OR TWO-DAY SIX FLAGS TICKET.
Coupon valid only at Six Flags; Main Gate ONLY. This coupon valid for up to five (5) members of your immediate family. Discount applies only to regular ticket price, tax not included. Cannot be combined with any other discount offer.
Expires November 1, 1981. 18104

Chesapeake Bay Maritime Museum
Maritime museum in 16-acre waterfront complex.
With this voucher:
50¢ discount.
Summer open daily; winter open daily except on Mon., Jan.-Feb. open weekends only.
Closed Christmas.
Voucher limit: 4 family members.

Amish Acres
Restored 80-acre Amish farm; tours, horse-drawn rides, family-style restaurant.
With this voucher:
$1.00 discount.
Closed Dec.-Apr.
Voucher limit: 4 family members.

Frontier Town
Authentic log and rock village.
With this voucher:
One free admission with one paid admission.
Voucher limit: 1 person.

Bay State-Spray & Provincetown Cruises
Look for the RED TICKET OFFICE, 1/2 way down Long Wharf.
With this voucher:
$2.00 off Provincetown cruise.
Sailing Apr.-Oct. For schedules and information call (617) 723-7800.
Voucher limit: 5 family members.

Conner Prairie Pioneer Settlement
With this voucher:
One adult admission free with each paid adult admission.
Open Apr. 1-Dec. 13. Closed Easter and Thanksgiving Weekend. Closed Mon. in May-Oct.; closed Mon. and Tues. in Apr., Nov.-Dec.

Medora Musical
Upper Midwest's finest family variety show. Adjacent to Theodore Roosevelt Nat. Park.
With this voucher:
One free admission per family.
Open mid-June thru 1st week in Sept.
Voucher limit: 1 person.

The Farmers' Museum
With this voucher:
$1.00 off adult admission. Voucher may not be used for combination tickets with other museums. Open daily May 1-Oct. 31; closed Mon. Nov. 1-Apr. 30. Closed New Year's, Thanksgiving and Christmas.
Voucher limit: 2 family members.

Museum of the Fur Trade
Restored trading house: trade goods, Indian garden.
With this voucher:
One adult half price for each paid adult admission. Daily June-Labor Day. (Closed except by appointment rest of year.)
Voucher not to be used with any other discount program.

Red River & No. Plains Regional Museum
Pioneer days in the village of Bonanzaville, U.S.A.
With this voucher:
25% discount. Voucher not valid during Fair Week or Pioneer Days.
Museum open all year. Closed Memorial Day.
Voucher limit: 4 family members.

Fenimore House
With this voucher:
$1.00 off adult admission. Voucher may not be used for combination tickets with other museums. Closed Jan.-Mar. Closed New Year's, Thanksgiving and Christmas.
Voucher limit: 2 family members.

Lake Mead Yacht Tours
Tours cruise along portions of Lake Mead's 550 miles of shoreline.
With this voucher:
50¢ discount.
Closed Christmas.
Voucher limit: 2 persons.

Cincinnati Zoo
Features nation's first and only Insect World Building, new outdoor Gorilla exhibit and rare white tigers.
With this voucher:
$1.00 discount. Voucher valid everyday of the year except Sundays and holidays.
Open all year.
Voucher limit: 1 person.

The New York Experience
Multiple screens and unique special effects recreate the sights and sounds of New York City. Spectacular.
With this voucher:
50¢ off regular adult admission/25¢ off regular child's admission. Prices subject to change.
Open all year; showings every hour on the hour.
Voucher limit: 2 family members.

Strawbery Banke
A ten-acre historic waterfront neighborhood.
With this voucher:
$1.00 discount.
Daily Apr. 15-Nov. 15.
Guided tours by appointment, Nov.-May.
Voucher limit: 2 family members.

— — — Clip along dashed lines — — — —

Pro Football Hall of Fame
With this voucher:
20% discount on adult and children admissions only. Voucher may not be used in combination with any other discount program. Prices subject to change without notice.
Closed Christmas.
Voucher limit: entire party.

Philipsburg Manor
With this voucher:
One free admission with purchase of same type of regular admission.
Closed New Year's, Thanksgiving and Christmas.

M/V "Mt. Washington"
50-mile, 3 1/4-hour excursion cruise on Lake Winnipesaukee.
With this voucher:
50¢ off adult fare, children 5-12, 25¢ off. Discount applies only to M/V "Mt. Washington." Open Memorial Day weekend-Oct. 15. Phone (603) 366-5531.
Voucher limit: family members.

TSA-LA-GI
Recreated Cherokee village; Trail of Tears drama; Cherokee National Museum and Rural Museum Village.
With this voucher:
Adults 10% discount. Village open May 5-Aug. 23. Closed Mon. Open weekends through Sept. 1. Theatre open June 20-Aug. 22. Closed Sunday. Museum open year round (except Mon. during winter).
Voucher limit: 6 persons.

Rockefeller Center Tour and Observation Roof
One-hour tour including history, artwork, private roof gardens. Radio City Music Hall and Observation Roof on 70th floor.
With this voucher:
30¢ discount.
Closed New Year's, Christmas.
Voucher limit: 6 persons.

U.S.S. Ling - World War II Submarine
With this voucher:
25% discount.
June-Sept., 10 a.m.-6 p.m.; Oct.-May, 10 a.m.-5 p.m. Visiting hours subject to change.
Closed New Year's, Easter, Thanksgiving, Christmas.
Voucher limit: family members.

Wildlife Safari
Over 600 African and Asian animals and birds roam in habitats similar to those of their native lands. Restaurant, gift shop, children's zoo. Walking Safaris by reservation.
With this voucher:
One adult admission for $1.00 less than the regular admission price. This certificate may not be used in conjunction with any other discount offers. Open daily all year.
Voucher limit: 1 person per car.

Sunnyside
With this voucher:
One free admission with purchase of same type of regular admission.
Closed New Year's, Thanksgiving and Christmas.

Bronx Zoo
Largest urban zoo in the United States; over 3,400 animals representing 630 species of wildlife.
With this voucher:
20% off discount on general admission, Safari tour, children's zoo, World of Darkness, Wild Asia Skyfari and animal ride (children only) purchased as a combination ticket. Many attractions closed during winter; zoo open all year. Valid Fri.-Mon. before 1:30 p.m.
Voucher limit: family members.

— — — Clip along dashed lines — — — —

The Franklin Institute Science Museum
America's foremost science museum and planetarium, featuring hands-on exhibitry, planes, ships, trains, Benjamin Franklin National Memorial.
With this voucher: 50¢ off regular Museum admission. Voucher may not be used in combination with any other discount program. Closed New Year's, July 4, Nov. 27-Dec. 24-25.
Voucher limit: 4 persons.

Van Cortlandt Manor
With this voucher:
One free admission with purchase of same type of regular admission.
Closed New Year's, Thanksgiving and Christmas.

Circle Line Sightseeing Yachts
Cruises around Manhattan Island.
With this voucher:
Adults, $1.00 off current rate.
Daily Apr. thru mid-Nov. (Closed rest of year.)
Voucher limit: 6 family members.

Hershey Gardens
23 colorful acres of floral displays; world-renowned roses; 6 themed areas. Free parking.
With this voucher:
25¢ discount.
Open Apr. 15-Oct. 31. (Closed rest of year.)
Voucher limit: family members.

Old Salem
Restored 18th-century Moravian Congregation Town.
With this voucher:
20% discount on price of admission.
Closed Christmas.
Voucher limit: 6 family members.

Crossroads Sightseeing
Variety of bus tours of New York City.
With this voucher:
$1.00 off any regular daytime tour.
Closed Christmas.

Hershey Museum of American Life
Pennsylvania German, American Indian collections. Stiegel glassware, multimedia introduction to museum.
With this voucher: 50¢ discount.
Open daily. Closed New Year's, Thanksgiving and Christmas.
Voucher limit: 1 person.

Tweetsie Railroad
With this voucher:
$1.00 off current rate.
Daily Memorial Day weekend-Labor Day weekend. After Labor Day, open weekends only.
Voucher limit: immediate family members.

Day Line
All-day excursion up Hudson River to Bear Mountain, West Point and Poughkeepsie.
With this voucher:
$1.00 off adult admission, children 2-11, 50¢ off. Daily Memorial Day thru mid-Sept.
Voucher limit: 6 family members.

25¢ off **National Tower** Gettysburg, Pennsylvania AMERICAN EXPRESS TRAVEL & LEISURE ROAD ATLAS	**10% off** **Black Hills Passion Play** Spearfish, South Dakota AMERICAN EXPRESS TRAVEL & LEISURE ROAD ATLAS	**50¢ off** **Shelburne Museum** Shelburne, Vermont AMERICAN EXPRESS TRAVEL & LEISURE ROAD ATLAS
20% off **Pennsylvania Dutch Farm** Grange Rd. Mt. Pocono, Pennsylvania AMERICAN EXPRESS TRAVEL & LEISURE ROAD ATLAS	**$AVING$** **Confederama** 2742 Tennessee Ave. Chattanooga, Tennessee AMERICAN EXPRESS TRAVEL & LEISURE ROAD ATLAS	**50¢ off** **Historic Michie Tavern Museum** Charlottesville, Virginia AMERICAN EXPRESS TRAVEL & LEISURE ROAD ATLAS

Clip along dashed lines

$1.00 off **Poconos' Magic Valley and Winona 5 Falls** U.S. 209 Bushkill, Pennsylvania AMERICAN EXPRESS TRAVEL & LEISURE ROAD ATLAS	**20% off** **Tommy Bartlett's WATER CIRCUS** U.S. 441 Pigeon Forge, Tennessee 37863 AMERICAN EXPRESS TRAVEL & LEISURE ROAD ATLAS	**$AVING$** **Skyline Caverns** U.S. 340 Front Royal, Virginia AMERICAN EXPRESS TRAVEL & LEISURE ROAD ATLAS
50¢ off **Quiet Valley Living Historical Farm** Bus. U.S. 209, 3½ miles south of Stroudsburg, Pennsylvania AMERICAN EXPRESS TRAVEL & LEISURE ROAD ATLAS	**50¢ off** **Alamo Village Vacation & Movie Land** Brackettville, Texas AMERICAN EXPRESS TRAVEL & LEISURE ROAD ATLAS	**10% off** **Hatfields & McCoys— Honey in the Rock** Grandview State Park Beckley, West Virginia AMERICAN EXPRESS TRAVEL & LEISURE ROAD ATLAS
$1.00 off **Roadside America** I-78, Shartlesville Exit Shartlesville, Pennsylvania AMERICAN EXPRESS TRAVEL & LEISURE ROAD ATLAS	**$1.00 off** **Aquarena Springs** Aquarena Springs Dr. San Marcos, Texas AMERICAN EXPRESS TRAVEL & LEISURE ROAD ATLAS	**20¢ off** **Circus World Museum** Baraboo, Wisconsin AMERICAN EXPRESS TRAVEL & LEISURE ROAD ATLAS

Clip along dashed lines

$1.00 off **International Tennis Hall of Fame** 194 Bellevue Ave. Newport, Rhode Island AMERICAN EXPRESS TRAVEL & LEISURE ROAD ATLAS	**20% off** **"Remember the Alamo" Theatre/Museum** 315 Alamo Plaza San Antonio, Texas 78205 AMERICAN EXPRESS TRAVEL & LEISURE ROAD ATLAS	**$1.00 off** **Green Bay Packer Hall of Fame** 1901 S. Oneida Green Bay, Wisconsin AMERICAN EXPRESS TRAVEL & LEISURE ROAD ATLAS
25¢ off **Cypress Gardens** U.S. 52 Charleston, South Carolina AMERICAN EXPRESS TRAVEL & LEISURE ROAD ATLAS	**50¢ off** **Texas Ranger Hall of Fame and Museum** I-35 Waco, Texas AMERICAN EXPRESS TRAVEL & LEISURE ROAD ATLAS	**20% off** **La Crosse Queen Paddlewheel Boat** Riverside Park La Crosse, Wisconsin AMERICAN EXPRESS TRAVEL & LEISURE ROAD ATLAS

Clip along dashed lines

50¢ off **Magnolia Plantation and Gardens** S.C. 61 Charleston, South Carolina AMERICAN EXPRESS TRAVEL & LEISURE ROAD ATLAS	**$AVING$** **Hansen Planetarium** 15 S. State St. Salt Lake City, Utah AMERICAN EXPRESS TRAVEL & LEISURE ROAD ATLAS	**$1.00 off** **Jackson Hole Aerial Tram** Teton Village, Wyoming AMERICAN EXPRESS TRAVEL & LEISURE ROAD ATLAS
50¢ off **Black Hills Caverns** S. Dak. 44 West Rapid City, South Dakota AMERICAN EXPRESS TRAVEL & LEISURE ROAD ATLAS	**50¢ off** **Santa's Land** U.S. 5 Putney, Vermont AMERICAN EXPRESS TRAVEL & LEISURE ROAD ATLAS	**10% off** **Mad River Boat Trips** 153 N. Cache (State Stop Shops) Box 2222, Jackson, Wyoming 83001 AMERICAN EXPRESS TRAVEL & LEISURE ROAD ATLAS

Shelburne Museum
35 buildings: Americana, steamboat, miniature circus, quilts, tools, railroad, toys. Picnic area; cafeteria; gift shop.
With this voucher:
50¢ discount.
Daily mid-May thru late Oct. (Limited schedule through winter.)
Voucher limit: 6 persons.

Black Hills Passion Play
Outdoor drama depicting the life of Christ.
With this voucher:
10% off all adult admissions June, July, Aug. For reservations write Box 469, Spearfish, S. Dak. 57783.
Voucher limit: 6 family members.

National Tower
307-ft. observation tower with dramatic sound program. Enclosed and open decks, air conditioned, elevators, gift shops, picnic area.
With this voucher:
25¢ discount.
Open daily. Weekends only Dec.-Feb. Closed New Year's and Christmas.
Voucher limit: family members.

Historic Michie Tavern Museum
A famous tavern of the 1700 s; "Inn of the Presidents."
With this voucher:
50¢ off adult admission.
Closed New Year's and Christmas.
Voucher limit: family members.

Confederama
See Confederama first to better appreciate Chattanooga. Dynamic exhibit and dioramas; gifts and souvenirs. "Civil War - The Battle of Chattanooga - That Sealed the Fate of the Confederacy."
Closed Thanksgiving and Christmas.
With this voucher:
One free admission with one paid admission of the same type.

Pennsylvania Dutch Farm
Amish exhibit home, animals, seasonal horse-drawn rides, "Frontier Shop."
With this voucher.
20% off regular admission.
Voucher limit: entire party.

Skyline Caverns
With this voucher:
$1.00 off adult admission, children 50¢ off.
Closed Christmas Day.

Tommy Bartlett's WATER CIRCUS
A two-hour live show featuring world champion water skiers performing near to impossible feats; the most outrageous stage show in the country; and from the sky a tremendous helicopter trapeze act you won't want to miss!
With this voucher:
20% off general admission price. Open May 22-Sept. 27. Voucher may not be used in combination with any other discount program.
Voucher limit: entire party.

Poconos' Magic Valley and Winona 5 Falls
The Poconos' only amusement park. Working craftspeople, thrilling rides.
With this voucher:
$1.00 off per person. Cannot be combined with any other discount program. No cash value.
Open Apr.-Oct. (Closed rest of year.)
Voucher limit: 5 family members. Expires Oct. 1982.

— — — Clip along dashed lines — — — —

Hatfields & McCoys—Honey in the Rock
Outdoor musical theater played in repertory.
With this voucher:
10% off all tickets.
June 20-Aug. 30. (Not valid on Sats.) Closed Mondays.

Alamo Village Vacation & Movie Land
Authentic recreation of Old San Antonio for John Wayne's movie "The Alamo."
With this voucher:
50¢ discount.
Voucher limit: family members.

Quiet Valley Living Historical Farm
Guided tour of a living historical Pennsylvania Dutch farm dating from 1765.
With this voucher:
50¢ off adult admission.
Daily June 20-Labor Day. (Closed rest of year.)
Voucher limit: 4 adults.

Circus World Museum
With this voucher:
20% off regular admission.
Open mid May-mid Sept.
Voucher limit: entire party (except family rate).

Aquarena Springs
Complete tour including glass-bottom boat ride, submarine theatre, Texana Village, Sky Spiral.
With this voucher:
$1.00 discount on complete tour ticket.
Closed Christmas.
Voucher limit: family members.

Roadside America
The world's greatest indoor miniature village.
With this voucher:
$1.00 discount.
Voucher limit: 2 adults per coupon.

Green Bay Packer Hall of Fame
Multimedia presentations, exhibits, memorabilia and life-size figures tell the story of the Green Bay Packers from 1919 to the present.
With this voucher:
$1.00 off adult admission, children 50¢ off.
Closed Christmas.
Voucher limit: 1 family member.

"Remember the Alamo" Theatre/Museum
Authentic historic MultiVision presentation of the siege of the Alamo; across from the Alamo.
With this voucher:
20% discount.
Closed New Year's, Thanksgiving, Christmas.

International Tennis Hall of Fame
With this voucher:
$1.00 discount.
Closed New Year's Day and Christmas.

— — — Clip along dashed lines — — — —

La Crosse Queen Paddlewheel Excursion Boat
"La Crosse Queen" Authentic paddlewheel excursion boat on the mighty Mississippi River.
With this voucher:
20% discount on 1-hr. trips only. Daily Memorial Day weekend-Labor Day. Charter only Apr.-May, Sept.-Oct. (Closed rest of year.)
Voucher limit: 6 family members.

Texas Ranger Hall of Fame and Museum
With this voucher:
50¢ discount.
Closed New Year's, Thanksgiving, Christmas.
Voucher limit: 1 person.

Cypress Gardens
With this voucher:
25¢ discount.
Daily Feb. 15-Apr. 30. (Closed rest of year.)
Voucher limit: 6 persons.

Jackson Hole Aerial Tram
With this voucher:
$1.00 discount.
Open late May-late Sept.
Voucher limit: 1 person.

Hansen Planetarium
With this voucher:
One free admission with one paid adult regular admission.
Closed holidays.
Voucher limit: 1 person.

Magnolia Plantation and Gardens
17th-century gardens by the Ashley River; nature trails, bike and canoe rentals. 500-acre wildlife preserve, petting zoo, mini horse ranch; snack shop, gift shop; wildlife observation tower; 16th-century maze. Plantation House additional.
With this voucher:
50¢ off for all visitors. Open all year.

Mad River Boat Trips
With this voucher:
10% discount. Open mid-May thru Sept. All trips include round-trip transportation from Jackson and all river gear. Overnight trip provides all camping gear. Waterproof storage for cameras. Coupon invalid if trip reservations are through any agency.
Voucher limit: family members.

Santa's Land
Four-season family attraction; Santa's home May to Christmas. Winter schedule: sleigh rides, ski touring, ice skating. Pancake House, Christmas shops.
With this voucher:
50¢ discount.
Closed Thanksgiving, Christmas.
Voucher limit: family members.

Black Hills Caverns
Crystal-lined wind caverns; snack bar, gift shop, hiking trail and picnic area. 1-hour, 1/2-hour, candlelight and spelunking tours offered.
With this voucher:
50¢ discount. May 15-Oct. 1. (Closed rest of year.)
Voucher limit: family members.

Alphabet Games

These games are for two or more people. The more players, the merrier. Parents and children can both join in the fun.

Shopping

This game will test your memory. The first player may say "I bought an airplane." The next player must repeat the airplane and add something beginning with a B, such as "I bought an airplane and a ball." Then the next player may say, "I bought an airplane, a ball and a cuckoo clock."

Each player must repeat, in alphabetical order, all the things others have bought and then add his own. Crazy things add to the fun. If a player forgets something, he is out of the game.

Alphabet Stories

The first player starts a story using three "A's," like this: "My name is Allen. I like apples. I am going to Arizona to get some." The next player uses three "B's" like this: "My name is Betsy. I live in Boston. I have a pet bear."

I See

Use whatever you see in the car or along the road to run through the alphabet. The first player starts by looking for "A's." He may say, "I see an automobile." The next player may say, "I see a B in that sign 'Buffalo, 20 miles.'" Then, "I see a chimney."

The game can be played by each person in turn, or by all players looking for the next letter regardless of turn. There are some hard letters, such as J, Q, X, Y and Z. Allow lots of time.

City sCrAmbLe

How many famous cities can you unscramble? Set a time limit, such as ten minutes, and score your answers. See who can unscramble the most city names in the allotted time.
Hint: The capital letter indicates the first letter of the city name.

1. cogCahi _____ Chicago _____
2. laDsal _____
3. noniMeapsli _____
4. xePonih _____
5. tsoBon _____
6. wekMaeliu _____
7. fufBola _____
8. laSt keLa iCyt _____
9. neneCehy _____
10. amhOa _____
11. danlorO _____
12. sDe onMsei _____
13. akloOhma tiyC _____
14. lasesThalea _____
15. toreDit _____
16. tacronSn _____
17. donhiRcm _____
18. Cnadveell _____
19. reGne yBa _____
20. pmeMsih _____
21. xuoSi yCti _____
22. agriNaa lsFal _____
23. Bseio _____
24. weN kYro _____
25. Mmiia _____

1. Chicago 2. Dallas 3. Minneapolis 4. Phoenix 5. Boston 6. Milwaukee 7. Buffalo 8. Salt Lake City 9. Cheyenne 10. Omaha 11. Orlando 12. Des Moines 13. Oklahoma City 14. Tallahassee 15. Detroit 16. Scranton 17. Richmond 18. Cleveland 19. Green Bay 20. Memphis 21. Sioux City 22. Niagara Falls 23. Boise 24. New York 25. Miami.

States You Will Visit On Your Trip

Color or lightly pencil all the states you have already visited, using different colors for each state if you wish. Then color or pencil more darkly the states you expect to visit on this trip. This will show you the states you will see for the first time.

Sports Quiz

Here are questions to test your knowledge of American sports. After you have answered these questions, try coming up with your own to stump your traveling companions.

1. What major-league baseball team has won more World Series than any other?_____
2. What award is given annually to the champions of the National Hockey League?_____
3. Name the location of the 1980 Winter Olympic Games. _____ The summer games. _____
4. Who holds the record as the all-time leading touchdown scorer in professional football?_____
5. Who was the winner of the 1976 Olympic decathlon with a record 8,618 points?_____
6. What pro basketball player holds the record for most points scored in a career? _____

Name the sport in which these people participate or have participated. Think of more names, and ask your traveling companions to name the sport.

7. Gordie Howe	16. Bob Cousy
8. Chris Evert-Lloyd	17. Bobby Orr
9. Ty Cobb	18. Dorothy Hamill
10. Nancy Lopez Melton	19. Jack Dempsey
11. Sugar Ray Leonard	20. Pete Rose
12. Lou Brock	21. Billy Jean King
13. Mario Andretti	22. Larry Czonka
14. Tom Watson	23. Steve Cauthen
15. Mark Spitz	24. Joe Namath

1. New York Yankees	13. Auto Racing
2. Stanley Cup	14. Golf
3. Lake Placid, N.Y., Moscow	15. Swimming
4. Jim Brown	16. Basketball
5. Bruce Jenner	17. Hockey
6. Wilt Chamberlain (30,335 pts.)	18. Ice Skating
7. Hockey	19. Boxing
8. Tennis	20. Baseball
9. Baseball	21. Tennis
10. Golf	22. Football
11. Boxing	23. Horse Racing
12. Baseball	24. Football

Games for the Car

Twenty Questions

This game is for two or more players. Someone starts by thinking of something he sees in the car (or along the road). Then the others take turns asking him questions which can only be answered by *yes* or *no*, until someone correctly guesses the object. If the object is guessed in less than 20 questions, the player who guessed correctly picks the next object. If the players use the 20 questions without guessing correctly, the original selector picks again. The game may also be played by thinking of places or things, but you must tell the other players if the subject chosen is animal, vegetable or mineral.

Tall Tales

A game for two or more players. The first player starts a story with something he sees along the road, such as "*I see a restaurant.*" Then each person in turn continues the story by adding crazy sentences, such as "*Two monkeys are having lunch in the restaurant,*" followed by "*The monkeys like to eat the paper napkins with whipped cream.*"

Town Names

A game for one person or any number of players. Write the name of a city or town you see along the road. Then see how many words you can form from the letters. For example, you pass through Madison. From this name you can make *mad, is, son* and *on*. For another example, by mixing the letters in Springfield you can make *spring, field, sing, ring, din, pin, den* and others.

Buzz

"Buzz" is a counting game played by three or more players. The players count numbers, but each time a seven, a multiple of seven, or a number that has a 7 in it is reached, the word *Buzz* is substituted. Thus: 1, 2, 3, 4, 5, 6, *Buzz*, 8, 9, 10, 11, 12, 13, *Buzz*, 15, 16, *Buzz*, and so on.

At the start of the game, the players take their turns in a clockwise order. When a *Buzz* number is called, the order reverses to counterclockwise until another *Buzz* number is reached. Thus: Player A, Player B, Player C until *Buzz* is called, then backwards to Player B, Player A, Player C, and so on.

A player who hesitates too long when it is his turn or who does not call *Buzz* at the correct time, is out of the game. The last player to be eliminated is the winner.

Spot the Vehicles

The first player to see a vehicle points it out to other players and marks his lettered box. After setting a time limit, total the scores.

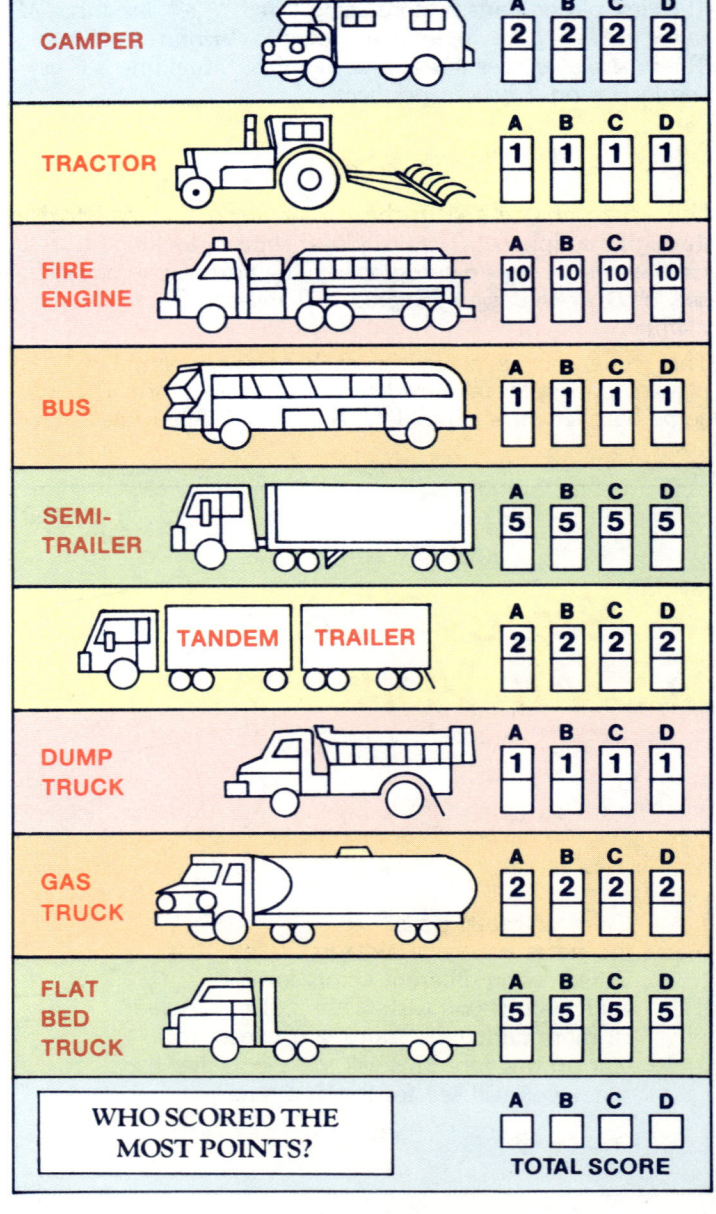

		A	B	C	D
CAMPER		2	2	2	2
TRACTOR		1	1	1	1
FIRE ENGINE		10	10	10	10
BUS		1	1	1	1
SEMI-TRAILER		5	5	5	5
TANDEM TRAILER		2	2	2	2
DUMP TRUCK		1	1	1	1
GAS TRUCK		2	2	2	2
FLAT BED TRUCK		5	5	5	5
WHO SCORED THE MOST POINTS?		A	B	C	D

TOTAL SCORE

Did you know that America's shortest river, the D River in Oregon, is only 440 feet long at low tide—while the longest, the Mississippi-Missouri, stretches more than 3,700 miles. Below are more unusual U.S. Facts and Figures that you and your family may find interesting as you travel.

Geographic center of the conterminous United States
Near the town of Lebanon in Smith County, Kansas (39°50′N, 98°35′W).

Largest states
1. Alaska (589,757 sq. mi.).
2. Texas (267,338 sq. mi.).
3. California (158,693 sq. mi.).

Smallest states
1. Rhode Island (1,214 sq. mi.).
2. Delaware (2,057 sq. mi.).
3. Connecticut (5,009 sq. mi.).

Largest county
San Bernardino County, California (20,119 sq. mi.).

Smallest county
New York, New York (23 sq. mi.).

Northernmost city
Barrow, Alaska (71°17′N).

Southernmost city
Hilo, Island of Hawaii (19°43′N).

Easternmost city
Eastport, Maine (66°59′W).

Westernmost city
Nome, Alaska (165°25′W).

Highest city
Leadville, Colorado (10,200 ft.).

Lowest town
Calipatria, California (-184 ft.).

Oldest national park
Yellowstone National Park, Wyoming, Montana, Idaho (1872).

Longest natural bridge
Landscape Arch, in Arches National Park near Moab, Utah, spans 291 ft. and rises 100 ft. above the canyon floor.

Highest waterfall
Yosemite Falls, California (total in three sections: 2,425 ft.): Upper Yosemite Fall (1,430 ft.), Cascades in middle section (675 ft.), Lower Yosemite Fall (320 ft.).

Longest rivers
1. Mississippi-Missouri (3,710 mi.).
2. Rio Grande (1,885 mi.).
3. Yukon (1,770 mi.).

Shortest river
D River, Oregon. Connects Devil's Lake to Pacific Ocean and is 440 ft. long at low tide.

Fastest rapids
The fastest rapids ever navigated are the Lava Falls on the Colorado River in Arizona. At times of flood they attain a speed of 30 m.p.h.

Highest mountains
1. Mount McKinley in Alaska (20,320 ft.).
2. St. Elias in Alaska (18,008 ft.).
3. Foraker in Alaska (17,400 ft.).

Lowest point
Death Valley, California (-282 ft.).

Deepest lake
Crater Lake, Oregon (1,932 ft.).

Largest underground lake
Lost Sea, 300 ft. underground in Craighead Caverns, Sweetwater, Tennessee. Covers an area of 4½ acres.

Longest cave system
Mammoth Cave National Park, Kentucky, has a length of 141.77 mi.

Rainiest spot
Mt. Waialeale, Hawaii (annual average rainfall 440 inches).

Largest gorge
Grand Canyon, Colorado River, Arizona (217 miles long, 4 to 18 miles wide, 1 mile deep).

Deepest gorge
Hells Canyon, Snake River, Idaho (7,900 ft.).

Highest temperature
On June 10, 1930, a temperature of 134°F. was recorded in Death Valley, California.

Lowest temperature
On January 23, 1971, a temperature of -80°F. was recorded in Prospect Creek, Alaska.

Strongest surface wind
Mount Washington, New Hampshire (recorded 1934—231 m.p.h.).

Fastest land speed
739.6 m.p.h., attained by Stan Barrett on Dec. 17, 1979, Edwards Air Force Base, California in a rocket-powered car.

Fastest air speed
2,194 m.p.h. in a Lockheed SR-71, attained by Capt. Eldon Joersz and Maj. George Morgan, Jr. on July 28, 1976 over Beale Air Force Base, California.

Biggest dam
New Cornelia Tailings, Ten Mile Wash, Arizona (274,026,000 cu. yards of materials used).

Largest building
Boeing 747 assembly plant, Everett, Washington (205,600 cu. ft.; covering 47 acres).

Tallest structures
1. TV tower, Blanchard, North Dakota (2,063 ft.).
2. B.R.E.N. Tower, Nevada (1,527 ft.).
3. American Electric Power Co. smokestack, West Virginia (1,206 ft.).

Largest lakes
1. Superior (31,820 sq. mi.).
2. Huron (23,010 sq. mi.).
3. Michigan (22,400 sq. mi.).

Highest bridge
Royal Gorge, Colorado (1,053 ft. above water).

Longest suspension bridges
1. Verrazano-Narrows, Lower New York Bay (4,260 ft.).
2. Golden Gate, San Francisco Bay, California (4,200 ft.).
3. Mackinac, Straits of Mackinac, Michigan (3,800 ft.).
4. George Washington, Hudson River, New York-New Jersey (3,500 ft.).

Tallest buildings
1. Sears Tower, Chicago (110 stories—1,454 ft.).
2. World Trade Center, New York City (110 stories—1,350 ft.).
3. Empire State Building, New York City (102 stories—1,250 ft.).

Longest vehicular tunnels
1. Brooklyn-Battery, East River, New York City (1.7 mi.—1950).
2. Holland, Hudson River, New York-New Jersey (1.6 mi.—1937).
3. Lincoln, Hudson River, New York-New Jersey (1.5 mi.—1937).

Oldest living tree
Bristlecone Pine named Methuselah; age 4,600 years. Located on the California side of the White Mountains.

Largest living thing
Sequoia tree named General Sherman stands 272 feet 4 inches tall, girth of 79.1 feet, 2,145 tons. Found in Sequoia National Park, California.

Tallest tree
Howard Libbey Redwood tree is 366.2 feet tall. Located in Creek Grove, California.

Largest aquarium
John G. Shedd Aquarium in Chicago, Illinois. Total capacity 450,000 gallons.

Largest oceanarium
Marineland of the Pacific, Palos Verdes Peninsula, California. The total capacity is 2,500,000 gallons.

Largest deserts
1. Mojave, in California (15,000 sq. mi.).
2. Great Salt Lake, in Utah (10,000 sq. mi.).

Largest island
Kodiak, Alaska (5,363 sq. mi.).

Tallest active geyser
Steamboat Geyser in Yellowstone National Park, Wyoming erupted to a height of 380 ft.

Most frequently erupting geyser
"Old Faithful" in Yellowstone National Park, Wyoming, which erupts on the average of every 66 minutes.

Largest ocean
Pacific Ocean, representing 45.8% of the world's total oceans.

STATE	CAPITAL	NICKNAME	STATE TREE	STATE FLOWER	SQ. MILES (RANK)	POPULATION (RANK) *	SALES TAX	TIME ZONES
Alabama	Montgomery	Heart of Dixie	Southern pine	Camellia	51,609 (29)	3,870,251 (22)	4%	Central
Alaska	Juneau	None	Sitka spruce	Forget-me-not	589,757 (1)	400,142 (50)	—	Pacific/Yukon/ Alaska-Hawaii/ Bering
Arizona	Phoenix	Grand Canyon State	Paloverde	Saguaro cactus blossom	113,909 (6)	2,719,225 (29)	5%	Mountain
Arkansas	Little Rock	Land of Opportunity	Pine	Apple blossom	53,104 (27)	2,284,037 (33)	3%	Central
California	Sacramento	Golden State	California redwood	Golden poppy	158,693 (3)	23,545,061 (1)	6%	Pacific
Colorado	Denver	Centennial State	Colorado blue spruce	Rocky Mtn. columbine	104,247 (8)	2,882,061 (28)	3%	Mountain
Connecticut	Hartford	Constitution State	White oak	Mountain laurel	5,009 (48)	3,096,454 (24)	7½%	Eastern
Delaware	Dover	First State	American holly	Peach blossom	2,057 (49)	594,779 (47)	—	Eastern
Florida	Tallahassee	Sunshine State	Sabal palmetto palm	Orange blossom	58,560 (22)	9,579,963 (7)	4%	Cen./East.
Georgia	Atlanta	Peach State	Live oak	Cherokee rose	58,876 (21)	5,404,384 (13)	3%	Eastern
Hawaii	Honolulu	Aloha State	Candlenut	Hibiscus	6,450 (47)	964,680 (39)	4%	Alaska-Hawaii
Idaho	Boise	Gem State	White pine	Syringa	83,557 (13)	943,629 (41)	3%	Pac./Mtn.
Illinois	Springfield	Prairie State	White oak	Violet	56,400 (24)	11,355,062 (5)	4%	Central
Indiana	Indianapolis	Hoosier State	Tulip poplar	Peony	36,291 (38)	5,461,103 (12)	4%	Cen./East.
Iowa	Des Moines	Hawkeye State	Oak	Wild rose	56,290 (25)	2,909,463 (27)	3%	Central
Kansas	Topeka	Sunflower State	Cottonwood	Sunflower	82,264 (14)	2,356,032 (32)	3%	Cen./Mtn.
Kentucky	Frankfort	Bluegrass State	Kentucky coffee tree	Goldenrod	40,395 (37)	3,642,795 (23)	5%	Cen./East.
Louisiana	Baton Rouge	Pelican State	Cypress	Magnolia	48,523 (31)	4,199,542 (18)	6%	Central
Maine	Augusta	Pine Tree State	Eastern white pine	White pine cone/tassel	33,215 (39)	1,123,670 (38)	5%	Eastern
Maryland	Annapolis	Old Line State	White oak	Black-eyed Susan	10,577 (42)	4,198,113 (19)	5%	Eastern
Massachusetts	Boston	Bay State	American elm	Mayflower	8,257 (45)	5,728,288 (11)	5%	Eastern
Michigan	Lansing	Great Lake State	White pine	Apple blossom	58,216 (23)	9,238,634 (8)	4%	Cent./East.
Minnesota	St. Paul	North Star State	Red pine	Pink & white lady's slipper	84,068 (12)	4,069,356 (21)	4%	Central
Mississippi	Jackson	Magnolia State	Magnolia	Magnolia	47,716 (32)	2,511,491 (31)	5%	Central
Missouri	Jefferson City	Show-Me State	Dogwood	Hawthorn	69,686 (19)	4,906,480 (15)	3⅛%	Central
Montana	Helena	Treasure State	Ponderosa pine	Bitterroot	147,138 (4)	783,698 (44)	—	Mountain
Nebraska	Lincoln	Cornhusker State	Cottonwood	Goldenrod	77,227 (15)	1,564,901 (35)	3%	Cent./Mtn.
Nevada	Carson City	Sagebrush State	Single-leaf pinon	Sagebrush	110,540 (7)	800,312 (43)	3%	Pacific
New Hampshire	Concord	Granite State	White birch	Purple lilac	9,304 (44)	918,959 (42)	—	Eastern
New Jersey	Trenton	Garden State	Red oak	Purple violet	7,836 (46)	7,342,164 (9)	5%	Eastern
New Mexico	Santa Fe	Land of Enchantment	Pinon	Yucca	121,666 (5)	1,295,474 (37)	3¾%	Mountain
New York	Albany	Empire State	Sugar maple	Rose	49,576 (30)	17,507,541 (2)	4%	Eastern
North Carolina	Raleigh	Tar Heel State	Pine	Dogwood blossom	52,586 (28)	5,847,788 (10)	3%	Eastern
North Dakota	Bismarck	Sioux State	American elm	Wild prairie rose	70,665 (17)	652,437 (46)	3%	Cent./Mtn.
Ohio	Columbus	Buckeye State	Buckeye	Scarlet carnation	41,222 (35)	10,772,342 (6)	4%	Eastern
Oklahoma	Oklahoma City	Sooner State	Redbud	Mistletoe	69,919 (18)	3,001,252 (26)	2%	Central
Oregon	Salem	Beaver State	Douglas fir	Oregon grape	96,981 (10)	2,618,126 (30)	—	Pac./Mtn.
Pennsylvania	Harrisburg	Keystone State	Hemlock	Mountain laurel	45,333 (33)	11,828,095 (4)	6%	Eastern
Rhode Island	Providence	Little Rhody	Red maple	Violet	1,214 (50)	945,835 (40)	6%	Eastern
South Carolina	Columbia	Palmetto State	Palmetto	Carolina jessamine	31,055 (40)	3,069,825 (25)	4%	Eastern
South Dakota	Pierre	Coyote State	Black Hills spruce	Pasque flower	77,047 (16)	688,217 (45)	4%	Cent./Mtn.
Tennessee	Nashville	Volunteer State	Tulip poplar	Iris	42,244 (34)	4,545,590 (17)	4½%	Cent./East.
Texas	Austin	Lone Star State	Pecan	Bluebonnet	267,338 (2)	14,173,876 (3)	4%	Cent./Mtn.
Utah	Salt Lake City	Beehive State	Blue spruce	Sego lily	84,916 (11)	1,459,010 (36)	4%	Mountain
Vermont	Montpelier	Green Mtn. State	Sugar maple	Red clover	9,609 (43)	511,299 (48)	3%	Eastern
Virginia	Richmond	Old Dominion	Dogwood	Dogwood blossom	40,817 (36)	5,323,412 (14)	4%	Eastern
Washington	Olympia	Evergreen State	Western hemlock	Rhododendron	68,192 (20)	4,114,738 (20)	5³⁄₁₀%	Pacific
West Virginia	Charleston	Mountain State	Sugar maple	Big rhododendron	24,181 (41)	1,930,787 (34)	3%	Eastern
Wisconsin	Madison	Badger State	Sugar maple	Wood violet	56,154 (26)	4,693,941 (16)	4%	Central
Wyoming	Cheyenne	Equality State	Cottonwood	Indian paintbrush	97,914 (9)	468,954 (49)	3%	Mountain

* Based on preliminary 1980 census. Source: U.S. Dept. of Commerce.

UNITED STATES

Names shown thus: Hollywood...90 B-5 refer to city inset on page 90 with grid reference location at B-5

ALABAMA

Map on page 22

Abbeville C-3
Albertville B-1
Alexander City C-2
Aliceville A-2
Andalusia B-3
Anniston C-2
Arab B-1
Ariton C-3
Ashland C-2
Ashville B-2
Athens B-1
Atmore B-3
Attalla B-1
Auburn C-2
Banks C-3
Bay Minette B-3
Bessemer B-2
Birmingham B-2
Boaz B-1
Brantley B-3
Brent B-2
Brewton B-3
Bridgeport C-1
Brundidge C-3
Butler A-2
Calera B-2
Camden B-2
Carbon Hill B-2
Castleberry B-3
Catherine B-2
Cedar Bluff C-1
Centre C-1
Centreville B-2
Chatom A-3
Childersburg B-2
Citronelle A-3
Clanton B-2
Clayton C-3
Clio C-3
Collinsville B-1
Columbia C-3
Columbiana B-2
Cullman B-1
Dadeville C-2
Decatur B-1
Demopolis B-2
Dothan C-3
Double Sprs. B-1
E. Brewton B-3
Elba B-3
Enterprise B-3
Eufaula C-3
Eutaw B-2
Evergreen B-3
Fairhope B-3
Fayette B-2
Flomaton B-3
Florala B-3
Florence B-1
Foley B-3
Fort Deposit B-3
Ft. Payne C-1
Franklin B-3
Frisco City B-3
Gadsden C-1
Geneva C-3
Goodwater B-2
Greenville B-3
Grove Hill B-3
Guin B-2
Guntersville B-1
Haleyville B-1
Hamilton A-1
Harpersville B-2
Hartford C-3
Hartselle B-1
Hayneville B-2
Headland C-3
Heflin C-2
Homewood B-2
Huntsville B-1
Hurtsboro C-2
Jackson B-3
Jacksonville C-2
Jasper B-2
Lafayette C-2
Lanett C-2
Leeds B-2
Linden B-2
Livingston A-2
Luverne B-3
Maplesville B-2
Marion B-2
Midway B-3
Midway C-2
Millport A-2
Mobile A-3
Monroeville B-3
Montevallo B-2
Montgomery B-2
Moulton B-1
Mountain Brook 76 F-3
Mt. Vernon B-3
New Brockton C-3
Northport B-2
Oak Hill B-3
Oakman B-2
Oneonta B-2
Opelika C-2
Opp B-3
Oxford C-2
Ozark C-3
Pell City B-2
Phenix City C-2
Phil Campbell B-1
Piedmont C-2
Plantersville B-2
Prattville B-2
Prichard B-3
Red Bay A-1
Reform B-2
Repton B-3
Roanoke C-2
Rockford B-2
Rogersville B-1
Russellville B-1
Safford B-2
Scottsboro B-1
Seale C-2
Selma B-2
Sheffield B-1
Stevenson C-1
Sulligent A-2
Sumiton B-2
Sylacauga B-2
Talladega B-2
Tallassee B-2
Theodore A-3
Thomasville B-3
Town Creek B-1
Troy B-3
Tuscaloosa B-2
Tuscumbia B-1
Tuskegee C-2
Union Sprs. C-2
Uniontown B-2
Uriah B-3
Vernon A-2
Wadley C-2
Wagarville B-3
Warrior B-2
Wetumpka B-2
Winfield B-2
Woodstock B-2
York A-2

ALASKA

Map on page 52

Adak B-4
Anchorage B-2
Attu A-3
Barrow B-1
Bethel A-2
Circle C-1
Cordova B-2
Delta Junc. C-2
Eagle C-2
Fairbanks B-2
Gambell A-1
Haines C-2
Homer B-2
Juneau C-2
Kenai B-2
Ketchikan D-3
Kodiak B-2
Kotzebue B-1
Livengood B-1
Nome A-1
Palmer B-2
Pt. Hope B-1
Seward B-2
Sitka C-3
Tok C-2
Umnak C-4
Valdez B-2

ARIZONA

Map on page 46

Aguila B-4
Ajo B-4
Alpine D-4
Apache D-5
Apache Jct. C-4
Ash Fork B-3
Avondale B-4
Benson C-5
Bisbee D-5
Black Canyon City B-3
Bouse B-4
Bowie D-4
Buckeye B-4
Cameron C-3
Camp Verde C-3
Casa Grande B-4
Cashion 100 A-3
Chambers D-3
Chandler C-4
Chinle D-2
Chino Valley B-3
Clarkdale B-3
Clifton D-4
Cochise D-4
Colorado City B-2
Concho D-3
Congress B-3
Continental C-5
Coolidge C-4
Cortaro 108 D-1
Davis Dam A-3
Douglas D-5
Duncan D-4
El Mirage 100 A-2
Eloy C-4
Fairbank C-5
Flagstaff C-3
Florence C-4
Fort Apache D-4
Ft. Defiance D-3
Fort McDowell 100 F-1
Ft. Thomas C-4
Fountain Hills 100 F-2
Fredonia B-2
Ganado D-3
Gila Bend B-4
Gilbert 100 E-4
Glendale B-4
Globe C-4
Goodyear 100 A-3
Grand Canyon B-2
Gray Mtn. C-3
Guadalupe 100 D-4
Heber C-3
Higley 100 F-4
Holbrook C-3
Humboldt B-3
Jacob Lake B-2
Jaynes 108 D-2
Jeddito C-3
Jerome B-3
Joseph City C-3
Kayenta C-2
Kingman B-3
Komatke 100 B-4
L. Havasu City A-3
Laveen 100 B-4
Liberty B-4
Litchfield Park 100 A-2
Lukachukai D-2
Mammoth C-4
Marana C-4
Marble Canyon C-2
Maryvale 100 B-3
Mayer B-3
McNary D-3
Mesa C-4
Miami C-4
Mohawk B-4
Morristown B-4
Navajo D-3
Nogales C-5
North Rim C-2
Nortons Corner 100 E-4
Nutrioso D-4
Oracle C-4
Oracle Jct. C-4
Oro Valley 108 E-1
Page C-2
Paradise Valley 100 D-2
Parker A-3
Patagonia C-5
Payson C-3
Peach Springs B-3
Pearce D-5
Peoria 100 B-2
Phoenix C-4
Picacho C-4
Pima D-4
Pine C-3
Portal D-5
Prescott B-3
Punkin Center C-4
Quartzsite A-4
Rainbow Lodge C-2
Red Lake B-3
Red Rock D-2
Rillito 108 D-1
Roosevelt C-4
Safford D-4
Sahuarita C-5
St. David C-5
St. Johns D-3
Salome B-4
San Carlos C-4
Sanders D-3
San Simon D-4
Santa Maria 100 B-3
Sasabe C-5
Scottsdale 100 D-3
Sedona C-3
Seligman B-3
Sells C-5
Sentinel B-4
Shonto C-2
Show Low D-3
Sierra Vista C-5
Snowflake C-3
Solomon D-4
Sonoita C-5
South Phoenix 100 C-3
South Tucson 108 E-2
Springerville D-3
Sun City B-4
Sunnyslope 100 C-2
Superior C-4
Surprise 100 A-1
Tempe B-4
Thatcher D-4
The Gap C-2
Tolleson 100 A-3
Tombstone D-5
Tonalea C-2
Tonopah B-4
Topock A-3
Tubac C-5
Tuba City C-2
Tucson C-4
Tumacacori C-5
Valentine B-3
Valle B-3
Vicksburg B-4
Wellton A-4
Why B-4
Wickenburg B-4
Wikieup B-3
Willcox C-4
Williams B-3
Window Rock D-3
Winkelman C-4
Winona C-3
Winslow C-3
Young C-3
Youngtown 100 A-2
Yucca A-3
Yuma A-4

ARKANSAS

Map on page 24

Altheimer B-2
Arkadelphia A-2
Arkansas City B-3
Ashdown A-3
Atkins A-2
Augusta B-2
Bald Knob B-2
Barton B-2
Batesville B-2
Bearden A-3
Beebe B-2
Benton A-2
Bentonville A-1
Berryville A-1
Blytheville B-2
Booneville A-3
Bradford B-2
Brinkley B-2
Calion A-3
Camden A-3
Cammack Village 91 H-1
Cave City B-2
Charleston A-3
Chidester A-3
Clarendon B-2
Clarksville A-2
Clinton A-2
Conway A-2
Corning B-1
Cotter A-1
Cove A-2
Crossett A-3
Danville A-3
Dardanelle A-3
DeQueen A-3
Dermott B-3
Des Arc B-2
DeWitt B-2
Dierks A-2
Dover A-2
Dumas B-3
Earle B-2
El Dorado A-3
Emerson A-3
Eudora B-3
Eureka Sprs. A-1
Evening Shade B-1
Fayetteville A-1
Fordyce A-3
Forrest City B-2
Fort Smith A-2
Fouke A-3
Genevia 91 M-3
Glenwood A-3
Gould B-3
Grady B-2
Gravette A-1
Green Forest A-1
Greenwood A-2
Gurdon A-3
Hamburg B-3
Hampton A-3
Hardy B-1
Harrisburg B-2
Harrison A-1
Hazen B-2
Heber Sprs. B-2
Helena B-2
Hope A-3
Hot Springs A-2
Hoxie B-1
Humphrey B-2
Huntington A-2
Imboden B-1
Jacksonville B-2
Jasper A-2
Jonesboro B-2
Judsonia B-2
Junction City A-3
Kingsland A-3
Kirby A-3
Lake Village B-3
Leachville B-2
Leslie A-2
Lewisville A-3
Little Rock A-2
Lockesburg A-3
Lonoke B-2
Louann A-3
Luxora C-2
Magnolia A-3
Malvern A-2
Mammoth Spr. B-1
Manila B-2
Mansfield A-2
Marianna B-2
Marked Tree B-2
Marshall A-2
Marvell B-2
McCrory B-2
McGehee B-3
McNeil A-3
Mena A-2
Monette B-2
Monticello B-3
Montrose B-3
Morrilton A-2
Mountain Home A-1
Mountain View A-2
Mount Ida A-3
Mulberry A-2
Murfreesboro A-2
Nashville A-3
Newport B-2
N. Little Rock A-2
Ola A-2
Osceola C-2
Ozark A-2
Pangburn B-2
Paragould B-1
Paris A-2
Parkin B-2
Piggott B-1
Pine Bluff B-2
Pocahontas B-1
Portland B-3
Prairie Grove A-2
Prescott A-3
Rison A-3
Rogers A-1
Russellville A-2
St. Joe A-1
Salem B-1
Searcy B-2
Sheridan A-2
Siloam Sprs. A-1
Smackover A-3
Sparkman A-3
Springdale A-1
Stamps A-3
Star City B-3
Stephens A-3
Strong A-3
Stuttgart B-2
Swifton B-2
Texarkana A-3
Thornton A-3
Trumann B-2
Tuckerman B-2
Van Buren A-2
Waldo A-3
Waldron A-2
Walnut Ridge B-1
Warren A-3
W. Memphis B-2
Wilmot B-3
Winslow A-2
Wynne B-2
Yellville A-1

CALIFORNIA

Map on pages 50, 51

Adelanto F-3
Adin B-1
Agnew 104 D-8
Alameda 104 C-6
Alamo 104 D-5
Albany 104 B-5
Alhambra 90 C-5
Alpine F-3
Altadena 90 C-4
Altamont 104 E-6
Alturas B-1
Alum Rock 104 E-8
Alviso 104 D-8
Anaheim F-4
Anderson A-1
Angels Camp B-2
Antioch 104 E-5
Apple Valley F-3
Arbuckle A-2
Arcadia 90 D-5
Arcata A-1
Arroyo Grande B-4
Artesia 90 D-7
Atascadero B-4
Atherton 104 C-7
Atolia F-3
Atwood 90 E-7
Auburn B-2
Avalon E-4
Avon 104 D-4
Azusa 90 E-5
Baker F-3
Bakersfield E-3
Baldwin Hills 88 D-6
Baldwin Park 90 D-5
Banning F-3
Barstow F-3
Bassett 90 D-5
Beaumont F-3
Bel Air 88 B-3
Belden B-1
Bell 90 C-6
Bellflower 90 C-7
Bell Gardens 90 C-6
Belmont 104 C-7
Belvedere 104 B-5
Belvedere Gdns 89 K-5
Benicia 104 C-4
Berkeley A-3
Bethel Island 104 F-4
Beverly Glen 88 B-2
Beverly Hills E-3
Bieber B-1
Big Bear Lake F-3
Big Pine E-2
Big Sur B-3
Bishop E-2
Black Point 104 B-4
Blairsden B-2
Blue Lake A-1
Blythe G-4
Bolinas 104 A-5
Boyle Heights 89 H-5
Brawley G-4
Brea 90 E-6
Brentwood 90 A-5
Brentwood 104 E-5
Bridgeport E-1
Brisbane 104 B-6
Broderick 103 G-1
Buellton B-4
Buena Park 90 D-7
Buena Vista 104 D-3
Burbank E-3
Burlingame 104 C-7
Burney B-1
Byron 104 F-5
Calexico G-4
Calipatria G-4
Calistoga A-2
Cambria B-4
Canby B-1
Carmel A-3
Carson 90 B-7
Castella A-1
Castro Valley 104 D-6
Castroville A-3
Cathedral City F-4
Cayton B-1
Cedarville B-1
Centerville 104 D-7
Century City 88 C-4
Cerritos 90 D-7
Chester B-1
Chico B-2
China Lake F-3
Chino Hills 90 F-6
Cholame B-4
Chowchilla B-3
Chula Vista F-3
City Terrace 89 J-4
Clairemont 105 K-4
Claremont 90 F-5
Clayton 104 D-5
Clearlake Oaks A-2
Cloverdale A-2
Clyde 104 D-4
Coachella F-4
Coalinga B-3
Coarsegold E-2
Coleville E-1
Collinsville 104 E-4
Colma 104 B-6
Coloma B-2
Colton F-3
Colusa B-2
Commerce 90 C-6
Compton 90 C-7
Concord 104 D-5
Corning A-2
Corona F-4
Coronado F-4
Corte Madera 104 B-5
Costa Mesa 90 E-8
Covina 90 E-5
Crescent City A-1
Cudahy 90 C-6
Culver City 90 B-6
Cummings A-2
Cupertino 104 D-8
Cypress 90 D-7
Daly City 104 B-6
Danville 104 D-5
Davis Creek B-1
Death Valley F-2
Death Valley Junction F-2
Delano E-3
Del Loma A-1
Del Mar F-4
Desert Center G-4
Diablo 104 D-5
Diamond Bar 90 E-6
Dorris B-1
Downey 90 C-6
Downieville B-2
Doyle B-1
Duarte 90 D-5
Dublin 104 D-6
Dunsmuir A-1
Eagle Rock 89 H-2
Earlimart E-3
East Irvine 90 E-8
East Los Angeles 90 C-6
East Palo Alto 104 D-7
East Pasadena 89 M-1
El Cajon F-4
El Centro G-4
El Cerrito 104 C-5
El Dorado B-2
El Granada 104 B-7
El Modeno 90 E-7
El Monte 90 D-5
El Portal E-2
El Segundo 90 B-6
El Sereno 89 J-3
Emeryville 104 C-5
Emigrant Gap B-2
Encanto 105 M-6
Escondido F-4
Eureka A-1
Exeter E-2
Fairfax 104 A-5
Fairfield B-2
Fairmead B-3
Fall River Mills B-1
Famoso E-3
Fillmore E-3
Florence 89 G-7
Fontana F-3
Forest Knolls 140 A-4
Ft. Bragg A-2
Ft. Ross A-2
Fortuna A-1
Foster City 104 C-7
Fountain Valley 90 D-8
Fowler E-2
Freeman F-3
Fremont 104 D-7
Fresno E-2
Fullerton 90 D-7
Garberville A-1
Gardena 90 B-7
Garden Grove 90 D-7
Geyserville A-2
Giant 104 B-4
Gilroy B-3
Glendale E-3
Glendora 90 E-5
Gonzales B-3
Granada Hills 90 A-4
Grantville 105 M-5
Grass Valley B-2
Greenbrae 104 A-5
Greenville B-1
Gridley B-2
Groveland B-3
Gustine B-3
Hacienda Hts. 90 D-6
Half Moon Bay A-3
Hanford E-2
Hawaiian Gdns. 90 D-7
Hawthorne 90 B-6
Hayward 104 D-6
Hayfork A-1
Healdsburg A-2
Hercules 104 C-4
Hermosa Beach 90 B-7
Highland Park 89 J-2
Highway Highlands 90 B-4
Hillsborough 104 B-7
Hinkley F-3
Hollister B-3
Hollywood 90 B-5
Holtville G-4
Hopland A-2
Hornbrook A-1
Huntington Beach E-4
Huntington Park 90 C-6
Hyde Park 88 E-7
Idyllwild F-4
Ignacio 104 B-4
Imperial G-4
Imperial Beach 105 L-8
Independence F-2
Indian Falls B-1
Indio F-4
Industry 90 D-6
Inglewood 90 B-6
Inyokern F-3
Irvine 90 E-8
Irvington 104 D-7
Jackson B-2
Johnstonville B-1
Keeler F-2
Kettleman City E-3
King City B-3
Kingsburg E-2
Klamath A-1
Kyburz B-2
La Canada 90 C-4
La Crescenta 90 C-4
Lafayette 104 C-5
Laguna Beach F-4
Lagunitas 104 A-4
La Habra 90 D-6
La Honda 104 C-8
La Jolla 105 K-4
Lake Isabella E-3
Lakeport A-2
Lakewood 90 C-7
La Mesa F-4
La Mirada 90 D-6
Lancaster F-3
La Palma 90 D-7
La Puente 90 D-6
Larkspur 104 A-5
La Verne 90 E-5
Lawndale 90 B-7
Laytonville A-2
Lebec E-3
Lee Vining E-2
Lennox 90 B-6
Likely B-1
Lincoln B-2
Lincoln Acres 105 M-6
Linda Vista 105 L-4
Lindsay E-2
Little Lake F-3
Livermore B-3
Livingston B-3
Lodi B-2
Lomita 90 B-7
Lompoc B-4
Lone Pine E-2
Long Beach E-4
Los Alamitos 90 D-7
Los Altos 104 D-8
Los Altos Hills 104 C-8
Los Angeles E-4
Los Banos B-3
Los Encinos 90 A-5
Los Gatos B-3
Los Molinos B-1
Los Serranos 90 F-6
Lost Hills E-3
Ludlow F-3
Lynwood 90 C-6
Mac Doel A-1
Madeline B-1
Madera B-3
Manhattan Bch. 90 B-7
Manteca B-3
Maricopa E-3
Marina 104 C-1
Marin City 104 B-5
Mariposa E-2
Martinez 104 C-4
Marysville B-2
Maywood 90 C-6
McCloud A-1
McKittrick E-3
Mecca G-4
Mendota B-3
Menlo Park 104 C-8
Merced B-3
Middletown A-2
Midway City 90 D-8
Milford B-1
Millbrae 104 B-7
Mill Valley 104 A-5
Milpitas 104 E-8
Mineral B-1
Miramar 104 B-7
Miranda A-1
Mission Beach 105 K-5
Mission San Jose 104 D-7
Mission Village 105 L-5
Modesto B-3
Mojave F-3
Monrovia 90 D-5
Montara 104 B-7
Montclair 90 F-5
Montebello 90 C-6
Monterey A-3
Monterey Park 90 C-5
Montgomery Cr. B-1
Montrose 90 C-4
Moraga 104 D-5
Morgan Hill B-3
Morningside Pk. 88 E-7
Morro Bay B-4
Moss Beach 104 B-7
Mountain View 104 D-8
Mount Baldy 90 F-4
Mount Hamilton 104 F-8
Mt. Shasta A-1
Napa A-2
National City 105 M-6
Needles G-3
Nestor 105 M-8
Nevada City B-2
Newark 104 D-7
Newberry Springs F-3
Newman B-3
Newport Beach E-4
Nicasio 104 A-4
Niles 104 D-7
Nob Hill 104 D-2
North Beach 104 E-1
North Hollywood 90 B-5
Northridge 90 A-4
Norwalk 90 D-7
Novato 104 A-4
Oakdale B-3
Oakland A-3
Oakley 104 E-4
Ocean Beach 105 K-5
Ocean Park 88 A-6
Oceanside F-4
Oilfields B-3
Ojai E-3
Olancha E-2
Olinda 90 E-6
Ontario F-4
Orange 90 E-7
Orick A-1
Orinda 104 C-5
Orland A-2
Oroville B-2
Oxnard E-3
Pacheco 104 D-5
Pacifica 104 B-7
Pacific Beach 105 K-4
Palm City 105 M-8
Palmdale E-3
Palms 88 C-5
Palm Springs F-4
Palo Alto A-3
Palos Verdes Estates 90 B-7
Panorama City 90 A-4
Paramount 90 C-7
Park La Brea 88 D-4
Pasadena E-3
Paso Robles B-4
Patterson B-3
Perris F-4
Petaluma A-2
Pico Rivera 90 C-6
Piedmont 104 C-5
Piercy A-2
Pine Valley F-3
Pinole 104 C-4
Pittsburg B-3
Placentia 90 E-7
Placerville B-2
Playa del Rey 88 B-8
Pleasant Hill 104 C-5
Pleasanton 104 E-6
Pomona 90 F-5
Pt. Arena A-2
Porterville E-2
Portola Valley 104 C-8
Princeton-by-the-Sea 104 B-7
Pulga B-2
Quincy B-2
Rafael Village 104 A-4
Ravendale B-1
Red Bluff A-1
Redding A-1
Redlands F-3
Redondo Beach 90 B-7
Redwood City A-3
Reseda 90 A-4
Richmond 104 B-3
Richmond (P.O.) 104 B-5
Riverside F-4
Rodeo 104 C-4
Rolling Hills 90 B-7
Rolling Hills Estates 90 B-7
Rosamond F-3
Rosemeade 90 D-5
Roseville B-2
Ross 104 B-5
Rossmoor 90 D-7
Rowland Hts. 90 E-6
Russian Hill 104 D-2
Sacramento B-2
St. Helena A-2
Salinas B-3
San Andreas B-2
San Anselmo 104 A-5
San Bernardino F-3
San Bruno 104 B-7
San Carlos 104 C-7
San Clemente F-4
San Diego F-4
San Dimas 90 E-5
San Fernando 90 B-4
San Francisco A-3
San Gabriel 90 C-5
San Geronimo 104 A-4
San Gregorio 104 B-8
San Jose B-3
San Juan Bautista B-3
San Juan Capistrano F-4
San Leandro 104 C-6
San Lorenzo 104 C-6
San Lucas B-3
San Luis Obispo B-4
San Luis Rey F-4
San Marino 90 C-5
San Mateo A-3
San Miguel B-4
San Pablo 104 B-5
San Pedro 90 B-8
San Rafael A-3
San Ramon 104 D-6
San Simeon B-4
Santa Ana F-4
Santa Barbara E-3
Santa Clara 104 D-8
Santa Cruz A-3
Santa Fe Springs 90 D-6
Santa Maria B-4
Santa Monica E-4
Santa Paula E-3
Santa Rosa A-2
Santa Venetia 104 B-5
Sattley B-2
Sausalito A-3
Scotia A-1
Seal Beach 90 C-8
Selma E-2
Shandon B-4
Sherman Oaks 88 B-1
Shingletown B-1
Shore Acres 104 D-4
Shoshone F-3
Sierra City B-2
Sierra Madre 90 D-5
Signal Hill 90 C-7
Silver Lake F-3
Smith River A-1
Soledad B-3
Sonora B-3
South El Monte 90 D-5
South Gate 90 C-6
S. Lake Tahoe B-2
South Pasadena 90 C-5
S. San Francisco 104 B-6
S. San Gabriel 89 L-5
Standish B-1
Stanton 90 D-7
Stinson Beach 104 A-5
Stockton B-3
Stove Pipe Wells D-2
Studio City 88 C-1
Sun City F-4
Sunland 90 B-4
Sunnyvale 104 D-8
Suno 104 D-7
Sunset Beach 90 D-8
Sun Valley 90 B-4
Susanville B-1
Sutter Creek B-2
Taft E-3
Tahoe City B-2
Tassajara 104 E-6
Tehachapi E-3
Telegraph Hill 104 E-2
Temecula F-4
Temple City 90 D-5
Terra Linda 104 A-4
Three Rivers E-2
Tiburon 104 B-5
Tipton E-2
Torrance 90 B-7
Tracy B-3
Trinidad A-1
Trona F-3
Truckee B-2
Tujunga 90 B-4
Tulare E-2
Tule Lake B-1
Turlock B-3
Tustin 90 E-8
Twentynine Palms G-3
Ukiah A-2
Union City 104 D-7
Universal City 90 B-5
Upland F-3
Upper Lake A-2
Vacaville A-2
Vallejo A-2
Van Nuys 90 A-5
Venice 90 A-6
Ventura E-3
Vernon 89 H-6
Victorville F-3
Vidal G-3
View Park 90 B-6
Visalia E-2
Vista F-4
Walnut 90 E-6
Walnut Creek 104 D-5
Walnut Park 89 H-7
Warm Springs 104 D-7
Wasco E-3
Watsonville A-3
Watts 89 H-8
Wawona E-2
Weaverville A-1
Weed A-1
Weldon E-3
Weott A-1
Westchester 88 C-7
West Covina 90 E-5
Western Addition 104 D-2
West Hollywood 88 D-3
West Los Angeles 88 B-5
Westminster 90 D-7
Westmorland G-4
West Pittsburg 104 D-4
Westwood (P.O.) B-1
Westwood 88 B-4
Wheatland B-2
Wheeler Ridge E-3
Whittier 90 D-6
Williams A-2
Willits A-2
Willow Creek A-1
Willows A-2
Wilmington 90 C-7
Woodacre 104 A-4
Woodfords E-1
Woodland B-2
Woodside 104 C-8
Yorba Linda 90 E-6
Yreka A-1
Yuba City B-2

COLORADO

Map on page 45

Aguilar C-4
Akron D-2
Alamosa B-4
Antonito C-4
Arlington D-3
Aspen B-3
Ault C-2
Aurora C-3
Axial B-2
Bailey C-3
Bayfield B-4
Blanca C-4
Boulder C-2
Branson D-4
Brighton C-3
Broadmoor 80 E-3
Brush D-2
Buena Vista B-3
Burlington D-3
Calhan C-3
Campo D-4
Canon City C-3
Castle Rock C-3
Cedaredge B-3
Central City C-3
Cheyenne Wells D-3
Colorado Sprs. C-3
Commerce City 81 J-1
Cope D-3
Cortez A-4
Cotopaxi C-3
Cowdrey B-2
Craig B-2
Creede B-4
Cripple Creek C-3
Crook D-2
Deer Trail C-3
Del Norte B-4
Delta A-3
Denver C-3
Dillon B-3
Dinosaur A-2
Divide C-3
Dove Creek A-4
Durango A-4
Eads D-3
Eagle B-3
Eaton C-2
Eckley D-2
Elk Springs A-2
Englewood 81 H-4
Estes Park C-2
Fairplay C-3
Flagler D-3
Fleming D-2
Florence C-3
Fort Collins C-2
Fort Garland C-4
Fort Morgan D-2
Fowler C-3
Fraser C-3
Fruita A-3
Georgetown C-3
Glendale 81 H-3
Glenwood Springs B-3
Golden C-3
Granada D-3
Granby B-3
Grand Jct. A-3
Grand Lake B-2
Grand Valley A-3
Greeley C-2
Gunnison B-3
Gypsum B-3
Hamilton B-2
Haswell D-3
Haxtun D-2
Hayden B-2
Hebron B-2
Holyoke D-2
Hot Sulphur Springs B-2
Hudson C-2
Hugo D-3
Idaho Sprs. C-3
Idalia D-3
Iliff D-2
Ivywild 80 E-3
Julesburg D-2
Kim D-4
Kit Carson D-3
Kremmling B-2
La Junta D-4
Lake City B-3
Lamar D-3
Las Animas D-3
Last Chance D-3
Leadville B-3
Limon D-3
Longmont C-2
Loveland C-2
Lyons C-2
Manitou Sprs. C-3
Maybell A-2
Meeker B-2
Merino D-2
Minturn B-3
Monte Vista B-4
Montrose A-3
New Castle B-3
Olathe A-3
Ordway D-3
Otis D-2
Ouray B-4
Pagosa Springs B-4
Palisade A-3
Palmer Lake C-3
Parlin B-3
Platteville C-2
Poncha Sprs. B-3
Pueblo C-3
Rangely A-2
Raymer D-2
Rifle B-3
Rio Blanco A-3
Rocky Ford D-3
Rustic C-2
Saguache B-3
Salida C-3
San Luis C-4
Sargents B-3
Sedgwick D-2
Seibert D-3
Sheridan 81 G-4
Sheridan Lake D-3
Silverton B-4
Simla C-3
South Fork B-4
Springfield D-4
Steamboat Springs B-2
Sterling D-2
Stratton D-3
Thatcher D-4
Timpas D-4
Toponas B-2
Trinidad C-4
Vail B-3
Victor C-3
Villa Grove C-3
Walden B-2
Walsenburg C-4
Wiggins C-2
Wolcott B-3
Woodland Park C-3
Wray D-2
Yampa B-2
Yuma D-2

CONNECTICUT

Map on page 14

Avon A-3
Bridgeport A-3
Bristol A-3
Canaan A-2
Clinton A-3
Colchester A-3
Cornwall Bridge A-3
Danbury A-3
Danielson A-3
Deep River A-3
Derby A-3
Granby A-3
Greenwich A-3
Guilford A-3
Hartford A-3
Jewett City A-3
Lakeville A-3
Litchfield A-3
Madison A-3
Manchester A-3
Meriden A-3
Middletown A-3
Milford A-3
Mystic A-3
New Britain A-3
New Haven A-3
New London A-3
New Milford A-3
Norfolk A-3
Norwalk A-3
Norwich B-3
Old Saybrook A-3
Putnam A-3
Rockville A-3
Sharon A-3
Stafford Sprs. A-3
Stamford A-3
Stonington B-3
Thomaston A-3
Thompsonville A-2
Torrington A-3
Wallingford A-3
Waterbury A-3
Watertown A-3
W. Willington A-3
Willimantic A-3
Windsor Locks A-3
Winsted A-3

DELAWARE

Map on page 19

Bellefonte 112 F-1
Bethany Beach E-2
Bridgeville E-2
Delmar E-2
Dover E-2
Elmhurst 112 D-2
Elsmere 112 D-2
Georgetown E-2
Harrington E-2
Laurel E-2
Lewes E-2
Milford E-2
Odessa E-2
Rehoboth Beach E-2
Seaford E-2
Smyrna E-2
Wilmington E-2
Wilmington Manor 112 D-3

DISTRICT OF COLUMBIA

Map on page 19

Washington D-2

FLORIDA

Map on page 25

Alfred C-2
Apalachicola A-2
Apopka C-2
Arcadia C-3
Archer B-2

FLORIDA (continued)

Arlington	86 C-6	Medley	93 A-3
Atlantic Beach	C-1	Melbourne	C-3
Auburndale	C-3	Melrose Park	92 K-3
Avon Park	C-3	Miami	C-3
Baldwin	C-2	Miami Beach	C-3
Balentine Manor	106 D-7	Miami Lakes	93 A-1
Bal Harbour	93 E-2	Miami Shores	93 C-2
Bartow	C-3	Miami Springs	93 A-3
Bay Harbor Is.	93 E-2	Milton	A-3
Bayshore Gdns.	106 D-7	Mims	C-2
Bayview	106 E-3	Miramar	92 J-6
Beach Park	106 B-8	Monticello	B-1
Belair	106 D-1	Moore Haven	C-3
Belleair	106 C-2	Mt. Dora	C-2
Belleair Beach	106 C-2	Naples	C-3
Belleair Bluffs	106 C-2	Newberry	B-2
Belleair Shores	106 C-2	New Pt. Richey	B-2
Belle Glade	C-4	New Smyrna Beach	C-2
Belle Haven	106 B-2	Niceville	B-3
Belle Isle	98 F-6	N. Bay Village	93 C-2
Belleview	B-2	N. Lauderdale	92 K-1
Beverly Terrace	106 E-7	N. Miami	93 C-2
Biltmore	A-6	N. Miami Beach	92 K-7
Biscayne Park	93 C-1	N. Redington Beach	106 D-2
Blountstown	B-3	Norwood	92 J-7
Boca Raton	C-3	Oak Hill	C-2
Bonita Sprs.	C-3	Oakland Park	92 L-2
Bowden	86 C-7	Ocala	C-2
Bowling Green	C-3	Okeechobee	C-3
Boynton Beach	C-3	Old Town	B-2
Bradenton	B-3	Oneco	106 E-6
Bradenton Beach	106 C-6	Opa-locka	93 B-2
Branford	B-2	Orlando	C-2
Bristol	A-1	Orlovista	98 D-5
Bronson	B-2	Ormond Beach	C-2
Brooksville	B-2	Ortega	86 B-7
Browns Village	93 C-1	Otter Creek	B-2
Bunche Park	93 C-1	Pahokee	C-3
Bunnell	B-2	Palatka	C-2
Bushnell	B-2	Palma Sola	106 D-6
Callahan	C-1	Palma Sola Pk.	106 D-6
Canal Point	C-3	Palm Beach	C-3
Cape Canaveral	C-2	Palmdale	C-3
Cape Coral	B-3	Palmetto	B-3
Capps	B-1	Palm River	108 B-5
Carlouel	106 C-2	Panama City	B-3
Carol City	92 H-7	Pass-a-Grille Beach	106 D-5
Carrabelle	A-2	Pembroke Park	92 L-6
Caryville	B-3	Pembroke Pines	92 J-6
Cedar Key	B-2	Pensacola	A-3
Chattahoochee	A-1	Perry	B-1
Chiefland	B-2	Pine Castle	98 F-6
Chipley	B-3	Pine Hills	98 D-4
Clarcona	98 D-5	Pinellas Park	106 D-3
Clearwater	B-3	Plantation	92 H-3
Clearwater Bch.	106 C-2	Plant City	B-3
Clermont	C-2	Pompano Beach	C-3
Clewiston	C-3	Ponte Vedra Beach	C-1
Coachman	106 D-1	Port Charlotte	B-3
Cocoa	C-2	Port St. Joe	B-3
Coconut Grove	93 C-5	Punta Gorda	C-3
Cooper City	92 H-4	Quincy	A-1
Coral Gables	C-3	Redington Shores	106 C-2
Cortez	106 C-6	Riverview	86 B-5
Crawfordville	A-1	Riviera Beach	C-3
Crescent City	C-2	Safety Harbor	106 E-1
Crestview	B-3	St. Augustine	C-2
Cross City	B-2	St. Cloud	C-2
Crystal Lake	B-3	St. Petersburg	B-3
Crystal River	B-2	St. Petersburg Beach	106 D-4
Dade City	B-2	Samoset	106 E-6
Dame Point	86 C-6	Sanford	C-2
Dania	92 J-4	San Jose	86 C-7
Davie	92 J-4	Sarasota	B-3
Daytona Beach	B-2	Sebastian	C-3
DeFuniak Sprs.	B-3	Sebring	C-3
DeLand	C-2	Seminole	106 D-3
Delray Beach	C-3	Seminole Park	106 E-4
Dinsmore	86 A-5	Snug Harbor	106 F-3
Doctor Phillips	98 D-6	South Bay	C-3
Dunedin	B-3	S. Jacksonville	86 C-7
Dunnellon	B-2	South Miami	93 A-6
Eastport	86 C-5	S. Pasadena	106 E-4
Eatonville	98 E-4	Starke	B-2
Edgewood	98 F-6	Stuart	C-3
Ellenton	106 E-5	Sulphur Sprs.	108 B-4
El Portal	93 D-3	Sunny Isles	93 E-1
Elwood Park	106 E-6	Sunrise	92 H-2
Eustis	C-2	Surfside	93 E-2
Everglades City	C-3	Taft	98 F-6
Fernandina Beach	C-1	Tallahassee	A-1
Fern Crest Village	92 J-4	Tallevast	106 F-7
Flagler Beach	C-2	Tamarac	92 H-1
Ft. Lauderdale	C-3	Tampa	B-3
Fort Meade	C-3	Tangelo Park	98 E-6
Fort Myers	C-3	Tarpon Springs	B-2
Fort Pierce	C-3	Temple Terr.	108 A-5
Fort Walton Beach	B-3	Tierra Verde	106 E-5
Freeport	B-3	Titusville	C-2
Fruitville	106 F-8	Treasure Is.	106 D-4
Gainesville	B-2	Ulmerton	106 D-2
Garden City	86 B-4	Valparaiso	B-3
Gilmore	86 D-6	Venetia	86 B-8
Golden Beach	92 M-7	Venice	B-3
Goodbys Lake	86 C-8	Verna	106 F-8
Gotha	98 E-5	Vero Beach	C-3
Graceville	B-3	Vineland	98 F-8
Green Cove Sprs.	B-2	Virginia Gdns.	93 B-4
Greenville	B-1	Wabasso	C-3
Groveland	C-2	Wakulla	A-1
Gulfport	106 E-4	Waldo	B-2
Hacienda Village	92 K-4	Walsingham	106 D-2
Haines City	C-3	Wauchula	C-3
Hallandale	92 L-6	Wesconnett	86 B-7
Havana	A-1	W. Hollywood	92 K-6
Hawthorne	B-2	West Miami	93 A-5
Hernando	B-2	W. Palm Beach	C-3
Hialeah	C-3	Wewahitchka	B-3
High Point	106 B-2	White Sprs.	B-1
High Sprs.	B-2	Whitfield Ests.	106 D-7
Holly Hill	C-2	Wildwood	C-2
Hollywood	C-3	Williston	B-2
Holopaw	C-2	Wilton Manors	92 L-2
Homestead	C-3	Windermere	98 C-6
Homosassa	B-2	Winter Garden	C-2
Hosford	B-1	Winter Park	C-2
Immokalee	C-3	Yankeetown	B-2
Indian Cr. Vil.	93 D-2	Ybor City	108 B-5
Indian Rocks Beach	106 C-2	Yeehaw Jct.	C-3
Indian Shores	106 C-2	Yukon	86 B-8
Inverness	B-2	Yulee	C-1
Island Estates	106 D-1	Zephyrhills	B-2
Jacksonville	C-1	Zolfo Sprs.	C-3
Jacksonville Beach	C-1		
Jasper	B-1		
Jupiter	C-3		
Kenneth City	106 E-3		
Key Biscayne	93 D-6		
Key Largo	C-3		
Key West	C-4		
Kissimmee	C-2		
La Belle	C-3		
Lake Buena Vista	98 G-7		
Lake Butler	B-2		
Lake Cain Hills	98 D-6		
Lake City	B-2		
Lakeland	B-2		
Lake Wales	C-3		
Lake Worth	C-3		
Largo	106 D-2		
Lauderdale Lks.	92 K-2		
Lauderhill	92 K-3		
Lazy Lake	92 L-2		
Leesburg	C-2		
Live Oak	B-1		
Longboat Key	106 C-6		
Madeira Beach	106 D-4		
Madison	B-1		
Marianna	B-3		
Mayo	B-1		

GEORGIA
Map on pages 22, 23

Abbeville	D-3	Blue Ridge	C-1
Adel	D-2	Bremen	C-1
Adrian	D-2	Brunswick	E-3
Alapaha	D-2	Buena Vista	C-2
Albany	D-1	Buford	C-1
Alma	D-3	Butler	C-2
Americus	C-2	Cairo	D-1
Arlington	D-1	Calhoun	C-1
Ashburn	D-2	Camilla	D-1
Athens	D-2	Canton	C-1
Atlanta	C-1	Carrollton	C-1
Augusta	C-2	Cartersville	C-1
Austell	C-1	Cedartown	C-1
Avondale Estates	74 E-7	Chamblee	74 E-6
Bainbridge	D-1	Chatsworth	B-1
Barnesville	C-2	Clarkesville	C-1
Baxley	D-3	Clarkston	74 D-1
Blackshear	D-3	Claxton	D-3
Blairsville	B-1	Clayton	B-1
Blakely	D-1	Cochran	D-2
		College Park	74 D-2
		Colquitt	D-1
		Columbus	C-2
		Comer	D-1
		Commerce	C-1
		Conley	74 E-8
		Constitution	74 E-7
		Cordele	D-2
		Cornelia	D-1
		Covington	C-2
		Crawfordville	D-2
		Cumming	C-1
		Cusseta	C-2
		Cuthbert	D-1
		Dahlonega	C-1
		Dallas	C-1
		Dalton	C-1
		Darien	E-3
		Dawson	D-1
		Dawsonville	C-1
		Decatur	C-2
		Donalsonville	D-1
		Doraville	74 E-6
		Douglas	D-2
		Douglasville	C-1
		Dublin	D-2
		Eastman	D-2
		East Pt.	C-1
		Eatonton	C-2
		Elberton	D-1
		Ellaville	C-2
		Ellijay	C-1
		Emory	74 E-7
		Fairburn	C-1
		Fairmount	C-1
		Fair Oaks	74 C-6
		Fargo	E-2
		Fitzgerald	D-2
		Folkston	E-3
		Forest Park	74 E-8
		Forsyth	C-2
		Fort Gaines	D-1
		Fort Valley	C-2
		Franklin	C-1
		Gainesville	C-1
		Geneva	C-2
		Glennville	E-3
		Gray	C-2
		Greensboro	D-2
		Greenville	C-2
		Griffin	C-2
		Hapeville	74 D-8
		Harlem	D-2
		Hartwell	D-1
		Hawkinsville	D-2
		Hazlehurst	D-3
		Henderson	D-2
		Hiawassee	B-1
		Hogansville	C-2
		Homerville	D-2
		Irwinton	D-2
		Jackson	C-2
		Jasper	C-1
		Jefferson	D-1
		Jeffersonville	D-2
		Jesup	E-3
		Jonesboro	C-2
		Kingsland	E-3
		LaFayette	C-1
		La Grange	C-2
		Lake City	74 E-8
		Lakeland	D-2
		Lavonia	D-1
		Lawrenceville	C-1
		Leary	D-1
		Leslie	D-2
		Lexington	D-2
		Lincolnton	D-2
		Louisville	D-2
		Ludowici	E-3
		Lumber City	D-3
		Lumpkin	C-2
		Lyons	D-3
		Mableton	74 C-6
		Macon	C-2
		Madison	D-2
		Manchester	C-2
		Marietta	C-1
		McCaysville	B-1
		McDonough	C-2
		McRae	D-2
		Meigs	D-1
		Metter	D-3
		Midway	E-3
		Milledgeville	D-2
		Millen	D-3
		Monroe	C-1
		Montezuma	C-2
		Monticello	C-2
		Morrow	74 E-8
		Moultrie	D-2
		Mt. Vernon	D-3
		Nahunta	E-3
		Nashville	D-2
		Newnan	C-1
		Newton	D-1
		Norcross	74 F-6
		North Atlanta	D-5
		Ocilla	D-2
		Oglethorpe	C-2
		Patterson	D-3
		Pearson	D-2
		Pelham	D-1
		Pembroke	D-3
		Perry	D-2
		Phenix City	78
		Pine Mtn.	C-2
		Plains	C-2
		Port Wentworth	D-3
		Quitman	D-2
		Ray City	D-2
		Red Oak	74
		Reidsville	D-3
		Richland	C-2
		Roberta	C-2
		Rochelle	D-2
		Rockmart	C-1
		Rome	C-1
		Rossville	C-1
		Roswell	74
		Royston	D-1
		St. Simons	E-3
		Sandersville	D-2
		Sandy Springs	74 D-6
		Savannah	D-3
		Scottdale	74 E-7
		Sea Island	E-3
		Senoia	C-2
		Smithville	C-2
		Smyrna	74 D-6
		Soperton	D-3
		Sparta	D-2
		Statesboro	D-3
		Stockton	D-2
		Stone Mtn.	74
		Stonewall	74 D-7
		Summerville	C-1
		Swainsboro	D-2
		Sylvania	D-3
		Sylvester	D-3
		Talbotton	C-2
		Tallapoosa	C-1
		Tate	C-1
		Thomaston	C-2
		Thomasville	D-1
		Thomson	D-2
		Tifton	D-2
		Toccoa	C-1
		Trion	C-1
		Tucker	74 F-6
		Twin City	D-2
		Tybee Island	E-3
		Unadilla	D-2
		Union Pt.	D-2
		Valdosta	D-2
		Vidalia	D-3
		Villa Rica	C-1
		Wadley	D-2
		Warm Sprs.	C-2
		Warner Robins	D-2
		Warrenton	D-2
		Washington	D-2
		Waycross	D-3
		Waynesboro	D-3
		West Pt.	C-2
		Willacoochee	D-3
		Winder	D-1
		Wrens	D-2
		Wrightsville	D-2
		Zebulon	C-2

HAWAII
Map on page 52

Aiea	A-5	Maunaloa	B-5
Ewa	A-5	Naalehu	D-5
Hilo	D-5	Nanakuli	A-5
Honokaa	D-5	Pahala	D-5
Honolulu	C-4	Pahoa	D-5
Kaaawa	A-5	Wahiawa	C-4
Kailua	C-4	Waialua	C-4
Kaneohe	A-4	Waianae	A-5
Kapaa	B-4	Wailuku	C-5
Kaunakakai	B-5	Waimanalo	A-5
Kekaha	B-4		
Laie	A-5		
Lanai City	B-5		
Lihue	B-4		

IDAHO
Map on page 42

Aberdeen	D-4	Rexburg	E-4
American Falls	D-4	Richfield	C-4
Arco	D-4	Riddle	B-4
Arimo	D-4	Rigby	D-4
Ashton	E-3	Riggins	B-3
Athol	B-2	Ririe	D-4
Baker	D-3	Roberts	D-4
Bancroft	D-4	Rogerson	C-4
Bellevue	C-4	Rupert	D-4
Blackfoot	D-4	St. Anthony	E-4
Bliss	D-5	St. Maries	B-1
Boise	B-4	Salmon	C-3
Bonners Ferry	B-1	Sandpoint	B-1
Bruneau	C-4	Shelley	D-4
Buhl	C-2	Shoshone	C-4
Burke	C-2	Soda Springs	D-4
Burley	D-4	Spencer	D-3
Caldwell	B-4	Spirit Lake	B-1
Cambridge	B-3	Springfield	D-4
Carey	C-4	Stanley	C-3
Cascade	B-3	Sun Valley	C-4
Challis	C-3	Swan Valley	E-4
Clark Fork	B-1	Tensed	B-2
Clayton	C-3	Teton	D-4
Coeur d'Alene	B-1	Tetonia	E-4
Copeland	B-1	Thatcher	D-4
Cottonwood	B-2	Twin Falls	C-2
Council	B-3	Victor	E-4
Craigmont	B-2	Wallace	B-2
Downey	D-4	Wayan	D-4
Driggs	E-4	Weiser	B-3
Dubois	D-3	Wendell	C-4
Eastport	B-1	Winchester	B-2
Eden	C-4		
Emida	B-2		
Emmett	B-4		
Fairfield	C-4		
Filer	C-2		
Genesee	B-2		
Gibbonsville	C-3		
Glenns Ferry	C-4		
Gooding	C-4		
Grace	D-4		
Grangeville	B-2		
Greer	B-2		
Hagerman	C-4		
Hailey	C-4		
Harrison	B-2		
Headquarters	B-2		
Holbrook	D-4		
Hollister	C-2		
Howe	D-4		
Idaho City	B-4		
Idaho Falls	D-4		
Inkom	D-4		
Irwin	D-4		
Jerome	C-4		
Kamiah	B-2		
Kellogg	B-2		
Ketchum	C-4		
Kimama	D-4		
Kooskia	B-2		
Lapwai	B-2		
Lava Hot Sprs.	E-4		
Leadore	D-3		
Lewiston	B-2		
Mackay	D-4		
Malad City	D-4		
Malta	D-4		
McCall	B-3		
McCammon	D-4		
Meridian	B-4		
Minidoka	D-4		
Montpelier	E-4		
Moreland	D-4		
Moscow	B-2		
Mountain Home	C-4		
Movie Sprs.	C-1		
Mullan	B-2		
Murphy	B-4		
Naples	B-1		
New Meadows	B-3		
New Plymouth	B-4		
Oakley	D-4		
Orofino	B-2		
Paris	E-4		
Parma	B-4		
Payette	B-4		
Pierce	B-2		
Plummer	B-2		
Pocatello	D-4		
Post Falls	B-1		
Potlatch	B-2		
Preston	D-4		
Priest River	B-1		
Rathdrum	B-1		

ILLINOIS
Map on page 31

Addison	79 G-5	Melrose Park	79 J-5
Albion	D-5	Mendota	B-2
Alpha	B-2	Meredosia	B-3
Alton	B-4	Metropolis	C-5
Annawan	B-2	Minonk	C-2
Ashley	C-4	Moline	B-2
Aurora	B-2	Momence	C-2
Barry	B-3	Monmouth	B-2
Beardstown	B-3	Morris	C-2
Bedford Park	79 J-7	Morrison	A-2
Belleville	B-4	Morton	C-2
Bellwood	79 J-5	Morton Grove	79 K-3
Belvidere	A-1	Mt. Carmel	C-5
Bensenville	79 H-4	Mt. Olive	C-4
Benton	C-5	Mt. Pulaski	C-3
Berkeley	79 H-5	Mt. Vernon	C-4
Berwyn	79 K-6	Murphysboro	C-4
Bloomington	C-3	Newton	C-4
Bridge View	79 K-7	Niles	79 K-3
Broadview	79 J-5	Norridge	79 K-4
Brookfield	J-6	Norris City	C-5
Brooklyn	106 C-4	Northlake	79 H-5
Burr Ridge	79 H-7	N. Riverside	79 J-6
Bushnell	B-2	Oak Brook	79 H-6
Cahokia	106 B-5	Oakbrook Terr.	79 H-6
Cairo	C-5	Oak Lawn	79 K-8
Canton	B-2	Oak Park	79 K-5
Carbondale	C-4	Oblong	C-4
Carlyle	C-4	Ogden	D-2
Carmi	C-5	Olney	C-4
Centralia	C-4	Oregon	A-1
Champaign	C-3	Ottawa	C-2
Charleston	C-4	Palisades	79 H-8
Chenoa	C-3	Pana	C-4
Chester	C-4	Paris	D-3
Chicago	D-2	Park Ridge	79 J-3
Chrisman	C-3	Paxton	C-3
Cicero	79 K-6	Pekin	C-2
Clarendon Hills	79 H-7	Peoria	C-2
Clinton	C-3	Petersburg	C-3
Countryside	79 J-7	Pittsfield	B-3
Crossville	C-5	Polo	A-1
Danville	D-2	Pontiac	C-3
Decatur	C-3	Princeton	B-2
Des Plaines	79 J-3	Quincy	B-3
Dixon	A-1	Rantoul	C-3
Downers Grove	79 G-7	Red Bud	C-4
Dupo	106 B-5	River Forest	79 K-5
DuQuoin	C-4	River Grove	79 J-5
Dwight	C-2	Riverside	79 J-6
E. Carondelet	106 B-5	Robinson	C-4
E. Peoria	C-2	Rochelle	B-1
E. St. Louis	106 B-4	Rock Island	B-2
Effingham	C-4	Roodhouse	B-3
Elgin	C-1	Rosemont	79 J-4
Elk Grove Vil.	79 H-4	Roseville	B-2
Elmhurst	79 H-5	Rushville	B-3
Elmwood Park	79 J-5	St. Charles	C-1
El Paso	C-3	Salem	C-4
Evanston	D-1	Sandoval	C-4
Evergreen Park	79 L-8	Sandwich	C-2
Fairfield	C-5	Santa Fe Park	79 J-8
Farmer City	C-3	Savanna	A-1
Farmington	B-2	Schiller Park	79 J-4
Flora	C-3	Shawneetown	C-4
Forest Park	79 K-7	Sheffield	B-2
Forest View	79 K-7	Shelbyville	C-4
Forrest	C-3	Skokie	79 K-3
Franklin Park	79 J-5	Springfield	C-3
Freeport	A-1	Sterling	A-1
Fulton	A-2	Stickney	79 K-6
Galena	A-1	Stone Park	79 J-5
Galesburg	B-2	Streator	C-2
Galva	B-2	Sullivan	C-4
Gibson City	C-3	Summit	79 K-7
Gilman	C-3	Taylorville	C-3
Good Hope	B-2	Tuscola	C-3
Grandview	102 F-7	Urbana	C-3
Grayville	C-5	Vandalia	C-3
Greenup	C-4	Vienna	C-5
Greenville	C-4	Villa Park	79 H-5
Hamilton	A-2	Virginia	B-3
Harrisburg	C-5	Watseka	C-3
Harvard	C-1	Waukegan	D-1
Harwood Hts.	79 K-4	Waverly	C-3
Havana	B-3	Wenona	C-2
Herrin	C-5	Westchester	79 J-6
Hickory Hills	79 J-8	Western Sprs.	79 H-7
Highland Hills	79 L-6	W. Frankfort	C-5
Hillside	H-7	Westmont	79 G-7
Hinsdale	79 H-7	White Hall	B-3
Hodgkins	79 H-7	Willowbrook	79 H-7
Hometown	79 L-8	Willow Springs	79 J-8
Hoopeston	D-3	Winchester	B-3
Indian Head Pk.	79 H-7	Windsor	C-4
Itasca	79 G-4	Wood Dale	79 H-4
Jacksonville	B-3	Woodstock	C-1
Jerseyville	B-3	York Center	79 G-6
Joliet	C-2		
Justice	79 J-8		
Kankakee	C-2		
Kansas	C-3		
Kewanee	B-2		
Kincaid	C-3		
Knoxville	B-2		
Lace	79 H-8		
La Grange	J-6		
La Grange Park	79 J-6		
La Harpe	B-2		
La Salle	C-2		
Lawrenceville	D-3		
Lewistown	B-2		
Lincoln	C-3		
Lincolnwood	79 K-4		
Litchfield	C-4		
Lombard	79 G-6		
Lyons	79 K-7		
Macomb	B-2		
Marion	C-4		
Marion Hills	79 H-7		
Mason City	C-3		
Mattoon	C-4		
Maywood	79 J-5		
Mc Cook	79 J-7		
McLeansboro	C-5		

INDIANA
Map on page 26

Akron	B-1	Fort Wayne	B-1
Albany	B-1	Fowler	A-1
Albion	B-1	Frankfort	A-1
Alexandria	B-1	Franklin	A-2
Anderson	B-1	French Lick	A-2
Angola	B-1	Fulton	A-1
Ardmore	102 B-7	Gary	A-1
Argos	B-1	Gas City	B-1
Attica	A-2	Goshen	B-1
Auburn	B-1	Greencastle	A-2
Augusta	86 B-1	Greenfield	B-2
Avilla	B-1	Greensburg	B-2
Bedford	A-2	Greenwood	A-2
Beech Grove	86 C-3	Hammond	A-1
Ben Davis	86 A-3	Hartford City	B-1
Berne	B-2	Haysville	A-3
Bicknell	A-2	Hebron	A-1
Bloomfield	A-2	Highwoods	86 B-2
Bloomington	A-2	Huntingburg	A-2
Boonville	A-3	Huntington	B-1
Bowling Green	A-2	Indianapolis	A-2
Brazil	A-2	Indian Village	102 D-7
Bremen	A-1	Jasper	A-2
Brookville	B-2	Jeffersonville	B-2
Brownstown	A-2	Kendallville	B-1
Bryant	B-1	Kentland	A-1
Burlington	B-1	Kirklin	A-1
Cambridge City	A-2	Knox	A-1
Cannelton	A-3	Kokomo	B-1
Carlisle	A-2	La Crosse	A-1
Charlestown	A-2	Lafayette	A-1
Cincinnati	A-2	Lagrange	B-1
Clarksville	90 B-1	La Porte	A-1
Clinton	A-2	Lawrence	86 D-2
Columbia City	B-1	Lawrenceburg	B-2
Columbus	A-2	Lebanon	A-1
Connersville	B-2	Liberty	B-2
Corydon	A-2	Ligonier	B-1
Covington	A-2	Linton	A-2
Crawfordsville	A-2	Logansport	A-1
Crown Point	A-1	Loogootee	A-2
Crows Nest	86 A-3	Lynhurst	86 A-3
Dale	A-2	Madison	B-2
Danville	A-2	Marion	B-1
Decatur	B-1	Marion Hts.	107 L-7
Deerfield	B-1	Mars Hill	86 B-4
Delphi	B-1	Martinsville	A-2
Demotte	A-1	Maywood	86 B-3
Dresser	107 L-7	Medaryville	A-1
Drexel Grds.	86 A-4	Meridian Hills	86 C-1
East Chicago	90 A-1	Merriam	B-1
Edgewood	86 C-4	Michigan City	A-1
Elkhart	B-1	Millersville	86 D-1
Elston	90 D-2	Mishawaka	B-1
Elwood	B-1	Mitchell	A-2
Etna Green	A-1	Monticello	A-1
Evansville	A-3	Montpelier	B-1
Farmland	B-1	Mooresville	A-2
Five Points	86 D-3	Morocco	A-1
Fortville	B-1	Muncie	B-1
		Nappanee	B-1
		Nashville	A-2
		New Albany	B-2
		New Augusta	86 A-1
		New Castle	B-2
		New Harmony	A-3
		Noblesville	B-1
		North Crows Nest	86 B-1
		N. Judson	A-1
		N. Vernon	B-2
		Notre Dame	102 D-7
		Oakland City	A-2
		Odell	A-1
		Oolitic	A-2
		Osgood	B-2
		Paoli	A-2
		Pendleton	B-1
		Peru	B-1
		Petersburg	A-2
		Plymouth	A-1
		Portland	B-1
		Princeton	A-2
		Ravenswood	86 C-1
		Redkey	B-1
		Remington	A-1
		Rensselaer	A-1
		Reynolds	A-1
		Richmond	B-2
		Rochester	A-1
		Rockport	A-3
		Rockville	A-2
		Rocky Ripple	86 B-1
		Roseland	102 D-7
		Rossville	A-1
		Rushville	B-2
		St. Meinrad	A-2
		Salem	A-2
		Sandborn	A-2
		San Pierre	A-1
		Santa Claus	A-3
		Scottsburg	B-2
		Sellersburg	B-2
		Seymour	A-2
		Shelbyville	A-2
		Shoals	A-2
		Shore Acres	86 C-5
		Silver Lake	B-1
		South Bend	B-1
		South Whitley	86 A-2
		Speedway	86 A-2
		Spencer	A-2
		Spring Hills	86 B-2
		Sullivan	A-2
		Sulphur	A-2
		Switz City	A-2
		Tell City	A-3
		Terre Haute	A-2
		Tipton	B-1
		Union City	B-1
		Valley Mills	86 A-4
		Valparaiso	A-1
		Veedersburg	A-2
		Versailles	B-2
		Vincennes	A-2
		Wabash	B-1
		Walkerton	A-1
		Wanatah	A-1
		Warren	B-1
		Warren Park	86 D-2
		Warsaw	B-1
		Washington	A-2
		Waterloo	B-1
		Westfield	B-1
		West Lafayette	90 E-2
		Winamac	A-1
		Winchester	B-1
		Wolcott	A-1
		Woodstock	86 B-2
		Worthington	A-2
		Wynnedale	86 B-2

IOWA
Map on page 37

Adel	E-2	Boone	F-1
Afton	F-2	Britt	E-1
Albia	F-2	Burlington	G-2
Algona	E-1	Calmar	E-1
Alta	E-1	Carroll	E-1
Alton	D-1	Centerville	F-2
Ames	F-2	Chariton	F-2
Anamosa	F-2	Charles City	E-1
Armstrong	E-1	Cherokee	E-1
Atlantic	E-2	Clarinda	F-2
Auburn	E-1	Clarion	E-1
Audubon	E-2	Clear Lake	E-1
Avoca	E-2	Clinton	G-2
Baldwin	F-2	Colo	F-1
Bedford	F-2	Corning	F-2
Belle Plaine	F-2	Correctionville	D-1
Bellevue	F-2	Corydon	F-2
Belmond	E-1	Council Bluffs	E-2
Bloomfield	F-2	Cresco	E-1
		Creston	F-2
		Davenport	G-2
		Decorah	E-1
		Denison	E-2
		Des Moines	F-2
		DeWitt	F-2
		Dows	E-1
		Dubuque	F-1
		Dunlap	E-2
		Early	E-1
		Eldora	F-1
		Eldridge	E-2
		Elkader	E-1
		Emmetsburg	E-1
		Estherville	E-1
		Fairfield	G-2
		Farmington	G-2
		Fonda	E-1
		Forest City	E-1
		Ft. Dodge	E-1
		Ft. Madison	G-2
		Glenwood	E-2
		Grand Jct.	E-1
		Greenfield	F-2
		Grinnell	F-2
		Grundy Cen.	F-1
		Guthrie Center	E-2
		Hampton	E-1
		Harcourt	E-1
		Harlan	E-2
		Hawarden	D-1
		Holstein	D-1
		Humboldt	E-1
		Ida Grove	E-1
		Independence	F-1
		Indianola	F-2
		Inwood	D-1
		Iowa City	G-2
		Iowa Falls	E-1
		Jefferson	E-1
		Keokuk	G-2
		Keosauqua	G-2
		Knoxville	F-2
		Lake City	E-1
		Lake Mills	E-1
		Lake View	E-1
		Lansing	E-1
		Laurens	E-1
		Le Mars	D-1
		Leon	F-2
		Lineville	F-2
		Logan	E-2
		Luxemburg	F-1
		Lyndon	F-2
		Lyons	D-2
		Macksville	D-2
		Madison	F-1
		Madrid	F-2
		Manchester	F-1
		Manly	E-1
		Mapleton	E-1
		Maquoketa	G-1
		Marengo	F-2
		Marshalltown	F-1
		Mason City	E-1
		McGregor	E-1
		Milford	E-1
		Missouri Valley	E-2
		Mondamin	D-2
		Montezuma	F-2
		Monticello	G-1
		Mt. Ayr	F-2
		Mt. Pleasant	G-2
		Mt. Vernon	G-2
		Muscatine	G-2
		Nashua	F-1
		Nevada	F-1
		New Hampton	E-1
		Newton	F-1
		Northwood	F-1
		Oakland	F-1
		Oelwein	E-1
		Ogden	F-1
		Osage	E-1
		Osceola	F-2
		Oskaloosa	F-2
		Ottumwa	F-2
		Panora	F-2
		Parkersburg	F-1
		Paulina	E-1
		Pocahontas	E-1
		Pomeroy	E-1
		Postville	E-1
		Primghar	D-1
		Red Oak	F-2
		Remsen	D-1
		Rock Rapids	D-1
		Rock Valley	D-1
		Rockwell City	E-1
		Sac City	E-1
		Sanborn	D-1
		Sheldon	D-1
		Shenandoah	F-2
		Sibley	D-1
		Sidney	F-2
		Sigourney	F-2
		Sioux Center	D-1
		Sioux City	D-1
		Sioux Rapids	E-1
		Sloan	D-2
		Spencer	E-1
		Spirit Lake	E-1
		Storm Lake	E-1
		Strawberry Pt.	F-1
		Stuart	F-2
		Tama	F-2
		Tipton	G-2
		Toledo	F-2
		Ute	E-2
		Villisca	F-2
		Vinton	F-1
		Wapello	G-2
		Washington	G-2
		Waterloo	F-1
		Waukon	E-1
		Waverly	F-1
		Webster City	E-1
		West Union	E-1
		Williamsburg	F-2
		Winterset	F-2
		Woodbine	E-2

KANSAS
Map on pages 38, 39

Abilene	E-1	Beloit	D-1
Admire	E-1	Bird City	C-1
Alma	E-1	Blue Mound	E-1
Almena	D-1	Blue Rapids	E-1
Alta Vista	E-1	Bucklin	E-1
Altamont	F-2	Buffalo	E-2
Anthony	F-1	Burden	E-2
Arkansas City	F-1	Burlingame	E-1
Arlington	E-1	Burlington	E-2
Ashland	F-1	Caldwell	F-1
Atchison	E-2	Cedar Vale	F-2
Atwood	D-1	Chanute	E-2
Augusta	E-1	Cherryvale	F-2
Baldwin	E-2	Chetopa	F-2
Baxter Sprs.	F-2	Cimarron	E-1
Belleville	E-1	Clay Center	D-1
		Clyde	E-1
		Coffeyville	F-2
		Colby	D-1
		Coldwater	E-1
		Columbus	F-2
		Concordia	D-1
		Council Grove	E-1
		Countryside	87 H-4
		Cunningham	E-1
		Dighton	D-1
		Dodge City	E-1
		Douglass	E-2
		Downs	D-1
		El Dorado	E-2
		Elkhart	F-1
		Ellinwood	E-1
		Ellis	D-1
		Ellsworth	E-1
		Emporia	E-2
		Englewood	F-1
		Erie	E-2
		Eskridge	E-1
		Eureka	E-2
		Fairview	E-1
		Fairway	87 J-4
		Florence	E-2
		Fort Scott	E-2
		Fowler	E-1
		Frankfort	E-1
		Franklin	F-2
		Fredonia	E-2
		Galena	F-2
		Garden City	E-1
		Garnett	E-2
		Girard	E-2
		Glasco	E-1
		Goodland	D-1
		Great Bend	E-1
		Greeley	E-2
		Greensburg	E-1
		Gypsum	E-1
		Halstead	E-1
		Hamilton	E-2
		Harper	E-1
		Haven	E-1
		Haviland	E-1
		Hays	D-1
		Haysville	E-1
		Herington	E-1
		Hiawatha	E-1
		Hill City	D-1
		Hillsboro	E-1
		Hoisington	E-1
		Holton	E-1
		Hope	E-1
		Horton	E-1
		Howard	E-2
		Hoxie	D-1
		Hugoton	F-1
		Hutchinson	E-1
		Independence	F-2
		Inman	E-1
		Iola	E-2
		Jetmore	E-1
		Jewell	D-1
		Johnson	E-1
		Junction City	E-1
		Kansas City	E-2
		Kensington	D-1
		Kingman	E-1
		Kinsley	E-1
		Kiowa	F-1
		La Crosse	E-1
		Lakin	E-1
		Larned	E-1
		Lawrence	E-2
		Leavenworth	E-2
		Leawood	87 J-5
		Lebanon	D-1
		Lenexa	87 G-5
		Leoti	D-1
		Liberal	F-1
		Lincoln	D-1
		Lindsborg	E-1
		Louisburg	E-2
		Lyndon	E-2
		Lyons	E-1
		Macksville	E-1
		Madison	E-2
		Manhattan	E-1
		Mankato	D-1
		Marion	E-1
		Marysville	E-1
		McDonald	D-1
		McPherson	E-1
		Medicine Lodge	F-1
		Merriam	87 H-4
		Minneola	F-1
		Mission	87 H-4
		Mission Hills	87 J-4
		Mission Woods	87 J-4
		Moline	E-2
		Montezuma	E-1
		Moran	E-2
		Mound City	E-2
		Mound Valley	F-2
		Moundridge	E-1
		Mulvane	E-1
		Neodesha	E-2
		Ness City	E-1
		Netawaka	E-1
		Newton	E-1
		Norton	D-1
		Oakland	112 E-6
		Oakley	D-1
		Oberlin	D-1
		Olathe	E-2
		Osage City	E-1
		Osawatomie	E-2
		Osborne	D-1
		Oswego	F-2
		Ottawa	E-2
		Overland Park	87 H-5
		Paola	E-2
		Parsons	F-2
		Peabody	E-1
		Phillipsburg	D-1
		Pittsburg	F-2
		Plains	F-1
		Plainville	D-1
		Pleasanton	E-2
		Prairie Village	87 H-4
		Pratt	E-1
		Protection	F-1
		Quinter	D-1
		Randolph	E-1
		Roeland Park	87 H-4
		Rush Center	E-1
		Russell	D-1
		Sabetha	E-1
		St. Francis	C-1
		St. John	E-1
		St. Marys	E-1
		Salina	E-1
		Scandia	E-1
		Scott City	D-1
		Sedan	F-2
		Selden	D-1
		Seneca	E-1
		Severy	E-2
		Sharon Sprs.	D-1
		Shawnee	87 G-4

Smith Center D-1
S. Haven E-2
Stafford D-2
Sterling D-1
Stockton D-1
Strong City E-1
Sublette C-2
Syracuse C-2
Topeka F-1
Tribune C-1
Troy F-1
Turner 87 H-3
Ulysses C-2
Valley Falls F-1
Victoria D-1
WaKeeney D-1
Wamego E-1
Washington E-1
Waterville E-1
Welborn 87 H-2
Wellington E-2
Westmoreland E-1
Westwood 87 J-4
Westwood Hills 87 J-4
Wichita E-2
Winfield E-2
Yates Center F-2

KENTUCKY

Map on pages 32, 33

Adairville B-2
Albany C-2
Arlington A-2
Ashland D-1
Athens 91 M-5
Avon 91 M-4
Barbourville D-2
Bardstown C-2
Bardwell A-2
Bedford C-1
Bellevue 80 E-7
Benton A-2
Berea C-2
Bowling Green B-2
Bracktown 91 L-4
Buechel 90 C-3
Burkesville C-2
Burnside C-2
Cadiz B-2
Campbellsville C-2
Carrollton C-1
Catlettsburg D-1
Cave City C-2
Central City B-2
Clay B-2
Clay City D-2
Clinton A-2
Cloverport B-2
Cold Spring 80 E-8
Coletown 91 L-5
Columbia C-2
Constance 80 C-8
Corbin C-2
Covington C-1
Crab Orchard C-2
Crescent Pk. 80 D-8
Crescent Sprs. 80 D-8
Cumberland D-2
Cynthiana C-1
Danville C-2
Dawson Sprs. B-2
Dayton 80 E-7
Devondale 90 D-2
Donerail 91 K-4
Drakesboro B-2
Eddyville A-2
Edmonton C-2
Elizabethtown B-2
Fairview B-2
Falmouth C-1
Faywood 91 K-4
Flemingsburg D-1
Florence C-1
Fort Mitchell 80 D-8
Fort Spring 91 K-5
Fort Thomas 80 E-8
Frankfort C-1
Franklin B-2
Fulton A-2
Garrett D-2
Georgetown C-1
Glasgow C-2
Grayson D-1
Greendale 91 L-4
Greensburg C-2
Greenup D-1
Guthrie B-2
Hardinsburg B-2
Harlan D-2
Harrodsburg C-2
Hartford B-2
Hawesville B-2
Hazard D-2
Henderson B-2
Hindman D-2
Hodgenville C-2
Hopkinsville B-2
Horse Cave C-2
Hutchinson 91 M-4
Hyden D-2
Indian Hills 90 C-2
Irvington B-2
Jackson D-2
Jenkins D-2
Jonestown 91 L-5
Keene 91 K-5
Lakeside Pk. 80 D-8
Lancaster C-2
Lawrenceburg C-1
Lebanon C-2
Leitchfield B-2
Lexington C-1
Liberty C-2
Little Georgetown 91 K-5
Little Texas 91 K-5
Livermore B-2
London C-2
Louisa D-1
Louisville C-1
Ludlow 80 D-7
Madisonville B-2
Manchester D-2
Marion A-2
Mattoxtown 91 L-4
Mayfield A-2
Maysville D-1
Middlesboro D-2
Midway 91 K-4
Millersburg C-1
Monticello C-2
Morehead D-1
Morganfield B-2
Morgantown B-2
Mt. Olivet C-1
Mt. Sterling D-1
Mt. Vernon C-2
Muir 91 M-4
Murray A-2
New Castle C-1
Newport C-1
Nicholasville C-2
Nortonville B-2
Nugents Cross Roads 91 K-4
Olive Hill D-1
Owensboro B-2
Owenton C-1
Paducah A-2
Paintsville D-2
Paris C-1
Parkers Lake C-2
Park Hills 80 D-8
Paynes Depot 91 K-4
Perryville C-2
Pikeville D-2
Pinckard 91 K-5
Pine Grove 91 M-5
Pineville D-2
Pisgah 91 K-4
Prestonsburg D-2
Princeton B-2
Providence (P.O.) B-2
Providence 91 L-5
Radcliff B-2
Richmond C-2
Russell D-1
Russell Sprs. C-2
Russellville B-2
St. Matthews 90 D-2
Salyersville D-2
Scottsville B-2
Sebree B-2
Shelbyville C-1
Shively 90 A-3
Silver Grove 80 F-8
Somerset C-2
S. Ft. Mitchell 80 D-8
Southgate 80 E-8
S. Shore D-1
Springfield C-2
Stanford C-2
Sturgis B-2
Uttingertown 91 M-4
Versailles C-1
Warsaw C-1
West Buechel 90 C-3
West Liberty D-2
Whitesburg D-2
Wickliffe A-2
Williamsburg C-2
Williamstown C-1
Winchester C-2

LOUISIANA

Map on page 24

Abbeville A-5
Alexandria A-4
Amite B-4
Arcadia A-3
Basile A-4
Bastrop B-3
Baton Rouge B-4
Bernice A-3
Berwick B-5
Bogalusa C-4
Bonita B-3
Bossier City A-3
Breaux Bridge B-4
Bunkie A-4
Buras C-5
Cameron A-5
Campti A-4
Chalmette C-5
Clarks A-3
Clinton B-4
Colfax A-4
Columbia B-3
Converse A-4
Coushatta A-3
Covington C-4
Crowley A-4
Delhi B-3
Denham Sprs. B-4
DeQuincy A-4
DeRidder A-4
Donaldsonville B-4
Dubach A-3
Eunice A-4
Farmerville A-3
Fat City 95 H-2
Ferriday B-4
Franklin B-5
Franklinton B-4
Gibsland A-3
Gibson B-5
Grand Chenier A-5
Grand Isle C-5
Greenwood A-3
Gretna C-5
Gueydan A-5
Hammond B-4
Harvey 95 K-5
Haynesville A-3
Hodge A-3
Homer A-3
Hornbeck A-4
Houma B-5
Jackson B-4
Jeanerette B-5
Jena A-4
Jennings A-4
Jonesboro A-3
Jonesville B-4
Kaplan A-5
Kenner B-5
Kentwood B-4
Kinder A-4
Lafayette B-4
Lake Arthur A-4
Lake Charles A-4
L. Providence B-3
Lecompte A-4
Leesville A-4
Logansport A-4
Longville A-4
Mamou A-4
Mangham B-3
Mansfield A-3
Mansura B-4
Many A-4
Marksville B-4
Mer Rouge B-3
Merryville A-4
Metairie 95 H-2
Minden A-3
Monroe B-3
Morgan City B-5
Morganza B-4
Napoleonville B-5
Natchitoches A-4
Newellton B-3
New Iberia B-4
New Orleans C-4
New Roads B-4
Oakdale A-4
Oak Grove B-3
Olla A-4
Opelousas A-4
Patterson B-5
Pecan Island A-5
Pineville A-4
Plain Dealing A-3
Plaquemine B-4
Pollock A-4
Port Allen B-4
Rayne A-4
Rayville B-3
Ruston A-3
St. Francisville B-4
St. Joseph B-4
St. Martinville B-4
Shreveport A-3
Simmesport B-4
Slidell C-4
Sondheimer B-3
Springhill A-3
Sulphur A-4
Tallulah B-3
Thibodaux B-5
Tullos A-4
Venice C-5
Vidalia B-4
Ville Platte A-4
Vinton A-4
Waterproof B-4
Welsh A-4
W. Monroe A-3
Westwego B-5
White Castle B-4
Winnfield A-4
Winnsboro B-3
Wisner B-3
Zwolle A-4

MAINE

Map on page 15

Abbot Village B-2
Alfred A-3
Allagash B-1
Ashland B-1
Auburn A-2
Augusta B-2
Aurora B-2
Bangor B-2
Bar Harbor B-2
Bath B-3
Belfast B-2
Bethel A-2
Biddeford A-3
Bingham B-2
Blue Hill B-2
Boothbay Hbr. B-3
Brewer B-2
Bridgewater C-1
Bridgton A-2
Brunswick B-3
Bucksport B-2
Calais C-2
Camden B-2
Caratunk B-2
Caribou B-1
Castine B-2
Cherryfield B-2
Clinton B-2
Cornish A-3
Damariscotta B-2
Danforth C-2
Deer Isle B-2
Dexter B-2
Dixfield A-2
Dixmont B-2
Dover-Foxcroft B-2
Eagle Lake B-1
E. Corinth B-2
Eastport C-2
East Waterboro A-3
Ellsworth B-2
Fairfield B-2
Falmouth 101 M-1
Farmington A-2
Ft. Fairfield B-1
Ft. Kent B-1
Freeport A-3
Fryeburg A-2
Gardiner B-2
Gorham A-3
Gray A-3
Greenville B-2
Guilford B-2
Hamlin C-1
Hampden B-2
Harmony B-2
Harrington C-2
Haynesville B-2
Houlton C-1
Island Falls B-1
Jackman A-2
Jonesport C-2
Kennebunk A-3
Kingfield A-2
Lagrange B-2
Lee B-2
Lewiston A-2
Limestone C-1
Lincoln B-2
Lisbon A-2
Livermore A-2
Livermore Falls A-2
Lubec C-2
Lynchville A-2
Machias C-2
Macwahoc B-2
Madawaska B-1
Madison B-2
Mars Hill C-1
Masardis B-1
Mattawamkeag B-2
Mexico A-2
Milbridge C-2
Millinocket B-2
Milo B-2
Monson B-2
Monticello C-1
Naples A-3
Newport B-2
Newry A-2
New Sharon A-2
Norridgewock B-2
N. Anson B-2
North Berwick A-3
Norway A-2
Oak Hill A-3
Old Orchard Beach A-3
Old Town B-2
Orono B-2
Patten B-1
Perry C-2
Phillips A-2
Pittsfield B-2
Popham Beach B-3
Portage B-1
Port Clyde B-3
Portland A-3
Presque Isle B-1
Princeton C-2
Rangeley A-2
Robbinston C-2
Rockland B-2
Rockwood B-2
Rumford A-2
Saco A-3
Sanford A-3
Sargentville B-2
Searsport B-2
Sherman B-2
Skowhegan B-2
Smyrna Mills B-1
Solon B-2
S. China B-2
S. Paris A-2
S. Portland 101 M-3
Stockton Sprs. B-2
Stonington B-2
Stratton A-2
Thomaston B-2
Topsfield C-2
Turner A-2
Union B-2
Unity B-2
Van Buren B-1
Vanceboro C-2
Waldoboro B-2
Waterville B-2
Welchville A-2
Wells A-3
Wesley C-2
Westbrook A-3
W. Enfield B-2
West Forks A-2
Whiting C-2
Wilton A-2
Winterport B-2
Winthrop A-2
Wiscasset B-2
Woodland C-2
York Village A-3

MARYLAND

Map on pages 18, 19

Aberdeen D-2
Annapolis D-2
Arbutus 78 B-4
Baltimore D-2
Bel Air D-2
Berlin E-2
Bethesda 109 J-6
Birchwood City 109 L-8
Bladensburg 109 L-6
Boonsboro D-2
Brentwood 109 L-6
Brooklandville 78 B-1
Brooklyn 78 C-4
Cabin John 109 J-6
Cambridge D-2
Capitol Hts. 109 L-7
Carney 78 D-1
Centreville E-2
Chevy Chase 109 K-6
Chillum 109 K-6
College Park 109 L-6
Colmar Manor 109 L-6
Coral Hills 109 L-7
Crisfield E-3
Cumberland C-2
Delmar E-2
Denton E-2
District Hts. 109 L-7
Dundalk 78 D-3
Easton D-2
Edmonston 109 L-6
Elkridge 78 A-4
Elkton E-2
Emmitsburg D-2
Essex 78 D-3
Fairmount Hts. 109 L-7
Forest Heights 109 K-8
Forestville 109 M-7
Frederick D-2
Frostburg C-2
Gaithersburg D-2
Glenarden 109 M-6
Glen Echo 109 J-6
Good Luck 109 M-6
Hagerstown D-2
Halethorpe 78 B-4
Hancock C-2
Havre de Grace D-2
Hillcrest Hts. 109 L-8
Hughesville D-2
Hyattsville 109 K-6
Kensington 109 K-5
Kentland 109 L-7
Keysers Ridge C-2
Landover Hills 109 L-6
Langley Park 109 L-6
Lanham 109 M-6
Lansdowne 78 B-4
La Plata D-2
Largo 109 M-7
Laurel D-2
Lexington Pk. D-2
Linthicum 78 B-4
Lochearn 78 A-2
Loch Raven 78 C-1
Middle River 78 D-2
Milford 78 A-2
Morningside 109 L-8
Mount Rainier 109 L-6
New Carrollton 109 L-6
Oakland C-2
Ocean City E-2
Overlea 78 D-2
Palmer Park 109 M-7
Parkville 78 C-2
Pikesville 78 A-1
Pocomoke City E-2
Pt. Lookout D-2
Potomac 109 H-5
Prince Frederick D-2
Princess Anne E-2
Providence 78 C-1
Pumphrey 78 B-4
Redhouse C-2
Reisterstown D-2
Relay 78 A-4
Riverdale 109 L-6
Rockville D-2
Rodgers Forge 78 B-1
Rosedale 78 D-2
Rossville 78 D-2
Ruxton 78 B-1
Salisbury E-2
Seat Pleasant 109 L-7
Shawsville D-2
Silver Hill 109 L-8
Silver Spring 109 K-6
Snow Hill E-2
Solomons D-2
Sparrows Point 78 D-4
Stoneleigh 78 C-1
Suitland 109 L-7
Takoma Park 109 K-6
Taneytown D-2
Thurmont D-2
Towson D-2
University Pk. 109 L-6
Waldorf D-2
Walker Mill 109 L-7
Westminster D-2
Williamsport D-2
Woodlawn 78 A-2
Woodmoor 78 A-2

MASSACHUSETTS

Map on page 14

Adams A-2
Aldenville 102 B-4
Amherst A-2
Athol A-2
Attleboro B-3
Auburn 112 B-8
Ayer B-2
Barnstable B-3
Barre A-2
Belchertown A-2
Bernardston A-2
Boston B-2
Brockton B-2
Cambridge 77 K-5
Cayenne 102 A-5
Charlestown 77 L-5
Chatham B-3
Chelsea 77 M-4
Chicopee 102 B-4
Chicopee Falls 102 B-4
Clinton B-2
Columbia Point 77 L-6
Concord B-2
Cummington A-2
Dalton A-2
Dedham B-2
East Boston 77 M-5
Easthampton A-2
East Millbury 112 C-7
East Springfield 102 C-4
Essex B-2
Everett 77 L-4
Fairlawn 112 C-7
Fall River B-3
Falmouth B-3
Fitchburg B-2
Framingham B-2
Gardner A-2
Gloucester B-2
Great Barrington A-2
Greenfield A-2
Haverhill B-2
Holyoke A-2
Huntington A-2
Hyannis B-3
Indian Orchard 102 C-4
Lawrence B-2
Lee A-2
Lenox A-2
Leominster B-2
Lexington B-2
Lowell B-2
Lynn B-2
Marblehead B-2
Marlborough B-2
Medford 77 K-4
Middleboro B-3
Milford B-2
Millbury 112 C-8
Nantucket B-3
New Bedford B-3
Newburyport B-2
N. Adams A-2
North Agawam 102 A-6
North Grafton 112 C-7
Northampton A-2
Northfield A-2
Oak Bluffs B-3
Orange A-2
Orient Hts. 77 M-4
Orleans B-3
Otis A-2
Palmer A-2
Pittsfield A-2
Plymouth B-3
Provincetown B-2
Quincy B-2
Revere 77 M-4
Roxbury 77 K-6
Salem B-2
Sandwich B-3
Shelburne Falls A-2
Somerville 77 K-4
South Boston 77 L-6
S. Deerfield A-2
Springfield A-2
Stockbridge A-2
Stoneville 112 A-8
Sturbridge B-2
Tatnuck 112 A-7
Taunton B-3
Vineyard Haven B-3
Waltham B-2
Ware A-2
Wareham B-3
Webster B-2
West Agawam 102 A-6
W. Boylston B-2
Westfield A-2
West Springfield 102 A-6
Williamstown A-2
Winchendon B-2
Woods Hole B-3
Worcester A-2
Woronoco A-2

MICHIGAN

Map on pages 28, 29

Adrian D-2
Alanson C-1
Albion C-2
Alger C-1
Algonac D-2
Allegan C-2
Allen Park 81 G-8
Alma C-2
Alpena D-1
Amasa A-1
Ann Arbor D-2
Atlanta C-1
Bad Axe D-2
Baldwin C-2
Bangor B-2
Baraga A-1
Battle Creek C-2
Bay City D-2
Bear Lake B-1
Benton Harbor B-2
Benzonia B-1
Bergland A-1
Berkley 81 H-4
Berrien Sprs. B-2
Bessemer A-1
Big Rapids C-2
Blaney Park C-1
Boyne City C-1
Brighton D-2
Bruce Crossing A-1
Cadillac C-1
Calumet A-1
Caseville D-2
Cassopolis C-2
Castle Hills 87 H-7
Cedar River B-1
Cedar Springs C-2
Cedarville C-1
Charlevoix C-1
Charlotte C-2
Cheboygan C-1
Chelsea C-2
Clare C-2
Clarkston D-2
Clinton C-2
Coldwater C-2
Coopersville C-2
Copper Harbor A-1
Covington A-1
Cross Village C-1
Crystal Falls A-1
Curran D-1
Davison D-2
Dearborn D-2
Decatur C-2
Delta Center 87 G-7
Delta Mills 87 G-6
DeTour Village D-1
Detroit D-2
Dowagiac B-2
Drummond D-1
Dundee D-2
E. Grand Rapids 84 C-5
East Jordan C-1
East Lansing 87 K-7
Eastport C-1
E. Tawas D-1
Eaton Rapids C-2
Eckerman C-1
Ecorse 81 H-8
Elberta B-1
Elk Rapids C-1
Elkton D-2
Emmett D-2
Empire B-1
Epoufette C-1
Escanaba B-1
Fairview C-1
Fenton D-2
Ferndale 81 H-5
Flint D-2
Fowlerville D-2
Frankfort B-1
Fremont B-2
Gaylord C-1
Germfask B-1
Gladstone B-1
Gladwin C-2
Grand Haven B-2
Grand Rapids C-2
Grayling C-1
Greenville C-2
Hamtramck 81 J-6
Hancock A-1
Harbor Beach D-2
Harbor Sprs. C-1
Harrison C-2
Harrisville D-1
Hart B-2
Hartford B-2
Hastings C-2
Hazel Park 81 J-5
Hesperia C-2
Highland Park 81 H-6
Holland B-2
Homer C-2
Houghton A-1
Howard City C-2
Howell C-2
Hudson C-2
Huntington Woods 81 H-4
Imlay City D-2
Interlochen C-1
Ionia C-2
Iron Mtn. A-1
Iron River A-1
Ironwood A-1
Ishpeming B-1
Ithaca C-2
Jackson C-2
Johnswood D-1
Jonesville C-2
Kalamazoo C-2
Kalkaska C-1
Kent City C-2
Lake City C-1
Lake Orion D-2
L'Anse A-1
Lansing C-2
Lapeer D-2
Lathrup Village 81 G-4
Laurium A-1
Leland C-1
Leslie C-2
Ludington B-2
Mackinaw City C-1
Madison Heights 81 J-4
Mancelona C-1
Manistee B-1
Manistique B-1
Manton C-1
Marion C-1
Marlette D-2
Marquette B-1
Marshall C-2
Mason C-2
Mass A-1
McBain C-1
McMillan C-1
Melvindale 81 G-8
Mendon C-2
Menominee B-1
Merrill C-2
Mesick C-1
Michigamme A-1
Middleville C-2
Midland C-2
Milan D-2
Millett 87 H-8
Mio C-1
Monroe D-2
Mottville C-2
Mt. Clemens D-2
Mt. Pleasant C-2
Munising B-1
Muskegon B-2
Nashville C-2
Naubinway C-1
Negaunee B-1
Newaygo C-2
Newberry C-1
New Buffalo B-2
Niles B-2
Northport C-1
Norway A-1
Oak Park 81 H-5
Onaway C-1
Onekama B-1
Ontonagon A-1
Oscoda D-1
Otsego Lake C-1
Owosso C-2
Oxford D-2
Paradise B-1
Paw Paw C-2
Pentwater B-2
Perry C-2
Petoskey C-1
Pigeon D-2
Pinconning D-2
Plainwell C-2
Pleasant Ridge 81 H-5
Pontiac D-2
Port Austin D-1
Port Huron D-2
Portland C-2
Port Sanilac D-2
Powers A-1
Raco C-1
Rapid River B-1
Reed City C-2
Remus C-2
Republic A-1
Richmond D-2
Richville D-2
River Rouge 81 H-8
Rogers City D-1
Romeo D-2
Roscommon C-1
Rose City C-1
Royal Oak D-2
Saginaw C-2
Sagola A-1
St. Charles C-2
St. Clair Shores D-2
St. Ignace C-1
St. James C-1
St. Johns C-2
St. Joseph B-2
St. Louis C-2
Saline D-2
Sandusky D-2
Saugatuck B-2
Sault Ste. Marie C-1
Scottville B-2
Sebewaing D-2
Seney B-1
Shelby B-2
Six Lakes C-2
Somerset C-2
Southfield D-2
South Haven B-2
Sparta C-2
Standish D-2
Stanton C-2
Stephenson A-1
Sterling Hts. D-2
Stockbridge C-2
Sturgis C-2
Suttons Bay C-1
Tawas City D-1
Taylor D-2
Tecumseh D-2
Tekonsha C-2
Three Rivers C-2
Traverse City C-1
Trenary B-1
Trout Lake C-1
Union City C-2
Unionville D-2
Vanderbilt C-1
Vassar D-2
Wakefield A-1
Walker 84 A-5
Warren D-2
Watersmeet A-1
Watervliet B-2
Wayland C-2
West Branch C-1
White Cloud C-2
Whitehall B-2
Williamston C-2
Wyoming C-2
Yale D-2
Ypsilanti D-2

MINNESOTA

Map on page 35

Ada E-2
Aitkin G-2
Albert Lea G-4
Alexandria F-3
Alvarado E-1
Anoka G-3
Appleton F-3
Arden Hills 91 K-6
Austin G-4
Bagley F-2
Barnesville E-2
Battle Lake F-2
Baudette F-1
Beaver Bay H-2
Belgrade F-3
Bemidji F-2
Benson F-3
Big Falls F-1
Blackduck F-2
Bloomington 91 K-8
Blue Earth F-4
Brainerd F-2
Breckenridge E-2
Brooklyn Center 91 K-6
Browns Valley E-3
Buffalo G-3
Caledonia H-4
Cambridge G-3
Canby E-3
Cannon Falls G-3
Cass Lake F-2
Chaska G-3
Chatfield G-4
Chisholm G-2
Clara City F-3
Cloquet G-2
Columbia Hts. 91 K-6
Cook G-2
Cotton G-2
Cromwell G-2
Crookston E-2
Crosby F-2
Dawson F-3
Deer River G-2
Detroit Lakes E-2
Dilworth 82 F-2
Donaldson E-1
Duluth H-2
E. Grand Forks E-2
Eden Valley F-3
Effie G-2
Elbow Lake E-3
Elk River G-3
Ely H-2
Erskine E-2
Fairfax F-3
Fairmont F-4
Falcon Heights 91 L-7
Faribault G-3
Farmington G-3
Fergus Falls F-2
Floodwood G-2
Forest Lake G-3
Fridley 91 K-6
Fulda F-4
Garrison G-2
Gaylord F-3
Glencoe F-3
Glenwood F-3
Graceville E-3
Grand Marais H-2
Grand Portage H-2
Grand Rapids G-2
Granite Falls F-3
Greenbush E-1
Hallock E-1
Halstad E-2
Hastings G-3
Herman E-3
Hibbing G-2
Hill City G-2
Hilltop 91 K-6
Hoffmans Corner 91 M-6
Hokah H-4
Hutchinson F-3
Illgen City H-2
International Falls G-1
Inver Grove Hts. 91 M-8
Isabella H-2
Ivanhoe E-3
Jackson F-4
Karlstad E-1
Kasson G-3
Kelliher F-2
Lake Benton E-3
Lake City G-3
Landfall 91 M-7
Lauderdale 91 K-7
Litchfield F-3
Little Canada 91 L-6
Little Falls F-3
Little Marais H-2
Long Prairie F-2
Luverne E-4
Madelia F-3
Madison E-3
Mahnomen F-2
Mahtomedi 91 M-6
Mankato G-3
Maplewood 91 M-6
Marshall F-3
McGregor G-2
Mendota 91 L-8
Mendota Heights 91 L-8
Milaca G-3
Minneapolis G-3
Montevideo F-3
Moorhead E-2
Moose Lake G-2
Mora G-3
Morris E-3
Morton F-3
Motley F-2
New Brighton 91 K-6
New London F-3
Newport 91 M-8
New Prague G-3
New Ulm F-3
North Branch G-3
Northfield G-3
Northome F-2
North St. Paul 91 M-7
Olivia F-3
Onamia G-2
Orr G-1
Ortonville E-3
Owatonna G-3
Park Rapids F-2
Paynesville F-3
Pine City G-3
Pipestone E-4
Preston G-4
Princeton G-3
Prosper H-4
Redlake F-2
Red Lake Falls E-2
Red Wing G-3
Redwood Falls F-3
Remer F-2
Richfield 91 K-8
Robbinsdale 91 J-6
Rochester G-4
Roseau F-1
Roseville 91 K-6
Rushford H-4
St. Anthony 91 K-6
St. Cloud F-3
St. James F-3
St. Paul G-3
St. Peter F-3
Sanborn F-3
Sandstone G-2
Sauk Centre F-3
Shakopee G-3
Shoreview 91 L-6
Silver Bay H-2
Slayton F-4
Sleepy Eye F-3
South St. Paul 91 L-8
Spring Valley G-4
Squaw Lake F-2
Stillwater G-3
Thief River Falls F-1
Tower G-2
Tracy F-3
Two Harbors H-2
Vadnais Heights 91 L-6
Virginia G-2
Wabasha G-3
Wadena F-2
Walker F-2
Warren E-1
Warroad F-1
Waseca G-3
Waterville G-3
West St. Paul 91 L-8
Wheaton E-3
White Bear Lake 91 M-6
Willernie 91 M-6
Williams F-1
Willmar F-3
Wilton F-2
Windom F-4
Winnebago G-4
Winona H-3
Winthrop F-3
Worthington F-4

MISSISSIPPI

Map on page 24

Aberdeen C-3
Ackerman C-3
Amory C-2
Anguilla B-3
Baldwyn C-2
Batesville B-2
Bay St. Louis C-4
Bay Springs C-4
Beaumont C-4
Belzoni B-3
Biloxi C-4
Booneville C-2
Brandon C-3
Brookhaven B-4
Brooksville C-3
Bude B-4
Byhalia C-2
Calhoun City C-3
Canton B-3
Carthage C-3
Centreville B-4
Clarksdale B-2
Cleveland B-3
Coffeeville C-3
Collins C-4
Columbia C-4
Columbus C-3
Corinth C-2
Crystal Sprs. B-3
D'Iberville 77 J-7
Drew B-3
Durant C-3
Ellisville C-4
Fayette B-4
Forest C-3
Fulton C-2
Gloster B-4
Greenville B-3
Greenwood B-3
Grenada B-3
Gulfport C-4
Hattiesburg C-4
Hazelhurst B-4
Hollandale B-3
Holly Springs C-2
Houston C-3
Indianola B-3
Iuka C-2
Jackson C-3
Kosciusko C-3
Landon 77 H-7
Lauderdale C-3
Laurel C-4
Leland B-3
Lexington B-3
Liberty B-4
Long Beach 77 G-8
Louisville C-3
Lucedale C-4
Lumberton C-4
Macon C-3
Magee C-4
Magnolia B-4
Marks B-2
Mathiston C-3
McComb B-4
Mendenhall C-4
Meridian C-3
Monticello B-4
Mt. Olive C-4
Natchez B-4
New Albany C-2
Newton C-3
North Gulfport 77 G-7
Okolona C-2
Oxford C-2
Pascagoula C-4
Philadelphia C-3
Picayune C-4
Pontotoc C-2
Poplarville C-4
Port Gibson B-4
Prentiss C-4
Quitman C-3
Raleigh C-3
Richton C-4
Ripley C-2
Rolling Fork B-3
Ruleville B-3
Sardis B-2
Scooba C-3
Senatobia C-2
Shannon C-2
Shaw B-3
Shelby B-3
Starkville C-3
Tchula B-3
Tremont C-2
Tunica B-2
Tupelo C-2
Tutwiler B-2
Tylertown B-4
Union C-3
Utica B-3
Vaiden C-3
Vicksburg B-3
Walnut C-2
Water Valley C-2
Waynesboro C-4
West Point C-3
Wiggins C-4
Winona C-3
Woodville B-4
Yazoo City B-3

MISSOURI

Map on page 37

Adrian E-3
Advance G-4
Affton 106 A-5
Albany E-2
Alton G-4
Anderson E-4
Appleton City E-3
Ash Grove F-4
Aurora F-4
Ava F-4
Avondale 87 K-2
Bella Villa 106 A-5
Bellefontaine Neighbors 106 B-2
Bel-Nor 106 A-3
Bel-Ridge 106 A-3
Berkeley 106 A-2
Bethany E-2
Birmingham 87 L-2
Black Jack 106 B-2
Blue Summit 87 L-3
Bolivar F-4
Bonne Terre G-4
Boonville F-3
Bowling Green G-3
Branson F-4
Brookfield F-3
Brunswick F-3
Buffalo F-4
Burlington Jct. E-2
Butler E-3
Cabool G-4
California F-3
Calverton Park 106 A-2
Camdenton F-3
Cameron E-3
Campbell G-4
Canton G-2
Cape Girardeau H-4
Carrollton F-3
Carthage E-4
Caruthersville H-4
Cassville E-4
Cement City 87 M-2
Charlack 106 A-3
Charleston H-4
Chillicothe F-3
Clarence F-3
Claycomo 87 K-1
Clayton 106 A-4
Clinton F-3
Cole Camp F-3
Collins F-4
Columbia F-3
Concordia F-3
Conway F-4
Cool Valley 106 A-3
Country Hills Club 106 B-3
Courtney 87 M-2
Craig E-2
Crane F-4
Cuba G-3
Dallas 87 J-5
Dellwood 106 B-2
DeSoto G-3
Dexter H-4
Dodson 87 K-5
Doniphan G-4
Drake G-3
Eagleville E-2
E. Kansas City 87 L-2
Edina F-2
Eldon F-3
El Dorado Springs E-4
Ellington G-4
Ellsinore G-4
Eminence G-4
Excelsior Sprs. E-3
Fair Play F-4
Farmington G-4
Ferguson 106 A-2
Festus G-3
Flat River G-4
Flordell Hills 106 B-3
Florissant 106 A-2
Forsyth F-4
Fort Bellefontaine 106 B-1
Fredericktown G-4
Fulton G-3
Gainesville F-4
Gallatin F-3
Galt F-2
Garden City E-3
Golden City E-4
Granby E-4
Grant City E-2
Grantwood 106 A-5
Green City F-2
Greenfield E-4
Greenville G-4
Halls Ferry 106 A-1
Hamilton F-3
Hanley Hills 106 A-3
Hannibal G-3
Harrisonville E-3
Hartville F-4
Hayti H-4
Hermann G-3
Hermitage F-4
Hickman Mills 87 K-5
Hillsdale 106 B-3
Holcomb H-4
Hollister F-4
Holmes Park 87 K-5
Houston G-4
Houston Lake 87 J-1
Huntsville F-3
Independence E-3
Ironton G-4
Jackson H-4
Jefferson City F-3
Jennings 106 B-3
Joplin E-4
Kahoka G-2
Kansas City E-3
Kennett G-4
Keytesville F-3
King City E-2
Kinloch 106 A-2
Kirksville F-2
Laclede F-3
Laddonia G-3
Lakeshire 106 A-5
Lamar E-4
Lancaster F-2
La Plata F-2
Lebanon F-4
Lewistown F-2
Lexington F-3
Liberty E-3
Licking G-4
Lincoln F-3
Louisiana G-3
Lucerne F-2
Lutesville G-4
Mackenzie 106 A-5
Macks Creek F-4
Macon F-3
Malden H-4
Mansfield F-4
Maplewood 106 A-4
Marceline F-3
Marionville F-4
Marlborough 106 A-5
Marshall F-3
Marshfield F-4
Maryville E-2
Memphis F-2
Mexico F-3
Milan F-2
Moberly F-3
Moline Acres 106 B-2
Monett F-4
Monroe City G-3
Montgomery City G-3
Mound City E-2
Mountain Grove F-4
Mountain View G-4
Mt. Moriah F-2
Neosho E-4
Nevada E-4
New London G-3

New Madrid H-4
Noel E-4
Normandy 106 A-3
Northern Hts. 87 J-1
N. Kansas City 87 K-2
Northmoor 87 J-1
Northwoods 106 B-3
Oaks 87 J-1
Oakview 87 J-1
Oakwood 87 J-1
Oakwood Manor 87 J-1
Oakwood Park 87 J-1
Oregon E-3
Osceola F-3
Owensville G-3
Ozark F-4
Pacific G-3
Pagedale 106 A-3
Palmyra G-3
Paris F-3
Parkville 87 H-1
Pasadena Hills 106 A-3
Pattonsburg E-2
Perry G-3
Perryville G-4
Piedmont G-4
Pine Lawn 106 B-3
Platte City E-3
Poplar Bluff G-4
Portageville H-4
Potosi G-4
Princeton F-2
Raytown 87 L-4
Rich Hill E-3
Richmond F-3
Richmond Hts. 106 A-4
Riverside 87 J-1
Riverview 106 B-2
Rockaway Bch. F-4
Rock Hill 106 A-4
Rock Port E-2
Rogersville F-4
Rolla G-3
St. Charles G-3
Ste. Genevieve G-3
St. George 106 A-5
St. James G-4
St. John 106 A-3
St. Joseph E-3
St. Louis G-3
Salem G-4
Salisbury F-3
Savannah E-3
Sedalia F-3
Seymour F-4
Shelbina G-3
Shrewsbury 106 A-4
Sikeston H-4
South Liberty 87 M-1
Spanish Lake 106 B-2
Springfield F-4
Stanberry E-2
Steele H-4
Steelville G-3
Sugar Creek 87 L-2
Sullivan G-3
Sycamore Hills 106 A-3
Tarkio E-2
Taylor G-3
Thayer G-4
Tipton F-3
Trenton F-2
Union G-3
Unionville F-2
Unity Village 87 M-5
University City 106 A-3
Urich F-3
Van Buren G-4
Vandalia G-3
Versailles F-3
Vienna G-3
Vinita Park 106 A-3
Warrensburg F-3
Warrenton G-3
Washington G-3
Waynesville F-4
Webb City E-4
Webster Groves 106 A-4
Wellston 106 B-3
Wentzville G-3
West Alton 106 B-1
West Plains F-4
Wilbur Park 106 A-5
Willow Sprs. F-4
Windsor F-3
Winona G-4
Winston E-3
Woodson Terrace 106 A-3

MONTANA

Map on pages 42, 43

Anaconda D-2
Arlee D-2
Ashland G-3
Augusta D-2
Babb D-1
Bainville H-1
Baker H-2
Belt E-2
Bigfork D-1
Big Sandy E-1
Big Timber F-3
Billings F-3
Black Eagle 84 C-1
Boulder D-2
Bozeman E-3
Bridger F-3
Broadus H-3
Browning D-1
Butte D-2
Cascade E-2
Chester E-1
Chinook F-1
Choteau D-2
Circle H-2
Clinton D-2
Clyde Park E-3
Cohagen G-2
Columbia Falls C-1
Columbus F-3
Conrad D-1
Cooke City F-3
Crane H-2
Crow Agency G-3
Culbertson H-1
Custer G-2
Cut Bank D-1
Darby C-2
Deer Lodge D-2
Dillon D-3
Divide D-3
Dodson F-1
Drummond D-2
Dupuyer D-1
Dutton E-2
East Glacier D-1
Ekalaka H-3
Elliston D-2
Elmo C-2
Ennis E-3
Eureka C-1
Fairfield E-2
Forsyth G-2
Fort Benton E-2
Fort Peck G-1
Frenchtown C-2
Froid H-1
Fromberg F-3
Gardiner E-3
Garrison D-2
Geyser E-2
Glasgow G-1
Glendive H-2
Grassrange F-2
Great Falls E-2
Hamilton C-2
Hardin G-3
Harlem F-1
Harlowton F-2
Havre F-1
Helena D-2
Hingham E-1
Hinsdale G-1
Hysham G-2
Ingomar G-2
Intake H-2
Jordan G-2
Judith Gap F-2
Kalispell C-1
Lame Deer G-3
Laurel F-3
Lavina F-2
Lewistown F-2
Libby C-1
Lima D-3
Lindsay H-2
Livingston E-3
Lodge Grass G-3
Lolo C-2
Loma E-2
Loring G-1
Malta G-1
Medicine Lake H-1
Melrose D-3
Miles City G-2
Missoula D-2
Monarch E-2
Moore F-2
Mosby G-2
Nashua G-1
Neihart E-2
Opheim G-1
Philipsburg D-2
Plains C-2
Plentywood H-1
Plevna H-2
Polson C-2
Poplar H-1
Ravalli C-2
Red Lodge F-3
Reedpoint F-3
Rock Springs G-2
Ronan D-2
Roundup F-2
Ryegate F-2
Saco G-1
St. Regis C-2
Scobey H-1
Shelby E-1
Sidney H-2
Stanford E-2
Sunburst D-1
Superior C-2
Sweetgrass D-1
Terry H-2
Thompson Falls C-2
Three Forks E-3
Townsend E-2
Troy B-1
Turner F-1
Vaughn E-2
Victor C-2
Virginia City D-3
Volborg H-3
Westby H-1
West Glacier C-1
W. Yellowstone E-3
Whitefish C-1
White Sulphur Springs E-2
Wibaux H-2
Wilsall E-3
Winnett F-2
Wisdom D-3
Wolf Creek E-2
Wolf Point H-1
Wyola G-3

NEBRASKA

Map on pages 36, 37

Ainsworth B-1
Albion D-2
Alliance A-1
Alma C-2
Ansley C-2
Arapahoe C-2
Arnold B-2
Arthur B-2
Ashland D-2
Atkinson C-1
Auburn E-2
Aurora D-2
Bartlett C-2
Bassett C-1
Bayard A-2
Beatrice D-2
Benkelman B-2
Big Sprs. A-2
Blair D-2
Blue Hill C-2
Boys Town D-2
Brady B-2
Bridgeport A-2
Broadwater A-2
Broken Bow C-2
Burwell C-2
Butte C-1
Cambridge B-2
Carter Lake 98 B-4
Central City D-2
Chadron A-1
Chappell A-2
Chester D-2
Clay Center C-2
Columbus D-2
Cozad C-2
Crawford A-1
Crete D-2
Culbertson B-2
Dakota City D-1
Dalton A-2
David City D-2
Deshler D-2
Dunning B-2
Elgin D-2
Ellsworth A-1
Elm Creek C-2
Ewing C-1
Fairbury D-2
Fairmont D-2
Falls City E-2
Franklin C-2
Fremont D-2
Friend D-2
Fullerton C-2
Geneva D-2
Genoa D-2
Gering A-2
Gordon A-1
Gothenburg C-2
Grand Island C-2
Grant B-2
Greeley C-2
Haigler B-2
Harrison A-1
Hartington D-1
Hastings C-2
Hay Springs A-1
Hebron D-2
Hemingford A-1
Holdrege C-2
Hooper D-2
Humboldt E-2
Hyannis B-2
Imperial B-2
Indianola B-2
Irvington 98 A-4
Kearney C-2
Kimball A-2
Lamar B-2
Laurel D-1
La Vista 98 A-6
Lewellen A-2
Lexington C-2
Lincoln D-2
Loup City C-2
Madison D-2
Maxwell B-2
Maywood B-2
McCook B-2
Merna C-2
Merriman B-1
Minden C-2
Mitchell A-2
Mullen B-1
Nebraska City D-2
Neligh D-1
Nelson C-2
Niobrara D-1
Norfolk D-1
N. Bend D-2
North Platte B-2
Oakland D-2
Ogallala B-2
Omaha D-2
O'Neill C-1
Ord C-2
Osceola D-2
Oshkosh A-2
Oxford C-2
Palisade B-2
Paxton B-2
Petersburg C-2
Pilger D-2
Plainview D-1
Plattsmouth D-2
Potter A-2
Ralston 98 A-5
Randolph D-1
Ravenna C-2
Red Cloud C-2
Rushville A-1
St. Edward D-2
St. Paul C-2
Sargent C-2
Schuyler D-2
Scottsbluff A-2
Scribner D-2
Seward D-2
Sidney A-2
Silver Creek D-2
Spencer C-1
Springview C-1
Stapleton B-2
Stratton B-2
Stromsburg D-2
Stuart C-1
Superior D-2
Sutherland B-2
Sutton D-2
Table Rock D-2
Taylor C-2
Tecumseh D-2
Tekamah D-2
Thedford B-2
Tilden D-1
Trenton B-2
Tryon B-2
Union D-2
Valentine B-1
Wahoo D-2
Walthill D-1
Wauneta B-2
Wayne D-1
Western D-2
West Point D-2
Winnebago D-1
Wisner D-1
Wymore D-2
York D-2

NEVADA

Map on page 49

Alamo C-3
Austin B-2
Baker C-2
Basalt A-2
Battle Mtn. B-1
Beatty B-3
Boulder City C-4
Caliente C-3
Carlin C-1
Carson City A-2
Coaldale A-2
Contact C-1
Currant C-2
Currie C-1
Dayton A-2
Denio A-1
Elko C-1
Ely C-2
Eureka C-2
Fallon A-2
Fernley A-2
Gabbs A-2
Glendale C-3
Golconda B-1
Goldfield B-3
Halleck C-1
Hawthorne A-2
Henderson C-4
Hiko C-3
Indian Springs C-3
Jackpot C-1
Jean C-4
Las Vegas C-3
Lovelock C-1
Lund C-2
Luning A-2
McDermitt B-1
McGill C-2
Mesquite C-3
Midas B-1
Mill City B-1
Mina A-2
Minden A-2
Moapa C-3
Montello C-1
Mountain City C-1
Nixon A-2
North Las Vegas C-3
Orovada B-1
Overton C-3
Owyhee B-1
Panaca C-3
Paradise C-3
Paradise Valley B-1
Pioche C-3
Preston C-2
Reno A-2
Schurz A-2
Searchlight C-4
Sparks A-2
Tonopah B-2
Tuscarora B-1
Valmy B-1
Virginia City A-2
Wabuska A-2
Warm Springs B-2
Wellington A-2
Wells C-1
Winnemucca A-1
Yerington A-2

NEW HAMPSHIRE

Map on page 14

Alton B-2
Ashland B-2
Berlin B-1
Bradford A-2
Bretton Woods B-1
Bristol B-2
Charlestown A-2
Chocorua B-2
Claremont A-2
Colebrook B-1
Concord B-2
Conway B-2
Danbury B-2
Derry B-2
Dover B-2
Epping B-2
Errol B-1
Exeter B-2
Farmington B-2
Franklin B-2
Glen B-1
Goffstown B-2
Gorham B-1
Groveton B-1
Hampton Beach B-2
Hanover A-2
Haverhill B-1
Henniker B-2
Hillsboro A-2
Keene A-2
Laconia B-2
Lancaster B-1
Lebanon A-2
Lisbon B-1
Littleton B-1
Manchester B-2
Meredith B-2
Milford B-2
Moultonborough B-2
Nashua B-2
Newport A-2
N. Conway B-1
N. Stratford B-1
Northwood B-2
N. Woodstock B-1
Peterborough B-2
Pittsfield B-2
Plymouth B-2
Portsmouth B-2
Rindge B-2
Rochester B-2
Rye Beach B-2
Sanbornville B-2
Troy A-2
Twin Mtn. B-1
Warren B-2
W. Ossipee B-2
W. Stewartstown B-1
Whitefield B-1
Winchester A-2
Wolfeboro B-2
Woodsville B-1

NEW JERSEY

Map on page 19

Asbury Park E-1
Atlantic City E-2
Audubon 80 C-3
Audubon Park 80 B-3
Barnegat E-2
Beideman 99 M-6
Bellmawr 80 B-3
Branchville E-1
Bridgeton E-2
Brooklawn 80 A-3
Buena E-2
Burlington E-1
Camden E-2
Cape May E-2
Cape May C.H. E-2
Carneys Point 112 F-2
Cliffside Park 96 C-3
Collingswood 80 C-2
Cramer Hill 80 C-1
Cuthbert Manor 80 C-2
Deepwater 112 F-3
Dover E-1
Edgewater 96 D-2
Egg Harbor City E-2
Elizabeth E-1
Fairview 80 B-2
Fairview (P.O.) 96 C-3
Fish House 99 M-6
Flemington E-1
Fort Lee 96 D-2
Franklin E-1
Freehold E-1
Gloucester City 80 B-3
Grantwood 96 C-2
Guttenberg 96 C-4
Hackettstown E-1
Haddon Heights 80 C-3
Hammonton E-2
Hasbrouck Hts. 96 A-1
Highland Park 80 B-3
Highlands E-1
Hightstown E-1
Hillcrest 80 C-1
Hoboken 96 B-5
Hudson Heights 96 C-3
Jersey City E-1
Lakehurst E-2
Lakewood E-1
Lambertville E-1
Leonia 96 C-1
Little Ferry 96 B-2
Long Branch E-1
Malaga E-2
Manahawkin E-2
Marsemere 96 C-2
Mays Landing E-2
Millville E-2
Moonachie 96 A-2
Morristown E-1
Mount Ephraim 80 B-3
Netcong E-1
Newark E-1
New Brunswick E-1
New Durham 96 B-4
Newton E-1
Oaklyn 80 B-3
Ocean City E-2
Palisade 96 D-2
Palisades Park 96 C-2
Parkside 80 B-2
Paterson E-1
Pavonia 80 C-1
Penns Grove E-2
Perth Amboy E-1
Phillipsburg E-1
Plainfield E-1
Pt. Pleasant E-1
Pompton Lakes E-1
Princeton E-1
Red Bank E-1
Ridgefield 96 C-2
Ridgefield Park 96 B-2
Rio Grande E-2
Salem E-2
Seaville E-2
Secaucus 96 A-4
Ship Bottom E-2
Somerville E-1
S. Dennis E-2
Sussex E-1
Teterboro 96 A-1
Toms River E-2
Trenton E-1
Tuckahoe E-2
Tuckerton E-2
Union City 96 B-4
Vineland E-2
Weehawken 96 B-4
Wellwood 99 M-6
W. Collingswood 80 B-2
West New York 96 B-4
Wildwood E-2
Woodlynne 80 B-2
Woodstown E-2

NEW MEXICO

Map on page 47

Abiquiu B-2
Alamogordo B-4
Albuquerque B-3
Anthony B-4
Apache Creek A-4
Arenal 74 A-2
Armijo 74 A-3
Artesia C-4
Atrisco 74 A-2
Aztec B-2
Belen B-3
Bernal C-3
Bernalillo B-3
Bernardo B-3
Blanco B-2
Blanco Trading Post B-2
Bloomfield B-2
Bluewater B-3
Buckhorn A-4
Caballo B-4
Capitan C-4
Caprock D-4
Capulin C-2
Carlsbad C-4
Carrizozo B-4
Cebolla B-2
Central B-4
Chama B-2
Cimarron C-2
Clayton D-2
Cliff A-4
Clines Corners C-3
Cloudcroft C-4
Clovis D-3
Columbus B-5
Corona C-3
Correo B-3
Costilla C-2
Crownpoint A-3
Cuba B-2
Cuervo C-3
Datil B-3
Deming B-4
Des Moines D-2
Dexter C-4
Dilia C-3
Dora D-4
Duran C-3
Eagle Nest C-2
Elida D-4
El Morro A-3
Encino C-3
Espanola B-3
Estancia C-3
Eunice D-4
Farmington A-2
Five Points 74 A-2
Folsom C-2
Ft. Sumner C-3
Fruitland A-2
Gallup A-3
Glenwood A-4
Glorieta C-3
Grady D-3
Gran Quivira C-3
Grants B-3
Grenville D-2
Hagerman C-4
Hatch B-4
Hillsboro B-4
Hobbs D-4
Hondo C-4
Hope C-4
Horse Springs A-4
Hurley B-4
Jal D-4
Lake Arthur C-4
Las Cruces B-4
Las Vegas C-3
Logan D-3
Lordsburg A-4
Los Alamos B-3
Los Candelarias 74 A-1
Los Duranes 74 A-1
Los Lunas B-3
Los Ojos B-2
Loving D-4
Lovington D-4
Lumberton B-2
Luna A-4
Malaga D-4
Maxwell C-2
Mayhill C-4
Melrose D-3
Mescalero C-4
Mills C-2
Milnesand D-4
Mora C-2
Moriarty B-3
Mosquero D-3
Mountainair B-3
Mountain Park C-4
Mt. Dora D-2
Nara Visa D-3
Newcomb A-2
Newkirk C-3
Newman B-5
Orogrande C-4
Pastura C-3
Pecos C-3
Portales D-3
Pueblo Bonito B-2
Quemado A-3
Questa C-2
Ramah A-3
Ramon C-3
Ranchos de Taos C-2
Raton C-2
Red River C-2
Regina B-2
Reserve A-4
Rodeo A-5
Roswell C-4
Roy C-3
Ruidoso C-4
San Antonio B-4
San Fidel B-3
San Jon D-3
San Jose 74 B-2
San Lorenzo B-4
Santa Barbara 74 B-1
Santa Fe C-3
Santa Rita A-4
Santa Rosa C-3
San Ysidro B-3
Scholle B-3
Separ A-4
Shiprock A-2
Silver City A-4
Socorro B-3
Springer C-2
Stanley B-3
Taos C-2
Tatum D-4
Texico D-3
Thoreau A-3
Three Rivers B-4
Tierra Amarilla B-2
Tohatchi A-3
Tres Piedras B-2
Trujillo C-3
Truth or Consequences B-4
Tucumcari D-3
Tularosa B-4
Vaughn C-3
Velarde C-2
Wagon Mound C-3
Watrous C-3
Whites City C-4
Willard C-3
Yeso C-3
Zuni A-3

NEW YORK

Map on pages 16,17

Adams D-2
Afton D-2
Albany E-2
Albertson 97 M-4
Albion B-2
Alder Creek D-2
Alexandria Bay D-1
Alton C-2
Amenia E-3
Amsterdam D-2
Andover C-2
Antwerp D-1
Aqueduct 97 G-8
Arcade B-2
Astoria 96 E-5
Attica B-2
Auburn C-2
Ausable Chasm E-1
Ausable Forks E-1
Avon C-2
Bainbridge D-2
Baldwinsville C-2
Ballston Spa. D-2
Batavia B-2
Bath C-2
Baxter Estates 97 L-2
Bay Shore E-3
Bayside 97 J-5
Beach Ridge 76 E-5
Beacon E-3
Beacon Hill 97 L-2
Beechhurst 97 H-4
Belfast B-2
Bellaire 97 K-6
Bellerose 97 K-6
Bellerose Terrace 97 K-6
Belmont C-2
Bergholtz 76 D-4
Binghamton D-2
Blue Mtn. Lake D-2
Boonville D-2
Brasher Falls D-1
Brewster E-3
Bridgewater D-2
Brighton 102 E-2
Broadlawn Hbr. 97 K-3
Broadway 97 H-5
Brockport B-2
Brocton B-2
Bronx 96 F-2
Bronx Beach 97 H-3
Brownsville 96 E-8
Buffalo B-2
Bushwick 96 E-7
Cairo D-2
Callicoon D-3
Cambria Heights 97 K-7
Cambridge E-2
Camden D-2
Canajoharie D-2
Canandaigua C-2
Canastota D-2
Candor C-2
Caneadea B-2
Canisteo C-2
Canton D-1
Cape Vincent C-1
Carthage D-2
Catskill D-2
Cattaraugus B-2
Cazenovia D-2
Central Square C-2
Champlain E-1
Chateaugay E-1
Chaumont C-1
Chautauqua B-2
Chestertown E-2
Chittenango D-2
City Island 97 H-2
Clayton C-1
Cleveland Hill 76 F-7
Cobleskill D-2
Cohocton C-2
Cohoes 75 J-3
College Point 97 G-4
Colonial Vil. 76 D-4
Colonie 75 H-4
Comstock E-2
Constantia D-2
Cooperstown D-2
Corning C-2
Corona 96 F-5
Cortland C-2
Cuba B-2
Dannemora E-1
Dansville C-2
Defreestville 75 K-4
Delhi D-2
Delmar 75 J-5
Deposit D-2
DeRuyter D-2
Dexter C-1
Douglas Manor 97 K-4
Douglaston 97 J-4
Downsville D-2
Doyle 76 F-8
Dryden C-2
Dunkirk B-2
Eagle Mills 75 K-3
E. Aurora B-2
East Elmhurst 96 F-5
East Hampton E-3
East New York 96 F-7
Eastwood 105 M-2
Edgewater 76 D-5
Edgewater Park 97 H-2
Edwards D-1
Eggertsville 76 F-7
Elizabethtown E-1
Ellenville D-3
Elmhurst 96 F-5
Elmira C-2
Elmont 97 L-7
Elsmere 75 J-5
Endicott C-2
Essex E-1
Fillmore B-2
Fine D-1
Floral Park 97 L-6
Floral Park Cen. 97 L-6
Flower Hill 97 L-3
Flower Hill Estates 97 M-3
Flushing 97 H-5
Fonda D-2
Forest Hills 97 G-6
Fort Covington D-1
Ft. Plain D-2
Fosterdale D-3
Franklin Square 97 M-6
Franklinville B-2
Fredonia B-2
Freeport E-3
Fresh Meadows 97 J-5
Fulton C-2
Garden Bay Manor 96 F-4
Garden City Park 97 M-5
Garden City S. 97 M-6
Gates 102 D-2
Geneseo C-2
Geneva C-2
Getzville 76 F-6
Glendale 96 F-7
Glen Oaks Vil. 97 K-5
Glens Falls E-2
Gloversville D-2
Goshen D-3
Gouverneur D-1
Gowanda B-2
Grand Gorge D-2
Grand Island 76 D-7
Grandyle 76 D-6
Granville E-2
Great Neck 97 K-3
Great Neck Estates 97 K-4
Great Neck Gardens 97 K-3
Great Neck Plaza 97 K-4
Great Valley B-2
Greene D-2
Green Island 75 J-3
Greenpoint 96 D-6
Greenport E-3
Greenwich E-2
Guilderland 75 G-5
Hague E-2
Hamburg B-2
Hamilton D-2
Hamlin C-2
Hammondsport C-2
Hancock D-3
Hannibal C-2
Harbor Acres 97 L-2
Harbor Hills 97 K-4
Harrisville D-1
Hartwick D-2
Hempstead E-3
Henderson C-2
Herkimer D-2
Herricks 97 M-5
Highland D-3
Hillside Heights 97 M-5
Hoffman 76 E-5
Hollis 97 H-6
Homer C-2
Hoosick Falls E-2
Hornell C-2
Horseheads C-2
Howard Beach 97 G-8
Hudson E-2
Hudson Falls E-2
Huntington E-3
Hyde Park E-3
Indian Lake D-2
Inland 76 D-6
Inlet D-2
Ithaca C-2
Jackson Heights 96 F-5
Jamaica 97 H-6
Jamaica Estates 97 H-6
Jamestown B-2
Jasper C-2
Jay E-1
Johnson City C-2
Johnstown D-2
Karner 75 H-4
Keene E-1
Keeseville E-1
Kenilworth 76 E-7
Kenmore 76 E-7
Kennedy B-2
Kennilworth 97 K-2
Kensington 97 K-4
Kerhonkson D-3
Kew Gardens 97 G-6
Kew Gardens Hills 97 H-6
Kings Point 97 K-3
Kingston D-3
LaFayette C-2
Lake George E-2
Lake Placid D-1
Lake Success 97 L-4
Lakeville Estates 97 L-5
Lancaster B-2
Latham 75 J-3
Laurelton 97 K-8
Le Roy C-2
Lewiston B-2
Lexington D-2
Liberty D-3
Little Falls D-2
Little Neck 97 K-4
Liverpool 105 K-1
Livingston Manor D-3
Lockport B-2
Long Beach E-3
Long Island City 96 D-5
Long Lake D-2
Loudonville 75 J-4
Lowville D-2
Lyon Mtn. D-1
Lyons C-2
Malba 97 G-4
Malone D-1
Malverne 97 M-8
Manhasset 97 L-3
Manhattan 96 C-5
Manorhaven 97 K-2
Mapleton 76 F-4
Margaretville D-2
Martinsville 76 E-5
Maspeth 96 E-6
Massena D-1
Mayfield D-2
Mayville B-2
Maywood 75 G-4
McKownville 75 H-5
Mechanicville E-2
Medina B-2
Menands 75 J-4
Middleburg D-2
Middletown D-3
Middle Village 96 F-6
Middleville D-2
Millbrook E-3
Millerton E-3
Mohawk D-2
Moira D-1
Monroe D-3
Montauk F-3
Monticello D-3
Morristown D-1
Munsey Park 97 L-3
Munson 97 M-7
Naples C-2
Narrowsburg D-3
Nassau E-2
Nassau Boulevard 97 M-6
Newark C-2
New Berlin D-2
Newburgh D-3
New Hyde Park 97 L-6
New Hyde Park Manor 97 L-5
New Salem 97 M-3
Newtonville 75 H-4
New York E-3
Niagara Falls B-2
Niskayuna 75 G-3
Norgate 97 L-4
North Creek D-2
North Hills 97 L-4
N. Lynbrook 97 L-8
N. New Hyde Pk. 97 L-5
N. Strathmore 97 L-3
N. Tonawanda 76 E-6
N. Valley Stream 97 L-7
Northville D-2
Norwich D-2
Norwood D-1
Nunda C-2
Oakland Gardens 97 J-5
Ogdensburg D-1
Olcott B-2
Old Forge D-2
Olean B-2
Oneida D-2
Oneonta D-2
Onondaga Val. 105 L-3
Orchard Beach 97 H-1
Ossining E-3
Oswego C-2
Ovid C-2
Owego C-2
Oxford D-2
Ozone Park 97 G-7
Painted Post C-2
Palenville D-2
Parkchester 97 G-2
Patchogue E-3
Paul Smiths D-1
Pavilion B-2
Peekskill E-3
Penn Yan C-2
Perry C-2
Petersburg E-2
Pine Hill 76 F-7
Plandome 97 L-3
Plandome Hts. 97 L-3
Plandome Manor 97 L-3
Plattsburgh E-1
Poland D-2
Port Chester E-3
Port Henry E-1
Port Jefferson E-3
Port Jervis D-3
Portville B-2
Port Washington 97 L-2
Port Washington North 97 L-2
Port Washington Terrace 97 L-1
Potsdam D-1
Poughkeepsie E-3
Pulaski C-2
Queens 97 G-6
Queens Village 97 K-6
Ravena E-2
Red Hook E-2
Rego Park 96 F-6
Rensselaer 75 K-5
Rhinebeck E-3
Richfield Sprs. D-2
Richmond Hill 97 H-7
Richmondville D-2
Ridgewood 96 E-7
Riverhead E-3
Riverview D-1
Rochester C-2
Rockville Cen. E-3
Rome D-2
Roscoe D-3
Rosedale 97 K-8
Roslyn Estates 97 M-3
Rouses Pt. E-1
Russell Gardens 97 K-4
Sackets Harbor C-2
Saddle Rock 97 K-3
Saddle Rock Estates 97 K-4
Sag Harbor E-3
St. Albans 97 J-7
St. Johnsburg 76 D-5
Salamanca B-2
Salem E-2
Sanborn 76 E-4
Sands Point 97 K-2
Sangerfield D-2
Saranac Lake D-1
Saratoga Sprs. D-2
Saugerties D-2
Sawyer 76 E-5
Schenectady D-2
Schroon Lake E-2
Schuylerville E-2
Sea Cliff 97 M-2
Searingtown 97 M-4
Seneca Falls C-2
Sevey D-1
Shakers 75 H-4
Shandaken D-2
Sharon D-2
Shawnee 76 E-4
Sheds D-2
Shelter Island E-3
Sherburne D-2
Sidney D-2
Silver Beach 97 H-3
Silver Creek B-2
Skaneateles C-2
Sloan 76 F-8
Sloansville D-2
Snyder 76 F-7
Solvay 105 K-2
Southampton E-3
S. Floral Park 97 L-6
S. Ozone Park 97 H-8
S. Valley Stream 97 K-8
Speculator D-2
Spencer C-2
Split Rock 105 K-3
Springfield Gardens 97 K-7
Springville B-2
Stamford D-2
Stewart Manor 97 L-6
Stillwater E-2
Strathmore 97 L-4
Suffern D-3
Sunnyside 96 E-5
Syracuse C-2
Taunton 105 K-2
Thomaston 97 K-4
Ticonderoga E-2
Tonawanda 76 D-6
Trenton D-2
Trout River D-1
Troy E-2
Trumansburg C-2
Tupper Lake D-1
University Gdns. 97 K-4
Utica D-2
Valley Stream 97 L-8
Van Etten C-2
Varysburg B-2
Vischer Ferry 75 G-3
Waddington D-1
Walden 76 F-8
Walmore 76 D-4
Walton D-2
Warrensburg E-2
Warsaw B-2
Waterloo C-2
Watertown D-2
Watervliet 75 J-3
Watkins Glen C-2
Waverly C-3
Wayland C-2
Webster C-2
Weedsport C-2
Wells D-2
Wellsville C-2
West Albany 75 J-4
Westfield B-2
Westport E-1
West Williston 97 M-5
Wevertown D-2
Whitehall E-2
White Plains E-3
Whitestone 97 H-4
Whitney Pt. C-2
Williamsburg 96 D-6
Williston 97 M-5
Windsor Park 97 J-5
Winfield 96 F-6
Wolcott C-2
Woodhaven 97 G-7
Woodside 96 E-5
Wurlitzer 76 E-5
Wurtsboro D-3
Wynantskill 75 K-3
Yonkers E-3
Youngstown B-2

NORTH CAROLINA

Map on pages 20, 21

Aberdeen D-3
Ahoskie E-2
Airlie 112 C-5
Albemarle C-3
Andrews B-3
Apex D-3
Asheboro D-3
Asheville B-3
Atlantic E-3
Atlantic Beach E-3
Barco E-2
Bat Cave B-3
Beaufort E-3
Belhaven E-3
Bennettsville D-3
Benson D-3
Bethel E-3
Biscoe D-3
Black Mtn. B-3
Blowing Rock C-2
Bolton D-3
Boone C-2
Bragtown 81 M-6
Brevard B-3
Bryson City B-3
Burgaw E-3
Burlington D-2
Burnsville B-3
Candor D-3
Canton B-3
Caraleigh 102 B-3
Carolina Beach E-3
Carthage D-3
Chapel Hill D-3
Charlotte C-3
Cherokee B-3
Chocowinity E-3
Clarkton D-3
Clinton D-3
Columbia E-3
Concord C-3
Deep Gap C-2
Dunn D-3
Durham D-3
Eden D-2
Edenton E-2
Elizabeth City E-2
Elizabethtown D-3
Elkin C-2
Elk Park B-2
Ellerbe D-3
Enfield E-2
Engelhard F-3
Faison D-3
Farmville E-3
Fayetteville D-3
Forest City C-3
Franklin B-3
Franklinton D-2
Fremont E-3
Fuquay-Varina D-3
Gastonia C-3
Goldsboro D-3
Graham D-2
Granite Falls C-3
Grantsboro E-3
Greensboro D-3
Greenville E-3
Hamlet D-3
Hatteras F-3
Hayesville B-3
Henderson D-2
Hendersonville B-3
Hickory C-3
Highlands B-3
High Point C-3
Hillsborough D-2
Hobucken E-3
Hot Springs B-3
Jacksonville E-3
Jefferson C-2
Kannapolis C-3
Kenansville E-3
Kings Mtn. C-3
Kinston E-3
Kitty Hawk F-2
Lake Lure B-3
Laurinburg D-3
Lenoir C-3
Lexington C-3
Liberty D-3
Lillington D-3
Lincolnton C-3
Linville C-2
Littleton E-2
Louisburg D-2
Lumberton D-3
Maiden C-3
Manns Hbr. F-3
Manteo E-3
Marion C-3
Marshall B-3
Masonboro 112 C-5
Maxton D-3
Mayodan C-2
Maysville E-3
Minnesott Bch. E-3
Mocksville C-3
Monroe C-3
Mooresville C-3
Morehead City E-3
Morganton C-3
Mt. Airy C-2
Mt. Holly C-3
Mt. Olive E-3
Murfreesboro E-2
Murphy B-3
Myrtle Grove 112 B-6
Nags Head F-3
New Bern E-3
Newton C-3
Norlina D-2
N. Wilkesboro C-2
Norwood C-3
Ocracoke E-3
Old Fort B-3
Oriental E-3
Oxford D-2
Pilot Mtn. C-2
Pinehurst D-3
Pinetops E-3
Pittsboro D-3
Plymouth E-3
Raeford D-3
Raleigh D-3
Ramseur D-3
Red Springs D-3
Reidsville D-2
Richlands E-3
Rich Square E-2
Roanoke Rapids E-2
Robersonville E-3
Rockingham D-3
Rocky Mount E-3
Rosman B-3
Rowland D-3
Roxboro D-2
Rutherfordton B-3
St. Pauls D-3
Salisbury C-3
Sandy Point E-3
Sanford D-3
Scotland Neck E-2
Seagate 112 C-5
Shelby C-3
Siler City D-3
Smithfield D-3
Snow Hill E-3
Southern Pines D-3
Southport D-4
Sparta C-2
Spring Hope D-2
Spruce Pine B-3
Statesville C-3
Supply D-3
Swan Quarter E-3
Swansboro E-3
Sylva B-3
Tabor City D-3
Tarboro E-3
Taylorsville C-3
Thomasville C-3
Topton B-3

Troy D-3
Tryon B-3
Twin Oaks C-2
Vanceboro E-3
Vilas C-2
Wadesboro C-3
Wake Forest D-2
Wallace E-3
Walnut Cove C-2
Warrenton D-2
Warsaw D-3
Washington E-3
Waynesville B-3
Whiteville D-3
Williamston E-3
Wilmington E-3
Wilson D-3
Windsor E-3
Winston-Salem C-2
Winter Park 112 B-5
Winton E-2
Wrightsville 112 C-4
Wrightsville Beach E-3
Yadkinville C-2
Yanceyville D-2
Zebulon D-3

NORTH DAKOTA

Map on pages 34, 35

Alexander B-2
Amidon B-2
Anamoose C-2
Ashley D-2
Beach B-2
Belfield B-2
Beulah C-2
Binford D-2
Bismarck C-2
Bottineau C-1
Bowbells B-1
Bowman B-2
Brocket D-1
Cando D-1
Carrington D-2
Casselton E-2
Cavalier E-1
Center C-2
Church's Ferry D-1
Clyde D-1
Cooperstown E-2
Crosby B-1
Dawson D-2
Devils Lake D-1
Dickinson B-2
Drake C-2
Dunseith D-1
Edgeley D-2
Elgin C-2
Ellendale D-2
Enderlin E-2
Fairmount E-2
Fargo E-2
Fessenden D-2
Finley E-2
Forman E-2
Fortuna B-1
Fredonia D-2
Garrison C-2
Glen Ullin C-2
Goldenvalley B-2
Grafton E-1
Grand Forks E-2
Granville C-1
Grassy Butte B-2
Halliday B-2
Hamilton E-1
Harvey C-2
Hazelton C-2
Hazen C-2
Hettinger B-2
Hillsboro E-2
Hurdsfield D-2
Jamestown D-2
Kenmare C-1
Killdeer B-2
Lakota D-1
La Moure D-2
Langdon D-1
Larimore E-2
Lehr D-2
Linton C-2
Lisbon E-2
Mandan C-2
Manvel E-1
Marmarth B-2
Mayville E-2
McClusky C-2
Medora B-2
Milnor E-2
Minnewaukan D-1
Minot C-1
Mohall C-1
Mott B-2
Napoleon D-2
New England B-2
New Rockford D-2
New Salem C-2
New Town B-1
Noonan B-1
Northwood E-2
Oakes E-2
Parshall B-2
Pembina E-1
Pingree D-2
Portal B-1
Ray B-1
Reeder B-2
Rhame B-2
Richardton B-2
Rock Lake D-1
Rolla D-1
Rugby D-1
Ryder C-2
St. Thomas E-1
Selfridge C-2
Stanley B-1
Starkweather D-1
Steele C-2
Tioga B-1
Tower City E-2
Towner C-1
Underwood C-2
Valley City E-2
Velva C-1
Verona E-2
Wahpeton E-2
Washburn C-2
Watford City B-2
Westhope C-1
Williston B-1
Wilton C-2
Wishek D-2
Wyndmere E-2

OHIO

Map on pages 26, 27

Aberdeen C-2
Akron D-1
Alliance D-1
Amberly 80 F-6
Andover D-1
Antwerp B-1
Ashland C-1
Ashtabula D-1
Athens C-2
Attica C-1
Bainbridge C-2
Barberton D-1
Bellaire D-2
Bellefontaine C-1
Bellevue C-1
Belpre D-2
Bexley 78 F-7
Blanchester B-2
Bloomingburg C-2
Blue Ash 80 F-6
Bluffton C-1
Bowling Green C-1
Bridgeport D-1
Bridgetown 80 C-7
Brooklyn 77 J-3
Brooklyn Hts. 77 L-3
Bryan B-1
Bucyrus C-1
Cadiz D-1
Caldwell D-2
Cambridge D-1
Canfield D-1
Canton D-1
Cardington C-1
Carey C-1
Carrollton D-1
Carthage 80 E-6
Catawba Island C-1
Celina B-1
Chagrin Falls D-1
Chardon D-1
Chauncey C-2
Chesapeake C-2
Cheviot 80 D-7
Chillicothe C-2
Cincinnati B-2
Circleville C-2
Clarington D-2
Cleveland D-1
College Hill 80 D-6
Columbus C-2
Columbus Grove B-1
Conneaut D-1
Coolville D-2
Corning C-2
Cortland D-1
Coshocton C-1
Creston D-1
Crooksville C-2
Cumberland D-2
Cuyahoga Hts. 77 L-3
Dayton B-2
Deer Park 80 F-6
Defiance B-1
Delaware C-1
Delphos B-1
Dover D-1
Dresden D-1
E. Liverpool D-1
Eaton B-2
Elmwood Place 80 E-6
Elyria C-1
Euclid D-1
Fairborn C-2
Fairfax 80 F-7
Fayetteville C-2
Findlay C-1
Fort Recovery B-1
Fostoria C-1
Franklin B-2
Fredericktown C-1
Fremont C-1
Friendship C-2
Galion C-1
Gallipolis C-2
Garfield Hts. 77 M-3
Geneva D-1
Georgetown B-2
Grandview Hts. 78 D-7
Granville C-1
Greenfield C-2
Greenville B-1
Greenwich C-1
Groesbeck 80 D-6
Hamilton B-2
Harrison B-2
Hebron C-2
Hicksville B-1
Hillsboro C-2
Holgate B-1
Huron C-1
Ironton C-2
Jackson C-2
Jacksontown C-2
Johnstown C-1
Kenton C-1
Kenwood 80 F-6
Kettering B-2
Kingsville D-1
Kinsman D-1
Lafayette C-2
Lakewood D-1
Lancaster C-2
Lebanon B-2
Leesburg C-2
Lima B-1
Lime City 108 B-3
Lisbon D-1
Locust Grove C-2
Lodi C-1
Logan C-2
London C-2
Lorain C-1
Loudonville C-1
Madisonville 80 F-7
Manchester C-2
Mansfield C-1
Marblehead C-1
Mariemont 80 F-7
Marietta D-2
Marion C-1
Martins Ferry D-1
Marysville C-1
Massillon D-1
Maumee C-1
Mayfield Hts. D-1
McArthur C-2
McConnelsville D-2
Mechanicsburg C-1
Medina D-1
Mercer B-1
Miamisburg B-2
Middleport C-2
Middletown B-2
Milford B-2
Millersburg C-1
Millwood C-1
Minerva D-1
Moline 108 C-3
Mount Airy 80 D-6
Mt. Gilead C-1
Mount Healthy 80 D-6
Mount Lookout 80 F-7
Mt. Sterling C-2
Mt. Vernon C-1
Napoleon B-1
Navarre D-1
Nelsonville C-2
Newark C-1
Newburgh Hts. 77 L-3
Newcomerstown D-1
New Concord D-1
New Lexington C-2
New Matamoras D-2
New Philadelphia D-1
New Richmond B-2
New Vienna C-2
Niles D-1
North Baltimore C-1
N. College Hill 80 D-6
Northwood 108 C-3
Norton 75 J-2
Norwalk C-1
Norwood 80 E-6
Oak Hill C-2
Oakley 80 F-7
Oakshade B-1
Oakwood 81 L-3
Oberlin C-1
Oregon 108 C-2
Orwell D-1
Ottawa C-1
Ottawa Hills 108 A-1
Oxford B-2
Painesville D-1
Parkman D-1
Parma D-1
Paulding B-1
Perrysburg 108 B-3
Phillipsburg B-2
Piedmont D-1
Piketon C-2
Piqua B-1
Plain City C-1
Pleasant Ridge 80 F-6
Point Place 108 C-1
Pomeroy C-2
Port Clinton C-1
Portsmouth C-2
Powhatan Point D-2
Price Hill 80 D-7
Proctorville 85 M-5
Put-in-Bay C-1
Ravenna D-1
Reading 80 F-6
Ripley C-2
Riverside 80 D-7
Riverside 81 M-1
Rossford 108 B-2
Roundhead C-1
Russellville C-2
St. Bernard 80 E-6
St. Clairsville D-1
St. Marys B-1
Salem D-1
Sandusky C-1
Sawyerwood 75 M-2
Sidney B-1
Silverton 80 F-6
Sinking Spring C-2
Somerset C-2
Springfield C-2
Steubenville D-1
Strasburg D-1
Strongsville D-1
Struthers 112 F-8
Sullivan C-1
Sunbury C-1
Sylvania C-1
Tiffin C-1
Toledo C-1
Toronto D-1
Troy B-1
Uhrichsville D-1
Union City B-1
Upper Sandusky C-1
Urbana C-1
Utica C-1
Vandalia B-2
Van Wert B-1
Versailles B-1
Waco 78 F-2
Walbridge 108 C-3
Waldo C-1
Wapakoneta B-1
Warren D-1
Washington C.H. C-2
Wauseon B-1
Waverly C-2
Waynesburg D-1
Waynesville B-2
Wellington C-1
Wellsville D-1
Westerville C-1
W. Jefferson C-2
W. Liberty C-1
West Union C-2
Westwood 80 D-7
White Oak 80 D-6
Williamsfield D-1
Williamstown C-1
Willoughby D-1
Wilmington C-2
Wilmot D-1
Woodsfield D-2
Woodville C-1
Wooster D-1
Wyoming 80 E-6
Xenia C-2
Youngstown D-1
Zanesville D-2

OKLAHOMA

Map on pages 38, 39

Ada E-3
Allen E-3
Altus D-3
Alva D-2
Anadarko D-3
Antlers F-3
Apache D-3
Arapaho D-3
Ardmore E-3
Arnett D-2
Atoka F-3
Barnsdall E-2
Bartlesville E-2
Beaver C-2
Bethany E-3
Blanchard E-3
Boise City B-2
Boswell F-4
Boynton F-3
Bristow E-3
Broken Bow F-3
Buffalo D-2
Burbank E-2
Burlington D-2
Cache D-3
Caddo E-3
Calvin E-3
Carmen D-2
Chandler E-3
Checotah F-3
Cherokee D-2
Chester D-2
Cheyenne D-3
Chickasha E-3
Chouteau F-2
Claremore F-2
Clayton F-3
Cleo Sprs. D-2
Cleveland E-2
Clinton D-3
Coalgate E-3
Collinsville F-2
Comanche E-3
Cordell D-3
Crowder F-3
Cushing E-2
Davidson D-3
Davis E-3
Delaware F-2
Dewey F-2
Drumright E-3
Duke D-3
Duncan D-3
Durant E-4
Elk City D-3
El Reno D-3
Enid E-2
Erick D-3
Eufaula F-3
Fairview D-2
Forgan C-2
Ft. Gibson F-3
Fort Supply D-2
Ft. Towson F-3
Frederick D-3
Gate C-2
Geary D-3
Goodwell C-2
Grove F-2
Guthrie E-3
Guymon C-2
Haileyville F-3
Healdton E-3
Heavener F-3
Hennessey E-2
Henryetta E-3
Hobart D-3
Holdenville E-3
Hollis D-3
Hominy E-2
Hooker C-2
Hugo F-3
Idabel F-4
Jay F-2
Jet D-2
Kingfisher D-3
Kiowa E-3
Konawa E-3
Lamont E-2
Laverne C-2
Lawton D-3
Madill E-3
Mangum D-3
Marietta E-4
Marlow E-3
May D-2
McAlester F-3
Medford E-2
Meeker E-3
Miami F-2
Midwest City E-3
Minco E-3
Morris F-3
Muskogee F-3
Nash E-2
Newkirk E-2
Nichols Hills 98 A-6
Norman E-3
Nowata F-2
Oilton E-2
Okarche D-3
Okemah E-3
Oklahoma City D-3
Okmulgee E-3
Pauls Valley E-3
Pawhuska E-2
Pawnee E-2
Perkins E-3
Perry E-2
Ponca City E-2
Pond Cr. E-2
Poteau F-3
Prague E-3
Pryor F-2
Purcell E-3
Randlett D-3
Ratliff City E-3
Ravia E-3
Ringling E-3
Rocky D-3
Roff E-3
Roosevelt D-3
Rosston D-2
Rush Springs E-3
Ryan E-3
Salina F-2
Sallisaw F-3
Sand Sprs. E-2
Sapulpa E-2
Sayre D-3
Seiling D-2
Seminole E-3
Shattuck D-2
Shawnee E-3
Skiatook E-2
Snyder D-3
Spiro F-3
Springlake Park 98 B-7
Stigler F-3
Stillwater E-2
Stilwell F-3
Stroud E-3
Sulphur E-3
Tahlequah F-3
Talihina F-3
Taloga D-2
Texhoma C-2
Tishomingo E-3
Tonkawa E-2
Tulsa F-2
Turpin C-2
Tyrone C-2
Valliant F-4
Vian F-3
Vici D-2
Vinita F-2
Walters D-3
Warner F-3
Watonga D-3
Waurika E-3
Wayne E-3
Waynoka D-2
Weatherford D-3
Weleetka E-3
Westville F-2
Wetumka E-3
Wewoka E-3
Wilburton F-3
Wilson E-3
Wister F-3
Woodward D-2
Wynnewood E-3
Yukon E-3

OREGON

Map on page 48

Adel C-4
Albany A-3
Alicel D-3
Aloha 100 B-7
Alsea A-3
Amity A-3
Arlington B-3
Ashland A-4
Astoria A-2
Azalea A-4
Baker C-3
Bandon A-4
Barton 100 E-8
Bay City A-3
Beatty B-4
Beaverton 100 B-7
Bend B-3
Bethany 100 B-6
Biggs B-3
Blue River A-3
Bly B-4
Boardman C-3
Boring 100 F-7
Brogan D-3
Brookings A-4
Brooklyn 100 D-7
Brothers B-4
Burlington 100 B-5
Burns C-4
Camas Valley A-4
Canyon City C-3
Canyonville A-4
Carver 100 E-8
Cascade Locks B-3
Cave Jct. A-4
Cedar Hills 100 B-6
Central Point A-4
Chemult B-4
Clackamas 100 D-7
Clatskanie A-2
Condon B-3
Coos Bay A-4
Coquille A-4
Cornelius 100 A-6
Cornelius Pass 100 B-6
Corvallis A-3
Cottage Grove A-4
Dallas A-3
Damascus 100 E-7
Dayville C-3
Drain A-4
Dufur B-3
Durham 100 C-8
Durkee D-3
Eagle Creek 100 F-8
East Portland 100 D-6
Elgin C-3
Elkton A-4
Enterprise D-3
Eugene A-4
Fairview 100 E-6
Farmington 100 B-7
Fishers Mill 100 E-8
Florence A-4
Forest Grove A-3
Ft. Klamath A-4
Fossil B-3
Garden Home 100 C-7
Gardiner A-4
Garibaldi A-3
Gilchrist B-4
Gladstone 100 D-8
Gold Beach A-4
Gold Hill A-4
Grand Ronde A-3
Grants Pass A-4
Grass Valley B-3
Gresham 100 E-7
Halfway D-3
Halsey A-3
Hampton B-4
Happy Valley 100 E-7
Hardman C-3
Harrisburg A-3
Hazeldale 100 B-7
Hebo A-3
Heppner C-3
Hermiston C-3
Hillsboro 100 A-6
Hood River B-3
Huntington D-3
Ione C-3
Ironside C-3
Jennings Lodge 100 D-8
John Day C-3
Johnson City 100 D-8
Jordan Valley D-4
Junction City A-3
Juntura D-4
Kenton 100 C-6
King City 100 B-7
Klamath Falls A-4
La Grande D-3
Lake Oswego 100 C-7
Lakeview B-4
Langlois A-4
La Pine B-4
Lebanon A-3
Lents 100 D-7
Lexington C-3
Lincoln City A-3
Linnton 100 C-6
Logan 100 E-8
Madras B-3
Mapleton A-3
Maupin B-3
McDermitt D-4
McMinnville A-3
Medford A-4
Merrill B-4
Midway 100 A-8
Mill City A-3
Milton-Freewater C-3
Milwaukie A-3
Mitchell B-3
Modoc Point B-4
Monmouth A-3
Montavilla 100 D-6
Moro B-3
Mt. Vernon C-3
Multnomah 100 C-7
Myrtle Creek A-4
Myrtle Point A-4
Neskowin A-3
Newberg A-3
Newport A-3
North Bend A-4
North Plains 100 A-6
North Powder D-3
Nyssa D-4
Oak Grove 100 D-7
Oakridge A-4
Ontario D-3
Oregon City A-3
Orenco 100 B-6
Orient 100 F-7
Paisley B-4
Park Place 100 D-8
Parkrose 100 D-6
Pendleton C-3
Philomath A-3
Piedmont 100 D-6
Pilot Rock C-3
Portland A-3
Port Orford A-4
Prairie City C-3
Prineville B-3
Progress 100 C-7
Prospect A-4
Rainier A-2
Raleigh 100 C-7
Redland 100 E-8
Redmond B-3
Reedsport A-4
Reedville 100 B-7
Richland D-3
Riley C-4
River Grove 100 C-8
Rockwood 100 E-6
Roseburg A-4
Rose City Park 100 D-6
Russellville 100 D-6
St. Helens A-3
St. Johns 100 C-6
Salem A-3
Scappoose A-3
Scholls 100 A-8
Scottsburg A-4
Seaside A-3
Sellwood 100 D-7
Seneca C-3
Service Creek B-3
Shaniko B-3
Sherwood 100 B-8
Silver Lake B-4
Sisters B-3
Six Corners 100 B-8
Somerset West 100 B-6
Spray C-3
Springfield A-3
Summer Lake B-4
Sutherlin A-4
Sweet Home A-3
Sylvan 100 C-6
The Dalles B-3
Tigard 100 C-7
Tillamook A-3
Toledo A-3
Troutdale 100 F-6
Tualatin 100 C-8
Ukiah C-3
Umatilla C-3
Unity C-3
University Pk. 100 C-6
Vale D-4
Valley Falls B-4
Wagontire B-4
Waldport A-3
Wallowa D-3
Wankers Corner 100 C-8
Warrenton A-2
Wasco B-3
West Linn 100 D-8
Weston C-3
West Portland 100 C-7
W. Portland Pk. 100 C-7
West Slope 100 C-7
West Union 100 B-6
Wheeler A-3
Willamina A-3
Winston A-4
Witch Hazel 100 B-7
Woodburn A-3
Wood Village 100 E-6

PENNSYLVANIA

Map on pages 18, 19

Abbottstown D-2
Abington 99 L-1
Albion B-1
Allentown E-1
Allison Park 101 J-2
Altoona C-1
Amity Hall D-1
Andorra 99 H-3
Ardsley 99 K-1
Arlingham 99 H-1
Armagh C-1
Ashbourne 99 K-3
Ashland D-1
Ashmead Village 99 L-3
Aspinwall 101 J-3
Avalon 101 G-3
Avoca 108 F-7
Baederwood 99 K-1
Bairdford 101 J-1
Bakerstown 101 J-1
Bala-Cynwyd 99 H-5
Baldwin 101 J-5
Barnesboro C-1
Barren Hill 99 H-2
Bauerstown 101 H-3
Beaver B-1
Beaver Falls B-1
Bedford C-2
Bellefonte D-1
Belle Valley 82 C-2
Bellevue 101 H-3
Belmont Hills 99 H-4
Ben Avon 101 G-3
Berlin C-2
Berwick D-1
Bethayres 99 M-1
Bethel Park 101 H-5
Bethlehem E-1
Bingen 74 E-5
Blairsville C-1
Blakely 102 B-7
Blawnox 101 K-3
Bloomsburg D-1
Blossburg D-1
Braddock 101 K-5
Bradford C-1
Bradford Woods 101 G-1
Breezewood C-1
Brentwood 101 H-5
Bressler 84 C-8
Bridesburg 99 M-5
Bridgeville 101 G-5
Brockway C-1
Brookville C-1
Broughton 101 H-6
Brownsville C-1
Burholme 99 M-3
Butler C-1
Cambridge Sprs. C-1
Camp Hill 84 A-8
Camptown D-1
Canton D-1
Carbondale E-1
Carlisle D-1
Carnegie 101 G-4
Castle Shannon 101 H-5
Catasauqua 74 C-3
Cedarbrook 99 K-2
Centerville B-1
Cessna C-1
Chambersburg D-2
Cheltenham 99 L-3
Chelten Hills 99 K-2
Chesney Downs 99 H-2
Chester E-2
Chestnut Hill 99 H-2
Chinchilla 102 A-7
Churchill 101 K-4
Clairton 101 J-6
Clarion C-1
Clarks Summit E-1
Claysburg C-1
Clearfield C-1
Clymer C-1
Coatesville E-2
Cold Point 99 G-1
Colonial Park 84 C-6
Columbia D-1
Colwyn 99 G-7
Confluence C-2
Conneaut Lake B-1
Connellsville C-1
Corry C-1
Coudersport D-1
Crafton 101 H-4
Cranberry C-1
Crescentville 99 L-3
Curry 101 H-6
Curtis Park 99 G-8
Curwensville C-1
Danville D-1
Darby 99 G-7
Del. Water Gap E-1
Dickson City 102 B-7
Dormont 101 H-5
Dorneyville 74 B-5
Dorseyville 101 J-2
Downingtown E-1
Doylestown E-1
Dravosburg 101 J-5
Driftwood C-1
Du Bois C-1
Duncansville C-1
Dunmore E-1
Dupont 108 F-7
Duquesne 101 K-5
Dushore D-1
Eaglehurst 82 A-2
Eagles Mere D-1
East Brady C-1
East Falls 99 J-4
East Lansdowne 99 G-6
Easton E-1
Eastwick 99 H-8
Ebensburg C-1
Edge Hill 99 J-1
Edgemont 84 C-7
Edgewood 101 J-4
Elizabethtown D-1
Elkins Park 99 L-2
Elmwood 99 H-7
Emmaus E-1
Emporium C-1
Emsworth 101 G-3
Enfield 99 J-1
Enhaut 84 C-8
Enola 84 A-7
Ephrata D-1
Erdenheim 99 H-2
Erie B-1
Etna 101 J-3
Everett C-1
Exeter 108 E-7
Fairview B-1
Feltonville 99 L-4
Fern Rock 99 K-3
Fernwood 99 G-7
Fitzwatertown 99 K-1
Florence B-1
Flourtown 99 H-1
Forest Hills 101 K-4
Forty Fort 108 E-7
Fountain Hill 74 D-4
Fox Chapel 101 K-3
Fox Chase 99 M-2
Frackville D-1
Frankford 99 L-4
Franklin C-1
Franklin Park 101 G-2
Freemansburg 74 E-4
Freeport C-1
Friedensville 74 D-5
Fullerton 74 C-3
Galeton D-1
Gap D-2
Gauff Hill 74 D-4
Georgetown 108 E-8
Germantown 99 J-3
Gettysburg D-2
Gibsonia 101 J-1
Girard B-1
Girard Point 99 J-8
Glassport 101 J-6
Glendale 101 G-5
Glen Rock D-2
Glenshaw 101 J-3
Glenside 99 K-2
Glenwood 82 B-2
Gold D-1
Grampian C-1
Greenawalds 74 B-3
Greencastle D-2
Greensburg C-1
Greentree 101 H-4
Greenville B-1
Greenwich Point 99 K-8
Hallstead E-1
Hamburg E-1
Hanover D-2
Harkness Point 99 H-8
Harlansburg B-1
Harrisburg D-1
Hawley E-1
Hazleton E-1
Heidelberg 101 G-5
Hellertown 74 E-5
Hershey D-1
Highland 99 H-3
Highland 101 H-2
Highland Park 84 B-8
Hillcrest 99 J-2
Hilldale 108 E-7
Hollidaysburg C-1
Hollywood 99 L-2
Homer City C-1
Homestead 101 J-4
Honesdale E-1
Hughestown 108 F-7
Hughesville D-1
Huntingdon D-1
Huntingdon Val. 99 M-1
Indiana C-1
Indianola 101 K-2
Ingomar 101 G-2
Ingram 101 G-4
Irwin C-1
Ivy Hill 99 J-2
Jamestown B-1
Jenkintown 99 L-2
Jersey Shore D-1
Johnsonburg C-1
Johnstown C-1
Juniata Park 99 L-4
Kane C-1
Kearsarge 82 B-3
Kensington 99 K-5
Kingston D-1
Kittanning C-1
Kutztown E-1
Lafayette Hill 99 G-2
Laflin 108 E-7
Lakewood 82 A-2
La Mott 99 K-3
Lancaster D-1
Lancasterville 99 H-1
Laporte D-1
Laurel Run 108 E-8
Laverock 99 J-2
Lawndale 99 L-3
Lawrence Park 82 C-1
Lawrenceville D-1
Lebanon D-1
Leeper C-1
Lehighton E-1
Leithsville 74 E-5
Lemoyne 84 B-8
Lenker Manor 84 C-7
Lenox E-1
Lewisburg D-1
Lewistown D-1
Liberty (P.O.) D-1
Liberty 101 K-6
Ligonier C-1
Littlestown D-2
Lock Haven D-1
Logan 99 K-4
Luzerne 108 E-8
Lykens D-1
Lynnewood Gdns 99 K-2
Mahanoy City D-1
Manayunk 99 H-4
Mansfield D-1
Marienville C-1
Mayfair 99 M-4
McConnellsburg C-2
McKees Rocks 101 G-4
McKinley 99 L-2
McKnight Vil. 101 H-3
Meadowbrook 99 M-1
Meadville B-1
Melrose Park 99 K-3
Mercer B-1
Mercersburg C-2
Merion 99 G-5
Meyersdale C-2
Middleburg D-1
Middletown D-1
Mifflinburg D-1
Mifflintown D-1
Milesburg D-1
Milford E-1
Millbourne 99 G-6
Millersburg D-1
Millvale 101 H-3
Millville D-1
Milton D-1
Miquon 99 G-3
Monroeton D-1
Montoursville D-1
Montrose (P.O.) E-1
Montrose 101 K-3
Morris D-1
Mount Airy 99 J-3
Mt. Jewett C-1
Mount Lebanon 101 H-5
Mount Oliver 101 H-4
Mt. Pocono E-1
Mount Union D-1
Muncy D-1
Munhall 101 J-5
Myerstown D-1
Narberth 99 G-4
New Alexandria C-1
New Bethlehem C-1
New Castle B-1
New Cumberland 84 B-8
New Kensington C-1
New Milford E-1
New Stanton C-1
Nicetown 99 K-4
Noble 99 K-2
Norristown E-1
North Braddock 101 K-4
North East C-1
North Hills 99 J-1
Northumberland D-1
Oak Lane 99 K-3
Oakleigh 84 C-7
Oberlin 84 C-8
Ogontz 99 L-2
Oil City C-1
Olney 99 K-4
Olyphant 102 B-7
Orbisonia C-1
Oreland 99 J-1
Overbrook 99 H-5
Oxford E-2
Paoli E-1
Paper Mill Glen 99 J-2
Paxtang 84 C-7
Paxtang Manor 84 C-7
Paxtonia 84 C-6
Penbrook 84 C-7
Penfield C-1
Penrose Park 99 H-7
Perrysville 101 H-3
Philadelphia E-1
Philipsburg C-1
Phoenixville E-1
Pittock 101 G-3
Pittsburgh B-1
Pittsfield C-1
Pittston E-1
Plains 108 E-8
Plainsville 108 E-7
Pleasant Hills 101 J-5
Pleasantville C-1
Point Breeze 99 H-7
Point Marion C-2
Polk C-1
Port Allegany C-1
Portland E-1
Port Matilda C-1
Port Richmond 99 L-5
Port Vue 101 K-5
Potters Mills D-1
Pottstown E-1
Pottsville D-1
Progress 84 C-7
Punxsutawney C-1
Quakertown E-1
Rankin 101 J-4
Reading E-1
Renovo D-1
Reynoldsville C-1
Rhawnhurst 99 M-3
Richmond 99 L-5
Ridgewood 108 E-8
Ridgway C-1
Rimersburg C-1
Rochester B-1
Rockledge 99 L-2
Roseglen 99 G-3
Roslyn 99 K-1
Rowland Park 99 L-3
Roxborough 99 H-4
Rydal 99 L-1
Ryers 99 L-3
St. Marys C-1
Sayre D-1
Scottdale C-1
Scranton E-1
Seidersville 74 D-4
Selinsgrove D-1
Shamokin D-1
Sharon B-1
Sharpsburg 101 J-3
Shawmont 99 H-3
Sheffield C-1
Shenandoah D-1
Shickshinny D-1
Shippensburg D-1
Shiremanstown 84 A-8
Slatington E-1
Slippery Rock C-1
Smethport C-1
Somerset C-1
State College D-1
Steelton 84 C-8
Stewartstown D-2
Stoystown C-1
Stroudsburg E-1
Summerdale 84 A-6
Sunbury D-1
Sunnybrook 99 J-1
Swissvale 101 J-4
Swoyersville 108 E-7
Sykesville C-1
Talley Cavey 101 J-2
Tamaqua E-1
Tarentum C-1
Taylor 102 A-8
Throop 102 B-7
Tioga (P.O.) D-1
Tioga 99 K-5
Tionesta C-1
Titusville C-1
Towanda D-1
Troy D-1
Tunkhannock D-1
Tyrone C-1
Undercliff 101 J-3
Union City C-1
Uniontown C-2
Upper Darby 99 G-6
Upper Talley Cavey 101 J-2
Valencia 101 H-1
Valley Falls 99 M-1
Valley Green 99 H-1
Vandergrift C-1
Walbert 74 A-4
Walnut Hill 99 M-2
Warren C-1
Warrendale 101 G-1
Washington B-1
Water Street C-1
Wattsburg C-1
Waynesboro D-2
Waynesburg B-2
Weldon 99 K-1
Wellsboro D-1
Wescosville 74 A-5
Wesleyville 82 C-1
West Chester E-2
West Fairview 84 B-7
West Homestead 101 J-4
West Mifflin 101 J-5
West Mill Creek 82 A-2
West Oak Lane 99 K-3
West Pittston 108 E-7
West View 101 H-3
West Wyoming 108 E-7
Wexford 101 H-1
Whitehall (P.O.) 74 B-3
Whitehall 101 H-5
Whitemarsh 99 H-1
Whitemarsh Vil. 99 J-2
Wilcox C-1
Wildwood 101 J-2
Wilkes-Barre D-1
Wilkinsburg 101 K-4
Willamsport D-1
Windber C-1
Wind Gap E-1
Wissinoming 99 M-4
Wister 99 K-4
Woodward D-1
Wormleysburg 84 B-7
Wyalusing D-1
Wyncote 99 K-2
Wyndmoor 99 J-2
Wynnefield 99 H-5
Wynnewood 99 G-5
Wyoming 108 E-7
Yatesville 108 F-7
Yeadon 99 G-7
York D-2
Youngsville C-1
Zelienople B-1

RHODE ISLAND

Map on page 14

Block Island B-3
Bristol B-3
Chepachet B-3
Jamestown B-3
Narragansett B-3
Newport B-3
Pawtucket B-3
Providence B-3
Westerly B-3
Wickford B-3
Woonsocket B-3

SOUTH CAROLINA

Map on page 23

Abbeville D-1
Aiken E-2
Allendale E-2
Anderson D-1
Andrews F-2
Arcadia 105 G-2
Arkwright 105 H-3
Bamberg E-2
Barnwell E-2
Batesburg E-2
Beaufort E-2
Beech Island E-2
Bennettsville F-1
Bethune E-1
Bishopville E-1
Blackville E-2
Branchville E-2
Camden E-1
Charleston F-2
Cheraw E-1
Chesnee E-1
Chester E-1
Chesterfield E-1
City View 84 D-1
Clemson D-1
Clinton D-1
Columbia E-1
Conway F-2
Cowpens E-1
Darlington E-1
Denmark E-2
Dillon F-1
Drayton 105 J-2
Easley D-1
Edgefield D-2
Effingham F-1
Enoree D-1
Estill E-2
Fairfax E-2
Fair Forest 105 G-2
Florence F-1
Folly Beach F-2
Fort Mill E-1
Fountain Inn D-1
Gaffney E-1
Gantt 84 D-3
Gardens Cor. E-2
Georgetown F-2
Great Falls E-1
Greenville D-1
Greenwood D-1
Greer D-1
Hampton E-2
Hardeeville E-2
Hartsville E-1
Hilton Head I. E-2
Jacksonboro E-2
Johnston E-2
Jonesville E-1
Kershaw E-1
Kingstree F-2
Lake City F-2
Lancaster E-1
Latta F-1
Laurens D-1
Lexington E-2
Liberty D-1
Little River F-2
Loris F-1
Manning E-2
Marion F-1
McBee E-1
McColl F-1
McCormick D-2
Moncks Corner E-2
Mt. Pleasant F-2
Mullins F-1
Myrtle Beach F-2
Neeses E-2
Newberry E-1
North E-2
N. Myrtle Beach F-2
Olanta F-2
Orangeburg E-2
Pageland E-1
Pickens D-1
Princeton D-1
Ridgeland E-2
Rock Hill E-1
St. Andrews 80 A-5
St. George E-2
St. Matthews E-2
Saluda E-2
Saxon 105 H-2
Seneca D-1
Society Hill F-1
Southern Shops 105 H-1
Spartanburg E-1
Summerton E-2
Summerville E-2
Sumter E-2
Swansea E-2
Timmonsville E-1
Turbeville F-2
Union E-1
Valley Falls 105 H-1
Walhalla D-1
Walterboro E-2
Ware Shoals D-1
Westminster D-1
Whitmire E-1
Whitney 105 H-2
Williston E-2
Winnsboro E-1
Woodruff E-1
Yemassee E-2
York E-1

SOUTH DAKOTA

Map on pages 34, 35

Aberdeen D-3
Alexandria E-4
Arlington E-3
Artesian E-4
Avon D-4
Barnard D-3
Batesland C-4
Belle Fourche B-3
Beresford E-4
Billsburg C-3
Bison B-3
Blunt D-3
Bonesteel D-4
Bowdle D-3
Bridgewater E-4
Britton E-3
Brookings E-3
Buffalo B-3
Canton E-4
Carter C-4
Chamberlain D-4
Clark E-3
Clear Lake E-3
Colman E-4
Colome D-4
Custer B-4
Davis E-4
Deadwood B-3
Dell Rapids E-4
De Smet E-3
Doland D-3
Dupree C-3
Eagle Butte C-3
Edgemont B-4
Elk Pt. E-4
Eureka D-3
Faith B-3
Faulkton D-3

Flandreau E-3
Fort Pierre C-3
Ft. Thompson D-3
Frederick D-3
Freeman E-4
Gann Valley D-3
Gettysburg D-3
Gregory D-4
Groton D-3
Hayes C-3
Hecla D-3
Henry E-3
Herreid D-3
Highmore D-3
Hill City B-4
Hot Springs B-4
Howard E-3
Howes B-3
Huron D-3
Interior C-4
Ipswich D-3
Isabel C-3
Kadoka C-4
Kennebec D-4
Keystone B-4
Kimball D-4
Lake Andes D-4
Lake City E-3
La Plant C-3
Lead B-3
Lemmon B-3
Leola D-3
Longvalley C-4
Madison E-3
Martin C-4
McIntosh C-3
McLaughlin C-3
Menno E-4
Midland C-3
Milbank E-3
Miller D-3
Mission C-4
Mitchell D-4
Mobridge C-3
Morristown C-3
Mound City D-3
Murdo C-4
Newell B-3
Oelrichs B-4
Oglala B-4
Onida D-3
Philip C-3
Pierre C-3
Pine Ridge B-4
Plankinton D-4
Presho C-4
Rapid City B-3
Redfield D-3
Roscoe D-3
Salem E-4
Scenic B-4
Selby D-3
Seneca D-3
Sioux Falls E-4
Sisseton E-3
Spearfish B-3
Stephan D-3
Sturgis B-3
Timber Lake C-3
Tripp E-4
Tyndall E-4
Vermillion E-4
Vivian C-4
Wagner D-4
Wall B-3
Watertown E-3
Waubay E-3
Webster E-3
Wessington D-3
Wessington Sprs. D-3
White Owl B-3
White River C-4
Winner D-4
Wolsey D-3
Woonsocket D-3
Yankton E-4

TENNESSEE

Map on pages 32, 33

Alcoa D-3
Athens C-3
Bartlett 94 D-1
Bean Sta. D-2
Bells A-3
Berry Hill 98 B-3
Bolivar A-3
Bordeaux 98 A-1
Bristol D-2
Brownsville A-3
Bruceton A-2
Camden A-2
Carthage C-2
Caryville C-2
Celina C-2
Centerville B-3
Chattanooga C-3
Clarksville B-2
Cleveland C-3
Collierville A-3
Columbia B-3
Cookeville C-2
Covington A-3
Cowan C-3
Crossville C-3
Cumberland Gap D-2
Dayton C-3
Decatur C-3
Decaturville A-3
Dickson B-2
Dover B-2
Dresden A-2
Ducktown C-3
Dunlap C-3
Dyersburg A-2
East Ridge 78 B-8
Elizabethton D-2
Erin B-2
Erwin D-2
Etowah C-3
Fayetteville B-3
Franklin B-3
Gallatin B-2
Gatlinburg D-3
Goodlettsville B-2
Greeneville D-2
Greenfield A-2
Halls A-3
Harriman C-3
Hartsville B-2
Henderson A-3
Hohenwald B-3
Humboldt A-3
Huntingdon A-3
Jackson A-3
Jamestown C-2
Jasper C-3
Jellico C-2
Johnson City D-2
Kingsport D-2
Knoxville D-3
La Follette D-2
Lake City C-2
Lawrenceburg B-3
Lebanon B-2
Lenoir City C-3
Lewisburg B-3
Lexington A-3
Linden B-3
Livingston C-2
Lookout
Mountain 78 A-8
Loudon C-3
Madisonville C-3
Manchester C-3
Martin A-2
Maryville D-3
McKenzie A-2
McMinnville C-3
Memphis A-3
Milan A-3
Millington A-3
Monteagle C-3
Monterey C-2
Morristown D-2
Mountain City E-2
Mt. Pleasant B-3
Murfreesboro B-3
Nashville B-2
Newbern A-2
Newport D-3
Norris C-2
Oak Ridge C-2
Obion A-2
Oneida C-2
Paris A-2
Parsons A-3
Pikeville C-3
Portland B-2
Pulaski B-3
Raleigh 94 C-1
Red Bank 78 B-6
Red Boiling Sprs. C-2
Ridgeside 78 C-8
Ripley A-3
Rockwood C-3
Rogersville D-2
Savannah A-3
Selmer A-3
Sevierville D-3
Sewanee C-3
Shelbyville B-3
Signal Hills 78 A-6
Signal Mtn. C-3
Smithville C-3
Soddy-Daisy C-3
Somerville A-3
S. Pittsburg C-3
Sparta C-3
Spencer C-3
Spring City C-3
Springfield B-2
Sweetwater C-3
Tazewell D-2
Tiftonia 78 A-8
Tiptonville A-2
Trenton A-3
Troy A-2
Tullahoma B-3
Union City A-2
Vonore C-3
Watertown B-2
Wauhatchie 78 A-8
Waverly B-2
Waynesboro B-3
Westmoreland B-2
Whitwell C-3
Winchester B-3
Woodbury B-3

TEXAS

Map on pages 40, 41

Abernathy D-1
Abilene E-1
Addicks 85 G-3
Alamo Heights 103 L-2
Alba G-1
Albany E-1
Aldine 85 K-1
Alice F-4
Alief 85 H-4
Alpine C-2
Alto G-2
Alvarado F-1
Alvin G-3
Alvord F-1
Amarillo B-4
Amherst C-1
Andrews C-1
Angleton G-3
Anson E-1
Anthony A-1
Aransas Pass F-4
Archer City E-1
Arlington 83 G-7
Asherton E-3
Aspermont D-1
Athens G-1
Atlanta G-1
Austin F-2
Bacliff 85 G-5
Baird E-1
Balch Springs 83 M-7
Balcones Hts. 103 K-2
Ballinger E-2
Balmorhea C-2
Bandera E-3
Barnhart D-2
Bastrop F-2
Bay City G-3
Baytown G-3
Bayview 85 G-5
Beaumont G-2
Beaumont Place 85 M-3
Bedford 83 G-6
Beeville F-3
Bellaire 85 J-4
Bellevue E-1
Bells F-1
Bellville F-3
Belton F-2
Benavides E-4
Benbrook 82 C-8
Benjamin E-1
Beverly Hills 109 G-2
Big Lake D-2
Big Spring D-1
Big Wells E-3
Bishop F-4
Blackwell D-1
Blanco E-2
Boerne E-3
Bogata G-1
Bonham F-1
Borger B-4
Bowie F-1
Brackettville D-3
Brady E-2
Breckenridge E-1
Bremond F-2
Brenham F-2
Bridgeport F-1
Bronte D-2
Brownfield C-1
Brownsville F-4
Brownwood E-2
Bryan F-2
Buckingham 83 L-5
Buffalo G-2
Bunker Hill 85 H-3
Burkburnett E-1
Burnet E-2
Caldwell F-2
Calvert F-2
Cameron F-2
Canadian B-4
Canton G-1
Canyon B-4
Carrizo Sprs. D-3
Carrollton 83 J-4
Carthage G-1
Castle Hills 103 L-1
Catarina E-3
Celeste F-1
Center G-2
Centerville G-2
Charlotte E-3
Childress B-4
Chillicothe C-4
Christoval D-2
Cisco E-1
Clairemont D-1
Clarendon B-4
Clarksville G-1
Claude B-4
Cleburne F-1
Cleveland G-2
Cloverleaf 85 M-3
Coahoma D-1
Cockrell Hill 83 J-7
Coleman E-2
College Sta. F-2
Colmesneil G-2
Colorado City D-1
Columbus F-3
Comanche E-2
Comfort E-3
Commerce G-1
Comstock D-3
Concan E-3
Conroe G-2
Coolidge F-2
Cooper G-1
Copperas Cove F-2
Corpus Christi F-4
Corrigan G-2
Corsicana F-1
Cotulla E-3
Crane C-2
Crockett G-2
Crosbyton D-1
Cross Plains E-1
Crowell E-1
Crystal City E-3
Cuero F-3
Daingerfield G-1
Dalhart A-3
Dallas F-1
Dalworthington
Gardens 82 F-8
Dayton G-2
Decatur F-1
Deepwater 85 M-4
DeKalb G-1
Del Rio D-3
Denison F-1
Denton F-1
Devine E-3
D'Hanis E-3
Diboll G-2
Dickens D-1
Dilley E-3
Dimmitt A-4
Donna F-4
Dryden C-2
Dublin E-1
Dumas A-4
Duncanville 83 J-8
Dyersdale 85 L-2
Eagle Pass D-3
East Houston 85 L-3
Eastland E-1
Eddy F-2
Eden E-2
Edgecliff 82 D-8
Edinburg F-4
Edna F-3
El Campo F-3
Eldorado D-2
Electra E-1
El Paso A-2
Encinal E-3
Encino E-4
Ennis F-1
Estelline B-4
Euless 83 G-6
Evant E-2
Fabens A-2
Fairbanks 85 H-2
Fairfield F-2
Falfurrias F-4
Fannin F-3
Farmers Branch 83 J-5
Farwell A-4
Fauna 85 M-2
Ferris F-1
Florence 83 H-8
Floresville F-3
Floydada D-1
Forest Hill 82 E-8
Fort Davis C-2
Fort Hancock B-2
Fort Stockton C-2
Fort Worth F-1
Franklin F-2
Frankston G-1
Fredericksburg E-2
Freeport G-3
Freer E-4
Friona A-4
Gail D-1
Gainesville F-1
Galena Park 85 L-4
Galveston G-3
Ganado F-3
Garden Oaks 85 K-3
Garden Villas 85 L-5
Garland 83 M-5
Garrison G-2
Gatesville F-2
Georgetown F-2
George West E-3
Giddings F-2
Gilmer G-1
Gladewater G-1
Glen Rose F-1
Golden Acres 85 M-4
Goldthwaite E-2
Goliad F-3
Gonzales F-3
Goodnight B-4
Graham E-1
Granbury F-1
Grandfalls C-2
Grand Prairie 83 H-7
Grand Saline G-1
Grapeland G-2
Grapevine 83 G-5
Greens Bayou 85 M-3
Greenville F-1
Gregory F-4
Groesbeck F-2
Groom B-4
Groveton G-2
Guthrie D-1
Hale Center D-1
Hallettsville F-3
Haltom City 82 E-6
Hamilton E-2
Hamlin D-1
Happy B-4
Harlingen F-4
Harper E-2
Harrisburg 85 L-4
Hartley A-4
Haskell E-1
Hearne F-2
Hebbronville E-4
Hedley B-4
Hedwig 85 J-3
Hempstead G-2
Henderson G-1
Henrietta E-1
Hereford A-4
Hermleigh D-1
Hico E-2
Higgins B-4
High Island G-3
Highland Park 83 K-6
Hillsboro F-1
Hilshire Village 85 J-3
Hockley G-2
Hondo E-3
Honey Grove G-1
Houmont Park 85 L-2
Houston G-3
Houston Gardens 85 L-3
Howellville 85 G-4
Hubbard F-2
Humble G-3
Hunters Creek 85 J-3
Huntington G-2
Huntsville G-2
Hurst 82 F-6
Hutchins 83 L-8
Iowa Park E-1
Iraan D-2
Irving 83 H-6
Italy F-1
Itasca F-1
Jacinto City 85 L-3
Jacksboro E-1
Jacksonville G-2
Jasper G-2
Jayton D-1
Jefferson G-1
Jersey Village 85 H-2
Jewett F-2
Johnson City E-2
Jourdanton E-3
Junction E-2
Karnes City E-3
Kaufman F-1
Kemp F-1
Kenedy F-3
Kennedale 82 E-8
Kent B-2
Kenwood 85 K-2
Kerens F-1
Kermit C-2
Kerrville E-2
Kilgore G-1
Killeen F-2
Kingsville F-4
Kirbyville G-2
Kosse F-2
Kountze G-2
LaGrange F-3
Lake Jackson G-3
Lakeview 83 H-7
Lake Worth 82 C-6
LaMarque 85 H-7
Lamesa C-1
Lampasas E-2
Langtry D-3
La Pryor E-3
Laredo E-4
Lawn E-1
Leakey E-3
Leonard F-1
Leon Valley 103 K-2
Levelland C-1
Liberty G-3
Linden G-1
Littlefield C-1
Livingston G-2
Llano E-2
Lockhart F-3
Lockney D-1
Lometa E-2
Lone Oak G-1
Longview G-1
Loraine D-1
Lorenzo D-1
Lott F-2
Lovelady G-2
Lubbock D-1
Lufkin G-2
Luling F-3
Mabank G-1
Mabelle E-1
Madisonville G-2
Magnolia Park 85 L-4
Malakoff F-1
Manchester 85 L-4
Marathon C-2
Marble Falls E-2
Marfa B-2
Marlin F-2
Marshall G-1
Mason E-2
Matador D-1
Mathis F-3
Maud G-1
McAllen E-4
McCamey C-2
McGregor F-2
McKinney F-1
McLean B-4
Meadow C-1
Meadowbrook 85 L-4
Memphis B-4
Menard E-2
Mercedes F-4
Meridian F-2
Mertzon D-2
Mesquite 83 M-7
Mexia F-2
Miami B-4
Midland D-2
Midlothian F-1
Milam G-2
Mineola G-1
Mineral Wells E-1
Mirando City E-4
Mission E-4
Monahans C-2
Morton C-1
Mt. Enterprise G-2
Mount Houston 85 L-2
Mt. Pleasant G-1
Mt. Vernon G-1
Muleshoe C-1
Munday E-1
Nacogdoches G-2
Navasota F-2
New Boston G-1
New Braunfels E-3
Newcastle E-1
Newman A-2
Newton G-2
New Waverly G-2
Nixon F-3
Nocona F-1
North Houston 85 J-2
North Shadydale 85 L-2
Oakwood G-2
Odessa C-2
O'Donnell D-1
Oklaunion E-1
Olmos Park 103 L-2
Olney E-1
Olton C-1
Orange G-2
Orla C-2
Ozona D-2
Paducah D-2
Paint Rock E-2
Palacios F-3
Palestine G-2
Palo Pinto E-1
Pampa B-4
Panhandle B-4
Pantego 82 F-7
Paris G-1
Park Place 85 L-4
Pasadena 85 M-4
Pearsall E-3
Pecos C-2
Perryton B-4
Pharr E-4
Pine Springs B-2
Piney Point 85 H-3
Pittsburg G-1
Plains C-1
Plainview D-1
Plano F-1
Pleasanton E-3
Port Aransas F-4
Port Arthur G-3
Port Bolivar 85 J-7
Port Isabel F-4
Port Lavaca F-3
Post D-1
Premont E-4
Presidio B-3
Pyote C-2
Quanah C-4
Quemado D-3
Quitman G-1
Ralls D-1
Ranger E-1
Rankin D-2
Raymondville F-4
Refugio F-3
Richardson 83 L-4
Richland F-2
Richland Hills 82 E-6
Richland Sprs. E-2
Richmond G-3
Ringgold E-1
Rio Grande City E-4
Rising Star E-1
River Oaks 82 D-7
River Oaks 85 J-3
Riviera F-4
Robert Lee D-2
Robstown F-4
Roby D-1
Rockdale F-2
Rockport F-3
Rocksprings D-3
Rockwall F-1
Rogers F-2
Roma-
Los Saenz E-4
Ropesville C-1
Roscoe D-1
Rosebud F-2
Rosenburg G-3
Rotan D-1
Round Rock F-2
Rusk G-2
Sabinal E-3
Saint Jo F-1
San Angelo D-2
San Antonio E-3
Sanatorium D-2
San Augustine G-2
San Benito F-4
Sanderson C-2
San Diego E-4
Sanger F-1
San Leon 85 H-6
San Marcos E-3
San Saba E-2
Sansom Park
Vil. 82 D-6
Santa Anna E-2
Saragosa C-2
Satsuma 85 G-2
Schulenburg F-3
Seagraves C-1
Sealy F-3
Seguin F-3
Seminole C-1
Seymour E-1
Shafter B-3
Shamrock B-4
Sharpstown 85 J-4
Sheffield D-2
Sherman F-1
Shiro G-2
Sierra Blanca B-2
Silsbee G-2
Sinton F-3
Skidmore F-3
Slaton D-1
Smith Point 85 K-5
Smithville F-3
Snyder D-1
Sonora D-2
South Bend E-1
South Houston 85 L-5
Southton 103 M-4
Spearman B-4
Springtown F-1
Spring Valley 85 H-3
Spur D-1
Stamford E-1
Stanton D-1
Stephenville E-1
Sterling City D-2
Stinnett B-4
Stockdale F-3
Stratford B-4
Strawn E-1
Sudan C-1
Sugar Land G-3
Sulphur Sprs. G-1
Sweetwater D-1
Taft F-4
Tahoka D-1
Taylor F-2
Teague F-2
Temple F-2
Tenaha G-2
Terrell F-1
Terrell Hills 103 M-2
Texarkana G-1
Texas City G-3
Texhoma B-3
Texline A-3
Thorndale F-2
Thornton F-2
Three Rivers E-3
Throckmorton E-1
Tilden E-3
Timpson G-2
Tivoli F-3
Toyah C-2
Toyahvale C-2
Trinity G-2
Tulia B-4
Turkey B-4
Tyler G-1
University Park 83 K-6
Uvalde E-3
Valentine B-2
Van Alstyne F-1
Van Horn B-2
Vega A-4
Vernon E-1
Victoria F-3
Waco F-2
Waelder F-3
Wallis F-3
Waxahachie F-1
Weatherford F-1
Weimar F-3
Wellington B-4
Wells G-2
Weslaco F-4
West F-2
W. Columbia G-3
Westover Hills 82 C-7
West University
Place 85 J-4
Westworth 82 C-7
Wharton G-3
Wheeler B-4
White Deer B-4
Whitesboro F-1
White
Settlement 82 C-7
Whitewright F-1
Wichita Falls E-1
Wills Point G-1
Windthorst E-1
Wink C-2
Winnsboro G-1
Winters E-2
Woodlake 85 H-4
Woodsboro F-3
Woodsdale 85 K-2
Woodville G-2
Wortham F-2
Yoakum F-3
Yorktown F-3
Zapata E-4
Zavalla G-2
Zipp City 83 M-7

UTAH

Map on page 44

American Fork C-2
Aurora B-3
Beaver B-3
Blanding D-4
Bluff D-4
Bountiful C-2
Brigham City B-2
Castle Dale C-3
Cedar City B-4
Centerfield C-3
Circleville B-3
Clearfield B-2
Cove Fort B-3
Delta B-3
Devils Slide C-2
Duchesne C-2
Dutch John D-2
Echo C-2
Elsinore B-3
Emery C-3
Enterprise B-4
Ephraim C-3
Eureka B-3
Fairview C-3
Fayette C-3
Ferron C-3
Fillmore B-3
Fountain Green C-3
Fremont Jct. C-3
Garden City C-2
Goshen B-3
Gouldings Trading
Post C-4
Granger 107 G-8
Grantsville B-2
Green River C-3
Gunnison B-3
Hanksville C-3
Hatch B-4
Heber City C-2
Helper C-3
Hinckley B-3
Holden B-3
Huntington C-3
Hurricane B-4
Jensen D-2
Junction B-3
Kanab B-4
Kanarraville B-4
Kanosh B-3
Kaysville B-2
Laketown C-2
Leamington B-3
Lehi C-2
Levan B-3
Loa C-3
Logan C-2
Lynndyl B-3
Manila D-2
Manti C-3
Marysvale B-3
Meadow B-3
Mexican Hat C-4
Milford B-3
Minersville B-3
Moab D-3
Modena B-4
Mona B-3
Monticello D-4
Morgan C-2
Moroni C-3
Mt. Carmel Jct. B-4
Mt. Pleasant C-3
Murray B-2
Myton D-2
Nephi B-3
Newcastle B-4
Ogden C-2
Orderville B-4
Orem C-2
Panguitch B-4
Paragonah B-4
Parowan B-4
Payson C-2
Pleasant Grove C-2
Price C-3
Provo C-2
Randolph C-2
Richfield C-3
Richmond C-2
Roosevelt C-2
Rosette B-2
St. George B-4
Salina C-3
Salt Lake City B-2
Santaquin C-3
Scipio B-3
Sevier B-3
Sigurd C-3
Smithfield C-2
Snowville B-2
Soldier Summit C-3
S. Salt Lake 107 H-8
Spanish Fork C-2
Spring City C-3
Springdale B-4
Springville C-2
Stockton B-2
Summit B-4
Thistle C-3
Thompson D-3
Tooele B-2
Toquerville B-4
Torrey C-3
Tremonton B-2
Vernal D-2
Vernon B-2
Veyo B-4
Wahsatch C-2
Wanship C-2
Wellington C-3
Wellsville C-2
Wendover A-2
Willard C-2
Woodruff C-2
Woodside C-3

VERMONT

Map on page 14

Addison A-1
Ascutney A-2
Barre A-1
Barton A-1
Bellows Falls A-2
Bennington A-2
Bethel A-2
Bradford A-2
Brandon A-2
Brattleboro A-2
Bridport A-2
Burlington A-1
Cambridge A-1
Charlotte A-1
Chester A-2
Dorset A-2
E. Berkshire A-1
Enosburg Falls A-1
Fair Haven A-2
Grand Isle A-1
Hardwick A-1
Island Pond B-1
Johnson A-1
Lowell A-1
Ludlow A-2
Lyndonville A-1
Manchester Cen. A-2
Middlebury A-1
Montpelier A-1
Morrisville A-1
Newport A-1
Norton B-1
Plainfield A-1
Plymouth A-2
Poultney A-2
Richford A-1
Rochester A-2
Rutland A-2
St. Albans A-1
St. Johnsbury A-1
Springfield A-2
Stockbridge A-2
Stowe A-1
Swanton A-1
Vergennes A-1
Wallingford A-2
Warren A-1
Waterbury A-1
Wells A-2
W. Danville A-1
Weston A-2
White River Jct. A-2
Wilmington A-2
Windsor A-2
Winooski A-1
Woodstock A-2

VIRGINIA

Map on pages 20, 21

Abingdon B-2
Adner E-2
Alexandria E-1
Altavista D-2
Amelia E-2
Amherst D-2
Annandale 109 H-8
Appalachia B-2
Appomattox D-2
Arlington E-1
Ashland E-2
Baileys
Crossroads 109 J-7
Bayside 98 E-2
Bedford D-2
Belleville 98 C-3
Berryville E-1
Big Island D-2
Big Stone Gap B-2
Blacksburg C-2
Blackstone D-2
Bluefield C-2
Boone 98 C-3
Bowers Hills 98 C-3
Bowling Green E-1
Bristol B-2
Broadway D-1
Brookneal D-2
Buchanan D-2
Buckroe Beach 98 D-1
Buena Vista D-2
Burkeville D-2
Cape Charles E-2
Central Garage E-2
Charlottesville D-2
Chase City D-2
Chatham D-2
Chesapeake E-2
Chincoteague F-2
Christiansburg C-2
Churchland 98 D-2
Churchville D-1
Clarksville D-2
Clifton Forge D-2
Clinchport B-2
Courtland E-2
Covington D-2
Crewe D-2
Crittenden 98 C-2
Crows C-2
Cuckoo E-2
Culpeper D-1
Damascus C-2
Danville D-2
Deep Creek 98 D-3
Dublin C-2
Eclipse 98 C-2
Elkton D-1
Ellerson 101 J-6
Emporia E-2
Exmore F-2
Falls Church 109 H-7
Fancy Gap C-2
Farmville D-2
Floyd C-2
Ft. Chiswell C-2
Fox Hill 98 D-1
Franklin E-2
Fredericksburg E-1
Front Royal D-1
Gainesville E-1
Galax C-2
Gate City B-2
Glasgow D-2
Glenns E-2
Gordonsville D-1
Goshen D-2
Grand View 98 D-1
Greenville D-2
Gretna D-2
Grundy B-2
Gum Spring D-2
Halifax D-2
Hampton E-2
Hanging Rock 101 K-6
Harrisonburg D-1
Hillsville C-2
Hopewell E-2
Hot Sprs. C-2
Independence C-2
Jonesville B-2
Kempsville 98 E-3
Keysville D-2
Kilmarnock E-2
Kiptopeke Bch. E-2
Lakeside 101 G-6
Land 98 E-3
Langley 109 J-6
Lawrenceville E-2
Lebanon B-2
Leesburg E-1
Lexington D-2
Lovingston D-2
Luray D-1
Lynchburg D-2
Lynnhaven 98 E-3
Madison D-1
Manassas E-1
Mapleton 98 E-3
Marion C-2
Marshall E-1
Martinsville D-2
McClean 109 J-6
McKenney E-2
Mechanicsville 101 J-6
Merrifield 109 H-7
Middleburg E-1
Monterey D-1
Morrison 98 C-1
New Market D-1
Newport News E-2
Norfolk E-2
Norton B-2
Ocean View 98 E-2
Old Point
Comfort 98 D-1
Onancock F-2
Orange D-1
Pearisburg C-2
Pennington Gap B-2
Petersburg E-2
Phoebus 98 D-1
Portsmouth E-2
Pulaski C-2
Quantico E-1
Radford C-2
Reedville E-2
Rich Creek C-2
Richlands C-2
Richmond E-2
Ridgeway D-2
Roanoke D-2
Rocky Mount C-2
Ruckersville D-1
St. Paul B-2
Salem C-2
Seven Corners 109 J-7
Smithfield E-2
South Boston D-2
South Hill D-2
South Norfolk 98 D-3
Sperryville D-1
Spring Grove E-2
Sprouses Cor. D-2
Staunton D-1
Stephens City D-1
Stony Creek E-2
Strasburg D-1
Stuart C-2
Suffolk E-2
Surry E-2
Tappahannock E-2
Tazewell C-2
Tysons Corner 109 H-7
Victoria D-2
Vienna 109 H-7
Vinton 101 M-7
Virginia Beach E-2
Wachapreague F-2
Warm Sprs. D-1
Warrenton E-1
Warsaw E-2
Waverly E-2
Waynesboro D-1
West Point E-2
West Norfolk 98 D-2
Williamsburg E-2
Winchester D-1
Windsor E-2
Woodstock D-1
Wytheville C-2
Yorktown E-2

WASHINGTON

Map on page 48

Aberdeen A-2
Alderton 107 L-5
Alderwood
Manor 107 L-1
Algona 107 L-5
Allyn 107 H-4
Anacortes A-1
Anatone D-2
Arcadia 107 H-5
Asotin D-2
Auburn A-2
Beaux Arts 107 L-2
Beaver A-1
Belfair 107 J-3
Bellevue B-2
Bellingham A-1
Berrydale 107 M-4
Bingen B-3
Blaine A-1
Bonney Lake 107 M-5
Boston Harbor 107 H-6
Bothell A-2
Bremerton A-2
Brewster C-1
Brinnon 107 H-2
Browns Point 107 K-4
Brownsville 107 J-2
Bryn Mawr 107 L-3
Buckley B-2
Buena B-2
Burien 107 L-3
Burlington A-1
Burton 107 K-4
Camas A-3
Carbonado 107 M-6
Cashmere B-2
Castle Rock A-2
Cathlamet A-2
Cedarhurst 107 K-3
Centralia A-2
Chehalis A-2
Chelan C-2
Chewelah D-1
Clarkston D-2
Cle Elum B-2
Clyde Hill 107 L-2
Coal Creek 107 M-3
Colby 107 K-3
Colchester 107 K-3
Colfax D-2
Colton D-2
Colville C-1
Connell C-2
Coulee City C-2
Coulee Dam C-1
Covington 107 M-4
Crosby 107 H-2
Daisy C-1
Dash Point 107 L-4
Davenport C-2
Dayton C-2
Deer Park D-2
Deming A-1
Des Moines 107 L-4
Dewatto 107 G-3
Dierienger 107 M-5
Dockton 107 K-4
Du Pont 107 J-6
Dusty D-2
Eastgate 107 M-3
Easton B-2
E. Port
Orchard 107 K-3
Edmonds 107 L-1
Elbe A-2
Eldon 107 G-3
Elgin 107 J-4
Ellensburg B-2
Ellisport 107 L-4
Elma A-2
Enetai 107 K-2
Entiat B-2
Enumclaw B-2
Ephrata C-2
Everett A-1
Evergreen 100 E-5
Federal Way 107 L-4
Fern Prairie 100 E-5
Fife 107 L-5
Fircrest 107 K-5
Fisher 100 E-6
Forest Beach 107 J-4
Forks A-2
Fox Island 107 J-5
Garfield D-2
Gig Harbor 107 K-4
Gilberton 107 K-2
Glencove 107 J-4
Glenwood 107 J-4
Gold Bar B-2
Goldendale B-3
Gorst 107 J-3
Graham 107 L-6
Grand Coulee C-1
Grandview B-2
Granger B-2
Grapeview 107 J-4
Harper 107 K-3
Hartstene 107 H-5
Holly 107 H-3
Home 107 J-5
Hoquiam A-2
Humptulips A-2
Hunters C-1
Hunts Point 107 L-2
Illahee 107 K-2
Ilwaco A-2
Indianola 107 K-1
Ione D-1
Issaquah 107 M-3
Juanita 107 L-2
Kalama A-2
Keller C-1
Kelso A-2
Kenmore 107 L-1
Kennewick C-2
Kent 107 M-4
Kettle Falls D-1
Keyport 107 K-2
Kingston 107 K-1
Kirkland 107 L-2
Lacey 107 J-6
Lakebay 107 J-5
Lake Forest
Park 107 L-1
Lakeview 107 K-5
Lakewood 107 K-5
Lakota 107 L-4
Laurier C-1
Leavenworth B-2
Lilliwaup A-2
Lind C-2
Lisabuela 107 K-4
Lofall 107 J-1
Longbranch 107 J-5
Longview A-2
Lynnwood 107 L-1
Maltby 107 M-1
Manchester 107 K-3
Manzanita 107 K-4
Maplewood 107 M-3
Marcus D-1
Maryhill B-3
McMillin 107 M-5
Medina 107 L-2
Mercer Island 107 L-3
Metaline Falls D-1
Midland 107 L-5
Milton 107 L-5
Monroe B-2
Montesano A-2
Monte Vista 107 K-5
Morton A-2
Moses Lake C-2
Mossyrock A-2
Mountlake
Terr. 107 L-1
Mt. Vernon A-1
Naches B-2
Navy Yard City 107 J-3
Neah Bay A-1
Neilton A-2
Newcastle 107 M-3
Newhalem B-1
Newport D-1
Nisqually 107 J-6
Normandy Park 107 L-3
Oakesdale D-2
Oakville A-2
Odessa C-2
Okanogan C-1
Olalla 107 K-4
Olympia A-2
Omak C-1
Opportunity D-2
Orchards 100 D-5
Orondo B-2
Oroville C-1
Orting 107 M-6
Pacific 107 L-5
Palouse D-2
Parkland 107 L-6
Pasco C-2
Pe Ell A-2
Pomeroy D-2
Ponders Corner 107 K-6
Portage 107 K-4
Port Angeles A-1
Port Blakely 107 K-2
Port Madison 107 K-2
Port Orchard A-2
Port Townsend A-1
Poulsbo 107 K-1
Proebstel 100 E-5
Prosser C-2
Pullman D-2
Purdy 107 J-4
Puyallup A-2
Quilcene A-2
Quincy B-2
Raymond A-2
Reardan D-2
Redmond 107 M-2
Redondo 107 L-4
Renton A-2
Republic C-1
Retsil 107 K-3
Richland C-2
Richmond
Beach 107 K-1
Richmond
Highlands 107 L-1
Ritzville C-2
Riverton Hts. 107 L-3
Rosalia D-2
Rosedale 107 J-4
Ruston 107 K-4
St. John D-2
Scandia 107 J-2
Seabeck 107 J-2
Seabold 107 K-2
Seahurst 107 L-3
Seattle A-2
Seattle Hts. 107 L-1
Sedro Woolley A-1
Sekiu A-1
Sequim A-1
Shelton A-2
Sifton 100 D-5
Silverdale 107 J-2
Skykomish B-2
Snohomish B-2
Snoqualmie B-2
Soap Lake C-2
South Bay 107 H-6
South Bend A-2
South Colby 107 K-3
South Prairie 107 M-6
Southworth 107 K-3
Spanaway 107 K-6
Spangle D-2
Spokane D-2
Sprague D-2
Springdale D-1
Steilacoom 107 K-5
Stevenson B-3
Sumas A-1
Sumner 107 L-5
Sunnydale 107 L-3
Sunnyside C-2
Suquamish 107 K-1
Tacoma A-2
Tahlequah 107 K-4
Tahuya 107 G-4
Thompson
Place 107 J-6
Thrift 107 L-6
Tillicum 107 K-6
Tonasket C-1
Toppenish B-2
Tracyton 107 J-2
Tukwila 107 L-3
Twisp B-1
Union 107 G-4
Union Mills 107 J-6
Vancouver A-3
Vashon 107 K-3
Vashon Heights 107 K-3
Vaughn 107 J-4
Victor 107 J-4
View Park 107 K-3
Waitsburg C-2
Walla Walla C-2
Wallula C-2
Walnut Grove 100 D-5

Warren 107 J-4
Washougal 100 F-6
Washtucna C-2
Waterman 107 K-2
Waterville C-2
Wauna 107 J-4
Wenatchee B-2
West Fork C-1
Westport A-2
Wilbur C-2
Wildcat Lake 107 J-2
Wilkeson 107 M-6
Winslow 107 K-2
Wollochet 107 K-5
Woodinville 107 M-1
Woodland A-3
Woodway 107 L-1
Yakima B-2
Yarrow Point 107 L-2
Yoman 107 J-5

WEST VIRGINIA

Map on pages 18,19

Alderson B-3
Ansted B-2
Baker C-2
Beckley B-3
Belington B-2
Berkeley Sprs. C-2
Bethlehem 109 K-2
Bluefield B-3
Bridgeport B-2
Bruceton Mills C-2
Buckhannon B-2
Burnsville B-2
Cameron B-2
Cedar Grove B-2
Chapmanville A-3
Charleston B-2
Charles Town D-2
Chester B-1
Clarksburg B-2
Clay B-2
Clendenin B-2
Edgewood 80 E-4
Elkins C-2
Fairmont B-2
Fayetteville B-2
Fort Gay A-2
Franklin C-2
Frost C-2
Gassaway B-2
Gauley Br. B-2
Gilbert B-3
Glenville B-2
Grafton C-2
Harman C-2
Harpers Ferry D-2
Harrisville B-2
Hinton B-3
Hundred B-2
Huntington A-2
Huttonsville C-2
Iaeger B-3
Ireland B-2
Ivydale B-2
Junction C-2
Kenova A-2
Kermit A-3
Keyser C-2
Keystone B-3
Kimball B-3
Lewisburg B-3
Linn B-2
Logàn B-3
Madison B-2
Mannington B-2
Marlinton B-2
Marmet B-2
Martinsburg C-2
Mill Point B-2
Millstone B-2
Milton A-2
Minnehaha Sprs. C-2
Moorefield C-2
Morgantown C-2
Moundsville B-2
Mt. Hope B-3
Mullens B-3
New Martinsville B-2
Norton B-2
Oak Hill B-3
Oakvale B-3
Paden City B-2
Parkersburg B-2
Parsons C-2
Pennsboro B-2
Petersburg C-2
Philippi C-2
Point Pleasant A-2
Princeton B-3
Racine B-2
Rainelle B-3
Ravenswood B-2
Richwood B-2
Ripley B-2
Rock Cave B-2
Romney C-2
Ronceverte B-3
St. Albans B-2
St. Marys B-2
Salem B-2
Seneca Rocks C-2
Shady Spr. B-3
Shinnston B-2
Sistersville B-2
Smithville B-2
South Hills 80 E-5
Spencer B-2
Summersville B-2
Sutton B-2
Sweet Sprs. B-3
Thomas C-2
Thornwood C-2
Valley Head C-2
Wardensville C-2
Wayne A-2
Webster Sprs. B-2
Weirton B-1
Welch B-3
Wellsburg B-1
W. Hamlin A-2
Weston B-2
West Union B-2
Wheeling B-1
White Sulphur Sprs. B-3
Williamson A-3
Williamstown B-2
Winfield B-2
Yukon B-3

WISCONSIN

Map on page 30

Abbotsford B-3
Adams C-4
Algoma D-3
Alma B-3
Aniwa C-3
Antigo C-3
Appleton C-3
Ashland B-2
Augusta B-3
Baldwin A-3
Baraboo C-4
Barron B-3
Bayfield B-2
Beaver Dam C-4
Beloit C-4
Berlin C-4
Black Creek C-3
Black River Falls B-3
Bloomer B-3
Boscobel B-4
Brown Deer 94 B-4
Bruce B-3
Cable B-2
Cadott B-3
Cameron B-3
Cashton B-4
Chetek B-3
Chilton C-3
Chippewa Falls B-3
Clintonville C-3
Coloma C-3
Columbus C-4
Cornell B-3
Crandon C-3
Cudahy 94 C-8
Cumberland A-3
De Pere D-3
Dickeyville B-4
Dodgeville C-4
Durand B-3
Eagle River C-3
Eau Claire B-3
Edgerton C-4
Elkhorn C-4
Ellsworth A-3
Evansville C-4
Fairchild B-3
Fennimore B-4
Florence C-3
Fond du Lac C-4
Ft. Atkinson C-4
Fox Point 94 C-4
Galesville B-3
Gillet C-3
Gills Rock D-3
Glendale 94 B-5
Gotham B-4
Green Bay D-3
Greendale 94 B-8
Greenfield 94 A-8
Green Lake C-4
Hales Corners 94 A-8
Hayward B-3
Hudson A-3
Hurley B-2
Iron River B-2
Janesville C-4
Jefferson C-4
Kenosha D-4
Kewaunee D-3
Kiel D-4
LaCrosse B-4
Ladysmith B-3
Lake Geneva C-4
Lancaster B-4
Laona C-3
Madison C-4
Manitowoc D-3
Marinette D-3
Marshfield B-3
Mauston B-4
Mayville C-4
Medford B-3
Mellen B-2
Menasha C-3
Menomonie A-3
Merrill C-3
Milwaukee D-4
Minocqua C-3
Minong B-2
Mondovi B-3
Monico C-3
Monroe C-4
Montreal B-2
Mountain C-3
Necedah B-3
Neenah C-3
Neillsville B-3
Nelson B-3
New London C-3
New Richmond A-3
Niagara C-3
Oconomowoc C-4
Oconto D-3
Ojibwa B-3
Oshkosh C-4
Osseo B-3
Owen B-3
Park Falls B-3
Pembine C-3
Phillips B-3
Plainfield C-3
Platteville B-4
Plymouth C-4
Portage C-4
Pt. Washington D-4
Pound C-3
Prairie du Chien B-4
Prentice B-3
Princeton C-4
Racine D-4
Readstown B-4
Reedsburg B-4
Rhinelander C-3
Rice Lake B-3
Richland Center B-4
Ripon C-4
River Hills 94 B-4
St. Croix Falls A-3
St. Francis 94 D-7
Sauk City C-4
Schofield C-3
Shawano C-3
Sheboygan D-4
Shorewood 94 C-5
Siren A-3
Sister Bay D-3
Solon Springs A-2
Sparta B-4
Spooner A-3
Stanley B-3
Stevens Pt. C-3
Sturgeon Bay D-3
Superior B-2
Sun Prairie C-4
Theresa C-4
Tomah B-4
Tomahawk C-3
Turtle Lake A-3
Two Rivers D-3
Viroqua B-4
Washburn B-2
Watertown C-4
Waukesha C-4
Waupaca C-3
Waupun C-4
Wausau C-3
Wausaukee C-3
Wautoma C-3
Wauwatosa 94 A-6
West Allis 94 B-7
West Bend C-4
Westby B-4
West Milwaukee 94 B-7
Whitefish Bay 94 C-5
Whitehall B-3
Wisconsin Dells C-4
Wisconsin Rapids B-3
Wittenberg C-3

WYOMING

Map on pages 42, 43

Afton E-4
Alcova G-4
Allendale 78 D-6
Alpine Jct. E-4
Baggs G-5
Basin F-3
Beulah H-3
Big Piney E-4
Bosler H-5
Boulder F-4
Buffalo G-3
Canyon E-3
Casper G-4
Cheyenne H-5
Chugwater H-5
Clearmont G-3
Cody F-3
Cokeville E-4
Daniel E-4
Deaver F-3
Douglas H-4
Dubois F-4
Eden F-4
Emblem F-3
Evanston E-5
Farson F-4
Fort Bridger E-5
Fort Laramie H-4
Ft. Washakie F-4
Four Corners H-3
Freedom E-4
Garland F-3
Gillette H-3
Glendo H-4
Glenrock H-4
Granger E-5
Green River F-5
Greybull F-3
Guernsey H-4
Hawk Springs H-5
Hiland G-4
Hudson F-4
Jackson E-4
Jay Em H-4
Kaycee G-4
Kemmerer E-5
Kirby F-4
La Barge E-4
Lander F-4
Laramie H-5
Lingle H-4
Little America F-5
Lost Springs H-4
Lovell F-3
Lusk H-4
Lyman E-5
Manderson G-3
Manville H-4
Medicine Bow G-5
Meeteetse F-3
Midwest G-4
Mills 78 C-5
Moneta G-4
Moorcroft H-3
Newcastle H-4
Opal E-5
Orin H-4
Osage H-4
Pinedale F-4
Pine Tree H-4
Point of Rocks F-5
Powder River G-4
Powell F-3
Ranchester G-3
Rawlins G-5
Red Desert F-5
Riverton F-4
Rock River H-5
Rock Springs F-5
Saratoga G-5
Shell G-3
Sheridan G-3
Shoshoni G-4
Sinclair G-5
Smoot E-4
Split Rock G-4
Sundance H-3
Ten Sleep G-3
Thermopolis G-4
Tie Siding H-5
Torrington H-4
Ucross G-3
Uptin H-3
Van Tassell H-4
Walcott G-5
Waltman G-4
Wamsutter F-5
Wheatland H-4
Worland G-4

MEXICO

Map on pages 70-73

Names shown thus: Iztacalco...116 E-7 refer to city inset on page 116 with grid reference location at E-7

Abasolo (Gto.) C-7
Abasolo (Tam.) H-4
Acambaro C-7
Acambay D-8
Acaponeta A-6
Acapulco C-9
Acatlán (Pue.) E-8
Acatlán (Oax.) E-8
Acatzingo E-8
Acayucan F-9
Aconchi B-2
Actopan D-7
Acuitzio C-8
Acula F-8
Aculco D-7
Acultzingo E-8
Agrícola Oriental 116 F-7
Agua Blanca (Hgo.) D-7
Agua Blanca (Q.R.) G-7
Agualeguas H-3
Agua Nueva G-4
Agua Prieta C-1
Aguascalientes B-7
Aguilera 116 C-5
Aguililla B-8
Ahomé C-4
Ahuacatlán A-7
Ahualulco A-7
Ahuehuetes 116 C-6
Ahuichila F-4
Ajacuba D-7
Ajalpan E-8
Ajijic B-7
Ajuchitlán C-8
Akumal H-6
Alacrán G-8
Alamo E-7
Alamos (D.F.) 116 D-7
Alamos (Son.) C-3
Alazán E-7
Alchichica E-8
Aldama (Chi) E-2
Aldama (Tam.) E-6
Alicante F-3
Allende (Coah) G-2
Allende (N.L.) H-4
Almoloya C-8
Altamira E-6
Altar B-1
Altata C-4
Altotonga E-8
Alvarado F-8
Amatepec C-8
Amatlán A-7
Amealco C-7
Ameca A-7
Amecameca D-8
America 116 B-7
Anahuac G-3
Angel Albino Corzo G-10
Antiguo Morelos D-6
Angangueo C-8
Antonio Rayón D-6
Anzures 116 C-6
Apan D-8
Apaseo C-7
Apatzingan B-8
Apitpac G-9
Apizaco E-8
Aquila B-8
Aquiles Serdán (Chi.) E-2
Aquiles Serdán (D.F.) 116 E-5
Aramberri H-4
Arandas B-7
Arcelia C-8
Argentina 116 B-5
Ario C-8
Arista G-5
Arizpe B-1
Arriaga G-9
Arteaga (Coah.) G-4
Arteaga (Mich.) B-8
Atasta G-8
Atenango del Río D-8
Atequiza B-7
Atitalaquia D-7
Atlacomulco D-8
Atlixco D-8
Atolinga B-7
Atotonilco (Hgo.) D-7
Atotonilco (Jal.) B-7
Atoyac B-8
Atoyac de Alvarez C-9
Atzimba C-8
Autlán A-8
Aviacion Civil 116 F-6
Axotla 116 C-8
Ayutla (Gro.) D-9
Ayutla (Jal.) A-7
Babicora D-2
Babanoyaba D-2
Bacalar G-7
Bachiniva D-2
Bacoachi C-1
Badiraguato D-4
Bahía de Tortugas A-4
Bahía Kino A-2
Bahía San Carlos B-2
Balancán H-9
Balsas D-8
Banderilla E-8
Bañon F-5
Barra de Cazones E-7
Barra de Navidad A-8
Barra de Tamilco E-7
Barrilaco 116 B-6
Barroterán G-3
Batopilas D-3
Bayas E-5
Becal G-6
Becanchén G-6
Bejucos C-8
Belisario Dominguez 116 E-5
Bella Unión G-4
Bellavista A-7
Bellote G-8
Benavidez F-2
Benito Juárez H-6
Benjamin Hill B-1
Bermejillo F-4
Bernal D-7
Boca del Río (Sin.) C-4
Boca del Río (Ver.) F-8
Bochil G-9
Bolaños B-7
Bolonchenticul G-6
Boguillas del Carmen F-2
Buenaventura D-2
Buenavista B-2
Buendia E-3
Buenos Aires E-5
Bufalo E-3
Bustamante (N.L.) G-3
Bustamante (Tam.) H-5
Burgos H-4
Cabo San Lucas C-5
Caborca A-1
Cacahuatepec D-9
Cacama 116 D-8
Cadereyta (N.L.) G-4
Cadereyta (Qto.) D-7
Caimanero D-5
Cajurachic C-2
Calera (Victor Rosales) F-5
Calkini G-6
Calnali D-7
Calpulalpan D-8
Calvillo B-7
Camacho F-4
Camargo H-3
Campeche H-8
Campo Aníbal D-4
Cananea B-1
Canatlán E-4
Cancún H-6
Candela G-3
Candelaria H-8
Cañitas de Felipe Pescador F-5
Cansancab G-6
Capacuaro B-8
Carapan B-8
Carbó B-2
Cardel E-8
Cárdenas (S.L.P.) D-6
Cárdenas (Tab.) G-8
Cardonal D-7
Carmen 116 C-8
Carrillo F-3
Casa Blanca (Mich.) B-8
Casa Blanca (N.L.) G-3
Casas H-5
Casas Alemán 116 E-5
Castaños G-3
Caté G-9
Catemaco F-8
Catorce G-5
Ceballos F-3
Cebollas E-5
Cedillos E-1
Cedral G-5
Celaya C-7
Celestún F-6
Celulosa D-2
Cenotillo G-6
Cerralvo H-3
Cerritos C-6
Chaccheito G-7
Chahuites F-9
Chalchihuites F-5
Chalchijapan F-9
Chalco D-8
Chamela A-8
Champotón H-8
Chapala B-7
Chapuihuacán D-7
Charcas G-5
Charco de Peña E-2
Chazumba E-8
Chemax H-6
Chenalhó G-9
Chencán H-8
Chencoyí G-7
Cherán C-8
Chetumal H-7
Chiapa G-9
Chiautla D-8
Chicomucelo G-10
Chicontepec D-7
Chicxulub Puerto G-6
Chietla D-8
Chignahuapan D-8
Chihuahua D-2
Chikindzonot G-6
Chila E-9
Chilapa D-9
Chilpancingo D-9
China H-4
Chinatú D-3
Chiquilá H-6
Choix C-3
Cholula D-8
Chontla E-7
Chopo 116 C-5
Chumpón H-6
Ciénega de Flores G-4
Cieneguilla E-4
Cihuatlán A-8
Cintalapa G-9
Ciudad Acuña G-2
Ciudad Alemán E-8
Ciudad Altamirano C-8
Ciudad Camargo E-3
Ciudad del Carmen H-8
Ciudad del Maiz D-6
Ciudad Gúzman B-8
Ciudad Hidalgo (Chis.) G-10
Ciudad Hidalgo (Mich.) C-8
Ciudad Juárez D-1
Ciudad Madero E-6
Ciudad Mante H-5
Ciudad Mier H-3
Ciudad Miguel Alemán H-3
Ciudad Obregón B-3
Ciudad Pemex G-9
Ciudad Serdán E-8
Ciudad Valles D-7
Ciudad Victoria H-5
Clavería 116 C-5
Coahuayana B-8
Coahuyutla C-8
Coalcomán B-8
Coatepec E-8
Coatlán E-9
Coatzacoalcos F-8
Cobá H-6
Cocula B-7
Cojumatlán B-7
Coixtlahuaca E-9
Colima B-8
Colonia Juárez C-1
Colonia Progreso F-5
Colonia Unesco G-9
Colonia Yucatan H-6
Colotlán B-6
Comalcalco G-8
Comanjilla C-7
Comitán G-9
Comonfort C-7
Compostela A-7
Concepción del Oro G-4
Concordia E-5
Copala E-5
Copatillo D-9
Córdoba E-8
Cortazar C-7
Cosalá D-4
Cosamaloapan E-8
Coscomatepec E-8
Cosmopolita 116 C-5
Cotija B-8
Coyame E-2
Coyotitán D-5
Coyuca de Benitez D-9
Coyuca de Catalan C-8
Cozón A-1
Cozumel H-6
Cristo Rey 116 B-7
Cruillas H-4
Cuajinicuilapa D-9
Cuatepin D-9
Cuatrociénegas F-3
Cuauhtémoc (Chi.) D-2
Cuauhtémoc (Chis.) H-10
Cuauhtemoc (D.F.) 116 C-6
Cuautémoc F-9
Cuautitlán D-8
Cuautla D-8
Cuchillo Parado E-2
Cuencamé F-4
Cuernavaca D-8
Cuicatlán E-9
Cuitzeo C-7
Cuitzeo de Hidalgo (Abasolo) C-7
Culiacán D-4
Cumpas C-1
Cunduacán G-8
Cuquio B-7
Cutzamala C-8
Cuyo H-6
Cuyoaco E-8
Cruillas H-4
Delicias E-2
Desemboque A-1
Diez de Octubre E-4
Dimas D-5
Dinamita F-4
Doctor Arroyo G-5
Doctor González H-4
Dolores Hidalgo C-7
Dorado D-4
Dos Aguas B-8
Durango E-4
Dzibalchén G-7
Dzilam Bravo G-6
Dzitás G-6
Ebano D-6
Edzná G-7
Ejido Insurgentes B-5
Ejido Río Frío C-9
Ejutla E-9
El Burrión C-4
El Capomo A-7
El Carmen G-5
El Casco E-4
El Fuerte C-3
El Grullo A-8
El Guaje F-3
El Largo C-2
El Manguito G-10
El Medano B-5
El Naranjo D-6
El Oasis B-2
El Oro C-8
Elota D-5
El Palmito E-4
El Pescadito C-7
El Reves F-3
El Rucio F-5
El Salto E-5
El Sifón 116 E-8
El Sueco D-2
El Temascal J-4
EL Tomaseño H-4
El Triunfo H-9
El Tule A-7
El Vergel D-3
El Zape E-4
El Zopilote A-7
Empalme B-2
Encarnación de Diaz B-7
Enramadas G-3
Ensenada A-3
Ent. F.C. del Sureste G-9
Escárgega H-8
Escuadrón 201 116 E-8
Escuinapa E-5
Escuintla G-10
Esperanza C-3
Espita H-6
Estación Creel D-3
Estación Don C-3
Estación Guzmán D-1
Estación Loreto C-3
Estación Sahuaro A-1
Estación San Juanito D-3
Estación Tefillas F-5
Estrella 116 E-5
Etchoropo C-3
Etla E-9
Etzatlán B-7
Eugenia 116 C-7
Federal 116 E-6
Felipe Cerrillo Puerto G-7
Fernandeño H-4
Filisola F-9
Flores Magón D-2
Fortín de las Flores E-8
Francisco I. Madeo 116 E-6
Francisco I. Madero (Coah.) F-4
Francisco I. Madero (Dgo.) E-4
Fresnillo F-5
Frontera G-8
Fronteras C-1
Fuente G-2
Galeana G-4
Gallego D-2
Garcia de la Cadena B-7
General Bravo H-4
General Cepeda G-4
General Simón Bolivar F-4
General Terán H-4
General Treviño H-3
General Trias D-2
Gertrudis Sanchez 116 E-5
Gogorrón C-7
Gómez Farias (Coah.) G-4
Gómez Farias (Tam.) H-5
Gómez Palacio F-4
Gonzáles D-6
Gonzáles Ortega F-4
Guachochic D-3
Guadalajara B-7
Guadalupe (Coah.) G-3
Guadalupe (Dgo.) E-3
Guadalupe (Zac.) F-5
Guadalupe Aguilera E-4
Guadalupe Bravos D-1
Guadalupe de la Joya H-4
Guadalupe Victoria E-4
Guamúchil C-4
Guanacevi D-4
Guanajuato C-7
Guasave C-4
Guaymas B-3
Guazarachic D-3
Guémez H-5
Guerrero (Chi.) D-2
Guerrero (Coah.) G-2
Guerrero (D.F.) 116 D-6
Guerrero (Zac.) F-5
Guerrero Negro B-4
Guevea F-9
Guimbalete F-3
Gutierrez Zamora E-7
Halachó G-6
Hampolol F-6
Hermanas G-3
Hermosillo B-2
Heroes de Churabusco 116 E-8
Hidalgo (N. L.) G-4
Hidalgo (Tam.) H-4
Hildago G-3
Hoctun G-6
Hopelchén G-6
Huajuapan de León E-9
Hualahuises H-4
Huamantla E-8
Huamuxtitlán D-9
Huandacareo C-8
Huatabampo C-3
Huatla E-8
Huatusco E-8
Huauchinango E-7
Huautla (Mor.) D-8
Huehuetoca D-8
Huejucar B-6
Huejuquilla B-6
Huejutla D-7
Huertecillas G-4
Huetamo C-8
Huichapan D-7
Huimanguillo G-9
Huitusi C-4
Huitzuco D-8
Huivulai B-3
Huixtla G-10
Huizache G-5
Hunucmá G-6
Icaiché G-7
Ignacio Allende F-4
Ignacio de la Llave F-8
Ignacio Zaragoza (Chi.) D-2
Ignacio Zaragoza (Tam.) D-6
Iguala D-8
Imuris B-1
Indé E-4
Industrial 116 D-5
Infiernillo B-8
Inguarán 116 E-5
Irapuato C-7
Isla F-9
Isla Aguada H-8
Iturbide H-4
Ixcamilpan D-8
Ixcaquixtla E-8
Ixcateopan D-8
Ixmiquilpan D-7
Ixtapa C-9
Ixtapan de la Sal D-8
Ixtapan del Oro C-8
Ixtepec F-9
Ixtlán B-7
Ixtlán del Rió A-7
Izamal G-6
Iztacalco 116 E-7
Iztapalapa 116 F-8
Izúcar de Matamoros D-8
Jacala D-7
Jacona B-8
Jala A-7
Jalapa (Tab.) G-9
Jalapa (Ver.) E-8
Jalcocolán A-7
Jalostotitlán B-7
Jalpa B-7
Jalpa de Méndez G-8
Jalpan D-7
Jaltenango G-10
Jaltipan F-8
Jamiltepec D-9
Janos C-1
Jaumave H-5
Jeréz F-5
Jesús Carranza F-9
Jiménez (Chi.) E-3
Jimenez (Coah.) G-2
Jiménez (Tam.) H-4
Jiménez del Teul F-5
Jimulco F-4
Jiquilpan B-8
Jocotepec B-7
Jojutla D-8
Jonuta G-8
Juan Aldama F-4
Juárez (Coah.) G-3
Juárez (N.L.) G-4
Juchatengo E-9
Juchipila B-7
Juchitán F-9
Juchitlán A-7
Julimes E-2
Justo Sierra 116 D-8
Juventino Rosas C-7
Kantunil G-6
Kantunil Kin H-6
Kinchil G-6
La Ascención (Chi.) C-1
La Ascensión (N.L.) H-4
La Barca B-7
La Barra E-7
La Bufa D-3
La Carbonera J-4
La Colorada B-2
La Coma H-4
La Cruz (Dgo.) A-6
La Cruz (Sin.) D-5
La Cuchilla F-4
La Cuesta F-2
La Gloria G-3
Lagos de Moreno C-7
Laguna Chapala A-4
Lagunas de Montebello H-9
Lagunillas D-7
La Huacana C-8
La Huerta A-8
La Isla D-1
La Laja E-7
La Loba H-4
Lampazos G-3
La Norteña C-2
La Ochoa F-4
La Paz C-5
La Perla E-2
La Pesca J-5
La Piedad C-7
La Placita B-8
La Poza B-4
La Rosa G-4
La Rosita G-2
Las Adjuntas E-5
Las Choapas F-9
Las Eutimias F-2
Las Margaritas H-9
Las Nieves E-3
Las Palmas 116 B-7
Las Palomas D-1
Las Rosas G-9
Las Trojes B-8
Las Varas A-7
La Trinitaria G-9
La Unión (Gro.) C-9
La Unión (Q.R.) G-7
Laurel B-6
La Ventura G-4
La Vibora F-3
La Zarca E-4
Lázaro Cárdenas B-9
Legaria 116 B-5
León C-7
León Fonseca C-4
Lerdo F-4
Liberación 116 C-5
Libra Unión G-6
Libres E-8
Liera H-5
Linares H-4
Llano Enmedio D-7
Llera H-5
Loma Bonita F-8
Loma de Plateros 116 B-8
Lomas Altas 116 A-7
Lomas Chapultepec 116 A-6
Lomas del Real E-6
Lomas de Sotelo 116 B-5
Lomas Hipodromo 116 A-6
Loreto (B.C.S.) B-4
Loreto (Zac.) C-6
Los Aldamas H-3
Los Azufres C-8
Los Corchos A-7
Los Mochis C-4
Los Muchachos G-4
Los Reyes B-8
Lourdes C-7
Luis Moya B-6
Macuspana G-9
Madera D-2
Magdalena B-1
Magdalena Mixhuca 116 D-7
Malpaso G-9
Mamantel H-8
Mano Marques E-9
Manuel D-6
Manzanillo A-8
Mapastepec G-10
Mapimi F-4
Maravatio C-8
Mariscala D-9
Martinez de la Torre E-7
Mascota A-7
Matamoros (Camp.) F-7
Matamoros (Coah.) F-4
Matamoros (Tam.) J-4
Matatan E-5
Matehuala G-5
Matias Romero F-9
Maxcaltzin D-6
Mazcanú G-6
Mazatán C-2
Mazatlán D-5
Medios Rios E-9
Méndez H-4
Meoqui E-2
Merida G-6
Mesa de Palotes C-7
Metepec D-7
Metzquititlán D-7
Metztitlán D-7
Mexcala D-9
Mexicali A-3
Mexicaltzingo 116 D-8
Mexico D-8
Mezquital E-5
Miahuatlán E-9
Miguel Auza F-4
Milpillas C-3
Minatitlán (Col.) A-8
Minatitlán (Ver.) F-9
Miwuihuana H-5
Misantla E-8
Mitla E-9
Mixcoac 116 B-8
Mocché H-6
Mocorito D-4
Moctezuma (D.F.) 116 E-6
Moctezuma (S.L.P.) G-5
Moctezuma (Son.) C-2
Moderna 116 D-7
Molango D-7
Monclova G-3
Monte Escobedo B-6
Montemorelos G-4
Montepio F-8
Monterrey G-4
Morelia C-8
Morelos (Coah.) G-2
Morelos (Sonora) B-2
Moroleón C-7
Motozintla G-10
Motul G-6
Moyahua B-7
Mulato F-9
Mulegé B-4
Muna G-6
Múzquiz G-3
Nabalam H-6
Nacimiento G-2
Naco B-1
Nacori Chico C-2
Nacozari C-1
Namiquipa D-2
Napoles 116 C-7
Naranjos E-7
Narvarte 116 D-7
Nátare C-2
Naucalpán 116 A-5
Nautla E-7
Nava G-2
Navojoa C-3
Navolato D-4
Nazas E-4
Necaxa E-7
Nejapa F-9
Nieves F-4
Niltepec F-9
Noche Buena 116 C-8
Nochistlán B-7
Nochixtlán E-9
Nocupetaro C-8
Nogales B-1
Nombre de Dios E-5
Nopala D-7
Novillero A-6
Novoava D-3
Nuariche G-4
Nueva Coahuila F-7
Nueva Colonia Agricola 116 E-7
Nueva Delicias D-2
Nueva Italia B-8
Nueva Reforma G-3
Nueva Rosita G-3
Nueva Santa María 116 C-5
Nuevas Tapias F-5
Nuevo Casas Grandes C-1
Nuevo Churumuco C-8
Nuevo Guerrero H-3
Nuevo Laredo H-3
Nuevo Morelos D-6
Nuevo Padilla H-4
Oaxaca E-9
Oballos G-3
Ocampo (Coah.) F-3
Ocampo (Gto.) C-7
Ocampo (Tam.) H-5
Oconi C-4
Ocosingo H-9
Ocotlán (Jal.) B-7
Ocotlán (Oax.) E-9
Ocozocoautla G-9
Octepec E-9
Ojinaga E-2
Ojocaliente B-6
Ojo Caliente (Ags.) B-7
Ojo Caliente (S.L.P.) C-7
Ojuelos C-7
Olinala D-9
Omealca E-8
Ometepec D-9
Opodepe B-2
Oriental E-8
Orizaba E-8
Ortiz B-2
Ostuacan G-9
Oxkutzcab G-6
Oxolotlán G-9
Ozuluama E-7
Pachuca D-7
Paila F-4
Pajapan F-8
Palau G-3
Palenque H-9
Palizada G-8
Palma de Cuautla A-6
Palmillas H-5
Panabá H-6
Pánuco D-6
Papaloapan F-9
Papantla E-7
Paracho B-8
Parácuaro B-8
Parás H-3
Paredón G-9
Parral E-3
Parras F-4
Parrita D-2
Paeo de Toro C-8
Pátzcuaro C-8
Pedriceña F-4
Penjamo C-7
Peñón Blanco E-4
Peñon de los Baños 116 F-6
Pensador Méxicano 116 E-5
Pensil 116 B-5
Peribán B-8
Pericos D-4
Perote E-8
Petatlán C-9
Petcacab H-7
Petlalcingo D-8
Peto G-6
Pich G-7
Pichucalco G-9
Piedras Negras (Coah.) G-2
Piedras Negras (Ver.) E-8
Piedritas F-2
Pihuamo B-8
Pijijiapan G-10
Pinotepa Nacional E-9
Pisté G-6
Pitiquito B-1
Placer H-7
Plateros F-5
Playa Azul B-8
Playa del Carmen H-6
Playa Vicente F-9
Plenitud 116 B-5
P. Natera G-5
Pochutla E-10
Polanco 116 B-6
Poncitlán B-7
Portezuela B-7
Porvenir D-1
Postal 116 D-7
Potam B-3
Pótrero de Gallegos B-6
Poza Rica E-7
Pozo de Higueras A-6
Pozos C-7
Presa del Baño A-7
Progreso (Coah.) G-3
Progreso (Yuc.) G-6
Pucté G-7
Puebla D-8
Puerto Angel E-10
Puerto Arista G-10
Puerto Arturo G-6
Puerto Ceiba G-8
Puerto Escondido E-10
Puerto Juárez H-6
Puerto Libertad A-2
Puerto Madero G-10
Puerto Marquéz C-9
Puerto Morelos H-6
Puerto Peñasco B-3
Puerto Real H-8
Puerto Vallarta A-7
Punta Abreojos B-4
Punta Allen H-6
Purépero C-8
Purificación A-8
Puruándiro C-7
Puruarán C-8
Pustunich G-7
Putla E-9
Queretaro C-7
Quila D-4
Quiriego C-3
Quiroga C-8
Quiviquinta E-5
Ramirez E-8
Ramos Arizpe G-4
Rancho Bizani A-1
Rancho Viejo D-7
Rayón (S.L.P.) D-7
Rayón (Son.) B-2
Reforma (Q.R.) G-7
Reforma (Sin.) D-4
Reforma Polanko 116 B-6
Reynosa H-3
Rincón de Romos B-6
Río Blanco 116 D-5
Río Bravo H-4
Río Grande F-5
Río Lagartos H-6
Río Verde C-7
Rodeo E-4
Roma 116 C-6
Romero Rudio 116 E-6
Romita C-7
Rosamorada A-6
Rosario (B.C.N.) A-3
Rosario (Sin.) E-5
Rosario de Tezopaco C-3
Rosarito B-4
Ruiz A-7
Sabancuy H-8
Sabinas G-3
Sabinas Hidalgo G-3
Sacramento G-3
Sahuaripa C-2
Sahuayo B-7
Sain Alto F-5
Salamanca C-7
Salina Cruz F-9
Salinas G-5
Salinas Guajardo J-4
Salinas Victoria G-4
Saltillo G-4
Salvatierra C-7
Samachic D-3
Samalayuca D-1
San Alvaro 116 C-5
San Andrés Tepepilco 116 D-8
San Andrés Tuxtla F-8
San Antonio 116 E-8
San Antonio de Campos F-6
San Augustin B-2
San Blas (Coah.) G-3
San Blas (Nay.) A-7
San Blas (Sin.) C-3
San Buenaventura G-3
San Carlos H-4
Sánchez Magallanes F-8
San Ciro de Acosta D-7
San Cristóbal de las Casas G-9
Sandia G-4
San Dimas E-4
San Diego C-7
San Esteban 116 A-5
San Felipe (B.C.N.) A-3
San Felipe (Gto.) C-7
San Felipe (Yuc.) H-6
San Fernando H-4
San Francisco del Oro D-3
San Fco. del Rincón B-7
San Gabriel Chilac E-8
San Ignacio (B.C.S.) B-4
San Ignacio (Sin.) D-5
San Juan de Aragón 116 F-5
San José de Llanates F-5
San José del Cabo C-5
San José Iturbide C-7
San José Purúa C-8
San Juan Capistrano A-6
San Juan de Guadalupe F-4
San Juan de Lima B-8
San Juan de los Lagos B-7
San Juan del Río (Dgo.) E-4
San Juan del Río (Gto.) C-7
San Juan Evangelista F-9
San Lorenzo F-4
San Luis B-3
San Luis Acatlán D-9
San Luis de la Paz C-7
San Luis del Cordero E-4
San Luis Potosí C-6
San Marcos (Gro.) D-9
San Marcos (Jal.) A-7
San Martín Chal. D-7
San Martin Hidalgo A-7
San Martin Texmelucán D-8
San Miguel F-2
San Miguel Allende C-7
San Miguel de Truces E-4
San Miguel el Alto B-7
San Miguel Regla D-7
San Miguel Zapotitlan C-4
San Pablo Huixtepec E-9
San Pedro G-8
San Pedro de la Canada C-7
San Pedro de las Colonias F-4
San Pedro del Gallo E-4
San Pedro de los Pinos 116 B-7
San Quintin A-3
San Rafael (Chi.) C-3
San Rafael (D.F.) 116 C-6
San Rafael (N.L.) G-4
San Rafael(Son.) B-2

San Roberto G-4
San Salvador el Seco .E-8
San Sebastián A-7
Santa Ana (Mex.) B-1
Santa Ana (Sonora) ..B-1
Santa Bárbara D-3
Santa Catarina C-6
Santa Clara (Dgo.) ...C-4
Santa Clara (Yuc.) ...G-6
Santa Cruz
Santa Cruz (Nay.) A-7
Santa Elena G-4
Santa Engracia H-5
Santa Fe ....116 .....A-8
Santa Julia ....116 ..C-6
Santa Lucía
  (D.F.) ....116 .....B-5
Santa Lucía (Zac.) ...A-6
Santa María
  (D.F.) ....116 .....B-6
Santa María (Dgo.) ...E-3
Santa María (Sin.) ...D-3
Santa María
  Camarones ....116 ..E-8
Santa María del Río ..C-7
Santa María Zaniza ...E-9
Santa Rosa A-6
Santa Rosa Juarequi ..C-7
Santa Rosalía B-2
Santa Teresa (Nay.) ..A-6
Santa Teresa (Tam.) ..H-4
Santiago (B.C.S.) C-5
Santiago (N.L.) G-4
Santiago Astata F-9
Santiago de la Peña ..E-7
Santiago Ixcuintla ...A-7
Santiago Juxtlahuaca .D-9
Santiago Papasquiaro .C-3
Santiago Tuxtla F-8
Santo Domingo G-4
Santo Domingo
  (S.L.P.) G-5
Santuario C-6
San Vicente (B.C.N.) .A-3
San Vicente (Tam.) ..H-5
Saric B-1
Sasabe A-1
Sauceda G-4
Saucillo E-2
Sauta A-7
Sayula (Jal.) B-7
Sayula (Ver.) F-9
Sayulita A-7
Sector Popular 116 ..E-8
Seguro Social ..116 ..H-6
Señor H-6
Seybaplaya H-8
Sierra Mojada D-3
Sihochac F-7
Silao C-7
Silvituc C-9
Simojovel G-9
Sinaloa D-2
Siqueiros D-5
Sisal G-6
Sola de Vega E-9
Soledad A-1
Soledad Diez
  Gutiérrez C-6
Sombrerete F-5
Sonoita A-1
Soteapan F-9
Soto la Marina H-5
Suchiapa G-9
Suchil G-7
Sumidero C-9
S. Urbina G-10
Symón F-4

Tacámbaro C-8
Tacuaba ....116 ....B-7
Tala B-7
Talismán J-4
Talpa de Allende ...A-7
Tamazula B-8
Tamazulapan E-9
Tamazunchale D-7
Tamiahua D-7
Tampak D-7
Tampico E-6
Tamuin D-7
Tancuayalab D-7
Tanque de
  Guadalupe G-4
Tanquián D-7
Tantoyuca D-7
Tapachula G-10
Tapanatepec F-9
Tasquillo D-7
Taxco D-8
Teacapán A-6
Teapa G-9
Tecalitlán B-8
Tecamachalco E-8
Tecolotlán A-7
Tecomán A-8
Tecoxautla E-9
Tecozautla D-7
Técpan C-9
Tecpatán G-9
Tecuala A-6
Tecuan A-7
Tehuacán E-8
Tehuantepec F-9
Tehuitzingo D-8
Tejupilco D-8
Tekax G-6
Tekit G-6
Telchac G-6
Teloloapan D-8
Temazcal F-9
Tempoal D-7
Tenacatita A-8
Tenango (Dgo.) D-8
Tenango (Hgo.) D-7
Tenango (Mex.) D-8
Tenosique H-9
Teocaltiche B-7
Teocelo E-8
Teocuitatlán B-7
Teopisca G-9
Tepatitlán B-7
Tepeaca E-8
Tepeguaje J-5
Tepehuanes E-4
Tepeji del Río D-8
Tepetonge B-6
Tepic A-7
Tequila B-7
Tequisistlán F-9
Tequisquiapan D-7
Tetela E-8
Texcoco D-8
Teziutlán E-8
Tezontepec D-7
Ticha A-7
Ticopo G-6
Ticul G-6
Tierra Blanca F-8
Tierra Colorada D-9
Tihuatlán E-7
Tijuana A-3
Tingambato C-8
Tiquicheo C-8
Tixtla D-9
Tizapán B-7

Tizayuca D-8
Tizimín H-6
Tlacoapa D-9
Tlacolula E-9
Tlacotepec E-9
Tlacotalpan F-8
Tlahualilo F-3
Tlalpan D-8
Tlaltizapán C-8
Tlaltenango C-7
Tlapa D-9
Tlapacoyan B-7
Tlaquepaque B-7
Tlaxcala D-8
Tlaxco D-8
Tlaxiaco E-9
Tlaxpala .....116 ..C-3
Tobarito C-2
Todos Santos C-4
Toluca D-8
Tomatlán A-8
Tomóchic C-2
Tonalá (Chis.) G-10
Tonalá (Jal.) B-7
Tonalá (Oax.) E-9
Tonila B-8
Topila D-7
Topolobampo C-4
Torreón E-3
Tototlán B-7
Totolapan E-9
Tres Estrellas 116 .E-5
Troncoso F-5
Tubutama B-1
Tuíta E-4
Tuito A-7
Tula (Hgo.) D-7
Tula (Tam.) H-5
Tulancingo D-7
Tultepec C-8
Tultitlan C-8
Tutupan E-4
Tutupec E-9
Tuxpan (Jal.) B-8
Tuxpan (Mich.) C-8
Tuxpan (Nay.) A-7
Tuxpan (Ver.) E-7
Tuxtepec E-8
Tuxtla Chico G-10
Tuxtla Gutiérrez ...G-9
Tuzantla C-8
Tzintzuntzan C-8
Tzucacab G-6
Umán G-6
Unión de Tula A-8
Unión Hidalgo F-9
Ures B-2
Uriangato C-8
Uruáchic C-3
Uruapan C-8
Valladolid H-6
Vallecillo G-3
Valle de Banderas ..A-7
Valle de Bravo C-8
Valle de Santiago ..C-7
Valle de Zaragoza ..D-3
Valle Gomez ...116 .C-5
Valle Hermoso (Q.R.) G-7
Valle Hermoso (Tam.) J-4
Valle Nacional E-8
Valparaíso B-7
Vega de Alatorre ...E-7
Veinte de
  Noviembre ...116 .E-6
Venado C-6
Venustiano Carranza G-9
Veracruz E-8
Vetagrande B-7
Vicam B-3
Vicente Guerrero
  (Cam.) G-7
Victor Rosales F-5
Viesca F-4
Vigia Chico H-6
Villa Ahumada D-1
Villa Aldama G-3
Villa Azueta F-9
Villa Carranza B-8
Villa Coronado C-3
Villa de Cos F-5
Villa Flores G-9
Villa Frontera G-3
Villagrán H-4
Villa Guerrero B-8
Villahermosa G-9
Villa Hidalgo (Dgo.) B-6
Villa Hidalgo (S.L.P.) D-7
Villa Juárez C-5
Villa Madero
  (D.F.) ......116 ..E-5
Villa Madero (Mich.) C-8
Villa Matamoros C-3
Villanueva B-6
Villa Ocampo C-3
Villa Unión (Coah.) .G-2
Villa Unión (Dgo.) .E-5
Villa Unión (Sin.) .D-5
Villa Victoria B-8
Virreyes .....116 ..B-8
Vista Hermosa de
  Neg. B-7
Volcán Ceboruco ...A-7

Walamo D-5

Xaltianguis D-9
X-cán H-6
Xiatil H-6
Xichu C-7
Xicoténcatl E-7
Xicotepec E-7
Xilitla D-7
Xmaben H-7
Xpujil G-7
Xul G-7

Yahualica B-7
Yanhuitlán E-9
Yautepec D-8
Yecora C-2
Yecuatla E-7
Yepáchic C-2
Yurécuaro B-7
Yuriria C-7

Zaachila E-9
Zacapoaxtla E-8
Zacapú C-8
Zacatal G-8
Zacatecas F-5
Zacatelco D-8
Zacatepec Mixes F-9
Zacatlán E-8
Zacatula C-8
Zacualtipán D-7
Zamora B-8
Zanatepec F-9
Zapopan B-7
Zapotiltic B-8
Zapotlán B-7
Zaragoza (Chi.) C-8
Zaragoza (Coah.) ...G-2
Zaragoza (Pue.) E-8
Zavalza D-7
Zihuatanejo C-9
Zimapán D-7
Zimatlán E-9
Zongolica E-8
Zumpango D-8
Zumpango del Río ...D-9

CANADA
Maps on pages 62-69

Names shown thus: York....114   C-6
refer to city inset on page 114 with
grid reference location at C-6

ALBERTA
Map on page 63

Alix K-3
Banff J-4
Bassano J-4
Beaverlodge H-5
Blairmore J-5
Bonnyville L-2
Bow Island K-5
Brooks K-4
Calgary J-4
Camrose K-3
Cardston K-5
Carway K-5
Castor K-3
Claresholm K-4
Consort L-3
Coronation L-3
Coutts L-5
Drumheller K-4
Edmonton J-3
Elk Point L-3
Entwistle J-3
Fairview H-2
Fort Macleod K-5
Fort Saskatchewan ..K-2
Grande Centre L-2
Grande Prairie H-2
Grimshaw H-2
Hanna K-4
Hardisty L-3
High Prairie H-2
High River J-4
Hinton J-3
Innisfail J-3
Jasper H-3
Killam K-3
Lacombe K-3
Lake Louise J-4
Leduc K-3
Legal K-3
Lethbridge K-5
Lloydminster L-3
Magrath K-5
Manning H-1
Mayerthorpe J-3
McLennan H-2
Medicine Hat L-4
Milk River K-5
Mundare K-3
Nanton K-4
Nordegg J-4
Olds J-4
Oyen L-4
Peace River H-1
Pincher Creek K-5
Provost L-3
Redcliff L-4
Red Deer K-3
Redwater K-3
Rimbey J-3
Rocky Mountain
  House J-3
Rycroft G-2
St. Paul L-2
Smoky Lake K-2
Stettler K-3
Stirling K-5
Stony Plain K-3
Strathmore J-4
Suffield L-4
Sylvan Lake J-3
Taber K-5
Two Hills L-3
Valleyview H-2
Vegreville K-3
Vermilion L-3
Viking K-3
Wainwright L-3
Waterton Park J-5
Westlock J-2
Wetaskiwin K-3
Whitecourt J-2
Wild Horse L-5
Winfield J-3
Youngstown L-4

BRITISH COLUMBIA
Map on pages 62, 63

Abbotsford E-5
Alert Bay C-4
Alexis Creek E-3
Anahim Lake D-3
Armstrong F-4
Ashcroft F-4
Balfour H-5
Barkerville F-3
Beatton River F-2
Belcarra ...115 ...K-2
Bella Coola C-3
Blue River F-3
Boston Bar F-5
Boundary Bay .115 .L-3
Bralorne E-4
Buick F-1
Burns Lake D-2
Cache Creek F-4
Campbell River D-4
Castlegar H-5
Chase F-4
Chetwynd F-2
Chilliwack F-5
Clearwater F-4
Clinton F-4
Cloverdale ..115 ..M-4
Courtenay D-4
Cranbrook H-5
Crescent Beach 115 H-5
Creston H-5
Dawson Creek F-2
Deep Cove ..115 ...K-1
Duncan D-5
Dundarave ..115 ..H-1
Eagle Harbour 115 .G-2
East Pine F-2
Elko J-5
Fairmont Hot Springs J-4
Fauquier H-5
Fernie J-5
Field J-4
Fort Fraser D-2
Fort St. James D-2
Fort St. John F-1
Glacier H-4
Golden H-4
Grand Forks H-5
Granisle C-2
Greenwood H-5
Hazelton C-2
Hixon E-3
Hollyburn ..115 ...H-1
Hope F-5
Horseshoe Bay E-5

Houston C-2
Ioco ....115 ...F-4
Kamloops F-4
Kaslo H-5
Kelowna G-5
Kelsey Bay C-4
Keremeos G-5
Kimberley H-5
Kispiox B-2
Kitimat B-2
Kootenay Bay H-5
Ladner .....115 ...M-4
Ladysmith D-5
Langley ....115 ...M-4
Lillooet F-4
Louis Creek F-4
Lumby G-5
Lytton F-4
Mackenzie E-2
Marguerite E-3
McBride E-3
McLeod Lake E-3
Merritt F-4
Monte Creek F-4
Nakusp H-4
Nanaimo D-5
Needles H-5
Nelson H-5
New Denver H-4
New Hazelton C-2
Newton .....115 ...L-3
New
  Westminster 115 .K-3
North Vancouver ...K-2
Ocean Falls C-3
Ocean Park .115 ...L-3
Oliver G-5
100 Mile House E-4
Ootsa Lake D-3
Osoyoos G-5
Parksville D-5
Penticton G-5
Pitt Meadows .115 .M-2
Port Alberni C-4
Port Alice C-4
Port Coquitlam 115 M-3
Port Hammond 115 ..M-3
Port Kells .115 ...M-3
Port Moody .115 ..K-2
Powell River D-4
Prespatou F-1
Prince George E-3
Prince Rupert B-2
Princeton F-5
Quesnel E-3
Radium Hot Springs J-4
Red Pass E-3
Revelstoke G-4
Roosville J-5
Rossland H-5
Savona F-4
Salmon Arm G-4
70 Mile House F-4
Sherman ....115 ...H-1
Sicamous G-4
Sidney E-5
Slocan H-5
Smithers C-2
Smithers Landing ..C-2
Sparwood J-5
Spences Bridge F-4
Spillimacheen H-4
Squamish E-5
Steveston ..115 ..M-4
Stewart B-2
Summit Lake B-3
Swanson Bay C-3
Takla Lake D-2
Takla Landing D-2
Tatla Lake D-4
Telkwa C-2
Terrace B-2
Topley C-2
Trail H-5
Tsawwassen L-4
Ucluelet D-5
Valemount E-3
Vancouver E-5
Vanderhoof D-2
Vernon G-4
Victoria E-5
Wadsley ....115 ..H-1
Wells E-3
Whalley ...115 ...L-3
White Rock .115 ..L-3
Williams Lake E-3
Windermere H-4
Winlaw H-5
Wonowon F-1
Yahk H-5

MANITOBA
Map on page 65

Berens River J-2
Birch River G-2
Birtle G-3
Boissevain H-4
Bowsman G-2
Brandon H-4
Camperville G-3
Carberry H-4
Carman H-4
Cowan G-3
Cranberry Portage .G-1
Dauphin H-3
Elkhorn G-4
Elm Creek H-4
Emerson J-4
Erickson H-3
Ethelbert G-3
Flin Flon G-1
Fort Garry .115 ..L-2
Gilbert Plains G-3
Gimli J-3
Gladstone H-3
Grand Rapids H-2
Grandview G-3
Killarney H-4
Lac du Bonnet J-3
Little Grand Rapids J-2
Lundar H-3
MacGregor H-4
Manitou H-4
Melita G-4
Minitonas G-2
Minnedosa H-3
Morden H-4
Morris J-4
Neepawa H-3
Ninette H-4
Oak Lake G-4
Overflowing River .G-2
Pilot Mound H-4
Pine Falls J-3
Portage la Prairie H-3
Rennie K-4
Reston G-4
Riverton J-3
Roblin G-3
Rosser ....115 ...K-6
Russell G-3
St. James-
  Assiniboia ..115 K-6
St. Laurent H-3
St. Vital ..115 ..M-8
Selkirk J-3
Sherridon G-1
Shoal Lake G-3
Snow Lake G-1
Souris G-4
Steinbach J-4
Swan River G-2
The Pas G-2
Treherne H-4
Victoria Beach J-3
Virden G-4
Wasagaming H-3
West Kildonan 115 L-6
Winnipeg J-4
Winnipeg Beach J-3
Winnipegosis H-3

NEW BRUNSWICK
Map on page 68

Allardville E-3
Alma E-4
Bath D-3
Bathurst E-3
Blackville E-3
Buctouche E-3
Campbellton D-2
Cape Tormentine ...F-3
Caraquet E-3
Chatham E-3
Dalhousie D-2
Doaktown D-3
Edmundston C-3
Fredericton D-3
Grand Falls D-3
Hampton D-4
Hartland D-3
Longs Creek D-4
McAdam D-4
Moncton E-3
Newcastle E-3
North Head E-4
Oromocto D-4
Perth-Andover D-3
Petitcodiac E-4
Port Elgin F-3
Richibucto E-3
Sackville E-4
St. Andrews D-4
St. George D-4
Saint John E-4
St. Leonard D-3
St. Quentin D-3
St. Stephen D-4
Shediac E-3
Shippagan E-3
Sussex E-4
Thomaston Cor. D-4
Tracadie E-3
Wilsons Beach D-4
Woodstock D-3

NEWFOUNDLAND
Map on page 69

Argentia K-3
Badger J-2
Baie Verte J-2
Bay Bulls L-3
Bay de Verde L-2
Belleoram J-3
Bishop's Falls K-2
Bonavista L-2
Botwood K-2
Buchans J-2
Burgeo J-3
Burin K-3
Carbonear L-3
Catalina L-2
Channel-Port aux
  Basques H-3
Clarenville L-2
Colinet L-3
Corner Brook J-2
Deer Lake J-2
Ferryland L-3
Forteau J-1
Gambo K-2
Gander K-2
Garnish K-3
Glovertown K-2
Goobies L-3
Grand Bank K-3
Grand Falls K-2
Harbour Breton K-3
Heart's Content L-3
Heart's Delight L-3
Holyrood L-3
Lewisporte K-2
Marystown K-3
Musgravetown L-2
Placentia K-3
Port Blandford L-2
Port Saunders J-1
Pouch Cove L-3
Roddickton K-1
St. Albans K-3
St. Anthony K-1
St. Lawrence K-3
St. Paul's J-2
Spaniard's Bay L-3
Springdale K-2
Stephenville J-2
Terrenceville K-3
Torbay L-3
Trepassey L-3
Trinity L-2
Trout River J-2
Wesleyville K-2

NOVA SCOTIA
Maps on pages 68, 69

Amherst E-4
Annapolis Royal ...E-4
Antigonish G-4
Baddeck G-3
Bass River F-4
Bridgetown E-4
Bridgewater E-4
Caledonia E-4
Canso G-3
Cape North G-3
Caribou F-4
Chester E-4
Chéticamp G-3
Clark's Harbour ...E-5
Dartmouth F-4
Digby E-4
Enfield F-4
Glace Bay H-3
Guysborough G-4
Halifax F-4
Ingonish Beach G-3
Inverness G-3
Kentville E-4
Liverpool E-4
Lockeport E-5
Louisbourg H-3
Lower Wedgeport ...D-5
Lunenburg E-4
Mahone Bay E-4
Margaree Forks G-3
Middleton E-4
Mulgrave G-4
Musquodoboit
  Harbour F-4
New Germany E-4
New Glasgow F-4
North Sydney G-3
Parrsboro E-4
Peggy's Cove F-4
Pictou F-4
Port Hawkesbury ...G-4
Port Hood G-3
Port Maitland D-4
Salmon River D-4
Saulnierville D-4
Sheet Harbour F-4
Shelburne E-5
Sherbrooke F-4
Springhill E-4
Stewiacke F-4
Sydney G-3
Tatamagouche F-4
Truro F-4
Weymouth E-4
Windsor E-4
Wolfville E-4
Yarmouth D-5

ONTARIO
Maps on pages 66, 67
•=See map on page 65

Algonquin Park G-4
Almonte H-4
Alta Vista .113 ..L-8
Arnprior H-4
Arthur F-4
Atikokan A-2
Bala G-4
Bancroft H-4
Barry's Bay H-4
Batchawana Bay D-3
Beardmore C-2
Beaverton G-4
Belleville H-4
Black Creek ..76 .D-7
Blenheim F-5
Blind River E-3
Bowmanville G-4
Bracebridge G-4
Brampton F-5
Brantford F-5
Brockville J-4
Burks Falls G-4
Burlington G-5
Byng Inlet F-3
Cambridge F-5
Capreol F-3
Cardinal
  Heights ...113 ..M-7
Carleton Place 113 L-8
Carlington .113 ..L-8
Carlingwood 113 ..L-7
Casummit Lake ....•L-3
Chapleau D-2
Chatham E-5
Chatsworth F-4
Cherrywood
  Acres .....113 ..H-7
Chippawa ...76 ...C-5
Clairville ..114 ..C-4
Clinton F-5
Cobden H-4
Cobocook G-4
Cobourg G-5
Cochrane E-2
Collingwood F-4
Combermere H-4
Confederation
  Heights ...113 ..L-8
Cornwall J-4
Corwin Crescent 113 D-8
Crescent Park 76 ..D-8
Crystal Beach 76 ..B-8
Cyrville ...113 ..M-7
Deep River H-3
Deux Rivières G-3
Dinorwic A-1
Drummond
  Heights ...113 ..H-8
Dryden ....•L-4
Dunnville G-5
Durham F-4
Dwight G-4
Eastway M-8
East York ..114 ..L-7
Eganville H-4
Elginfield F-5
Elliot Lake E-3
Elmvale Acres 113 M-8
Englehart E-2
English River A-2
Erie Beach .76 ...D-8
Etobicoke ..114 ..C-4
Falls View .113 ..H-8
Favourable Lake ..•L-2
Fenelon Falls G-4
Fergus F-5
Foleyet E-2
Fort Erie ..76 ...D-8
Fort Frances .....•L-4
French River F-3
Gananoque H-4
Geraldton C-2
Glebe ......113 ..L-8
Goderich F-5
Goldpines ....•L-3
Gore Bay E-4
Gravenhurst G-4
Greens Corners 113 G-7
Guelph F-5
Hamilton G-5
Havelock H-4
Hawkesbury J-4
Hawthorne
  Meadows ...113 ..M-8

Hearst E-2
Highland Park. 113 K-8
Hornepayne E-2
Huntsville G-4
Ignace A-2
Iroquois Falls E-2
Island Falls E-2
Jellicoe C-2
Kakabeka Falls B-2
Kaladar H-4
Kapuskasing E-2
Keewatin K-4
Kemptville J-4
Kenora ....•K-4
Kincardine F-4
Kingston H-4
Kirkland Lake E-2
Kitchener F-5
Lakefield G-4
Lancaster J-4
Larder Lake E-2
Laurentian
  View ....113 ..K-8
Leamington E-5
Lindsay G-4
Listowel F-4
Little Current F-4
London F-5
Longlac C-2
Madoc H-4
Manitouwadge D-2
Marathon C-2
Markdale F-4
Marmora H-4
Matachewan E-2
Matheson E-2
Mattawa G-3
Maynooth H-4
Meaford F-4
Midland G-4
Milton G-5
Minaki ....•K-4
Minden G-4
Mississauga G-5
Mitchell F-5
Morrisburg J-4
Napanee H-4
Nestor Falls .....•L-4
Newcastle G-4
New Edinburgh 113 L-7
New Liskeard E-2
Niagara Falls G-5
Nipigon B-2
North Bay G-3
North York .114 ..D-5
Oakville G-5
Onaping F-1
Orangeville F-4
Orillia G-4
Oshawa G-5
Ottawa J-4
Ottawa South 113 .L-8
Ottawa West .113 .L-7
Overbrook ...113 .L-7
Owensound F-4
Parry Sound G-4
Pembroke H-3
Penetanguishene ...F-4
Perth H-4
Peterborough G-4
Picton H-4
Port Burwell F-5
Port Elgin F-4
Port Hope G-5
Port Perry G-4
Queensway
  Gardens ...113 ..G-7
Rainy River ......•K-4
Red Lake ....•K-3
Renfrew H-4
Rideau Gardens 113 L-8
Rideau Park .113 ..L-8
Ridgeway ...76 ...E-7
Rockland J-4
Rolphton H-3
Round Lake H-4
St. Catherines G-5
St. Marys F-5
St. Thomas F-5
Sandy Hill .113 ..L-7
Sandy Lake ....•L-2
Sarnia E-5
Sault Ste. Marie ..D-3
Schreiber C-2
Shabaqua A-2
Shelburne F-4
Simcoe F-5
Sioux Lookout A-1
Sioux Narrows ...•K-4
Smiths Falls H-4
Smoky Falls E-1
Smooth Rock Falls .E-1
Southampton F-4
South Baymouth ...E-4
Stamford ...113 ..G-6
Stamford
  Centre .....113 ..H-7
Stirling H-4
Stratford F-5
Strathroy F-5
Sturgeon Falls G-3
Sudbury F-3
Sunderland G-4
Sundridge G-4
Terrace Bay C-2
Thamesville E-5
Thessalon D-3
Thunder Bay B-2
Tilbury E-5
Tillsonburg F-5
Timmins E-2
Tobermory F-4
Toronto G-5
Trenton H-4
Tweed H-4
Upsala A-2
Urbandale
  Acres .....113 ..M-8
Vanier .....113 ..L-7
Vermilion Bay ....•K-4
Wallaceburg E-5
Wawa D-2
Welland G-5
Wellington H-5
Whitby G-5
Whitney G-3
Wiarton F-4
Winchester J-4
Windsor E-5
Wingham F-4
Woodstock F-5
York ....114 ..C-6

PRINCE EDWARD ISLAND
Maps on pages 68, 69

Alberton F-3
Bedford F-3
Borden F-3
Charlottetown F-3
Kensington F-3
Montague F-3
N. Rustico F-3
Souris F-3
Summerside F-3
Tignish F-3

QUÉBEC
Map pages 68, 69
•=See map on page 67

Alma B-2
Amos ....•H-2
Amqui C-3
Andréville C-3
Anse-Pleureuse ...C-2
Baie Comeau B-2
Baie St.-Paul B-3
Baie Trinité B-2
Beaconsfield 113 ..H-6
Beattyville ......•H-2
Beauharnois 113 ..H-6
Berthierville J-4
Birch Manor .113 ..J-7
Boisbriand ..113 ..H-3
Bonaventure C-3
Brossard ...113 ..M-5
Cabano C-3
Cadillac ....•G-2
Cap-Chat C-2
Cap de la Madeleine •K-3
Cap des Rosiers ...F-2
Cap-St.-Jacques 113 G-6
Carleton C-3
Caughnawaga 113 ..K-6
Causapscal C-3
Chandler ....•J-2
Chapais ....•H-2
Chibougamau .....•H-2
Chicoutimi B-2
Chomedey ...113 ..J-4
Chute-des-Passes ..B-2
Coaticook B-4
Colombier C-2
Côte-St.-Luc 113 ..K-5
Dégelis C-3
Deux-Montagnes113 B-4
Disraëli B-4
Dolbeau B-2
Dollard-des-
  Ormeaux ...113 ..H-5
Dorion ....•J-4
Dorval .....113 ..J-6
Drummondville ...•K-4
Duparquet ....•G-2
Duvernay ...113 ..J-4
East Angus B-4
Fairville ..113 ..H-4
Forestville C-2
Ft. Coulonge ....•H-4
Gaspé C-2
Godbout B-2
Granby ....•K-4
Grande Rivière ...C-3
Grande Vallée C-2
Grand Remous ....•H-3
Greenfield Pk. 113 M-5
Hampstead ..113 ..K-5
Harrington Harbour H-1
Hauterive C-2
Havre Aubert G-3
Hébertville B-2
Hull ....113 ..G-5
Île-Bizard .113 ..G-5
Joliette ....•K-3
Jonquière B-2
Kazabazua ....•H-4
Kirkland ...113 ..H-6
La Baie B-2
Labelle ....•J-3
Lachine ....113 ..K-6
Lachute ....•J-4
Lac Mégantic B-4
Lakeside ...113 ..H-6
La Malbaie B-3
L'Annonciation ..•J-3
La Pérade K-3
Lapraire ...113 ..M-6
La Salle ...113 ..K-6
La Sarre ....•G-2
La Tuque ....•J-3
Laurentides .....•K-4
Laurier Sta. B-3
Laval ......113 ..J-3
Laval-des-
  Rapides ...113 ..J-4
Laval-Ouest 113 ..H-4
laval-sur-
  le-Lac ....113 ..J-4
Lemoyne ...113 ..M-5
Les Escoumins ...•B-1
Les Méchins C-2
Lévis B-3
Longueuil ..113 ..M-4
Macamic ....•G-2
Magog ....•K-4
Malartic ....•G-2
Maniwaki ....•H-3
Marieville ......•K-4
Marsoui C-2
Masson ....•J-4
Matagami ....•G-2
Matane C-2
Matapédia C-3
Mattawin ....•K-3
Montebello ......•J-4
Mont Joli C-2
Mont Laurier ....•J-3
Montmagny B-3
Montréal ..113 ..K-4
Montréal Ouest 113 K-6
Mont-Royal .113 ..K-5
Natashquan H-1
New Richmond C-3
Noranda ....•G-2
Oka ....•J-4
Ormstown ....•J-4
Outremont ..113 ..L-5
Parc de la
  Montagne ..113 ..K-7
Paspébiac C-3
Percé C-2
Pierrefonds 113 ..G-6
Plessisville B-3
Pointe au Père C-2
Pointe-Claire 113 H-5
Pont-Viau ..113 ..K-4
Port Cartier C-1
Port Daniel C-3
Port Menier D-2
Québec B-3
Richmond ....•J-4
Rigaud ....•J-4
Rimouski C-2
Rivière au Renard C-2
Rivière du Loup ..C-3
Rivière Ouelle B-3
Roberval B-2
Rock Island .....•K-4
Rouyn ....•G-2
Roxboro ...113 ..G-5
St. Adèle ....•J-4
Ste. Agathe des
  Monts ....•J-3

St. Jerome ......•J-4
St. Jovite ......•J-3
St-Lambert 113 ..M-5
St-Laurent .113 ..K-5
St.-Léonard 113 ..L-4
Ste.-Marie B-3
Ste.-Rose ..113 ..J-3
Ste. Siméon D-2
Sayabec C-3
Scott B-3
Sennetere ....•H-2
Sept-Îles D-1
Shawinigan ......•K-3
Sherbrooke B-4
Sorel ....•K-3
Strathmore 113 ..J-6
Tadoussac B-2
Taschereau ......•G-2
Temiscaming .....•G-3
Thetford Mines ...B-3
Thurso ....•J-4
Trois Pistoles C-3
Trois Rivières ..•K-3
Val-d'or ....•H-2
Vallée Jc. ......•K-3
Valleyfield .....•J-4
Verdun .....113 ..L-6
Victoriaville B-3
Ville Marie .....•G-3
Vimont ....113 ..J-3
Wakefield ....•H-4
Waterloo ....•K-4
Westmount ..113 ..L-5
Wrightville 113 ..K-7

SASKATCHEWAN
Map on pages 64, 65

Arcola F-4
Assiniboia E-4
Bengough E-4
Bienfait F-4
Biggar D-2
Blaine Lake D-2
Broadview F-3
Cabri C-3
Cadillac D-4
Canora F-2
Carlyle F-4
Carnduff F-4
Ceylon E-4
Chamberlain E-3
Chaplin D-3
Climax C-4
Cochin D-2
Corinne E-3
Craik E-3
Cudworth E-2
Dafoe E-3
Davidson E-3
Delisle D-2
Dinsmore D-3
Dorintosh C-1
Duck Lake D-2
Eastend C-4
Elrose D-3
Endeavour F-2
Estevan F-4
Foam Lake F-2
Fort Qu'Appelle ...E-3
Fox Valley C-3
Glaslyn D-2
Govenlock C-4
Green Lake D-1
Grenfell F-3
Gull Lake C-3
Herbert D-3
Hudson Bay F-2
Humboldt E-2
Imperial E-3
Indian Head F-3
Insinger F-2
Kenaston E-3
Kerrobert C-3
Kindersley C-3
Kyle D-3
Lafleche D-4
Langenburg F-3
Lanigan E-3
La Ronge E-1
Leader C-3
Leross F-3
Lloydminster C-2
Maidstone C-2
Maple Creek C-4
Meadow Lake D-1
Melfort E-2
Melville F-3
Midale F-4
Mistatim F-2
Monchy D-4
Montreal Lake E-1
Moose Jaw E-3
Moosomin F-3
Mossbank E-4
Naicam E-2
Nipawin E-2
Norquay F-2
North Battleford ..D-2
Northgate F-4
North Portal F-4
Ogema E-4
Outlook D-3
Oxbow F-4
Pelican Narrows ...F-1
Pinehouse Lake D-1
Ponteix D-4
Porcupine Plain ...F-2
Preeceville F-2
Prince Albert E-2
Qu'Appelle F-3
Radisson D-2
Raymore E-3
Regina E-3
Regway E-4
Rockglen E-4
Rosetown D-3
Rosthern D-2
Rouleau E-3
St. Louis E-2
Sandy Bay F-1
Saskatoon D-2
Shaunavon C-4
Shellbrook D-2
Snowden E-2
Southey E-3
Stockholm F-3
Stoughton F-4
Swift Current D-3
Tisdale E-2
Treelon C-4
Tribune F-4
Unity C-2
Val Marie D-4
Wadena F-2
Wakaw E-2
Wapella F-3
Waskesiu Lake D-1
Watrous E-3
West Poplar E-4
Weyburn F-4
Whitewood F-3
Wilkie D-2
Willowbrook F-2
Willow Creek F-2
Wolseley F-3
Wynyard E-2
Yellow Grass E-4
Yorkton F-3